Cambridge Studies in Biological and Evolutionary Anthropology 28

The Evolution and Genetics of Latin American Populations

The human genetic make-up of Latin America is a reflection of successive waves of colonization and immigration. To date there have been few works dealing with the biology of human populations at a continental scale, and while much information is available on the genetics of Latin American populations, most data remain scattered throughout the literature. This volume examines for the first time Latin American human populations in relation to their origins, environment, history, demography and genetics, drawing on aspects of nutrition, physiology and morphology for an integrated and multidisciplinary approach. The result is a fascinating account of a people characterized by a turbulent history, marked heterogeneity and unique genetic traits. Of interest to students and researchers of genetics, evolution, biological anthropology and the social sciences, this book will also appeal to anyone concerned with the multifaceted evolution of our species and constitutes an important volume not only for anthropological genetics, but also for Latin American research.

FRANCISCO M. SALZANO is Emeritus Professor of Genetics at the Federal University of Rio Grande do Sul in Porto Alegre, Brazil, and is one of the most eminent geneticists in Latin America working in the field of anthropology. While his main scientific interest is in human population genetics, his research also covers medical, animal and plant genetics. He has previously published a number of influential works in both English and Portuguese and has received several awards including the Almirante Alvaro Alberto National Prize and the Franz Boas High Achievement Award of the Human Biology Association.

MARIA CÁTIRA BORTOLINI is Associate Professor in the Genetics Department of the Federal University of Rio Grande do Sul in Porto Alegre, Brazil. Her research focuses on the genetic variability, at both protein and DNA levels, of African-derived and Amerindian populations of South America.

T0275618

Cambridge Studies in Biological and Evolutionary Anthropology

Series Editors

HUMAN ECOLOGY
C. G. Nicholas Mascie-Taylor, University of Cambridge
Michael A. Little, State University of New York, Binghamton
GENETICS
Kenneth M. Weiss, Pennsylvania State University
HUMAN EVOLUTION
Robert A. Foley, University of Cambridge
Nina G. Jablonski, California Academy of Science
PRIMATOLOGY
Karen B. Strier, University of Wisconsin, Madison

Consulting Editors
Emeritus Professor Derek F. Roberts
Emeritus Professor Gabriel W. Lasker

The Evolution and Genetics of Latin American Populations

FRANCISCO M. SALZANO

and

MARIA CÁTIRA BORTOLINI

CAMBRIDGE
UNIVERSITY PRESS

CAMBRIDGE UNIVERSITY PRESS
Cambridge, New York, Melbourne, Madrid, Cape Town, Singapore, São Paulo

Cambridge University Press
The Edinburgh Building, Cambridge CB2 2RU, UK

Published in the United States of America by Cambridge University Press, New York

www.cambridge.org
Information on this title: www.cambridge.org/9780521652759

First published 2002
This digitally printed first paperback version 2005

A catalogue record for this publication is available from the British Library

Library of Congress Cataloguing in Publication data

Salzano, Francisco M.
The genetics and evolution of Latin American populations/Francisco M.
Salzano and Maria Cátira Bortolini.
    p.   cm.
Includes bibliographical references and index.
ISBN 0 521 65275 8 (hbk.)
1. Human population genetics – Latin America.   2. Physical anthropology –
Latin America.   I. Bortolini, Maria Cátira, 1961–   II. Title.

QH455.S35 2001
599.93′5′098–dc21   2001025508

ISBN-13 978-0-521-65275-9 hardback
ISBN-10 0-521-65275-8 hardback

ISBN-13 978-0-521-02239-2 paperback
ISBN-10 0-521-02239-8 paperback

To Utopia, a world of freedom and love,
without oppressors and oppressed

To the Pioneers

Juan Comas

Darcy Ribeiro

James V. Neel

To our families, where everything started
and will eventually end

Contents

Preface

One of us (FMS) started to collect the information that was eventually incorporated in this book as early as 1956. In the ensuing years the compilation of the data continued, although not in a rigorously systematic way. Part of this information was used in the publication of a book about Brazilian populations in 1970, and the editing of another about Latin America in 1971. In more recent years the data collection became more systematic, and at the beginning of 1998 MCB started to help with the data analysis. In these three years in which we have been more specifically involved in the book's development much work was done, but with considerable enjoyment. Latin America's cultural, political and biologic histories make a fascinating subject, and we do not regret the decision, at least superficially, to describe and interpret them.

Naturally, the analysis of such a wide area and large population may have resulted in several gaps and involuntary omissions. Since our area of expertise is population genetics, the coverage of this aspect was much more complete than in the other subjects covered by the book, such as archeology, history, demography, biologic and social anthropology, or medicine. But since we believe that evolutionary studies in humans should be as comprehensive as possible, we decided to include them even at the risk of incurring such faults. Some friends helped by reading the different chapters and providing precious advice. They are as follows: 1. Origins: Marta M. Lahr (University of São Paulo, Brazil, and University of Cambridge, UK); 2. Environment and history: Claudia Wasserman (Federal University of Rio Grande do Sul, Brazil); 3. Socioeconomic indices, demography, and population structure: Juan Pinto (University of Valparaiso, Chile); 4. Ecology, nutrition and physiologic adaptation: Carlos E.A. Coimbra Jr and Ricardo V. Santos (Oswaldo Cruz Foundation, Rio de Janeiro, Brazil); 5. Morphology: Paulo A. Otto (University of São Paulo, Brazil); 6. Health and disease: Maria R. Passos-Bueno (University of São Paulo, Brazil); 7. Hemoglobin types and hemoglobinopathies: Marco A. Zago (University of São Paulo, Brazil); 8. Normal genetic variation at the protein and DNA levels: Marcos Palatnik (Federal University of Rio de Janeiro, Brazil); and 9. Gene dynamics: Rubén Lisker (Autonomous

National University of México, México). Of course, they are not responsible for any errors and omissions that still remain, particularly because not all of their suggestions have been adopted. Also as we write this preface, at the end of 2000, the literature concerning the subjects covered continues to appear at a steady pace. But everything has to come to an end; as we had a fixed deadline for the submission of the manuscript, only absolutely essential information was added at the final stages of the book's preparation.

During all these years we had the wonderful opportunity of enjoying the challenging and friendly environment provided by our colleagues at the Genetics Department of our university. Many of them provided, formally and informally, ideas and results that had been incorporated in many of the publications cited in this book. To all of them our heartful thanks. A special acknowledgment should be given to Laci Krupahtz, who was responsible for the preparation of the manuscript (text and figures) for the press. Her incredible work capacity and wonderful skill was essential for the completion of this work. She, Jaqueline Battilana and Patricia Koehler also helped us in the preparation of the author index, an important section for a reference book such as this one. Our research has been financed by many agencies along the years. We thank our present sponsors, the Programa de Apoio a Núcleos de Excelência (PRONEX), Conselho Nacional de Desenvolvimento Científico e Tecnológico (CNPq), Financiadora de Estudos e Projetos (FINEP), and Fundação de Amparo à Pesquisa do Estado do Rio Grande do Sul (FAPERGS) for precious help. Last but not least, acknowledgments are due to our many Latin American colleagues and friends who generously provided advice at many scientific meetings, as well as the bibliography on which this book was based; and to the subjects of our studies themselves, for their patience and collaboration.

Francisco M. Salzano
Maria Cátira Bortolini

1   *Origins*

> ... but the real reason why we study it [evolutionary biology] is that we
> are interested in origins. We want to know where we came from
>
> *John Maynard Smith and Eörs Szathmáry*

A diversity of sources

To understand the present biology of Latin American populations it is
important at the outset to emphasize the large diversity of their founding
stocks. The earliest migrants to the continent were those now called
Amerindians. The end of the fifteenth century witnessed the so-called
European discoveries, which set in motion a mass movement of people not
only from that continent, but also from Africa and Asia. In this chapter we
will present some background material for the characterization of these
migrants, needed for the evaluation of what occurred in the past 500 years.

Amerindians

Confusion at the first encounter

The denomination of 'Indian' to the people Christopher Columbus found
when he landed in America was due to a mistake, since he and his
companions imagined that they had arrived in India. This did not preclude
the quick dissemination of this generic designation, although some restric-
tions have been raised in relation to it (Maestri-Filho, 1994; Field, 1994).
America was named to honour Americus Vespucius, the Florentine navi-
gator, who, differently from Columbus, conceived the new lands as a New
World (Vespucio, 1951).

Controversies

There is much discussion about almost all aspects related to the arrival of
the earliest Americans. (Table 1.1 summarizes some of the questions, and
the evidence used to answer them.) In relation to their previous homeland

1

several options could be considered, but there is an almost unanimous consensus that they probably entered the American continent from Asia through the Bering Strait. A recent proposal to resurrect traces of Polynesian ancestry among these early Americans was shown to be unwarranted by Bonatto *et al.* (1996).

However, establishing an Asian origin is not enough. From where in Asia did they come from? A previous idea that they derived from groups inhabiting eastern Siberia was contradicted by evidence from mitochondrial DNA (mtDNA), since these groups lack a mtDNA haplogroup (B) that is well represented in Amerindians. The T cell lymphotropic virus type II, present in a large number of Amerindian groups, is absent in eastern Siberia. Both the virus and the mtDNA haplogroup B, however, do occur in the indigenous population of Mongolia, suggesting common ancestry between the present Amerindians and Mongolians (Neel *et al.*, 1994).

There is also much controversy about the date of arrival of these migrants, and the number of main waves of migration. The options are indicated in Table 1.1. There is increasing evidence that this date of entry is much earlier than has been supposed, but there are no undisputed indications about this from the archaeological and paleoanthropological material. Based on mtDNA data Bonatto and Salzano (1997*a,b*) suggested a model for the peopling of the Americas in which Beringia played a central role. This region would have been colonized by the Amerindian ancestors, and sometime after this colonization, they would have crossed the Alberta ice-free corridor and peopled the rest of the American continent. The collapse of this ice-free corridor during a few thousand years (about 14 000 to 20 000 years ago) isolated the people south of the ice sheets, giving rise to the bulk of North, Middle and South American Indians. The Na-Dene, Eskimo, and probably the Siberian Chukchi, would have originated from those who had stayed in Beringia, through a process of independent diversification.

This view, however, is not accepted by many scholars, who use other sources of evidence to question it. Here is not the place to examine all arguments in detail, and the reader is directed to the references given at the bottom of Table 1.1 for an appraisal of the most important aspects of these controversies. In relation to the number of migration waves, it would be wise to remember Brandon's (1961) assertion: 'There is no reason whatever to suppose that men of such times were consciously migrating, they were only living'.

Prehistoric development

Independently of what happened before, the fact is that a substantial number of people were present in what is now Latin America at the time of European arrival, by the end of the fifteenth century. But again there is much discussion about their precise numbers. The estimates generally relied on are: (a) prehistoric remains, (b) historical accounts, and (c) depopulation rates that occurred due to war, epidemics or other causes; and inferences from these factors are then made about the size of the putative original groups. All of these estimates are subject to errors, due to incomplete conservation of the prehistoric material, exaggerated or wrong testimonies, and local variation of the factors responsible for population decreases and eventual recoveries.

Selected estimates for the number of persons present in several areas, and in the total of Latin America, at the time of the European arrival are given in Table 1.2. For the whole region the number of people varies between 28 and 88 million, with a reasonable value, in our view, being 43 million. Their distribution through the continent, however, was uneven and related to the degree of socioeconomic development the different groups reached through time.

What are the main characteristics of these socioeconomic developments? The groups that first colonized the area were small, assembled as bands which relied mainly on hunting and gathering for their subsistence. They are generally classified under the generic name of Paleoindians, and archeological evidence of them is found all over the area, from Middle America to Tierra del Fuego. Some of the most important sites and cultures related to them are given at the bottom of Table 1.3.

Environmental changes and the extinction of the Pleistocene megafauna conditioned the development, initially in Middle America and the Andes, of agriculture. This occurred around 5000 BP, and led to a real revolution. By this time, the bands started to merge into larger groups or tribes. They could assemble in villages or stay dispersed in neighborhoods, but rarely exceeding, a few thousands of members. Fishing and the collection of mollusks was still an important means of subsistence in the coastal areas, and this strategy would continue together with the incipient agriculture.

The next stage in structural development was reached through the chiefdoms, with the integration of communities and the beginnings of stratification and hierarchies. This began around 2300 BP in different regions, including Middle America, the Intermediary Area (part of Central America, Ecuador and Colombia), as well as Caribbean, Andean and Amazonian territories. Two of the most sophisticated cultures of this period were the Middle American Olmec tradition, present in the La Venta site, and the Andean Chavín de Huántar (Table 1.3.).

Finally, in Middle America and the Andes, this type of social organization led to the development of states with incipient (Maya) or more developed urban centers (Toltec, Aztec, Inca). At the arrival of the Spaniards in Peru in 1532, the Inca Empire extended from Colombia to central Chile, from the Pacific Ocean to the eastern jungles – the largest political system created in the New World up to that time. Its capital, Cuzco, was as the Inca proclaimed 'the world's navel', and the Empire's total population is estimated to have been 12–14 million (Bethell, 1984).

This linear scheme of development is just an abstraction. Actually, the process was different in diverse regions, involved reversals, adaptation to local conditions, and in many places (as in the Amazon) the persistence of many groups with essentially the same way of life as they had at the time of colonization of the continent.

We can only guess about the conditions of life that these groups enjoyed under these varied circumstances. There is a long way from the basic egalitarian relationships prevailing in the hunter-gatherer bands to the Inca's complex hierarchy. The relationships between the center of the Empire and its periphery are also important. Neves and Costa (1998) considered this question with regard to the prehistoric people who inhabited the Atacama Desert of northern Chile and their association with the Tiwanaku empire. They used femur length as an indication of stature, and the latter as an assessment of standard of living. Femur length *per se*, and sexual dimorphism were evaluated (it is known that nutritional stresses affect males and females differentially, reducing sexual differences). Comparisons were made for three periods: before the Tiwanaku influence, during the period of the Empire's control, and post-Tiwanaku. Both indicators suggested a positive impact (better life conditions) during the associated Tiwanaku/Atacama period. Since the measures do not show any increase in variation in the people of this period, the authors concluded that the data do not support the hypothesis of a concentration of health in a few individuals during Tiwanaku times, but rather a generalized improvement of living conditions.

Europeans

Small but important

The small size of the European continent (it could be visualized as just a peninsula of the larger Asian continent) contrasts with its historical, cultural, political, and economic importance. Its history, in many respects, constitutes the essence of humankind's achievements. It was the birthplace of all occidental civilization, both in its material and spiritual aspects. The diversity of its peoples is only matched by their cultural variability. It was from there that, in the sixteenth century, the great adventure of the maritime expansion took place, considerably enlarging our view of ourselves.

Prehistory

Members of the genus *Homo* (*Homo erectus* and archaic *Homo sapiens*) were present in the region as early as 700 000 years ago. By the late Middle and early Upper Pleistocene (between 200 and 35 thousand years ago) the area was inhabited by a peculiar type of creature, the neanderthals. The appearance of modern humans (*Homo sapiens sapiens*) is documented only much more recently, that is, about 45 000 years ago. A heated debate has been going on for a long time now whether *Homo sapiens neanderthalensis* would have been completely substituted by *sapiens sapiens* without admixture, or whether such admixture did actually occur. Morphologic and genetic data were assembled and variously assessed (reviews in Stringer and Gamble, 1993; Trinkaus and Shipman, 1993). The direct examination of mitochondrial DNA sequences of the neanderthal-type specimen (Krings *et al.*, 1997), determined that they fall outside the variation of modern humans, suggesting that at least in relation to this organelle there was no overlap. These results, however, do not rule out the possibility that neanderthals contributed other genes to modern humans.

Cultural development

The cultural developments that took place in Europe from the time of *H. sapiens sapiens*' arrival up to around AD 900 are summarized in Table 1.4. It is a long history of achievements, from the simple stone artifacts of the Paleolithic to the Iron Age, reflected in its demographic structure which varied from single bands of hunter-gatherers to sophisticated empires. The interpretation of these facts is also subject to much discussion, which can be exemplified by the models advanced for the spread of agriculture.

Three models have been advanced for the spread of farming, a cultural development which started in the Near East. They are: (a) *demic diffusion:* a slow expansion of people from the Neolithic source population into

Europe who, due to their better suitability, either displaced or absorbed the less numerous Mesolithic hunter-gatherer populations (Ammerman and Cavalli-Sforza, 1984); (b) *cultural diffusion:* which did not involve intrusion of peoples, only of ideas, as well as the trade of crops (Dennell, 1983); and (c) *pioneer colonization:* migration of persons did occur, but was on a small scale, with much less influence in the genetic make-up of the populations involved (Zvelebil, 1986).

These models can be tested, considering the present genetic variability of European groups and that expected under the different alternatives proposed. Three selected examples will be presented here. Barbujani *et al.* (1995) developed five models of microevolution in European populations, considering also their relationships with Indo-European speakers. These were then compared with data from 26 nuclear protein genetic systems. The best correlations between observed and simulated data were obtained for two models in which dispersal depended basically on population growth, thus favoring the demic diffusion hypothesis. But diametrically different conclusions were reached by Richards *et al.* (1996), after an analysis of the mtDNA of 821 individuals from Europe and the Middle East. These authors found that the major extant European mtDNA lineages should have predated the Neolithic expansion, the conclusion being that the spread of agriculture was a substantially indigenous development. Fix (1996), on the other hand, using the HLA system as a model, made a computer simulation that indicated that clinal patterns similar to those observed in European populations could be due to temporal gradients in natural selection. These clines could have been influenced by domestication (specifically to animal husbandry) but not necessarily to the mechanism of demic diffusion. More recently Barbujani *et al.* (1998) and Chikhi *et al.* (1998) reexamined the problem both conceptually and using nuclear DNA markers, clearly favoring the demic diffusion model.

Factors responsible for the Maritime Epopee

More specific information about the events which shaped the lives of people living in the six European countries mainly involved with the Great Navigations of the sixteenth century is given in Table 1.5. They are characterized, in the period considered (AD 1000–1600) by an intense flux of peoples of different ethnic affiliations, wars, and the formation and dissolution of political units. The question that can be asked is: What factors were influential for the development of this great enterprise?

Undoubtedly, significant improvement in the art of navigation is the first point to be considered. In this connection, Portugal's Prince Dom Henrique (1394–1460) should be mentioned. He founded the School of Sagres, responsible for the training of a series of important Portuguese navigators, who successively, during one century, explored almost all regions of the world.

The structure of the European societies of that time should also be considered (reviews in Wehling and Wehling, 1994; Wasserman, 1996a; Wasserman and Guazzelli, 1996). With a density of 40–60 inhabitants per square kilometer, there was more physical contact among individuals, wider circulation of goods and ideas, and a more intensive exchange of experiences. The societies were structured in a rigid way, which generally involved the monarchy, the Church, the nobility, and the common people – bourgeois, artisans, and peasants. But this does not mean that regional differences did not exist. An important aspect in Portugal's and Spain's initial development must have been the Arab influence, with its characteristic absence of private property and of a pluralistic society in which there were opportunities irrespective of ethnic or religious affiliations. In Europe as a whole, however, the transition between feudalism and capitalism was actively under way. Within this context, the strengthening of the monarchies, the development of trade, the need for a stable monetary system, the crusading ideal that 'the others' should receive the word of Christ, and the Renaissance during the fourteenth to sixteenth centuries, were all important for the developments that took place at that time.

Africans

The cradle of mankind

There is consensus that the first species of our genus, *Homo habilis*, originated in Africa from an earlier genus *Australopithecus*, now extinct. It is thought that *H. habilis* first appeared some 2.5 million years ago (mya), and was replaced by *H. erectus* at about 1.5 mya. The latter spread to Asia and Europe. The next taxon in this phyletic line was *H. sapiens*, who appeared around 500 thousand years ago (kya) in Africa, and later dispersed to Europe and west Asia. Modern humans (*H. sapiens sapiens*) are first found in East and South Africa around 100 kya (Table 1.6).

Cultural development

Later developments in Africa are summarized in Table 1.6. People from three basic stocks are recognized as the ancestors of all African populations: Khoisanids, Negroids and Caucasoids. Descendants from the first today inhabit almost exclusively southern Africa, but were once found over a wider area. Present-day representatives are the Khoi and San, notable for characteristics such as steatopygia and languages with characteristic clicks.

Negroids would have given rise to the groups who live today in the tropical forest and in much of eastern and southern Africa, the pygmies probably representing a long-term adaptation to tropical forest. These would include Nilo-Saharan speakers, West Africans and Bantus.

Groups related to modern Caucasoid populations lived in North Africa. In the Maghreb a specific Paleolithic culture developed from 22 000 to 7500 BC, named Iberomarusian, typically of the Cro-Magnon type. At the end of the Pleistocene – early Holocene they coexisted with a pre-Neolithic culture, the Capsian. People from this culture were hunter-gatherers and fishermen who consumed large amounts of mollusks. Later they acquired pottery and sheep, but always retained Paleolithic characteristics.

Neolithic developments involved the appearance of agriculture, cattle domestication, and more efficient fishing economies. Written documents appear in Egypt about 3000 BC. The advent of the Metal Age furnished conditions for the organization of early states and empires (more details in Table 1.6).

State formation

Starting with the Egyptian and Ethiopian empires, a series of states originated in many parts of the African continent. Some of these are listed in Table 1.7, with the approximate epoch of their existence and location. However, no consensus has been formed for why they originated in so many parts of the African continent. Partial explanations could be: (a) the superiority of iron weaponry which, monopolized by a few, would lead to this development; (b) the increment of long-distance trade, determining the formation of centralized urban groups; (c) the production of economic surplus, bringing about specialization of labor and a restricted group of 'power' goods that could be kept in the hands of a few; (d) the need for social management of larger and larger groups, determined by demographic growth; and (e) conquest, or imitation of neighboring states

(Curtin *et al.*, 1978). It is probable that a combination of these factors may have been important, and that they differed in diverse circumstances.

The tragic commerce

Enslavement is part of the history of all ancient populations, but it never before or after reached the levels of the sixteenth to nineteenth centuries, with its tragic impact on African societies. As was emphasized by Curtin *et al.* (1978), it is ironic that maritime contact, which ended Africa's long isolation, should have led to a situation in which its own peoples became Africa's main export product. To understand clearly all aspects of the phenomenon it should be understood that: (a) the trade was uneven along the continent; areas such as the present Republic of South Africa and East Africa, from Tanzania to Ethiopia, were not involved in it; and (b) the Africans themselves were active agents in the process. For instance, there was a convention in West Africa (though not in Central Africa or Mozambique) that the European slave traders should stop at the waterside. Africans themselves would act as middlemen in the trade to the interior.

Details about the main regions where the slave trade was active, ports of exportation, African dealers, and the relative importance of these regions are presented in Table 1.8. The traffic occurred mainly along the West Coast and Mozambique, through several ports, and distinct states acted as dealers. The flow varied among regions during the four centuries of the trade. For instance, in Senegal and Gambia the peak of the traffic in absolute numbers occurred in the eighteenth century, while for Mozambique this happened a century later. The net result, however (e.g. the intercontinental transportation of about 80 000 people per year in the 1780s) was only matched by the intentional overseas European migration that took place in the nineteenth and twentieth centuries. Reader (1998) estimated as 9 million the number of slaves who were shipped across the Atlantic between 1451 and 1870.

What was the impact of the slave trade on the African populations? Some small ethnic groups were completely wiped out. Others suffered heavily for a time (Yoruba, Wolof), some were lightly involved (Benin), while still others (such as the Dahomey) may have profited at the expense of their neighbors. The sheer physical destructiveness of the trade was not high enough to produce differences in social health and progress between aggressors and victims. Its most serious damage to African society was the negative social imprinting that was established upon all the unfortunate victims, and their descendants, who suffered this unethical process (Curtin *et al.*, 1978).

Other contributors

The Asian connection

Migrants to Latin America came from almost everywhere. But in historical times and numerically, besides the groups already mentioned, some of the most important contributors (especially for Middle America, the northern part of South America, and southern Brazil) were people from East Asia (mainly Chinese and Japanese), as well as East Indians and Pakistani.

Asia is presently the most populous continent on earth, 1997 estimates indicating a population of 3.5 billion. More than half are concentrated in China (1.2 billion) and India (0.9 billion). The continent also has a rich prehistoric and historic past. Aspects related to the Mongolian and Siberian contributions to the formation of Latin American Amerindians were discussed earlier in this chapter. Here we shall concentrate on the main events that in the past shaped East Asian (Table 1.9) and East Indian (Table 1.10) populations.

A millenarian civilization

Pre-sapiens fossils and artifacts have been found in several places in the present Chinese territory, as well as early *Homo sapiens*, but no neanderthals. In what is now Japan evidence for human occupation near Tokyo 30 kya (the Sanrizuka site), and the oldest pottery in the world (12.7 kya) has been found.

Some of the main events and cultures identifiable in the prehistory or history of East Asia are listed in Table 1.9. The first written documents and urban life can be dated to the epoch of the Shang dynasty, 3.7 kya. The ensuing history is complicated because we are considering a very large territory, high population numbers, and tremendous fragmentation of these populations, with frequent migration from one region to another. However, few points deserve mention. Technical improvements in agriculture were mainly responsible for the population growth. With the end of the feudal system, farmers owned their land, while merchants and artisans became independent of the lords. A new elite also emerged, composed of administrators, teachers and philosophers. During the following years, a complex civilization was developed, which succeeded in developing a closely knit social and political structure that benefited from trade with its neighbors. Assimilation of other people (Turkic, Mongol, Tungusic) was also active, resulting in a closer approximation of northern Chinese to these populations and to Koreans rather than to the Chinese of the southern provinces.

Kaleidoscopic people and social structures

The Indian subcontinent provides us with a tremendous amount of social, cultural and ethnic diversity. Main aspects of its development are indicated in Table 1.10. The Indus valley was the scene of urban developments that were not as early as those of Mesopotamia, but which flourished giving rise to important structures such as the Harappan civilization, with its most important city, Moenjodaro, which, at its peak, may have had 40 000 inhabitants or more.

Migration from a wide set of peoples (Indo-Aryans, Greeks, Sakas, Kushanas, Huns, Arabs) and their admixture with the autochthonous populations led to a myriad biological and ethnic diversity. Superimposed on this a complex social structure was developed, including tribal clusters, castes, varnas, caste clusters, jatis, religious communities and sect clusters. Since marriage is regulated by the boundaries of these structures, they should be considered, together with geographic location and language, in any attempt at a delineation of an East Indian group.

The 'discovery'

Starting with one mistake

'During all night we heard the sounds of birds' wrote Cristopher Columbus in his diary on 9 October 1492 (Colombo, 1984). Three days afterwards and a little more than 2 months after they left the port of Palos (Spain), the three ships commanded by the Genoese admiral reached a small island called by the surprised and curious natives Guanahani, which was immediately rebaptized as San Salvador. Columbus experienced, in this first contact, a feeling of wonder. In his own words, 'and when I arrived here a perfume so good and gentle came from the flowers and trees that was the sweetest thing in earth'.

The ambitious and devout Columbus believed that he had reached the rich and unknown regions of the Far East, since he had the idea that traveling eastward inevitably would lead him to Asia and the Indies. Contouring the island of Cuba he even identified geographic details described by Marco Polo in his circumnavigation of the Indochina peninsula

(O'Gorman, 1986). In 1506, after being charged by the Inquisition, and poor and forgotten due to the fact that he had not discovered the riches that had stimulated his four expeditions, Christopher Columbus died refusing to accept the evidence that he had not reached the periphery of Kublai Khan's empire.

To a certain extent it is absurd to speak of the 'discovery' of a continent, which at the time of Columbus's arrival would have held more than 43 million people, magnificent civilizations, and cities larger than Madrid. In addition, it is possible that the continent had been visited by Vikings 500 years previously (Cavalli-Sforza *et al.*, 1994). But the fact is that America would never be the same after this event, its history frequently being divided into pre- and post-Columbian times.

Ideology and prejudices

The Europeans arrived in the New World at a climax of their cultural, military and religious force. This generated a strong Eurocentric feeling of superiority (Herrero, 1996). This imperative was rooted in the fact that the Christian culture and the written tradition were considered as superior. Consequently classified as inferior, or even excluded from the concept of culture altogether, were practices based mainly on oral, popular, and non-Christian traditions, such as those of the Indians and of the Africans brought to the Americas as slaves (Todorov, 1983; Bern, 1995). In addition, important scientific and technological developments gave such a material strength to the Europeans that they became persuaded of the superiority of their civilization.

These facts, and their ignorance of indigenous and African achievements, contributed to the colonizers conceiving of the New World as a place of degradation and exclusion; this as well as mercantile and other views on the possibilities for exploitation. This led to one of the greatest genocides in the history of mankind, with the relegation to the periphery a system of cultures that would not be in keeping with the European tradition. The resulting exploitation system contributed to the welfare of the economy and the wealth of the Iberian and Anglo-Saxon empires for centuries. But the strength of this hegemony was not enough to avoid biological admixture, cultural exchanges, and the blending of colonizers and colonized in a new synthesis. Concomitantly, independent nations were formed. The Latin American world emerged out of turbulence.

Table 1.1. *Controversies related to the arrival of the first Americans, and the types of evidence considered*

Questions	Sources of evidence
1. From where did they come?	1. Geologic
1.1. Siberia	2. Archeologic
1.2. Mongolia	3. Paleoanthropologic and morphologic
1.3. Somewhere else in Asia	4. Linguistic
1.4. Polynesia	5. Medical
1.5. Other places	6. Protein polymorphisms
2. When did they arrive?	7. DNA polymorphisms
2.1. Not earlier than 15 000 years ago	
2.2. Earlier (30 000–40 000 years ago)	
3. How many main waves of migration?	
3.1. Just one	
3.2. Three	
3.3. Four	

Selected references: 1. Geologic: Lemmen *et al.* (1994). 2. Archeologic: Hoffecker *et al.* (1993); Prous (1995); Roosevelt *et al.* (1996); Meltzer (1997). 3. Paleoanthropologic: Steele and Powell (1992); Munford *et al.* (1995); Lahr (1995); Neves *et al.* (1996). 4. Linguistic: Greenberg *et al.* (1986); Greenberg (1987); Diamond (1990). 5. Medical: Confalonieri *et al.* (1991); Neel *et al.* (1994). 6. Protein polymorphisms: Schanfield (1992); Cavalli-Sforza *et al.* (1994). 7. DNA polymorphisms: Horai *et al.* (1993); Torroni *et al.* (1994); Wallace (1995); Merriwether and Ferrell (1996); Forster *et al.* (1996); Bonatto *et al.* (1996); Bonatto and Salzano (1997*a,b*).

Table 1.2. *Estimated numbers for the population present in selected areas and the total for Latin America at the time of the European discovery*

Area	Estimated population (millions)		
	Minimum[a]	Maximum[b]	Most probable
Central México	11	25	14
Central America	0.9	6	5
Caribbean (including Venezuela)	1	10	4
Colombia	—	—	3
Central Andes	12	32	13
Araucanian territory	—	—	1
Brazil	1	7	2
Other	—	—	1
Total	28	88	43

[a]Adding 2 million to account for the missing estimates.
[b]Adding 8 million to account for the missing estimates.
Source: Bethell (1984); Verano and Ubelaker (1992).

Table 1.3. *Chronology of prehistoric and historic sites and cultures in Middle and South America*

Chronology	Middle America Highlands	Middle America Lowlands	Intermediary area	Carib	Amazon	Andes Coast	Andes Highlands	Patagonia and Tierra del Fuego
10000 BC S			Taima Taima			Monte Verde	Pikimachay	Los Toldos
C	*Clovis*							
9000 BC S			La Sueva Turrialba				El Inga	Cueva Fell
C								
8000 BC S	*Ajuereado*		Las Vegas					*Magallanes I*
C						Quereo	Guitarrero	*Magallanes II*
7000 BC S	Guila Naquitz		El Abra				Pachamachay	
6000 BC C	*El Riego*							
5000 BC S	Zohapilco		Cerro Mangote			Paloma		
4000 BC S				Banwari Trace		Quiani		Englefield Island
C								*Magallanes III*
3000 BC S			Puerto Hormiga					
C	*Coxcatlán*							
2500 BC S						Pampa		
C	*Abejas*							
2000 BC S			*Valdivia* Pastaza Real Alto		Tutishcainyo	Huaca Prieta Aspero	Kotosh	*Magallanes IV*
C								
1500 BC S		San Lorenzo				El Paraiso La Florida		
C	*Purrón*							
1000 BC S	Tlatilco					Cerro Sechin Moxeke	Wankarani	
C		*Olmec*	*Barrancoid Malambo*		*Ananatuba*			

500 BC	S	Cuicuilco	Nakbe, La Venta						
	C							*Chavín*	
AD/BC	S	Monte Albán	Tikal, El Mirador	La Tolita		Hupa-Iya		Pukara	
	C				*Saladoid*				
AD 500	C	*Izapa*	*Classic Maya*	*Saladoid*	*Ostionoid*	*Marajoara*	*Mochica*, *Nazca*		
	S	Teotihuacán		Manta, San Agustín			Sipán		
AD 1000	S	Tula, Cholula	Chichén Itzá, Palenque	Sitio Conte		Mojos, Manacupurú		Tiwanaku, Huari	
	C			*Quimbaya, Calima, Milagro*					
AD 1500	P	Aztec	Maya	Chibcha	Taino			Inca	Ohna, Yaghan
	S	Tenochtitlán	Mayapán			Several, Santarém	Chan Chan		
	C	*Toltec*			*Chicoid*	*Aruã*	*Chimú*		*Magallanes V*

P, People (ethnic groups); S, Sites; C, Cultures.
Source: Fiedel (1996).

Table 1.4. *Chronology of prehistoric and historic European development*

Chronology	Stages of development	Cultures	Places
700–500 kya	*H. erectus/H. antecessor*	Oldowan	Atapuerca, Ceprano
500–200 kya	'Archaic' *Homo sapiens*	Acheulean	Many sites
200–35 kya	Neanderthals	Mousterian	Almost everywhere at its peak of population development
40 kya	*Homo sapiens sapiens*		
	Paleolithic	Aurignacian, Gravettian	
		Lascaux	France, northern Spain
18 kya	Mesolithic	Microliths	Several places
10 kya	Neolithic	Cordial ware, Linear pottery	From Greece to all over Europe
8 kya	Middle Neolithic		
4 kya	Copper mining		Began in the Balkans
		Megalithic monuments	
		Stonehenge	England
3 kya		Bell beaker	Western and central Europe
		Globular amphora	Central, northern and eastern Europe
		Corded ware	
		Battleaxe	
			Ukraine and north of Caucasus mountains
	Pastoral nomadism	Minoan civilization	Crete
2 kya	Bronze age	Mycenean civilization	Greece
		'Urnfield'	Southern Germany, Czech Republic, Slovakia, Austria
		Hallstatt	Western Austria
1 kya	Iron age	La Tène	Northwestern Switzerland
		Etruscan	Italy
		Phoenician influence	Italy, Spain
		Greek empire	Eastern Mediterranean area
		Roman empire	Mediterranean, Balkans, France, England
AD 500	Middle Ages	'Barbarian' influence	All over Europe
AD 900		Magyar influence	Hungary
		Arabian influence	Italy, Spain
		Turkish influence	Balkans

kya: thousand years ago.
Source: Cavalli-Sforza *et al.* (1994).

Table 1.5. *Some key events during the first sixteen centuries of the Christian Era in the five countries mainly involved with the Great Navigations of the sixteenth century*

Chronology	Countries				
	France	Netherlands	Portugal	Spain	British Isles
Around AD 1	Celtic domination	Celtic domination	Roman empire	Roman Empire	Celtic domination
500	Frankish invasion	Frankish invasion	Visigoth invasion	Visigoth invasion	Anglo-saxon invasion
700			Arab invasion	Arab invasion	
800	Holy Roman–German empire	Holy Roman–German empire			Viking invasion
1000–1100			Christian conquest of Coimbra and Lisbon	Kingdoms of Navarra, León Castela and Aragón	Norman invasion
1200–1300			End of Arab occupation	End of Arab occupation	Magna Carta
1301–1500	One-hundred-years war	Burgundy domination	Avis dynasty		One-hundred-years war Wars of the Roses
1501–1600	Religious conflicts and peace	Spanish domination and independence in 1579	Spanish domination	Carlos I and the largest kingdom in Europe	Tudor dynasty and foundation of the Anglican Church

Source: Camargo (1998).

Table 1.6. *Chronology of prehistoric and historic African development*

Chronology	Stages of development	Peoples/cultures/events	Places
2.5 mya	*Homo habilis*		
1.7 mya	*Homo erectus*		
100 kya	*Homo sapiens sapiens*		
10 kya	Paleolithic	Khoisanids	Omo, Border caves, Klasies river mouth
		Negroids	East and southern Africa
		Caucasoids	West Africa
		Iberomarusian	North Africa
		Capsian	Maghreb
			Libya
7.5–5 kya	Neolithic		Egypt
7.0 kya	Fishing economy	Ishango	Lakes Mobutu and Turkana
4 kya	Agriculture		Sudan, Ethiopia
3 kya	Written documents	Egyptian civilization	Egypt
		Bubalis	Southern Algeria, Libya, northern Chad
		Round head	Egypt
2.3 kya		Incorporation of Egypt into the Hellenic World	
2.3 kya		Gwisho	Zambia
1.5 kya	Cattle domestication	Bovidian	
1 kya		Equidian	
		Bantu expansion	Central and southern Africa
0.6 kya	Metal Age	Kush kingdom	Meroe, Nubia
0.5 kya		Nigerians	Nok
		Lake Victorians	Urewe
AD 100		Roman influence	Northern Africa
700		Arab expansion	Egypt, East Africa
800		Phoenician influence	North Africa

mya: million years ago.
Source: Curtin *et al.* (1978); Cavalli-Sforza *et al.* (1994).

Table 1.7. *A list of some of the African states or empires which were formed in a period of 3.4 thousand years*

Chronology	African states or empires	Places
1500 BC	Egypt	Nile valley
600 BC	Ethiopia	Ethiopia
AD 100	Takrur	Senegal river
400	Ghana	Sahel
400	Aksum	Tigre plateau
800	Gao (a Songrai state)	Niger bend
800	Kanem and Borno	Lake Chad
900	Fatimid	Maghreb
1000	Masudi	Central Mozambique
1100	Almoravid	Maghreb
1100	Mwene Mutapa	Zambezi-Limpopo region
1200	Darfur	Chad
1200	Almohad	Maghreb
1200	Borno	Lake Chad
1300	Mali	Sahel
1500	Malawi	East African interior
1500	Hausa	Central Sudan
1500	Luba	Central Africa
1500	Lunda	Central Africa
1500	Kongo	South of the Zaire river
1600	Songrai	Mali, Senegal
1630	Kuba	South of Sankuru river
1700	Mossi	West of the Niger river
1700	Tio kingdom of Makoko	North of the Stanley Pool
1800	Merina	Madagascar
1850	Futaanke	Between the Niger and Senegal rivers
1890	Samori Ture	Northern Ivory Coast and Ghana

Source: Curtin *et al.* (1978).

Table 1.8. *The African slave traffic*

Region	Ports of exportation	Main dealers	Importance
Senegal/Gambia	St Louis, Rufisque, Joal, Cape Verde islands	Jahanke, Wolof, Bambara, Fuuta Tooro	Furnished about one-third of all the slaves exported from Africa before 1600. Peak in absolute numbers in the eighteenth century
Upper Guinea and Sierra Leone	Cacheu, Cape Verde islands, Conakry	Mane, Fube, Fuuta Jalo	One-third of the slaves exported in the sixteenth century. Total contribution less than 5% of the total
Windward Coast	Fragmented, Grand Lahou	Kuranko, Bambara, Juula, Asante	Never very important, spurts in the sixteenth and eighteenth centuries
Gold Coast	25 major stone forts	Akwamu, Asante, Accra	Up to 1660 small, main export product was gold. Afterwards, increase in the slave trade which reached 75% of total export value by 1680.
Bight of Benin	Whydah, Ajashe, Lagos	Dahomey, Yoruba, Oyo, Edo, Allada, Ouidah	The Benin state generally restricted or prohibited the export of male slaves, in contrast to the other states
Niger Delta and Cameroons	Bonny, Brass, Kalabari	Igo, Efik, Awka, Aro	Supply less numerous but steady
Mozambique	Mozambique, Kilwa, Zanzibar		Major source of slaves to Brazil in the nineteenth century

Source: Curtin *et al.* (1978).

Table 1.9. *Chronology of prehistoric and historic East Asian development*

Chronology	Stages of development	Peoples/cultures/events	Places
230–500 kya	*Homo erectus*	Earliest specimens	Lower caves of Zhoukoudian, near Beijing
300 kya	*Homo sapiens*	Oldest modern specimens	Dali, Jinniushan and Maba
67 kya			Liujiang
30 kya		Human occupation	Near Tokyo
12.7 kya	Neolithic	First pottery (oldest in the world)	Japan
10 kya		Jomon culture	Several places in Japan
8.5 kya		Pre-Yang-Shao culture	Several places, China
7.8 kya		Classical Yang-Shao	Shaanxi, Hebei, Hunan
7.0 kya	Agriculture		Shandong province, China
6.0 kya		Ta-Pen-Keng cultures	South coastal region
4.8 kya		Lung-Shan period	China
		Xia dynasty	China
3.7 kya	Writing documents, urban life	Shang dynasty	Center in Honan, northern China
3.1 kya	Feudal system	Chou dynasty	Capital: Shensi
2.5 kya		Division in several warring states	Northern China
2.4 kya		Qin dynasty	
2.2 kya		Yayoi period	Japan
AD 222		Han dynasty	China
	Beginning of history	'Six dynasties' period	China
400		Kofun period	Japan
500		Sui dynasty	China
900		Sung dynasty	China
1200		Mongol conquest	China
1300		Ming dynasty	China
1600		Ching dynasty	China

Source: Bowles (1984); Cavalli-Sforza *et al.* (1994).

Table 1.10. *Chronology of prehistoric and historic East Indian development*

Chronology	Stages of development	Peoples/cultures/events	Places
5.6 kya	Neolithic	Pre-Harappan cultures	Amri, Kot-Diji
5.0 kya		Burzahom	Kashmir valley
4.0 kya	Urban life	Harappan civilization	Moenjodaro, Harappa
3.5 kya	Iron Age	Painted Gray Ware	Gangetic region
2.5 kya		At least 15 kingdoms	Several places
2.0 kya		Indo-Aryan arrival	Migration from Central Asia through the Iranian plateau
2.4 kya		Greek and Saka migrations	Several places
1.5 kya	Entrance into history	Mauryan empire	All India except the extreme south
AD 100–200		Kushana and Hun invasions	Several places
400		Gupta dynasty	Northern India
800		Arab invasion	Northern India
1000		Turk invasion	Northern India
1526		Moghal empire	Almost all India

Source: Cavalli-Sforza *et al.* (1994); Papiha (1996).

2   *Environment and history*

Everything has a history   *J.B.S. Haldane*

A convenient geographic subdivision

Latin America can be conveniently divided in two main regions: Middle America and South America. Figure 2.1. shows a map of the area as a whole, with its main geographic coordinates and political units. The largest country in Middle America is Spanish-speaking México. Along the continental region several other Spanish-speaking countries can be discerned, while among the Caribbean islands there lives a highly diversified population, with a variety of languages. The largest country in South America is Portuguese-speaking Brazil, of continental size, while to the north, west and south are nations whose main language is Spanish.

A highly diversified environment

The region extends widely, from about 32° north to 60° south, and from 120° to 20° west of Greenwich. Within this range Middle America occupies from about 8°N to 32°N, with South America continuing to the continent's southern limit. A brief description of the geography of Middle and South America follows.

Middle America

Geologically Middle America can be divided in three regions. The first represents the southern continuation of characteristics of the western United States. At about latitude 20°N the region is separated by one of the world's greatest volcanic axes, with notable volcanoes in México. The second is constituted by east–west mountain forms that extend from the Pacific Coast of Central America eastward through the islands of Cuba, Hispaniola and Puerto Rico, to the Virgin Islands. Several platforms of coral limestone such as the peninsulas of Yucatan and Florida, and the Bahama Islands, can be distinguished. The third presents two strings

23

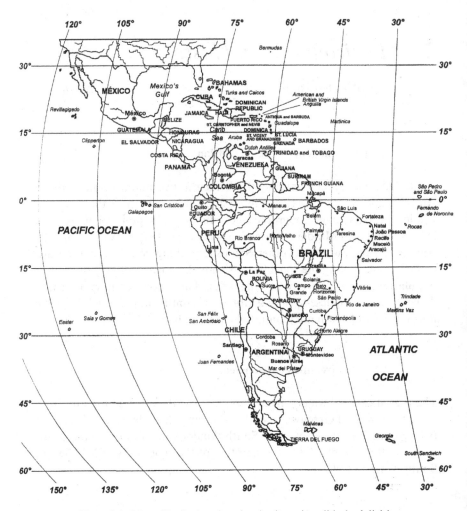

Figure 2.1. Map of Latin America, showing its main political subdivisions.

of volcanoes; one occurring from the lowland of Nicaragua to South America, and the other starting in the Virgin Islands and going from Trinidad to the southern continent.

Most of Middle America enjoys a tropical climate. The winds prevail from the east carrying a large amount of water vapor, thereby bringing heavy rains to the eastern sides of the Caribbean islands. Not uncommonly the region north of latitude 15° is ravaged by violent storms. On the Pacific coast there is high climate variability. In the Mexican northwest the climate is very dry, while in the south there is a typical monsoon, with summer rainfall and a winter dry season. A large part of the area consists of highlands, with mild temperatures, but the east of continental Central America is covered by tropical forests.

South America

The western side of the continent is dominated by the Andes mountains. To the east there are the great highland areas of the Guianas and the Brazilian coast. Three important river systems are present in between these highlands – in the north the Orinoco, along the equator the Amazon and, in the south, the Plata–Paraná–Paraguay system.

The greater part of South America has temperatures that are moderately high. In the equatorial and Atlantic coastal areas the climate is of the tropical rainy type supporting, in the Amazon, one of the largest tropical forests of the world. In Brazil's interior the climate is divided into a wet and dry season, while in the northeast severe droughts occur. The southern part of South America is in the middle latitudes, with the open ocean moderating the extremes of low and high temperatures. But stormy weather occurs in the southern portions of Argentina and Chile, although they seldom have much snow in the winter. A long narrow desert area crosses the continent diagonally; it starts at the Pacific coast near the border between Ecuador and Peru and the west coast is dry all the way to latitude 30°S in Chile, including the Atacama desert, one of the driest in the world. Also included in this complex are southern Peru and Bolivia, and northern Argentina. Patagonia is dry south to the Strait of Magellan.

Environment/human interaction

Today's most important challenge is to utilize this enormously varied environment without causing too much damage. This question has been variously considered by many specialists from Latin America and outside, for example, in the collective works edited by Hladik *et al.* (1993) and Nishizawa and Uitto (1995). All these authors emphasize the fragility of the tropical forests. Oldeman *et al.* (1993) asked 'Shall we eat the tropical forests or shall we eat from the tropical forest?' The dialectic relationships between Amazonia and the Brazilian northeast were also considered. In one we have plentiful water, while in the other there is an arid geography, with recurring droughts. Ideally, therefore, their environmental problems should be considered in a complementary fashion, avoiding Matsumoto's

(1995) prediction, which envisaged the woodland zone of the northeast with what would be 'Amazonia in the future after development!'

Power distribution: the two Americas

Independently of the uncertainties that existed at the time of the Great Navigations about the nature of the new lands, it was agreed that they should be explored, colonized and incorporated into the European kingdoms that had promoted the expeditions which led to their discoveries. Strengthened by the union the kingdoms of Castilla and Leon, and free from the Arab domination, Spain tried to obtain from the Vatican the property rights to the lands discovered by Columbus. At that time there were papal rights over regions occupied by Christians. Portugal, however, with its Christian tradition and history of fighting the unfaithful, and by virtue of also having financed maritime exploration and discovery well in advance of Spain, claimed its rights to some lands. After several rounds of negotiations, diplomatically, Pope Alexander VI divided the New World in two portions. In the Tordesillas treaty of 1494 everything that existed east of a meridian set 370 leagues west of the Cabo Verde archipelago would be Portuguese, while the other side would belong to Spain.

This decision met with opposition from the other European kingdoms, and was difficult to enforce due to the distance between Europe and the Americas, as well as the size of the territories under consideration. Therefore, numerous navigators of other nationalities, mainly British, French and Dutch, sponsored by their monarchs and financed by the emerging and increasingly rich mercantile bourgeoisie, sailed towards the new lands.

The Conquest

Spanish America (see Elliott, 1997a)

Two great arches of conquest swept from the Antilles to continental America. The first, from Cuba between 1516 and 1518, crossed México from 1519 to 1522, destroying the Aztec federation and then radiating outwards north and south from the Mexican Central Plateau. In 1524 the movement south had expanded from the present Guatemala and El Salvador, but it would still take another 20 years for the large Yucatán centers to fall under some type of Spanish control. The advance north was slower. Between 1529 and 1536 Nuño de Guzmán invaded México's north and west and established the kingdom of New Gallaecia; and only in 1562–1575 was a further region of the Mexican northwest, New Vizcaya, conquered (by Francisco de Ibarra).

The other arch, beginning in Panamá, extended for a brief period towards Nicaragua (1523–1524), and then took the Pacific route south for the conquest of the Inca Empire (1531–1533). From Peru the *conquistadores* turned north towards Quito (1534) and Bogotá (1536), where they met other groups who were going south along the coasts of Venezuela and Colombia. While an expedition under the command of Gonzalo Pizarro left Quito in 1541 to explore the Amazon basin, other groups went south in the direction of Chile, where Santiago was founded in 1542 by Pedro de Valdivia. The conquest of Chile, however, would be delayed by a long war with the Araucanian Indians. On the other side of the continent a European expedition tried without success to occupy a region on the La Plata river in 1535–1536, and afterwards left an advance post for the colonization of Paraguay. Buenos Aires was founded the first time in 1536, destroyed in 1541, and founded again in 1580, this time by groups from Asunción.

Some of the main events that took place in the two established vice-kingdoms of New Spain (México) and Peru are given in Tables 2.1 and 2.2, respectively. As can be noted, the conquest involved a network of alliances and fights, not only between Spaniards and the Indian empires, but within the two groups. In the end those better endowed technologically took control, although in the long-run the future political units would incorporate, in a complex way, values of both conquerors and conquered.

Portuguese America (see Johnson, 1997; Bueno, 1998b)

Brazil was discovered by Europe on 22 April 1500 by a fleet of 13 ships commanded by Pedro Alvares Cabral, a nobleman and member of Portugal's royal court. The first contacts with the Indians took place during the week the sailors spent in the region now known as Porto Seguro. Details of these first contacts were carefully documented in a letter from a member of the crew, Pero Vaz de Caminha, to the king Dom Manuel.

A second expedition, involving three ships under the command of Gonçalo Coelho, left Lisbon in May 1501. Several places were explored along the coast and the first specimens of pau-brasil (*Caesalpina echinata*), the tree that eventually served to name the country, was brought to Portugal.

A contract was then made between the court of Portugal and a commercial consortium, which financed at least two other expeditions to Brazil. However, the crown recovered the direct control of the commerce involving the new land, and in 1530 Dom João III decided that some type of a permanent colony should be established in Brazil. This was one of the objectives of Martim Afonso de Sousa's expedition (1532). His fleet of five ships transported 400 colonizers to the new lands, and he founded there the first sugar mill. He was also committed to the exploration of the area surrounding the Rio de la Plata, a task that he had difficulties in accomplishing. Other details about the first 30 years of Brazil's colonization are given in Bueno (1998*b*).

In 1534 Martim Afonso was still in Brazil and received news that the colony had been divided in 15 parts (Capitanias), whose administration had been given to 12 Lords, and that he had received two of them. These Lords had complete powers of administration and justice over their lands, although the commerce of pau-brasil remained a monopoly of the crown.

This decentralized system never functioned very well; contacts between Portugal and the Capitanias occurring more frequently than between the Capitanias themselves. Only 10 were colonized in the sixteenth century, and of these only two (São Vicente and Pernambuco) could be considered a success before 1550.

In contrast with Spanish America, the Portuguese did not have to fight against other empires in Brazil's colonization. The Indian groups had not attained such a degree of social complexity, and were much more dispersed through the territory. Episodes of confrontation were therefore much less common. On the other hand, Spanish and French expeditions to the Brazilian coast were dealt with mostly through diplomatic negotiations, which at least provisionally, prevented the danger of occupation by non-Portuguese colonizers.

Colonial period

Spanish America (see Osório, 1996; Elliott, 1997b; Brading, 1997)

To understand what happened in Spanish and Portuguese America, as well as in other places, during the Colonial Period, we should consider two factors: (a) the economic practices of the emerging European states, which gave rise to the so-called Colonial Mercantilist System; and (b) the contact and confrontation in the American continent between Iberian, Indian and African socioeconomic customs and structures.

The first point to be emphasized is the formation of *complementary* and *dependent* productive systems, which resulted in the supply of precious metals and tropical products to the Old World, which in turn promoted the production of industrial items and other products for export to the colonies. However, the monopoly enjoyed by the colonial states brought about a distorted and hypertrophied type of economy in the dependent colonies, atuned basically to the outside world, rather than their own needs.

To guarantee this control, a complex socioeconomic structure was established in Spanish America. The main events which determined this organization are listed in Table 2.3. In Spain the House of Contracts was responsible for all the dealings with the Colony, and this exchange was subjected to the rulings of a Council. In the New World, first two Vice-Kingdoms, and then two others were established. Additional controls were provided by governors, mayors, county councils, judges and other agents. Periodic visits by mandataries from the Crown also occurred.

The labor relationships were conditioned by, on one hand, the pre-Hispanic autocratic tradition, coupled with the Indians' collective use of the land; and on the other by the introduction of the European feudal system. The latter led to the institution of the *encomiendas*, which entitled chosen persons to force the performance of work or the supply of goods by the Indians. The *Leyes Nuevas* (New Laws), promulgated in 1542, prohibited this type of slavery, in part due to the frightening decrease of the Indian population that occurred after the Conquest. But a different system substituted it, the *repartimiento de indios*, that is, the compulsory performance of labor by those designated by their community. Labor was temporary (about 1 month in México; 1 year in Peru), and the workers would serve again only after a given period (7 years at the beginning; every 2 years at the Potosí mines of Peru). This form of compulsory work lasted until the eighteenth century in most Spanish colonies.

Mining was undoubtedly the main determinant of the type of economy developed in Spanish America. The difficult labor conditions prevailing at the mines led to the decimation of whole Indian populations (such as those of Santo Domingo, Puerto Rico and Cuba). But on the other hand it served as a promoter for the development of other economic sectors, due to the needs generated by it, and by the formation of centers which consumed agrarian products.

Progressively the Spaniards occupied the rural areas, introducing European plants and animals, and exploiting tropical products such as tobacco, cacao, and indigo. These products were already exploited at a commercial scale by the second half of the sixteenth century. The largest impact was produced by the introduction of cattle, which caused, in the most populous areas, the destruction of open plantations and/or their conversion to grassland. On the other hand, cattle raising made possible the colonization of previously uninhabited land.

The result was the appearance of the *haciendas*, large self-sufficient rural properties founded on the work of a small number or poor, enslaved servants, who would furnish meat, leather, and a small surplus of cultivated plants. For Spanish America as a whole this institution appears in the sixteenth century, acquired importance in the seventeenth century, and dominated the landscape during the last 100 years of the Colonial Period.

African slavery was present in Spanish America from the European arrival, but it was never as important as in Portuguese America and the Caribbean. Its main influence occurred in the eighteenth century, in the Carib coasts of New Granade and Venezuela (with the cacao cultivation), in the littoral of Ecuador and Peru, as well as in Cuba (in the latter case, associated with the system of sugar plantations).

The relationships between colonizers and colonized were never completely peaceful. The reforms introduced by José de Galvez (who became Secretary of the Council of the Indies in the period 1776–87), clearly favoring Spaniards from those born in the Colony, were received with discontent and unrest. The tobacco monopoly, expulsion of the Jesuits (1767) and other unpopular policies generated rebellions, one of the most important being that which occurred in south Peru, headed by a local chief named José Gabriel Condorcanqui who adopted the name of Tupae Amaru, the last Inca emperor. The movement extended from Cuzco to La Paz, and was controlled with difficulty. On the other hand, the local elite had developed sufficient power in terms of economic resources and military conditions to question the supremacy of the Empire. Therefore, with the occupation of Spain by Napoleon Bonaparte, the abdication of Carlos IV and the coronation of Joseph Bonaparte in 1808, the time was ripe to demand independence.

Portuguese America (see Wehling and Wehling, 1994; Costa, 1998)

As in Spanish America, the colonial economic history of Portuguese America can be understood as resting on three main factors: external dependence, latifundium, and slavery. All foreign commerce was controlled by Portugal, which in addition would impose heavy taxation on all products. Several commodities were the exclusive right of the Crown. The main monopolies were those of pau-brasil (abolished in 1823), whaling (1603–1798), tobacco (1642–1820), and salt (1658–1801). The latifundium involved mainly the sugar plantations of the northeast and cattle raising in many regions, especially in the south. The latter was connected with the territorial expansion into the interior. Slaves were mainly African and imported in millions during this period and the nineteenth century. Indian slavery was also employed, especially at times when the African traffic was interrupted or diminished. Its occurrence was especially marked in São Vicente (at the littoral of the present state of São Paulo) and Rio de Janeiro in the sixteenth and seventeenth centuries, as well as in Maranhão in the seventeenth and eighteenth centuries.

The main events in Portuguese America during the colonial period are summarized in Table 2.4. During the sixteenth century the littoral was occupied, and the main economic activities were of the extractive type. Several general governors tried to administer the colony with varied success. The French occupied the region around Rio de Janeiro for a period of 15 years, but were finally ousted in 1557. The main characteristic of the seventeenth century was geographical expansion, the main group responsible for penetration of the interior was the *bandeirantes*; these were adventurers financed from different sources who enslaved Indians, and searched for gold and diamonds. A further French invasion occurred, but in the north, and was of shorter duration. On the other hand, the Dutch stayed for 30 years in Pernambuco, in the northeast, and Spanish Jesuits tried twice to establish a kind of Indian Republic in the south. Also of note was the Palmares African-derived community, which resisted Portuguese attacks for almost a century. The seventeenth century closes with the discovery of large gold reserves in central Brazil, which affected all economic life for the following century. This was the time of the consolidation of the new frontiers, which more than doubled the territory originally assigned to Portugal by the Tordesillas treaty. On 13 January 1750, the Madrid treaty consecrated the *uti possidetis* principle, recognizing the Portuguese expansion.

As in Spanish America, the political situation in Europe led to conditions at the beginning of the nineteenth century for Brazil's independence, which occurred in 1822.

British, Dutch and French possessions (Camargo, 1998)

The three other naval powers of the fifteenth and sixteenth centuries also obtained their share of what is now Latin America. The regions conquered included the Caribbean area, north South America and islands located at the extreme south of the Atlantic Ocean.

The British colonization occurred from the sixteenth to the nineteenth centuries and involved most of the Caribbean and West Indies islands, Belize (formerly British Honduras), Guiana (former British Guiana), and in the far south the islands of Falklands, South Georgia and South Sandwich. English ships were almost everywhere in the New World by the end of the sixteenth century. But it was in the seventeenth century that the colonization of North America took place, and this served as a convenient base for the conquest, in this same century, of the area surrounded by the Carib Sea. The southern islands were occupied in the eighteenth century, while the present Guiana only became a British dominion by 1815.

The French colonization included the present Haiti, Guadeloupe and Martinique in Middle America, and French Guiana in northern South America. They also used their base in North America, which they had settled at the beginning of the seventeenth century, to extend their power to Middle and South America. French Guiana, however, was only confirmed as a French territory in 1817.

The Dutch possessions comprised what is now Aruba and Netherlands Antilles in the Carib Sea, plus Surinam, where the Dutch arrived at the end of the sixteenth century. British colonizers, however, also started to arrive there in 1630. In 1667 England traded its territory here with the Netherlands to obtain the city of New Amsterdam (present-day New York!) The country became officially Dutch in 1815.

The history of these territories (with the exception of the southern islands) was basically the same, independent of the colonizer. The region had been previously occupied by Arawak and Warao Indians, who ultimately were decimated or displaced to other areas. Massive importation of African slaves then occurred, to work in sugar and other types of tropical plantations. Significant numbers of East Indian migrants came to Trinidad and Tobago, Guiana and Surinam.

Some of these places still remain as territories of other countries, while others acquired independence only recently (for instance, St Kitts and Nevis in 1983). But Haiti was the first of the Latin American nations to proclaim its independence, in 1804.

The road to independence

Two main events occurred outside Latin America which decisively influenced the beginning of the independence process in the continent. The first was the French revolution of 1789–1799, with it's libertarian message to all peoples of the world. The second was a consequence of the reaction against the French republican regime, which led to Napoleon Bonaparte's seizing power in France and his campaigns to conquer other European nations, in particular Spain and Portugal. The consequent weakening of the Iberian empires was the starting signal for the colonies to launch the independence process.

However, conditions prevalent at the time in the different regions of Latin America were very diverse and therefore, the impetus of the process was also variable (its chronology is presented in Table 2.5). As was already mentioned, the people of Sainte Domingue were the first to proclaim independence, adopting the native name of Haiti. Within 30 years, between 1811 and 1841, 17 nations in Middle and South America declared independence. Another, more recent, decisive period was from 1962 to 1983, in which 13 nations in the Caribbean area and in northern South America obtained their autonomy. In between we have the special cases of Cuba and Panamá, which attained independence respectively in 1902 and 1903.

It is outside the scope of this book to give a detailed description of how these 33 nations obtained their independence. Some of the main events related to this process in some of the Middle and South American countries, as well as an indication of the people who mainly influenced them, are given in Tables 2.6 to 2.9.

Starting with Middle America, the main successes which led to the independence of México, Guatemala, Cuba, Haiti, the Dominican Republic and Costa Rica are listed in Table 2.6. While the Dominican Republic's declaration of independence occurred without a fight in 1821, the following year the country was occupied by Haitian troops, and the end of Haitian control occurred only 22 years afterwards. On the other side of the island, the main names to be mentioned in the rebellion against France are Toussaint Louverture, Jean Jacques Dessalines, Alexandre Pétion and Henri Christophe. In 1804 Dessalines was made Perpetual Governor, but 2 years later he was assassinated and a civil war started, which separated the nation into northern and southern parts. Reunification occurred only some years afterwards, under the leadership of Jean Pierre Boyer.

The proclamation of independence of Guatemala and México occurred almost simultaneously, the first on 15 September and the second on 28 September 1821. But a long history of struggle, which started at the beginning of the century, was necessary to reach this result. Miguel Hidalgo, José Inácio Rayón and José Maria Morelos should be remembered in these fights for Mexican independence.

Especially violent was Cuba's road to autonomy, with the 10 Year's War (1868–78), José Marti's death in 1895, and the general unrest that culminated with the United States' military intervention in 1898, which lasted for 4 years.

Costa Rica's independence was closely related to events in Central America as a whole. First becoming a member of the Mexican Empire (1821), the country afterwards participated in the Central American Federal Republic (1823–40). In 1821 Costa Rica was not so much a country, but rather a territory ruled by four towns (Cartago, Heredia, San José and Alajuela). Conflicts over which of them should take the leadership, as well as over the controversial leadership of Braulio Carrillo (1838–42) occurred. Finally San José emerged as the capital of the State, and then of the Republic of Costa Rica, proclaimed in 1848.

In terms of South America, while several names could be cited as heroes of the independence, the leading figure was undoubtedly Simon Bolivar. The main events of his life are presented in Table 2.7. He played a decisive role in the independence of Venezuela, Colombia, Ecuador, Peru and Bolivia, but his dreams of a continental-size Spanish-speaking nation never materialized. His last days, tormented by sickness, and with very few friends, were dramatically described by Márquez (1989).

The main events in the independence of southern South America (Paraguay, Argentina, Uruguay and Chile), are listed in Table 2.8. The independence of Argentina and Chile can be considered together, the decisive event being the battles of Chacabuco and Maipu in Chile, won by the Argentinian José de San Martin, who also participated in the independence of Peru. Another name to be remembered in Chile is Bernardo O'Higgins, head of the first Chilean government. Uruguay's independence occurred later (1828), in part due to the Argentinian and Brazilian interests in its territory. José Artigas and Juan Antonio Lavalleja were two of the key people who made the autonomy possible.

The events in Brazil (Table 2.9) were different in many ways from those that occurred in other Latin American countries. The process started with the transfer of the Portuguese royal family to Brazil, escaping from Bonaparte's invasion of Portugal. The change of the center of power from Europe to South America was accompanied by a series of developments favorable to Brazil, such as the opening of the ports to all friendly nations and the establishment of all the institutions needed for the proper functioning of the Empire. Schools of medicine were founded in Bahia and Rio de Janeiro (1808), as well as the Royal Printing House (1808), and Astronomical Observatory (1809), the Royal Military Academy, and the first Public Library (1810). The commercial relationships with England were strengthened, and in 1815 Brazil reached the status of vice-kingdom, together with Portugal and Algarves.

The Regent Dom João was solemnly crowned as Dom João VI at Rio de Janeiro in 1818, and 3 years later returned to Portugal. It is generally known that the king counseled his son Pedro de Alcântara that 'If Brazil separates from Portugal, you should put the crown in your head, before an adventurer takes the power'. Next year, pressed by the Court to return to Portugal, Dom Pedro stayed, and in response to other abusive determinations declared Brazil's independence in September 7, 1822. He was crowned emperor as Dom Pedro I on 1 December, and ruled until 1831 when, due to many political problems, he was forced to abdicate in favor of his son, and returned to Portugal.

Further developments

Many peculiarities related to different regions and nations could be presented here, but they would cover much more space than we have available. Therefore, we will just give a brief general outline of what happened with the Latin American nations after the struggles for independence, up to the present time. Since what happened in most cases was a revolution from the top, political instability occurred afterwards and is still present in many areas. To a large extent the unrest is a consequence of economic and social problems. As in colonial times, most of the now independent Latin American countries organized their economies with a view to the outside, rather than considering internal demands. Chronic structural problems also remained such as urban and rural poverty, conditions exacerbated by the oligarchic control of the land. Practically all these nations can be classified as belonging to the Third World (or worse!), and their interchanges with the First World were and are clearly characterized as being of an unequal nature. Previous loans and investments have led to enormous foreign debts, and the sheer size of the payments of interest seriously affects their budgets, preventing measures of a social nature, or new investment. Chronic cycles of dictatorships and more liberal regimes are the rule, sometimes influenced or determined from the outside. The present globalization process has not improved the situation, and on the contrary may act towards the indefinite prolongation of this unfortunate state of affairs, in which semi-feudal conditions are developing together with an early, savage form of capitalism. The socialist regime of Cuba could be viewed as an exception; but the smallness of the country's size and resources, coupled with a severe international boycott, has led to difficult times there. More details about these questions can be found in Wasserman (1996b), and Wasserman and Guazzelli (1996).

Present situation

Middle America

Tables 2.10 and 2.11 present selected information about the present Middle American countries and territories. Their total area amounts to 3.4 million km², in which a population of 164.74 million lives. They are distributed among 21 nations and 12 territories. Although ethnic classification varies among these units, in rough figures we can verify that more than half of the population can be classified as admixed. Amerindians occupy a second position (about 21%, with large numbers especially in México and Guatemala), while approximately 13% are classified as white, and a similar number as black. Descendants of East Indian migrants are especially frequent in Trinidad and Tobago, but their total prevalence is less than 1% in the total area. A similar frequency is observed from people who came from other Asian countries, especially China.

In Middle America the official language is most frequently English (19 nations or territories), followed by Spanish (10), French (3), Dutch (2) and Creole (2). Dominica, Haiti and Puerto Rico have adopted two of these languages as official ones. As for systems of government, monarchic or republican parliamentarianism is the most prevalent (present in 17 and 4 nations or territories, respectively), while the presidentialist system occurs in 10. Haiti is under a military regime, while Cuba is a socialist republic.

South America

Similar data pertaining to South America can be examined in Tables 2.12 and 2.13. Its total area (17.85 million km²) is five times larger than that of Middle America, and its total population (321.11 million) about twice as high. Twelve nations and three territories occur in the region. About an equal proportion (42–43%) of white and admixed subjects live there, blacks and Amerindians being less well represented (around 6% each). East Indian descendants occur mainly in Guiana and Surinam, the latter also including appreciable numbers of people of Javanese and Chinese origin. Descendants of Japanese migrants appear mainly in Brazil.

People from nations with just an official language speak Spanish (six countries), English (three nations or territories), Portuguese (Brazil), Dutch (Surinam) or French (French Guiana). Paraguay adopted two official languages (Spanish and Guarani), while Spanish, Quechua and Aymara are official languages in Bolivia and Peru. A presidential form of government exists in 10 countries, while parliamentary government (either republican or monarchic) obtains in five nations or territories.

An interpretative synthesis

Society formation and problems of development

Sociocultural development can operate in two distinct ways (Ribeiro, 1970). In one situation the people affected are the *agents* of their own development. The process then can be labeled as an *evolutionary acceleration*. Societies that have mastered a new technology are able to preserve their ethnic–cultural character while advancing socially. Conversely, when the people affected are the *recipients* of the cultural innovation, the process is *historical incorporation*. Societies in this situation suffer the impact of technologically more developed nations and are subjugated by them, losing their autonomy and sometimes having their ethnic character damaged or destroyed.

In Latin America, the historical configuration can be viewed as basically resulting from the second type of process. The former colonial domination and the deliberate population transfer that occurred over the years shaped the new societies in different ways. Obviously, after the European contact, the process was not one of mutual interchange between autonomous cultural entities. The populations of the continent, therefore, suffered an alienation process from which they are only now beginning to recover. Counteracting their efforts for a distinctive way of life is the present globalization process, with imposing policies which generally favor First World nations only.

Classification

The peoples of Latin America can be classified, according to Ribeiro (1970, 1977), as indicated in Table 2.14. Present-day societies could be classified as Witness, New or Transplanted Peoples, the fourth category, Emergent, being absent in the region. These categories reflect the peculiarities of the process that took place along the continent, giving rise to distinct biological and cultural units. They will be characterized in more detail in the next chapters.

Table 2.1. *Discovery and conquest of México*

Chronology	Event
1511	Occupation and colonization of Cuba under the command of Diego Velázquez
1517	First expedition towards México, under the direction of Hernández de Córdoba.
1518	Second expedition, commanded by Juan de Grijalva
1519	Third expedition, under the command of Hernán Cortés. 25 March: Battle of Centla; 21 April: arrival in Vera Cruz and negotiations with Montezuma, the Emperor of Tenochtitlán; July: rupture with Velázquez; August: Alliance with Cempoala Indians; fight and alliance with the Tlaxcaltecans; march towards the Mexican capital, where the Spaniards arrive on 8 November
1520	Cortés leaves México to fight Pánfilo de Narvaez. In his absence Pedro de Alvarado orders the massacre of a large part of the Mexican nobility (23 May). In the confrontation that follows Montezuma dies. Cortés and his allies are forced to leave the city in the *Noche Triste* (30 June)
1521	May: Siege of México; 13 August: the capital of Tenochtitlán surrenders to the Spaniards and their allies
1521–3	Conquest of the northeast. The Huastecans present a strong resistance to Cortés
1522	Submission of the Mixtecs of Tututepec
1524	Departure of Pedro de Alvarado to conquer Guatemala, and of Cristóbal de Olid to Honduras
1525	Execution of Cuauhtémoc, México's last sovereign. Huastecan rebellion
1529	Occupation of Michoacán by Nuño de Guzmán

Source: Bernand and Gruzinski (1997).

Table 2.2. *Discovery and conquest of Peru*

Chronology	Event
1532	Battle of Cajamarca and Inca Atahualpa's arrest by Francisco Pizarro
1533	Atahualpa's execution (29 August)
1534	March: Cuzco's foundation. Manco Inca is proclaimed emperor in accordance with the Spaniards
1535	Diego de Almagro leaves for Chile. Foundation of Lima (18 January)
1536	Siege of Cuzco by Manco Inca, who rebels against the Spaniards
1537	Paullu Inca is crowned in Cuzco by Almagro, who claims the city's government
1538	Las Salinas battle. Hernando and Gonzalo Pizarro defeat Almagro, who is killed
1539	Paullu Inca's alliance to the Pizarro brothers and confirmation in his post
1541	Assassination of Francisco Pizarro and his half-brother Francisco Martín de Alcántara
1542	Battle of Chupas with the victory of governor Cristóbal Vaca de Castro
1544	Gonzalo Pizarro is proclaimed Peru's governor and general captain. Manco's death
1544–8	Gonzalo Pizarro decides to sever connections with Spain
1546	Añaquito's battle, with the defeat of the vice-king Blasco Nuñez de Vela
1548	Xaquixaguana's battle. Defeat and death of Gonzalo Pizarro
1554	End of the civil wars

Source: Bernand and Gruzinski (1997).

Table 2.3. *Establishment of the socioeconomic structure of colonial Spanish America*

Chronology	Event
1503	Foundation of the Casa de Contratación de las Indias (House of Contracts for the Indies)
1510	Institution of governors as representatives of the Crown. Recognition of conquerors as governors (for instance, Vasco Nuñez de Balboa was designated governor of Darien in this year)
1524	Foundation of the Real y Supremo Consejo de las Indias (Royal Council for the Indies)
1535	Creation of the vice-kingdom of Nueva España (New Spain), with its capital in México City. First vice-king: Don Antonio de Mendoza (1535–49)
1543	Creation of the vice-kingdom of Peru, with its capital in Lima (first vice-king: Blasco Nuñez Vela)
1717	Creation of the vice-kingdom of Nueva Granada (New Grenade), with its capital in Santa Fé de Bogotá
1776	Creation of the vice-kingdom of Rio de la Plata, with its capital in Buenos Aires
Sixteenth to eighteenth centuries	Audiencias (audiences), organized to supervise the compliance of the royal dispositions. Established in New Spain: Santo Domingo (1511); México (1527); Guatemala (1543); Guadalajara (1548). Vice-kingdom of Peru: Panamá (1538); Lima (1543); Santa Fé de Bogotá (1548); Charcas (1559); Quito (1563); Chile (1563). At the local level, alcaldes (mayors) and cabildos (councils) were designated either by the Crown or by the vice-kings. The corregidores de indios would also act as agents of the Crown in the Indians' dealings with their encomenderos, who, by royal designation, could demand work and goods of the Indians. General visits (of which those headed by José de Gálvez in New Spain (1765), Peru (1776) and New Grenade (1778) were the most famous) were also held to verify that the regulations were being followed)

Source: Osorio (1996); Elliott (1997b); Brading (1997).

Table 2.4. *Main events which occurred in Portuguese America during the colonial period*

Chronology	Event
Sixteenth century	Occupation of the littoral, extractive economy, 100 000 people
1549–53	First governor, Tomé de Souza. Foundation of Salvador, fights with the Tupinambá
1553–7	Second governor, Duarte da Costa. French invasion of Rio de Janeiro (1555). Foundation of Guairá, by the Spaniards, in the south (1557)
1557–72	Third governor, Mem de Sá. Foundation of Rio de Janeiro (1565). French defeat (1567)
1571–8	Two separate administrations, from Salvador (Luis de Brito) and Rio de Janeiro (Antonio Salema)
1578–83	Governor Lourenço da Veiga. Colonization of the northeast
1583–7	Governor Manuel Teles Barreto
1587–91	Governing councils
1591–1602	Governor Francisco de Sousa
Seventeenth century	Geographical expansion (cycle of the *bandeirantes*), development of the cattle and sugar economy, 350 000 people
1612–15	The French in Maranhão. Foundation of São Luis
1616	Foundation of Belém
1624–54	The Dutch in Bahia and Pernambuco
1626–9	Jesuit missions in the south and their destruction
1682	Return of the Jesuits to southern Brazil
1600–94	The Palmares African-derived *quilombo*
1695	First signs of large gold reservoirs in Central Brazil
Eighteenth century	Consolidation of the new frontiers (Madrid Treaty, 1750), gold and diamond mining, 3.3 million people
1700–21	Colonization of the center of the country. Minas Gerais receives administrative autonomy
1737	Foundation of Rio Grande
1740–53	Colonization of the south
1756	The Jesuits are defeated at the Seven Peoples Mission
1763–1801	Spanish intervention in southern Brazil
1763	Transfer of the Brazilian capital from Salvador to Rio de Janeiro

Source: Wehling and Wehling (1994); Costa (1998).

Table 2.5. *Chronology of the achievement of independence by the Latin American countries*

Year	Areas and countries
Middle America	
Spanish America	
1821	Dominican Republic, Guatemala, México
1838	Honduras, Nicaragua
1841	El Salvador
1848	Costa Rica
1902	Cuba
1903	Panamá
The Caribbean	
1804	Haiti
1962	Barbados, Jamaica, Trinidad and Tobago
1973	Bahamas
1974	Grenada
1978	Dominica
1979	St Lucia, St Vincent and Grenadines
1981	Antigua and Barbuda, Belize
1983	St Kitts and Nevis
South America	
Spanish America	
1811	Paraguay, Venezuela
1816	Argentina
1818	Chile
1819	Colombia
1820	Peru
1825	Bolivia, Uruguay
1830	Ecuador
Others	
1822	Brazil
1966	Guiana
1975	Surinam

Source: Camargo (1998).

Table 2.6. *Main events related to the independence of some of Middle America's countries*

Year	Events
México	
1808	The vice-king, Don José de Iturrigaray, and the Royal Council, decided to challenge France's occupation of Spain
1809	Beginning of the war of independence. Miguel Hidalgo fights against the Spanish authorities, is defeated and executed in 1811
1811	José Inácio Rayón and José Maria Morelos maintain the fight for independence
1814	Re-establishment of Spanish power. The vice-king is Felix Maria Callejas
1821	Proclamation of independence
1822	Agustin Iturbide is proclaimed emperor
Guatemala	
1811–17	Rebellions against Spanish power, strongly opposed by the governor José Bustamante y Guerra
1817	Carlos de Urrutia y Montoya is named as new governor
1821	General Assembly and declaration of independence, written by José Cecilio del Valle. Gabino Gainza is maintained as leader of the nation
Cuba	
1868–78	Ten Years War of Independence
1880–95	Several attempts to overthrow Spanish rule. José Martí, one of the most important Cuban leaders, dies in battle on 19 May 1895
1895–8	General unrest and military action. United States intervention
1902	End of US military occupation
Haiti	
1800	Toussaint Louverture defeats André Rigaud, as well as the Spaniards from the Dominican Republic, proclaiming the island's autonomy within France
1801	An army of 20 000 soldiers is sent to the island to confirm French control. Toussaint Louverture is defeated and sent to France, where he dies in 1802
1802–3	Fights for independence. Main leaders: Jean Jacques Dessalines, Alexandre Pétion and Henri Christophe
1804	On 1 January the Saint-Domingue colony proclaims its independence taking the name of Haiti. Jean Jacques Dessalines is named Perpetual Governor and in October proclaims himself emperor
Dominican Republic	
1821	Proclamation of independence without resistance. José Núñez de Cáceres is designated head of the state
1822	Occupation by Haiti
1844	End of Haitian control. Proclamation of the Dominican Republic by Francisco del Rosario Sánchez and Ramón Mella. Juan Pablo Duarte returns to the country from exile and helps in the consolidation of the new nation

Table 2.6. (*cont.*)

Year	Events
Costa Rica	
1821	Independence of Central America
1823	Battle of Ochomogo; San José becomes capital of Costa Rica
1824	Founding of the Central American Federal Republic
1835	War of the League, won by San José
1838–42	Dictatorship of Braulio Carrillo
1848	Declaration of Republic of Costa Rica

Source: Zavalla (1964); Hall (1964); Carbonell (1964); Ureña (1964); Molina and Palmer (1998).

Table 2.7. *Main events in Simon Bolivar's life*

Year	Events
1783	Birth on July 24
1799	January 19, travels to Spain
1802	May 26, marriage with Maria Teresa Rodríguez del Toro.
	July 12, returns to Venezuela
1803	January 22, Maria Teresa dies in Caracas.
	October 23, returns to Spain
1810	April 18, confined in his hacienda, he does not participate in the first events related to the Venezuelan independence.
	June 9, diplomatic mission in London.
	December 5, returns to Venezuela
1811	March 2, First Venezuelan Congress.
	July 5, Venezuela's proclamation of independence.
	July 23, Bolivar's first experience of war in Valencia, under the command of Francisco de Miranda
1812	March 26, Caracas earthquake.
	July 6, Puerto Cabello's fortress, under his command, is lost due to a betrayal.
	September 1, arrival in Curaçao in his first exile.
	December 24, beginning of Magdalena's river campaign
1813	February 28, Cúcuta's battle.
	August 6, triumphant entry into Caracas.
	October 14, acclaimed as Captain-General and Liberator
1814	December 12, after several battles establishes the government in Bogotá
1815	May 10, voluntary exile in Jamaica.
	December 24, arrival in Haiti, where he establishes a friendship with president Alexandre Pétion
1816	March 31, Los Cayos' expedition for a new campaign for Venezuela's liberation
1819	February 15, he is elected Venezuela's president and starts the liberation campaign of Nova Granada.
	August 7, Boyacá's battle.

Table 2.7. (*cont.*)

Year	Events
	December 17, creation of Colombia's republic. Election as its president
1821	June 27, Carabobo's battle, assuring Venezuela's independence
1822	May 24, Pichincha's battle.
	June 16, triumphant entry into Quito, together with José Antonio Sucre. Meets Manuela Sánchez, who will accompany him for many years.
	July 11, arrival in Guayaquil and Ecuador's incorporation to Colombia.
	July 26–27, interview with José de San Martin
1823	March 17, an army of 3000 soldiers is sent by Bolivar to fight for Peru's independence.
	May 14, the Congress of Peru calls him to end the Peruvian civil war.
	September 1, arrival in Lima
1824	February 10, he is named Peru's dictator.
	August 6, Junin's battle.
	December 7, Lima's liberation.
	December 9, victory of José Antonio Sucre in Ayacucho
1825	February 18, Peru's Congress does not accept has abdication of the dictatorship.
	August 6, announcement of the creation of the Republic of Bolivia, liberated by José Antonio Sucre
1827	January 1, confirms José Antonio Páez as Venezuela's president.
	January 5, new request for abdication of Colombia's presidency.
	March 16, severs relationship with Francisco de Paula Santander.
	July 5, Colombia's Congress refuses to allow his resignation of the presidency and demands his presence to take the presidential oath.
	September 10, arrival in Bogotá; assumes the presidency against strong opposition
1828	September 25, attempt to assassinate him
1829	January 1, travels to Ecuador due to conflicts with Peru.
	October 29, returns to Bogotá
1830	March 1, leaves power to Domingo Caycedo, president of the government council and leaves for Fucha.
	May 8, leaves Bogotá for Cartagena, with the objective of leaving the country.
	June 4, José Antonio Sucre is murdered in Berruecos.
	Bolivar receives the news on July 1 and is deeply upset.
	December 1, arrives, gravely ill, at Santa Marta.
	December 17, dies in the San Pedro Alejandrino quinta, with very few friends in attendance

Source: V.R. Martínez, in Márquez (1989).

Table 2.8. *Main events related to the independence of southern South American countries*

Year	Events
Paraguay	
1810	Due to the events in Spain, the governor Bernardo de Velasco y Huidobro called the convocation of a junta, which decided to remain loyal to Fernando VII, but voted for autonomy in relation to the Buenos Aires junta
1811	Battle in Cerro Mbaé, later called Cerro Porteño, in which the Argentinian army was defeated. In Asunción, deposition of the governor. Power was taken by a junta, headed by Fulgencio Yegros. Other important leaders were Pedro Juan Caballero and José Gaspar de Francia. Treaty with Buenos Aires in which the Paraguayan autonomy was recognized
1813	Convocation of a congress and elaboration of the first constitution. Francia and Yegros were made consuls
1814	Francia becomes dictator
Argentina	
1810	May 25 revolution. Deposition of the vice-king, Baltazar Hidalgo de Cisneros. Unsuccessful attempt at a counter-revolution. Assumption of power by Cornélio de Saavedra. Another important leader was Mariano Moreno
1810–20	General unrest and fighting
1816	Declaration of independence of the Provincias Unidas de América del Sur
1817–18	Battles of Chacabuco and Maipu (Chile), won by José de San Martin.
1819	Alliance with Chile
Uruguay	
1811	Asencio declaration against the Spaniards. Leaders: Venancio Benavides and Pedro Vieira. Several battles and a Pacification Treaty. José Artigas is compelled to leave Uruguay, and takes with him thousands of civilians, an episode known as the exodus of the Oriental people
1814	The Argentinian army defeats the Spaniards and takes Montevideo
1817	Portuguese invasion and conquest of Montevideo. Head of state is Carlos Frederico Lecor
1820	Artigas's defeat in Tacuarembó. Sometime afterwards he takes refuge in Paraguay, where he dies 30 years later
1825	Juan Antonio Lavalleja and 32 other patriots disembark in La Agraciada and start Uruguay's liberation. Incorporation of the nation into the Rio de la Plata United Provinces
1828	Peace treaty between Argentina and Brazil, guaranteeing Uruguay's independence
Chile	
1810	Formation of a junta, presided over by Mateus de Toro y Zambrano
1811–13	Fights with the Spaniards. Bernardo O'Higgins is one of the Chilean leaders
1814	Lircay's treaty, recognizing Spanish domination
1817–18	Battle of Chacabuco and Maipu, won by José de San Martin. Bernardo O'Higgins is named as head of the Chilean government. Proclamation of Chile's independence
1823	Bernardo O'Higgins abdicates and leaves for Peru, where he dies in 1842

Source: Benítez (1964); Levene (1964); Heras (1964*a,b*).

Table 2.9. *Main events related to Brazil's independence*

Year	Events
1808	Transference of the Portuguese royal family to Brazil, accompanied by around 15000 others. The family's head, Dom João, has been regent since 1792 due to the illness of Dona Maria I, The Mad. Opening of ports to all friendly nations
1809	Occupation of French Guiana in retaliation to Bonaparte's invasion of Portugal
1810	Commercial treaties with England
1815	Proclamation of the United Kingdom of Portugal, Brazil and Algarves
1817	Revolution in Pernambuco, promptly suppressed by the central government
1818	Coronation of Dom João VI
1820	Constitutional revolution in Portugal
1821	Dom João VI leaves for Portugal
1822	The Portuguese court demands the return of Dom João VI's son, Dom Pedro de Alcântara, to Portugal. Counseled by José Bonifácio de Andrada e Silva and others he decides to stay in Brazil (January 9, 'Fico', or I stay), declaring the country's independence on September 7. He is crowned emperor as Dom Pedro I on December 1
1823	War of independence: Bahia, Maranhão, Pará, Cisplatina. Convocation and dissolution of the Constituent Assembly
1824	A constitution is imposed by force. New rebellion in Pernambuco (Equator Confederation), also suppressed by the government
1831	Abdication of Dom Pedro I in favor of his son, Dom Pedro II, and his return to Portugal

Source: Calmon (1964); Bueno (1998a).

Table 2.10. *Selected information about the present Middle American countries and territories*

Country/territory	Area (million km²)	Population		Ethnic distribution (%)	Official language	System of government
		Year	Number (millions)			
Nations						
Antigua and Barbuda	0.0004	1996	0.06	Black: 91 Mixed: 4 White: 2 Other: 3	English	Monarchic, parliamentary
Bahamas	0.01	1997	0.29	Black: 85 White: 12 Other: 3	English	Monarchic, parliamentary
Barbados	0.0004	1997	0.26	Black: 80 Mixed: 16 White: 4	English	Monarchic, parliamentary
Belize	0.02	1997	0.20	Black: 44 Creole: 30 Amer.: 18 White: 4 East Ind.: 4	English	Monarchic, parliamentary
Costa Rica	0.05	1997	3.60	White: 87 Mestizo: 8 Black: 2 Asian: 2 Amer.: 1	Spanish	Republic, presidential
Cuba	0.11	1997	11.10	Black: 62 White: 37 Asian: 1	Spanish	Socialist republic
Dominica	0.0007	1996	0.07	Black: 91 Mestizo: 6 Amer.: 2 Other: 1	English Creole	Republic, parliamentary

Table 2.10. (*cont.*)

Country/territory	Area (million km²)	Population		Ethnic distribution (%)	Official language	System of government
		Year	Number (millions)			
Dominican Republic	0.05	1997	8.10	Euro-Afr.: 74 White: 15 Black: 11	Spanish	Republic, presidential
El Salvador	0.02	1997	5.90	Euro-Amer.: 94 Amer.: 5 White: 1	Spanish	Republic, presidential
Grenada	0.0003	1996	0.10	Black: 82 Mestizo: 13 East Ind.: 3 White: 2	English	Monarchic, parliamentary
Guatemala	0.11	1997	11.20	Mestizo: 56 Amer.: 41 Asian: 3	Spanish	Republic, presidential
Haiti	0.03	1997	7.40	Black: 96 White: 3 Mixed: 1	French Creole	Military regime
Honduras	0.11	1997	6.00	Mestizo: 90 Amer.: 7 Black: 2 White: 1	Spanish	Republic, presidential
Jamaica	0.01	1997	2.50	Black: 75 Mixed: 13 East Ind.: 1 Other: 11	English	Monarchic, parliamentary
México	1.97	1992	94.30	Mestizo: 60 Amer.: 30 White: 9 Other: 1	Spanish	Republic, presidential

Nicaragua	0.13	1997	4.40	Mestizo: 69 / White: 17 / Black: 9 / Amer.: 5	Spanish	Republic, presidential
Panama	0.76	1997	2.70	Mestizo: 70 / Amer.: 20 / White: 10	Spanish	Republic, presidential
St Kitts and Nevis	0.0003	1996	0.04	Black: 97 / White: 3	English	Monarchic, parliamentary
St Lucia	0.0006	1996	0.14	Black: 96 / East Ind.: 3 / White: 1	English	Monarchic, parliamentary
St Vincent and Grenadines	0.0004	1996	0.11	Black: 96 / Amer.: 2 / White: 2	English	Monarchic, parliamentary
Trinidad and Tobago	0.005	1997	1.30	Black: 57 / East Ind.: 41 / White: 1 / Other: 1	English	Republican, parliamentary
Territories						
Anguilla	0.0001	1994	0.009	Black: 86 / Mixed: 11 / White: 3	English	Monarchic, parliamentary
Aruba	0.0002	1991	0.07	—	Dutch	Monarchic, parliamentary
Bermuda	0.00005	1991	0.06	Colored: 60 / Europ.: 40	English	Monarchic, parliamentary
Cayman	0.0003	1995	0.03	—	English	Monarchic, parliamentary
Dutch Antilles	0.0004	1981	0.17	Creole: 92 / White: 6 / Other: 2	Dutch	Monarchic, parliamentary

Table 2.10. (*cont.*)

Country/territory	Area (million km²)	Population		Ethnic distribution (%)	Official language	System of government
		Year	Number (millions)			
Guadeloupe	0.002	1993	0.41	Creole: 77 Black: 10 Mixed: 10 White: 2 Other: 1	French	Republican, parliamentary
Martinique	0.001	1993	0.38	Black: 94 White: 3 East Ind.: 2 Other: 1	French	Republican, parliamentary
Montserrat	0.0001	1996	0.01	Black: 93 Mixed: 6 White: 1	English	Monarchic, parliamentary
Puerto Rico	0.009	1996	3.70	White: 80 Black: 20	Spanish English	Republic, presidential
Turks and Caicos	0.0004	1995	0.01	—	English	Monarchic, parliamentary
Virgin Islands (American)	0.0003	1990	0.10	Black: 70 Mixed: 19 White: 11	English	Republic, presidential
Virgin Islands (British)	0.0001	1994	0.02	Black: 87 Mixed: 12 White: 1	English	Monarchic, parliamentary

Source: Camargo (1998).

Table 2.11. *Ethnic distribution in Middle America (in millions of population)*[a]

Country/territory	White	Black	Amer.	Creole/Mestizo/Mixed	East Ind.	Other Asians	Other
Antigua and Barbuda		0.05					0.01
Bahamas	0.03	0.25					0.01
Barbados	0.01	0.21					
Belize	0.01	0.09	0.03	0.04			
Costa Rica	3.13	0.07	0.04	0.06	0.01	0.07	
Cuba	4.11	6.88		0.29		0.11	
Dominica		0.06		0.01			
Dominican Republic	1.22	0.89		5.99			
El Salvador	0.06		0.29	5.55			
Grenada		0.08		0.01			0.01
Guatemala	0.22		4.59	6.27		0.34	
Haiti	0.06	7.11		0.07			
Honduras		0.12	0.42	5.4			
Jamaica		1.88		0.33	0.02		0.27
Mexico	8.49		28.29	56.58			0.94
Nicaragua	0.75	0.40	0.22	3.03			
Panama	0.27		0.54	1.89			
St Kitts and Nevis		0.04					
St Lucia		0.13			0.01		
St Vincent and Grenadines		0.10					0.01
Trinidad and Tobago	0.01	0.74			0.54		0.01
Anguilla		0.008		0.001			
Bermuda	0.02	0.04					
Dutch Antilles	0.01			0.16			
Guadeloupe	0.01	0.04		0.36			
Martinique	0.01	0.36			0.01		
Montserrat		0.01					
Puerto Rico	2.96	0.74					
Virgin Islands (American)	0.01	0.07	0.02				
Virgin Islands (British)		0.02					
Total[a]	21.39	20.39	34.42	86.06	0.59	0.52	1.26
(%)	12.99	12.38	20.91	52.28	0.36	0.32	0.76

[a]Total number considered: 164.63 million individuals. For three territories (110 000 people) there are no data on ethnic breakdown. This is a summary of the data shown in Table 2.10.

Table 2.12. *Selected information about the present South American countries and territories*

Country/territory	Area (million km²)	Population		Ethnic distribution (%)	Official language	System of government
		Year	Number (millions)			
Nations						
Argentina	2.78	1997	35.70	White: 85 Mestizo: 7 Other: 8	Spanish	Republic, presidential
Bolivia	1.10	1997	7.80	Quechua: 30 Aymara: 25 Mestizo: 15 White: 15 Other: 15	Spanish Quechua Aymara	Republic, presidential
Brazil	8.55	1996	157.08	White: 55 Mixed: 38 Black: 6 Other: 1	Portuguese	Republic, presidential
Chile	0.76	1997	14.60	Mestizo: 95 Amer.: 3 Other: 2	Spanish	Republic, presidential
Colombia	1.14	1997	37.10	Mestizo: 58 White: 20 Black: 18 Amer.: 1 Other: 3	Spanish	Republic, presidential
Ecuador	0.28	1997	11.90	Mestizo: 55 Amer.: 25 White: 10 Black: 10	Spanish	Republic, presidential
Guiana	0.21	1997	0.85	East Ind.: 51 Black: 30 Mestizo: 11 Amer.: 5 Other: 3	English	Republic, parliamentary

Paraguay	0.41	1997	5.10	Mestizo: 95 Amer.: 3 White: 2	Spanish Guarani	Republic, presidential
Peru	1.28	1997	24.40	Amer.: 45 Mestizo: 37 White: 15 Other: 3	Spanish Quechua Aymara	Republic, presidential
Surinam	0.16	1997	0.44	Black: 41 East Ind.: 37 Asian: 17 Amer.: 3 Other: 2	Dutch	Republic, parliamentary
Uruguay	0.18	1997	3.20	White: 88 Mestizo: 8 Black: 4	Spanish	Republic, presidential
Venezuela	0.91	1997	22.80	Mestizo: 67 White: 21 Black: 10 Amer.: 2	Spanish	Republic, presidential
Territories						
French Guiana	0.08	1993	0.14	Creole: 72 Asian: 6 Mixed: 6 Amer.: 4 Other: 12	French	Republican, parliamentary
Falkland	0.01	1996	0.002	White: 99 Other: 1	English	Monarchic, parliamentary
Georgia and South Sandwich	0.004	—	0.001	—	English	Monarchic, parliamentary

Source: Camargo (1998).

Table 2.13. *Ethnic distribution, South America (in millions of population)*[a]

Country/ territory	White	Black	Amer.	Creole/ Mestizo/ Mixed	East Ind.	Other Asians	Other
Argentina	30.34			2.50			2.86
Bolivia	1.17		4.29	1.17			1.17
Brazil	86.39	9.43		59.69			1.57
Chile			0.44	13.87			0.29
Colombia	7.42	6.68	0.37	21.52			1.11
Ecuador	1.19	1.19	2.97	6.55			
Guiana		0.26	0.04	0.09	0.43		0.03
Paraguay	0.10		0.15	4.85			
Peru	3.66		10.98	9.03			0.73
Surinam		0.18	0.01		0.16	0.08	0.01
Uruguay	2.81	0.13		0.26			
Venezuela	4.79	2.28	0.45	15.28			
French Guiana			0.01	0.11		0.01	0.01
Falkland	0.002						
Total[a]	137.87	20.15	19.71	134.92	0.59	0.09	7.78
(%)	42.94	6.27	6.14	42.02	0.18	0.03	2.42

[a]Total number considered: 321.11 million individuals. No ethnic breakdown is available for Georgia and South Sandwich (1000 people). This is a summary of data shown in Table 2.12.

Table 2.14. *Darcy Ribeiro's typology, as applied to Latin American populations*

Witness peoples
Modern representatives of highly civilized groups who felt the impact of the European expansion.
Examples: México, Guatemala, Bolivia, Peru, Ecuador.

New peoples
Those who have arisen from the conjunction, deculturation and fusion of African, European and Amerindian ethnic matrices.
Examples: Brazil, Venezuela, Colombia, Antilleans, some of the peoples of Central America, Chile, Paraguay.

Transplanted peoples
Modern nations created by the migration of Europeans to new parts of the world, re-establishing there ways of life similar to those of their nations of origin.
Examples: Costa Rica, Argentina, Uruguay.

Emergent peoples
Populations who have grown to nationhood from the tribal level.
Examples: Absent in Latin America, indicating the severity of domination by the conquerors and later by the local elites. The Araucanians of southern South America, because of their population size and ethos, could have become an Emergent people in the second half of the nineteenth century, but were decimated by Argentinians and Chileans, their survivors being confined to reservations.

Source: Ribeiro (1970, 1977).

3 *Socioeconomic indices, demography and population structure*

Births are the main cause of deaths   *Millôr Fernandes*

Biology, culture and the environment

The word population may have different meanings to distinct specialists. The point to be emphasized, however, is that it has *structure*. When the question is asked about how populations change in size over time (mainly a historical question), both intrinsic and extrinsic factors should be considered. In most general terms population behavior depends on individual behavior, that in turn is adjusted in a complex way by both genetic and environmental factors. The latter may comprise the physical environment, with its variables of temperature, humidity and general geologic conditions; or the socioeconomic system, developed by our cultural skills, which can also serve as a buffering layer that protects us from the rigors of the outside world.

Members of a Mendelian (sexual and cross-fertilizing) population have sex, and the proportion between males and females (the sex ratio) may significantly influence its future. People are born and die, and factors related to fertility, morbidity and mortality should also be considered. People also move (with the exception of those who already died!), and intra- and interpopulation mobility should be taken into consideration. In evolutionary terms, these different agents ultimately condition the opportunity for the action of natural selection, leading to the survival of the genetically best endowed and the elimination of those with lower 'fitness'.

Ideally, concepts from the several areas of knowledge mentioned above should closely interact, to obtain answers to the problems raised by different aspects of population fate. Unfortunately they seldom act this way. Good examples of such interactions, however, can be found in the works of Harrison and Boyce (1972), Ward and Weiss (1976), Harrison (1977), Adams *et al.* (1990), and Boëtsch *et al.* (1996).

55

Does Latin America exist?

According to Blakemore and Smith (1985*b*), the term 'Latin America' is a French invention which became common in the 1860s, when Napoleon III was trying to maintain Maximilian of Austria on a Mexican throne under French tutelage. The question, therefore, is pertinent, and was asked by Inglehart and Carballo (1997) when they were analyzing the data from the World Values Surveys. These were studies conducted on three different occasions, beginning in 1981, 1990, and 1995, respectively, and which involved a large set of questions answered, in the first survey, by almost 60 000 respondents in 43 societies. The items covered a wide range of topics, from religion to politics, to sexual norms, to attitudes toward science. These different orientations tended to go together in coherent patterns. Thus, in the 1990–91 survey a principal component analysis indicated that 51% of the cross-national variance could be explained by two main axes involving the dialectical relationships between traditional versus rational–secular authorities and survival versus well-being.

Four Latin American countries were included in this first survey (Argentina, Brazil, Chile and México), and they all grouped in a compact cluster in the statistical analysis made, reflecting the fact that in global perspective they have relatively similar value systems. The survey was repeated in Argentina and México with an interval of nine years (1981–90) with similar results, and the inclusion, in the third survey, of Venezuela, Dominican Republic and Puerto Rico, showed that they would fit into the Latin American cluster quite well. The picture is therefore coherent and relatively stable, justifying any investigation which considers this region and their people as a whole. Andrade (1991) provided a synthetic panorama of Latin America, stressing the similarities and differences between countries and the sociopolitical problems confronted by them. More extended treatments appeared in James (1969) and Blakemore and Smith (1985a). In sheer numbers, while in 1950 Latin Americans represented 49% of the hemisphere's population, this value jumped to 61% in 1995 (WHO, 1998). In world terms in 1992–95 the land area occupied by Latin America was 14.9%, its population represented 9.3%, and its gross economic product 8.3% of the world total (Huntington, 1997).

Micro- versus macroanalyses

Basically, any type of scientific question can be approached either at the micro or at the macro levels. The first tends to be related, at least in biology, to reductionist, and the latter to interactionist, approaches. However, this is not necessarily so. Both more restricted or wider subjects of enquiry can be studied giving emphasis to one or a few functional units, or to an array of interactive units.

Demography is not an exception. We can investigate population behavior over time (the subject of historical demography) at different levels of inclusiveness (local population, parish or county, state or province, nations, regional blocks, the whole of mankind). At the micro level it is possible to identify primary agents of change more easily, and even (especially with the advances in computer techniques), to construct pedigrees involving the whole community. A good Latin American example of this latter approach is the study performed by Castilla and Adams (1990) in Aicuña, Argentina, which led to the construction of a well-documented pedigree extending for 16 generations and including 8573 individuals.

Macroanalyses, on the other hand, try to establish general tendencies and generally involve large populations whose characteristics are evaluated at given time intervals, as in national censuses. Since many people are included in the collection, coding and analyses of the data, there is an obsessive preoccupation with the need to devise techniques that would correct for inaccuracies, which could lead to inability to detect real, unexpected factors of change.

Both approaches have advantages and disadvantages. The ideal, then, is to incorporate both of them in any given analysis, connecting the demographic with other systems, especially the socioeconomic one (Livi-Bacci, 1990).

Macrodemography and economics

Some basic socioeconomic information about the Middle and South American countries is given in Tables 3.1 and 3.2. Considering both areas in general (see the averages at the bottom of the tables) we can verify that there are not many differences among them, with the exception of the Gross Internal Product (GIP), that is 6.9 times higher in South than in Middle America (US$ 113041 million versus 16378 million). External debt, however, is 3.2 times higher in South America (31% of the GIP). In Middle America this variable is heavily influenced by México's high debt, which is 3.7 times higher then those of all the other nations considered together.

The figures show a marked departure from corresponding figures in the First World. Latin American populations are still growing at about 1.5%

per year, the average number of children per woman in the age interval between 15 and 49 years being 2.9. In contrast, life expectancy is still low (67–8 years for males, 73 years for females). These figures are 10 years lower than those obtained for Japan. Infant mortality at 28–35‰ is about 3.5 times higher than that found in the developed world. As a consequence of this and other parameters, the index of human development or IHD at 0.76–0.78 is lower than that of the First World, which is equal to or higher than 0.90. As for average GIP, Middle America's is only 2%, and that of South America only 10% of those observed in rich nations whose GIPs amount to values higher than one billion dollars.

There are heterogeneities within the two regions. In Middle America, Cuba shows a demographic increase of only 0.4% per year, while Grenada (− 1.1%) and St Kitts and Nevis (− 0.5%) show declining rates. On the other side of the range, in Guatemala the rate of increase is of an annual 2.8%. Contrasting values are also observed in South America (Uruguay, 0.6%, Paraguay 2.6%). Life expectancy and the IHD are low in Haiti (53 years for males, 56 for females, 0.34) and Bolivia (males 60 years and females 63 years, 0.59) but much higher in Barbados or Costa Rica (74 years and 79 years, respetively 0.91 and 0.89), as well as Chile (males 72 years and females 78 years, 0.89). Brazil's external debt is much higher than those of other South American countries, but its relationship with the GIP is much more favorable (24%) than is true for México (66%).

Other aspects about the standard of living of Latin Americans are presented in Tables 3.3 and 3.4. Poverty and extreme poverty are the pitiful conditions of respectively 37–38% and 14–16% of Latin Americans, although, as has been indicated, the situation varies between countries (72% of poor people in Bolivia, only 20% in Paraguay). It is also important to emphasize that definitions of poverty vary and the index may fluctuate widely in time. The numbers presented in Table 3.3 were from a study made in 1994–95 by the World Bank. Rocha (1996) made similar evaluations in six Brazilian metropolitan regions, and although she obtained, for 1994, a value similar to the one given in the table (38%), she estimated a significant decline 2 years later (29%), attributed to the relative stabilization of the Brazilian economy which occurred with the establishment of the Real Plan (*real* is the new Brazilian currency). On the other hand, the World Health Organization (WHO 1998) estimated that the number of people below the poverty level in Latin America increased from 197 million in 1990 to 209 million in 1994, of whom 65% lived in urban areas.

Within a country of continental size like Brazil, conditions may vary widely. Using an index of domestic comfort (which quantified the material goods present in a given home), Ribeiro (1997) examined the rural and urban conditions prevailing in 14 Brazilian counties. The extreme values were obtained in Santarém (state of Pará, in the north), with the lowest, and Ibirama (state of Santa Catarina, in the south) with the highest indices.

Numbers about wealth distribution are given in Table 3.4. While in 18 First World countries the poorest 20% of the population possess 6.3% of the total wealth, and the richest 20% possess 39.9%, the figures for Latin America are much more ill-balanced (3.6–4.0% and 56.7–57.2%, respectively). Gini's coefficient gives a quantitative estimate of wealth concentration. The distribution is relatively more even in Jamaica or Bolivia (0.40), but much worse in Brazil (0.57).

Multivariate analysis can also be used to determine in a comparative way similarities and dissimilarities in socioeconomic conditions among the different Latin American countries. Table 3.5 presents the result of one of such analysis. Based on 16 variables the situation in 23 Latin American nations is compared both within the block, as well as contrasting it with conditions that prevail in North America. Two principal components explain 75% of the variation generated by the 16 characteristics, and the first can be interpreted as a measure of the degree of development of the countries. Five variables (nos. 1, 2, 5, 11, 12) showed a marked discriminative power, and the results indicated the formation of four groups (designated as of maximum, average superior, average inferior and minimum development), as well as some subgroups.

As was expected, extremes of development appear in the USA and Canada (maximum) and Bolivia and Haiti (minimum). The other groups or subgroups occur independently of geographic location. Three of the nations with highest GIPs (Brazil, México and Venezuela) are placed in the same subgroup, but the other two (Argentina and Colombia) are separated. The highest level of development in Latin America (as assessed by the First Principal Component) occurs in small countries (Cuba, Trinidad and Tobago, Uruguay, Chile).

Historical demography

What is historical demography? As the name indicates it is a type of fusion between these two areas of study, namely the investigation of the variability in time and space of populations which lived in the past. Several lines of research can be identified within the discipline, such as: (a) family reconstitutions, (b) temporal series and population reconstitutions, (c) exploration of census data and other demographic statistics, and (d) family histories (Reher, 1997). On the other hand, Marcílio (1997) identifies three thematic lines: (a) nuptiality, family, concubinage and infancy, (b) structures and population dynamics, and (c) mortality and morbidity. An important aspect, not considered by these authors, is the relationship between these studies and genetic–evolutionary investigations. It is through the analysis of mating patterns, which are in part conditioned by mobility and migration, fertility, morbidity and mortality of past populations, that we can make genetic inferences about their present counterparts.

According to Bassanezi (1997) and Reher (1997), historical demography acquired an identity with the work of Louis Henry in the 1950s in France. In England Thomas Hollingsworth, David Glass, D.E.C. Eversley and John Hajnal should be remembered, while in the USA the Berkeley School (with key figures such as Sherbune Cook, Lesley Simpson and Woodrow Borah) comes to mind. The Berkeley School was responsible for the first historical demographic studies in Latin America (México). Almost simultaneously works in the field started to appear in Argentina (Nicolás Sánchez-Albornoz) and Brazil (Maria Luiza Marcílio), both influenced by the French investigators. Other centers of investigation appeared later, in México, Central America, Colombia, Chile and Uruguay, leading to the present burgeoning of the discipline (Marcílio, 1990).

As for the inferences which can be drawn about the present genetic variability by relying on the information provided by historical documents, pioneer work on this approach was made by Newton Freire-Maia, Oswaldo Frota-Pessoa and Pedro H. Saldanha in Brazil. Outside Latin America two key figures are Gabriel W. Lasker (USA) and Italo Barrai (Italy). Juan Pinto-Cisternas has been working extensively, for decades, on these problems, both in Chile and Venezuela, while in the latter country numerous studies have been undertaken by Alvaro Rodríguez-Larralde. Lorena Madrigal (now in the USA) developed a very detailed program of historical investigation in Escazú, Costa Rica.

Table 3.6 lists selected bibliographic sources for the study of historical demography in Latin America. This list should be considered as just a starting point for the voluminous literature on the subject. Research performed in 12 nations is reported in the indicated sources. Of special interest is the book edited by Nadalin *et al.* (1990), which gives part of the contributions presented at the First Congress on the History of Populations in Latin America, held in Ouro Preto, Brazil in July 1989. Five general areas of study were covered: (a) spatial distribution of historical populations in Latin America; (b) components of demographic growth; (c) comparative perspectives on nuptiality, family formation and fertility;

(d) slave populations in Latin America – special demographic characteristics; and (e) population and economy. These subjects were covered with data obtained in Argentina, Brazil, Chile, Costa Rica, Ecuador, México, Paraguay, Peru and Uruguay.

Sheer numbers are not important in reaching significant insights into biological problems from demographic data. Examples which demonstrate this statement can be presented from the information obtained in two small island populations of French extraction situated in the Caribbean Sea. A study by Dyke (1971) in the Northside community of St Thomas, Virgin Islands, addressed the question of the number of potential mates a given person has in a small population. This has obvious implications both for genetics and the social sciences. In the first case, this number will greatly influence the degree of genetic diversity that a community will display through generations. On the other hand, the choice of a mate depends on complex psychological, socioeconomic and cultural factors (such as the concept of beauty). Dyke (1971) took into consideration all individuals who would be available for marriage at any time between 1908 and 1966. After correcting for age, lineal relatives (parents, children, grandparents and grandchildren, great-grandparents and great-grandchildren) and the existence of people already married in the community, he arrived at the conclusion that the mean number of females for the average male over all years in the Northside mate pool is 71 before, and 41 after eliminating married individuals. Correlating this with inbreeding data, he surprisingly verified that those who marry endogamously should be more distantly related than those who do not, a conclusion that was confirmed by the actual data.

Leslie (1983) considered another problem, studying the population of St Barthélemy, near St Thomas, which was responsible for the formation of the latter through the emigration of one of its segments. The question addressed by Leslie was whether, in the calculation of potential mates, the use of cohorts would not introduce biases, since in humans the generations overlap. He verified that estimations of consanguinity expected under random mating differ minimally whether a cohort or a continuous model is used. But caution should be taken in the generalization of this finding when the population is being subjected to substantial demographic change, if the group is small, or the period of analysis differs from that considered in the investigation.

Sociobiological theory predicts that the higher variance in reproductive success of males may lead parents with more resources to prefer to raise sons, while parents with fewer resources would prefer daughters. This question was investigated by Brittain *et al.* (1988), also using

St Barthélemy as a test case. They verified that changes in the level of prosperity, as measured by sex ratios at birth and by the mean age of males at first marriage, appear to have no effect on the probability of dying during the first 5 years of life, for either sex, during the period of study. A characteristic of the Caribbean way of life may lead to no discrimination between a male or a female child. Thus, while males may provide stronger economic support, they tend to emigrate more easily, leaving daughters as the sole support of their parents in old age.

Mortality is a key parameter in demographic analysis, but the factors that influence this variable are seldom investigated in detail. Brittain (1992) evaluated 10 independent variables that could predict death before the first birthday for 4411 births that took place in St Barthélemy, from 1878 to 1976, issued by 978 women. Death of the mother was the most important factor in determining the death of the index child within the first year of life. Survival of the offspring, therefore, not surprisingly, is not independent of the mother's survival. This point, however, is generally not considered in demographic statistics. As Brittain (1992) has emphasized interview data, which of course has to rely on living women (unless we develop means to communicate with the dead!) may lead to underestimates of childhood mortality.

Migration

Introduction

Migration can involve long or short distances, people of the same or different ethnic extractions, can occur between and within countries, from rural to urban areas or vice-versa, and can be studied by focusing on a single community or a larger area. Our species is notable by its high mobility, and Latin Americans are no exception. However, the long-range migrations which shaped the present demographic structure of Latin American countries, and the type of migration which occurred within and between them, was highly variable. A summary of the main migratory movements that led to the formation of some of these nations, as well as of those that occurred within and between them, follows.

Spanish-speaking countries and Carib

Table 3.7 supplies information about migrations related to 11 Spanish-speaking nations, as well as to Guiana and the Caribbean islands. Immigration by European groups had varied importance in these countries, being especially marked in Argentina. In the Carib the African-derived influence was much more marked. Although not much is known about the place of origin of the people who came from Africa (except, in large outline, the contributing regions), for Europeans historical data are available. Besides Spaniards, the role of Italians, Portuguese, Germans, French and English is well documented.

Since colonial times there has been a constant flux of people within Latin American countries, conditioned by socioeconomic and environmental reasons. A special kind of this movement is that from rural to urban areas, leading to poles of attraction that formed huge conglomerates such as México City, Buenos Aires, Caracas, Lima, Bogotá, Santiago and Montevideo. The percentage of total population living in cities varies markedly in the nations surveyed, from a low of 36% in Guiana, to a high of 93% in Venezuela. This parameter is generally considered an indicator of development, but its relationship with standards of living is not linear. For instance, Costa Rica, with the highest index of human development in the region (cf. Table 3.1) has only 50% of its population living in cities.

A relatively recent phenomenon has been the increased amount of emigration of people from certain countries, mainly due to political reasons (Chile, Cuba, Paraguay, Uruguay), but also in search for better employment conditions (México).

Brazil

The amount of demographic information available for Brazil is considerable, due especially to two institutions: the Instituto Brasileiro de Geografia e Estatística (IBGE; Brazilian Institute of Geography and Statistics) and the Associação Brasileira de Estudos Populacionais (ABEP; Brazilian Association of Population Studies). Both have systematically produced a large number of documents, and ABEP is responsible for the publication of the *Revista Brasileira de Estudos de População* (*Brazilian Journal of Population Studies*). Selected information about these investigations is provided in Table 3.8, while Figure 3.1 gives the country's regions and states with their capitals, for the location of geographic references.

The first point to consider is the relative contribution of the European and modern Asiatic *immigration*, as compared with indigenous growth, in the formation of the present Brazilian population. It was suggested that while immigration played a key role in the growth of São Paulo and the states of the south, it was not important for the growth of the Brazilian population as a whole. Mortara (1947), examining the population growth

Figure 3.1. Map of Brazil showing its main geographic regions, states, and their capitals.

of some American countries between 1840 and 1940, concluded that immigration contributed directly or indirectly to their population growth in the following frequencies: Argentina, 58%; USA, 44%; Canada, 22%; and Brazil, 19%. Four years later (1951), using the same methodology, Mortara arrived at a lower contribution for Brazil (10%), a value that Levy (1974) considered nearer to reality. This question was recently reconsidered by Clevelario (1997), who found however that this contribution should oscillate between 12% and 24%, 18% being the most probable figure.

Both culturally and physically, the *Portuguese* were those who most markedly shaped Brazilian history. It is estimated that from the beginning of the nineteenth century up to the 1950s 80% of the Portuguese who left their country legally came to Brazil (Klein, 1989). The total number of Portuguese migrants to Brazil in the period 1820–1972 is calculated as 1.8 million. The peak of this immigration occurred in 1904–1914, with 412 607 people. Information about the destination of these immigrants in the period 1920–70 indicated the southeast as the most preferred region, especially Rio de Janeiro and São Paulo (Levy, 1974). The assimilation of these migrants into Brazilian life was gradual. Data on 25 074 unions in Rio de Janeiro during 1907–16 indicated 3.2 times more endogamous marriages than those expected if there was no choice regarding nationalities. The figure for São Paulo in 1934–46 is even more extreme: 7.7 times more endogamous unions in relation to those expected by chance, among 657 495 marriages (Klein, 1989).

Italians were the most frequent immigrants after the Portuguese (1.6 million during 1820–1972; Levy, 1974), and showed an even stronger tendency for endogamy, in the same two sets of data: 25.3 times more endogamous unions than those expected by chance in Rio de Janeiro, and 9.5 more in São Paulo, in disagreement with the assertions of Klein (1989).

In a previous review (Salzano, 1987), the figure of 3.6 million *African* immigrants was arrived at, with half of it occurring between 1701 and 1810. Historical statistics published in the same year (IBGE, 1987), and therefore unavailable at the time of the preparation of that review, suggest a somewhat higher figure of 4.0 million people, although the relative proportions in the three periods considered remained similar (1551–1701, 16% versus 15%; 1701–1810, 52% versus 48%; 1810–57, 32% versus 37%).

In terms of *internal mobility*, Taschner and Bógus (1986) characterized Brazil as having a highly mobile population. Their study indicated that more than 20% of the people changed residence from one county to another, and 7% changed residency from one state to another in the period covered. The rate of urbanization was high, with a tendency for the formation of a net of nine metropolitan areas (Belém, Fortaleza, Recife, Salvador, Belo Horizonte, Rio de Janeiro, São Paulo, Curitiba and Porto Alegre). Martine (1994), on the other hand, recognized three periods of frontier expansion, related to the conquest and development of rural areas. The first involved the west of São Paulo, north and west of Paraná, west of Santa Catarina, and southeast of Mato Grosso, and occurred from the 1930s to the 1960s; the second took place in the central portion of the country (Goiás, Mato Grosso do Sul, and Maranhão) from the 1940s to the 1960s; while the last phase involved the Amazonian region in the 1970s. For details about this phase see Droulers and Maury (1981). The contribution of these movements to an eventual change in the rural–urban flux was small in periods 1 and 2 (13% and 3%, respectively) and negligible in period 3. There was also a significant reduction in urban growth in the 1980s; but despite that the metropolitan regions had a growth of 8.3 million people – a number equivalent to the total population of Sweden! Presently these regions have 42.7 million inhabitants, which means that 3 in 10 Brazilians now live in a metropolitan city.

As is occurring in other Latin American countries, at present a significant number of Brazilians are now leaving the country in the search for better living opportunities. There is no systematic registration of movements of Brazilians across the borders, so that it is difficult to estimate the real magnitude of the *emigration* phenomenon. However, based on the censuses of 1970, 1980 and 1991, and on the assumption that the mortality rates in the 1970s and 1980s have been constant, Carvalho (1996) arrived at the surprising conclusion that the Brazilian population may have lost 1.0–2.5 million people due to emigration in the past decade. This agrees with estimates listed in Table 3.8, suggesting that the magnitude of the phenomenon is important enough to be worthy of consideration by the government agencies.

Community studies are important in several contexts. Table 3.9 gives information about four demographic methods of estimating migration that have been used by Brazilian geneticists to interpret their data. Note that the results already considered by Salzano and Freire-Maia (1970) are not included there, particularly a series of studies on mean matrimonial radius (the average of the distances between the birthplaces of the spouses and the place where they married), a measure that may be roughly similar to the parent–offspring distance. A relatively recent estimate of it for the city of Florianópolis yielded 131 km (Agostini and Meirelles-Nasser, 1986). Another measure of dispersion is isolate size, well covered in Frota-Pessoa (1971).

The results of Table 3.9 include 13 localities, situated in three regions. Those located in the north are distributed along the Amazon river, but no discernible gradient is observed in this west–east axis. Óbidos, a relatively small town situated in the central portion of this distribution, generally presents low values for all the four estimators, the opposite occurring with Santarém, the second largest city in the state of Pará, which generally shows two to five times higher figures than Óbidos. Two of the northeastern populations (Natal and Aracaju) present lower values than those of the north in two of the measures (individual migration, marital distance), intermediate for parent–offspring distance, but higher for the exogamy index. Thus, unions occur more frequently there with outsiders, but the localities involved are geographically closer than is true in the north. Two other studies concerned with northeasterners involved a small island community (Lençóis Island), and a sample of predominantly north-eastern migrants ascertained in São Paulo. Both generally presented low values for marital and parent–offspring distances.

Not much information is available for the southern communities, but the results do not depart greatly from those observed elsewhere, with the exception of the African-derived individuals studied in Curitiba, who generally show high values. This may be due to the fact that this city constitutes a pole of attraction for people looking for better working conditions, and persons of this ethnic extraction may have been more prone to try this option than European-derived individuals.

As expected, the numbers obtained in previous generations are about half those observed presently, with the exception of the exogamy index. In relation to the latter, extremely low values were found in the north for the grandparent generation.

There is not much more data outside Brazil, in relation to these measures of migration. Those that we could locate are presented in Table 3.10. Little difference between generations was found in St Barthélemy and Bluefields. The first is highly endogamous, while Bluefields, Dota, Fortín Lavalle and Valparaiso (both the city and the rural community of Casablanca situated nearby) are much more exogamic. In the Jujuy Province of Argentina there is a clear dichotomy between Puna/Quebrada and Valle/ Ramal, that is most probably conditioned by altitude. The first two regions (the more isolated) are located 2500 meters or more above sea level, while the other two are at a much lower altitude.

Inbreeding and isonymy

Patterns of mating are of obvious importance for the fate of a given gene in a population. For instance, if the gene is rare, deleterious and recessive, its effect will only appear in homozygosis, an event that would be made much more probable in the offspring of consanguineous marriages. This problem has been extensively investigated, especially in the 1960s and 1970s, as a component of the genetic load theory. In Brazil, N. Freire-Maia conducted a whole program of investigations which lasted for decades (reviews in Salzano and Freire-Maia, 1970; Freire-Maia, 1971, 1984). This author also verified that if all consanguineous marriages could be prevented, as much as 7% of postnatal (up to age 20) mortality and morbidity would be avoided in the Brazilian northeast, where such unions are particularly frequent (Freire-Maia, 1990).

The classical way of measuring the amount of inbreeding of a group is the calculation, using genealogies, of the number of marriages between people who are related (say, first cousins, first cousins once removed, etc.) that occurred in that group. Sewall Wright's *inbreeding coefficient* (*F*) can be defined as the probability that a child from a consanguineous marriage has in double dose any of the alleles present in a specific locus of common ancestors. Another related measure is Gustave Malécot's *coefficient of kinship* (*φ*), that can be defined as the probability that two alleles drawn from two individuals, not necessarily mates, are identical by descent. Thus the coefficient of inbreeding of a progeny is identical to the coefficient of kinship of their parents.

Kinship coefficients can be calculated using other alternatives besides pedigrees, namely genotypes or, due to the fact that they are also inherited, surnames. Isonymy (identity of surnames among individuals) is a relatively easy method of estimating relatedness and has been much employed with this purpose. The problem of estimating inbreeding by the classical method is that human memory is short. Therefore, remote consanguinity cannot be ascertained, and this may be important in the estimation of *F* for large populations (Salzano and Freire-Maia, 1970).

Other complications exist, and they have been considered using two Caribbean communities already mentioned, St Barthélemy and St Thomas, by Leslie *et al.* (1978, 1980, 1981). There are intricate relationships between probability of mating, avoidance and preference for consanguineous marriages, which were carefully examined through both empirical observations and mathematical modeling by these authors. They observed that in St Barthélemy nearly 60% of all individuals reaching mating age never reproduce, and that these persons tended to be more closely related than those who do reproduce. This results in nonrandom inbreeding, which reduces total pedigree inbreeding. On the other hand, consanguinity avoidance is compensated for by an apparent preference for more distant relationships, especially second cousins. Application of a model which predicts the effects of these factors on the genetic variability suggests that under some circumstances nonrandomness in celibacy and emigration may have a greater influence on genotype distributions than does consanguinity avoidance.

At least two reports exist about inbreeding in pre-Hispanic Latin America. Christensen (1998) developed a pedigree tracing all known relatives of Lord Eight Deer Jaguar Claw, a Mixtec ruler. The data involved the affairs of royal dynasties of various towns, located throughout the western portion of the modern state of Oaxaca, and was distributed for five (tenth to sixteenth) centuries. He arrived at an average *F* value of 0.0527, somewhat lower than that of first cousins (0.0625), but the pedigree included seven sibling marriages and several uncle–niece and aunt–nephew pairings. The pedigree drawn by Cruz-Coke (1965*a*), involving the Andean Inca dynasty, is even more striking. Considering information from 12 generations he arrived at the conclusion that about half of the genes of Huáscar, the last legitimate Inca, would have been derived from the first Inca, Mango Capac.

There is much information about inbreeding levels in Latin American populations, especially due to the efforts of Newton Freire-Maia, a Brazilian geneticist. Selected information, obtained through church and state legal records is presented in Table 3.11. The results include 15 Latin American countries, although the most extensive bodies of data derive from Brazil and Chile. For Middle America the *F* values (all multiplied by 100 000) for whole countries, discounting a small sample from Panamá, are distributed between 31 (México) and 142 (El Salvador). But a much higher prevalence (440) was obtained for Dota, a rural community in Costa Rica. As for South America, whole country estimates from the north of the continent (Venezuela, Colombia, Ecuador, Peru, Brazil) fluctuate around 200, while in the south (Bolivia, Argentina, Uruguay, Chile) the values are lower (around 55). Much higher values, however, were observed in small rural communities of Venezuela, Brazil, Argentina and Chile.

In all samples there are two clear trends towards reduction of the levels (a) with time and (b) from rural to urban communities. In Brazil there is a marked difference between the northeast (365) and south (81) regions. In Chile also, there are large differences within the country (14–654) that can, however, be ascribed to the two above-indicated trends.

Surnames are inherited, as occurs with genes, and therefore can be a valuable tool in the analysis of the genetic structure of populations. They can also be used as indicators of nationality or ethnic affiliation, and this type of data will be considered later in this book. Now we will review studies performed in Latin American populations using isonymy (the occurrence of the same surname between actual or potential mates) as an approach (Table 3.12).

The pioneering studies of Gabriel W. Lasker in this area are well recognized, and were performed in Peruvian and Mexican populations. The most extensive and systematic investigations developed to date in Latin America, however, are due to the efforts of Juan Pinto-Cisternas, Alvaro Rodríguez-Larralde, Italo Barrai, and associates. As is indicated in Table 3.12, their studies included almost all Venezuela, using a variety of methods of analysis. In this way they obtained inferences about the extent of isolation (and opportunities for mating) both at the micro and macro levels.

The advantages of the isonymy approach in relation to genetic markers (possibility of analysis of past, historical individuals or populations, facility in obtaining complete coverage of a given group) should not disguise that this approach also has limitations. Among them we could mention polyphyletism (the adoption of a given surname independently from inheritance), and illegitimacy. More subtle are questions related to the proportion of isonymous pairs and the inbreeding coefficient of children, as well as the requirement that the variances of sibship size for male and female parents are equal. These problems condition that the isonymy method will generally overestimate the level of inbreeding of a given population. Ideally, therefore, it should only be employed in groups that are well-known to the researcher, and should be complemented by other approaches. But as a rough indication of structure, and with proper care, it can be usefully employed.

Demographic trends

The speed of its population growth is the most marked characteristic of Latin America throughout the twentieth century. Thus, while Latin Americans represented only 5% of the world population in 1920, this number increased to 8% in 1975. During this period, the average rate of population growth for Middle and South America was approximately the same, but with marked differences among countries or territories (Tables 3.13 and 3.14). In Middle America the most marked difference occurred between Costa Rica (whose population increased 5.2 times in the period) and Barbados (1.7 times only). In South America the extremes occurred between Venezuela (5.3 times) and French Guiana (1.8 times).

This population growth was due, as elsewhere in the world, to declining mortality and constant birth rates. Marked changes, therefore, occurred in the age structure of these populations, that can be dramatically exemplified by the life expectancy numbers. Around 1920 life expectancy for the whole region was estimated as 27 years, while in 1975 this number jumped to 61 years (Marcílio, 1980)! Additional information about demographic trends in Costa Rica can be found in Fernández et al. (1976), and for Brazil in Wong et al. (1987), Saad (1991), and Martine and Camargo (1997/1998).

Population policies and trend reversals

It is clear that the time of almost unrestricted population growth is over for a large number of countries, in Latin America and elsewhere. The reasons for this are manifold. Perhaps more important than any government policy was the 'contraceptive revolution', that made possible to many the easy and relatively cheap access to several contraceptive methods. Some governments have also clearly established official policies of population control, and many international organizations developed aggressive programs with the same objective. The result is that population growth is being significantly reduced. Data about the use of contraceptives in nine Latin American countries were compared with Asiatic and African results in Morris *et al.* (1981).

Cruz-Coke (1971) furnished data about the effects of contraceptive campaigns that were performed in Santiago, Chile in the period 1960–70. The number of women using contraceptives increased exponentially, resulting in a decrease of the gross birth rate from 35 to 25 per thousand, and an increase in the proportion of primiparas from 25% to 35%. He argued that since it is well known that the first gestation always carries a higher risk of prematurity and of congenital malformations, that one of the effects of the campaigns would be the increase, in relative terms, of these events among newborns of that city.

Teenage pregnancies can lead to social, besides the biological problems, due to the fact that the mothers are generally not yet engaged in the economically active part of the population. Therefore, it is important to realize that Brazilian girls below 20 years of age are marrying and having children in about the same proportion as 25 years ago. According to Henriques *et al.* (1989), before their twentieth birthday 5 in each 10 girls are sexually active; almost 4 in 10 enter marriage, the same proportion conceives a baby; and 3 in 10 have a child. In rural areas these percentages are higher.

Table 3.15 furnishes a general view of the use of contraceptives by Brazilian women in the middle 1980s. In Brazil as a whole, at the time, only 60% of those who were married or lived in consensual unions used contraceptive methods. The proportion of those who did not use contraceptives was particularly high in the northeast, the highest figure (72.4%) occurring in the state of Maranhão. On the other hand, it is striking that for the country as a whole sterilization was more common than the contraceptive pill as a means of avoiding pregnancy (29.3% versus 22.9%). These proportions, however, varied widely among regions. The frequency of sterilization was particularly high (in proportional terms) in the north, northeast, and central-west regions. As many as 50.9% of the women interviewed in Goiás had been sterilized!

Additional information about these sterilizations was provided by Oliveira and Simões (1988). Around 62% of their total were performed after 1980, and while in the more developed areas about two-thirds of them were payed for, the inverse occurred elsewhere, with as many as 75% of them having been performed free of any payment in the northeastern states of Rio Grande do Norte and Alagoas. Since they generally occurred after Cesarean deliveries performed in state-owned or state-subsidized hospitals, it is clear that despite the fact that the Brazilian government does not officially endorse such policies, they are at least allowed or tolerated. Minella (1998) provided an extensive list of the studies related to contraception and sterilization in Brazil in the period 1986–96.

In another study on family planning among low-income women in Rio de Janeiro during 1984–85 Costa *et al.* (1990) verified that almost half (47%) had been sterilized. Compared with similar data obtained in Bogotá and other Colombian cities, it was clear that Colombian women used reversible methods in a much higher proportion (78%). Additionally, they were more varied than those employed in Brazil, where the contraceptive pill predominated.

Other aspects of reproductive health in México, Colombia, Brazil, Paraguay, Argentina and Chile were considered in Hardy (1998). They included questions of assisted reproduction (the other side of the coin – help to those who have biological problems in reproducing). In this regard, a report was presented which reflected a consensus achieved in a conference held in Reñaca, Chile in 1995, on the ethical and legal aspects of these procedures. Eleven Latin American countries were represented at the meeting, and their representatives approved the document. Another aspect considered was the question of male sterilization, provided by studies performed in Colombia and Brazil.

Halberstein (1980) developed a detailed biodemographic study of Bimini, a small Bahamian island, which included the examination of religious documents, censuses, statistical records, as well as an anthropological survey of its inhabitants. The island's population has stabilized since 1963, as a result of an intricate combination of biosocial factors which affected its patterns of fertility, mortality and migratory behavior. In-depth studies such as this one are important for the understanding of the phenomena which occur on a larger scale. Another investigation in which the demographic data were analyzed together with other sets of information was that conducted among the Tlaxcaltecans by a team coordinated by Crawford (1976).

The changing impact of mortality

As elsewhere, mortality rates have been decreasing in Latin America during this century, to a large extent due to the relative control achieved over the infectious diseases in the most developed areas. But the rate with which these changes are occurring varies a great deal, and deficient documentation make comparisons among countries difficult. To redress this, two studies were conducted, coordinated by the Panamerican Health Organization. The first was of adult mortality and included 10 Latin American cities, which were compared with Bristol, UK and San Francisco, CA, and was conducted in 1962–64 (Puffer and Griffith, 1967). The second was concerned with child (less than 5 years of age) mortality. It included urban and rural areas from eight Latin American countries, in the period 1968–70, which were contrasted with Sherbrooke in Canada and San Francisco, CA (Puffer and Serrano, 1973). A concise analysis of the results was presented by Laurenti et al. (1976), who also included other information on the total population and descendants of Japanese from São Paulo. As expected, wide differences were observed among these populations. For instance, the prevalence of deaths due to stomach cancer was twice as high in Bogotá than in São Paulo. Another calculation made concerned the number of child deaths that could have been prevented, if the conditions prevailing in São Paulo had been the same as those of Sherbrooke. Ferreira and Flores (1997/1998), on the other hand, compared the incidence of deaths before one day of life in São Paulo and Costa Rica, analyzing, in the former, endogenous and exogenous causes in the period 1930–78.

Four different approaches were applied by Victora and Vaughan (1985) to test the hypothesis that patterns of land tenure and agricultural production in Rio Grande do Sul, Brazil, were important infant mortality determinants. They concluded that social class, as defined by ownership in these rural lands, played a very important role in the determination of mortality and malnutrition differentials in that population.

Using a principal components analysis, which included five health indicators, Szwarcwald et al. (1997) proposed an index to evaluate quantitatively infant mortality rates in Brazil. Two indicators (proportion of deaths due to congenital malformations, and proportion of those dying due to diarrhea) can be considered as clear measures of socioeconomic development. They were able to identify three large groups of states within Brazil. The poorest set persists with a pattern that resembles that of India, while the most developed is so different as to resemble that prevalent in Belgium. Generally, the comparison with First World indices indicated that the decline in infant mortality in Brazil during the 1980s was small, suggesting that more specific interventions are needed.

Other selected examples of mortality studies in Brazil are presented in Table 3.16. As for studies in other countries, Johnston et al. (1989), after examining several sociodemographic variables, verified that educational attainment of the mother was significantly related to child mortality in a disadvantaged community located on the fringe of Guatemala City. Causes of death in the Puna de Atacama region in the period 1890–1950 were investigated by Bejarano *et al.* (1997). They found that the patterns were similar in Susques (Argentina) and San Pedro de Atacama (Chile), confirming that these populations are a part of the same highland ecosystem present there. García-Moro and Hernández (1997), on the other hand, investigated the patterns of mortality present in the Chilean population of Tierra del Fuego during 1890 and 1995.

Sex ratio and twinning

Two large studies (Feitosa and Krieger, 1992, 1993) analyzed the factors that may influence sex ratio at birth, both of liveborn and stillborn babies. The samples involved from one million to two million subjects, born in 11 Latin American countries (Costa Rica, Colombia, Venezuela, Ecuador, Brazil, Peru, Bolivia, Paraguay, Argentina, Uruguay and Chile), and were based on hospital records obtained through the Latin American Collaborative Study on Congenital Malformations (ECLAMC). The first study included the period 1982–86, and the second that between 1967 and 1986. Secular, spatial, biological and socioeconomic variables were found to influence the sex ratio and, notably, both among live or stillbirths, it was found that this ratio is decreasing with time in a parabolic fashion.

Beiguelman *et al.* (1995) investigated the annual variation in sex ratio in two southeastern Brazilian maternity hospitals from 1984 to 1993. They found that while the sex ratio variation among singletons was very small, that of twin births was extremely high. This may be a consequence of the variability occurring for male or female monozygotic twins.

The frequency of twinning in the state of Nuevo León, México during 1977 and 1978 was investigated by Garza-Chapa *et al.* (1984), while Carrera de Boscán and Rodríguez (1984) studied it in Caracas, Venezuela (year 1991) and Campana and Roubicek (1996) in the Province of Buenos Aires (period, 1982–95). A much rarer event, the birth of conjoined twins, was considered, taking into account 1.7 million births registered in the Latin American Study of Congenital Malformations (ECLAMC) during 1967–86 in 11 Latin American countries. The prevalence rate observed was 1/75 000 births (Castilla *et al.*, 1988). Baena de Moraes *et al.* (1989), on the other hand, tried to verify whether the declining twinning rates observed in several countries also occurred in Brazil. The population studied was in Campinas, in the State of São Paulo, and the period investigated was from

1965 to 1985; the answer was positive. The decline in dizygotic (DZ) twinning was attributed to the reduction of the fertility advantage of the more fecund DZ twin-prone women by the introduction of effective birth control, while other mechanisms could be causing the decline independently of zygosity. This problem was considered for a more extended and earlier period of time (1925–65) by Beiguelman and Villarroel-Herrera (1993) in the same population. They verified that the decline started before the introduction of anovulatories, due to a decrease in the proportion of African-derived parturients and a fall in the parity of the European-derived women. Finally, Beiguelman *et al.* (1997) could not find an association between twinning rates and social class in either European-derived or African-derived mothers from Campinas.

Demography and genetics

It was indicated in the first section of this chapter that population structure may significantly affect patterns of genetic variation. The factors involved may include migration (and data about it were presented in Tables 3.9 and 3.10), as well as mating preferences (Tables 3.11 and 3.12). Other types of demographic measures that can be used to infer genetic structure were described and the pertinent data reviewed in Salzano and Freire-Maia (1970). Again, Gabriel W. Lasker was among the first to point out the importance of these approaches (see, for instance, Lasker and Kaplan, 1964), and more recent studies covering different aspects of two isolated Brazilian communities have been performed by Freire-Maia and Cavalli (1978), Freire-Maia *et al.* (1978) and Souza and Culpi (1992).

A convenient way of integrating mortality and fertility studies is Crow's (1958) Index of Opportunity for Selection. In principle, the index would indicate how much change in mean population fitness may occur under the assumption that all variation in offspring number is genetically determined. Other factors, however, may influence mortality and fertility. Therefore, only relatively high values (or high differences between populations) may be of genetic significance (Adams and Smouse, 1985).

Bearing this in mind we can examine the data obtained in different Latin American populations concerning this index (Table 3.17). It can be separated into two components, one due to mortality (Im) and the other to fertility (If). Among the 21 estimates presented in Table 3.17, 62% presented higher Ifs than Ims, pointing to the importance of fertility differences among human populations. Total indices (Is) with values below, say 0.40, do not contradict the null hypothesis of no genetic variation in fertility, but much higher values (2.16, 2.46) were obtained in México or Chile (1.78, 1.79). All of these populations are agrarian communities, and the high opportunity for selection values are mainly due to high mortality components.

The future

In demographic and sociologic terms we should expect, for Latin America (always remembering the considerable heterogeneity that exists within the region) the following future trends. 1. Declines of mortality and fertility rates. 2. Increase in the percentage of old (above 65 years of age) people. For Brazil, it is expected that from the present 5% this value will increase to 11% by 2030 (Berquó, 1989; see also Moreira, 1998). As a consequence, specific policies should be implemented for this segment of the population (Branco, 1991). 3. Changing patterns in the relationships between the sexes: higher divorce rates, increased frequency of consensual unions, higher proportion of females without partners (in the population older than 65 years of age; in Brazil in recent times, 76% of males were married, against 32% of females; Berquó, 1989). Women live alone for several reasons including: they survive longer than men (and the latter prefer to marry younger females) also, increasingly women are choosing to remain alone, and to raise or adopt children independently of a male partner.

The Latin American economy was never very healthy, and suffered a lot with the Mexican crisis of 1994, as well as with the more recent (1998, 1999) Asiatic and Brazilian stock market problems. This leads to the staggering figure of the increase, in the region, of people falling into poverty at a rate of two per minute (Londoño, 1996)! The reversal of this pitiful situation is not easy. The World Bank estimated that an annual total of US$ 60 billion up to 2005 in investment (electricity, transportation, telecommunications, water and sanitation) would be necessary (Burki and Edwards, 1996). Another key problem is education. Presently the Latin American worker has, on average, just a little more than 5 years of instruction. For 2005 the need would be for an average of 7–8 years, depending on the rate of economic growth (Londoño, 1996).

A general formula for the easing of Latin American economic problems was given by Londoño (1996): fairer financing, pluralist and responsible organization, incentives to efficiency and quality, more competitive rendering of services, and consumers with greater and more informed choice. Granted that, rising investment in Latin American human potential could not only improve economic growth and social indices, but also provide a stronger basis for the advent of a representative democracy.

A key initiative that could assure success in these endeavors is the conjunction of efforts among nations. At the south of the continent Argentina, Brazil, Paraguay and Uruguay are developing a series of measures with the objective of unifying their economies through the MER-COSUL, a common market enterprise. In the north Brazil, Colombia, Ecuador, Guiana, Peru and Venezuela are devising ways to develop the Amazonian region with a minimum of environmental impact (Aragón and Imbiriba, 1989). Many dilemmas face Latin Americans in the road to development, as exemplified by the cultivation of sugar cane, which can be used either for food or fuel (Brown, 1980). Development having as its main target a more equitable and happy life for all should be the object of our efforts.

Table 3.1. *Selected socioeconomic information about Middle American countries*

Country	Annual demographic growth (%)	Life expectancy (years) at birth M/F[a]	Average no. children[b]	Infant (first year) mortality (‰)	Gross internal product (US$ million)	External debt (US$ million)	Index of human development[c]
Antigua and Barbuda	0.4	71/76	1.7	18	493	370	0.89
Bahamas	1.5	70/77	1.9	14	3460	—	0.89
Barbados	0.8	74/79	1.7	9	1742	574	0.91
Belize	2.5	73/76	3.7	30	578	261	0.81
Costa Rica	2.1	74/79	2.9	13	9233	3800	0.89
Cuba	0.4	74/78	1.5	9	16585	—	0.72
Dominica	0.8	67/71	2.3	34	227	93	0.87
Dominican Republic	1.7	69/73	2.8	37	11277	4259	0.72
El Salvador	2.2	66/72	3.1	36	9471	2583	0.59
Grenada	−1.1	68/73	3.8	—	276	113	0.84
Guatemala	2.8	65/70	4.9	44	14489	3275	0.57
Haiti	1.9	53/56	4.6	72	2043	807	0.34
Honduras	2.8	67/72	4.3	45	3937	4567	0.57
Jamaica	0.9	72/77	2.4	13	4406	4270	0.74
México	1.6	69/75	2.7	33	250038	165743	0.85
Nicaragua	2.6	66/71	3.8	46	1911	9287	0.53
Panama	1.6	72/76	2.6	23	7413	7180	0.86
St Kitts and Nevis	−0.5	63/69	2.4	30	225	56	0.85
St Lucia	0.7	62/72	2.9	17	556	128	0.84
St Vincent and Grenadines	0.6	71/74	2.3	19	256	205	0.84
Trinidad and Tobago	0.8	71/76	2.1	13	5327	2556	0.88
Average	1.3	68/73	2.9	28	16378	11059	0.76

[a]M, Males; F, Females.

[b]Average number of liveborn children per woman aged between 15 and 49 years.

[c]The index of human development combines life expectancy, degree of schooling, and *per capita* income variables in a single number. It varies from 0 to 1, the higher value representing better living standards.

Source: Camargo (1998), based in several international publications.

Table 3.2. *Selected socioeconomic information about South American countries*

Country	Annual demographic growth (%)	Life expectancy (years) at birth M/F	Average no. children[a]	Infant (first year) mortality (‰)	Gross internal product (US$ million)	External debt (US$ million)	Index of human development
Argentina	1.3	70/77	2.6	22	281 060	89 747	0.88
Bolivia	2.3	60/63	4.4	69	6131	5266	0.59
Brazil	1.4	64/71	2.4	42	749 000	178 200	0.78
Chile	1.4	72/78	2.4	12	67 297	25 562	0.89
Colombia	1.7	68/74	2.7	26	76 112	20 760	0.85
Ecuador	2.0	67/72	3.1	36	17 939	13 957	0.77
Guiana	1.4	61/68	2.3	58	595	2104	0.65
Paraguay	2.6	67/72	4.2	41	7743	2288	0.71
Peru	1.7	66/71	3.0	47	57 424	30 831	0.72
Surinam	0.7	69/74	2.4	24	333	12 390	0.79
Uruguay	0.6	70/76	2.2	18	17 847	5307	0.88
Venezuela	2.0	70/76	3.0	23	75 016	35 842	0.86
Average	1.6	67/73	2.9	35	113 041	35 188	0.78

[a] Average number of liveborn children per woman in the age interval between 15 and 49 years.
Source: Camargo (1998), based in several international publications.

Table 3.3. *Poverty in selected Latin American countries (%)*

Country	Poverty	Extreme poverty
Middle America		
El Salvador	38	10
Honduras	53	32
Jamaica	34	—
México	26	8
Nicaragua	50	19
Trinidad and Tobago	21	11
Average	37	16
South America		
Argentina	18	3
Bolivia	72	—
Brazil	43	—
Chile	24	9
Colombia	33	18
Guiana	43	28
Paraguay	20	3
Peru	54	21
Average	38	14

Source: Burki and Edwards (1996).

Table 3.4. *Wealth distribution in selected Latin American countries*

Country	Poorest 20%	Richest 20%	Gini coefficient[a]
Middle America			
Costa Rica	4.0	50.8	0.43
Dominican Republic	4.2	55.6	0.47
Guatemala	2.1	63.0	0.54
Honduras	2.7	63.5	0.54
Jamaica	6.0	48.4	0.40
México	4.1	55.9	0.47
Panama	2.0	59.8	0.52
Average	3.6	56.7	0.48
South America			
Bolivia	5.6	48.2	0.40
Brazil	2.1	67.5	0.57
Colombia	3.6	55.8	0.48
Chile	3.7	62.9	0.52
Peru	4.9	51.4	0.43
Average	4.0	57.2	0.48
18 First World countries			
Average	6.3	39.9	0.33

[a]This coefficient was estimated from World Bank figures and is an estimate of the relative concentration of wealth. A figure of 1.0 indicates a completely concentrated system, 0.0 a perfectly equitable one.
Source: Vandermeer (1996).

Table 3.5. *Socioeconomic conditions in 23 Latin American countries, compared with those prevailing in the USA and Canada: a multivariate comparison*

Rank in development	Countries	Average of the First Principal Component
1. Maximum	1.1. USA, Canada	−0.275
2. Average superior	2.1. Cuba, Trinidad and Tobago, Uruguay, Chile	−0.116
	2.2. Jamaica, Costa Rica, Panama	−0.070
	2.3. Venezuela, México, Brazil	−0.019
	2.4. Argentina, Guiana (isolated)	
3. Average inferior	3.1. Colombia, Dominican Republic	0.111
	3.2. El Salvador, Ecuador, Paraguay, Peru, Nicaragua, Honduras, Guatemala	(general)
4. Minimum	4.1. Bolívia, Haiti	0.240

The techniques employed were Principal Component Analysis and Cluster Analysis, considering the following variables: 1, mortality rates, less than 5 years of age, per 1000 livebirths; 2, same, less than 1 year of age; 3, annual births divided by total deaths in children 0–4 years; 4, *Per capita* Gross National Product; 5, life expectancy at birth; 6, literacy, percentage of persons above age 15 years who can read and write; 7, number of radio sets per 1000 inhabitants; 8, number of TV sets per 1000 inhabitants; 9, proportion of the population of age less than 5 years; 10, gross death rates; 11, gross birth rates; 12, number of children per woman past reproductive age; 13, percentage of the population living in urban areas; 14, average inflation (1980–87); 15, percentage of births assisted by trained health personnel; 16, rate of maternal mortality (deaths caused by gestation problems).
Source: Curi (1991).

Table 3.6. *Selected sources for the study of historical demography in Latin America*

Region or nation	Content	Authors
Latin America	Detailed historical analyses on populations of 9 countries	Nadalin *et al.* (1990)
Argentina	Mating preferences of the Spaniards who migrated to Buenos Aires, 1890–1900	Caratini *et al.* (1996)
Bahamas	Population size and basic demography, Bimini Island	Halberstein (1980)
Brazil	Inbreeding and mating patterns, many localities	Saldanha (1960, 1962*a*), Freire-Maia (1971), Frota-Pessoa (1971)
	Japanese immigration, 1908–1960	Guaraciaba (1966)
	General pattern of migration and admixture	Salzano and Freire-Maia (1970)
	Extensive analyses of the Afro-Brazilian population over time	ABEP (1988), Luna and Costa (1982), Costa (1983)
	General bibliographic review of Brazilian studies	Samara and Costa (1984)
	Historical statistics, economic, demographic, and social series, 1550–1985	IBGE (1987)
	An introduction and four contributions on concepts in historical demography, with emphasis on Brazil	Bassanezi (1997)
	Evolution of consanguineous marriages in Florianópolis, 1915–1980	Agostini and Meirelles-Nasser (1986)
Chile	Inbreeding in the Province of Valparaiso, 1917–1966	Lazo *et al.* (1970)
	Inbreeding in the community of Olmué, Valparaiso, 1904–1913, 1942–1954	Quezada and Barrantes (1973)
	Complete genealogy of the inhabitants of Easter Island	Englert and Cruz-Coke (1975)
	Mating structure, rural community of Puchuncaví	Pinto-Cisternas *et al.* (1977)
	Inbreeding in the Elqui Valley, 1875–1970	Zuñiga Ide (1980)
	Mating structure, community of Porvenir, Tierra del Fuego, 1897–1965	Hernández and García-Moro (1997)
Costa Rica	General review of African-derived persons	Melendez and Duncan (1974)
	Population structure and inbreeding in the Parish of Dota, 1888–1962	Barrantes (1975, 1978)
	Births, sex-ratio, twinning, mortality, inbreeding and isonymy in the parish of Escazú	Madrigal (1992, 1993, 1996, 1997) Madrigal and Ware (1997)
French Guiana	Historical information about the French colonization and its relationship with the Indians	Hurault (1989)

Table 3.6. (*cont.*)

Region or nation	Content	Authors
Guadeloupe	Mating pattern in St Barthélemy, a small island (25 km²), 1862–1961	Benoist (1966), Leslie (1983)
	Parent choice and child mortality in this same island	Brittain *et al.* (1988), Brittain (1992)
México	Evolution of the population with emphasis in the admixture between Indians and non-Indians	Romero Molina *et al.* (1976)
	Evolution of the population from prehistoric times to present, Tlaxcala	Crawford (1976)
	Temporal changes, 1900–1980 in a Zapotec-speaking community in the Valley of Oaxaca	Little and Malina (1989)
Peru	Six communities, general demographic information, isonymy 1908–1958	Lasker and Kaplan (1964), Lasker (1969)
	Six communities, Ichu River Valley, endogamy and exogamy rates, 1825–1914	Pettener *et al.* (1998)
Uruguay	General review of the colonization process	Sans (1992)
Venezuela	Inbreeding in Colonia Tovar, 1900–1977	Pineda *et al.* (1984)
	Inbreeding and isonymy, Birongo and La Sabana, 1900–1965	Castro de Guerra *et al.* (1990b), Pinto-Cisternas *et al.* (1990a)
	Isonymy and mating patterns, Quibor, 1866–1982	Pinto-Cisternas *et al.* (1990b)
Virgin Islands (USA)	Mating pattern in St Thomas, 1910–1960	Dyke (1971)

Table 3.7. *Selected information about migratory movements in Latin American countries or regions*

Country or region	Information	Authors
Argentina	The part played by immigration in its population growth is unique in the western hemisphere. Between 1881 and 1910 there was an annual net inflow of 2–3% added to the total population. In 1914 30% of the population were foreign-born. A large portion of them were from adjacent South American countries, but 3.4 million immigrants from overseas entered and remained there in the period 1857–1930. Italians and Spaniards (in the ratio 5:3) represented 80% of all immigrants	Crossley (1985)
	In 1936 the number of Argentinians of German origin was estimated as 210 000	Kunter (1987)
	African immigration: the slave population of Buenos Aires was estimated as 94 in 1726, but as 7802 in 1827	Caballero (1990)
	Spanish integration in Buenos Aires, 1890–1900: about equal number of endogamic (Spanish × Spanish) and exogamic (Spanish × non-Spanish) marriages	Caratini *et al.* (1996)
	Information about a borderline province (Jujuy): proportion of unions between persons locally born: 62% only, 97% of the foreign migrants were Bolivians	Alfaro and Dipierri (1996)
	Percentage of the urban population: 88	Camargo (1998)
Bolivia	Its population distribution has been changing, in part due to the rising importance of Santa Cruz. Percentages in four areas, in 1950–76, are as follows: Province of La Paz: 11/15; Altiplano: 42/33; Valle: 27/23; Oriente: 20/29	Smith (1985)
Caribbean Islands	About 4 million Africans were brought to the islands as slaves	Niddrie (1985)
	East Indians: 272 000 between 1838 and 1917; 200 000–400 000 Haitians now live in the Dominican Republic. The flux of people into, within, and out of the region is intense	Hoetink (1997)
	Percentage of the urban population: 61	Camargo (1998)
Chile	The country has *una loca geografía* (crazy geography; Benjamin Subercaseaux), being 4200 km in length, but having an average width of less than 200 km. The population concentrated mainly in the central provinces (in %: 1854, 70; 1884, 60; 1920, 55; 1940, 56; 1960, 59; 1982, 58) In the 1980s 39% of the country's population lived in Santiago's metropolitan region. Note that in the 1970s 300 000–700 000 persons left the country for political reasons	Blakemore (1985)

Table 3.7. (*cont.*)

Country or region	Information	Authors
	The number of Spanish migrants has been growing steadily from 1535 to 1960, the same occurring, but to a lesser extent, with Germans. African migrants entered the country in the sixteenth and seventeenth centuries, but in small numbers, and their genetic contribution was absorbed by those of other ethnic groups	Rothhammer (1987)
	Detailed examination of the migration into metropolitan Santiago	Elizaga (1970)
	Percentage of the urban population: 84	Camargo (1998)
Costa Rica	Good information about the European early settlers. Sixty percent of the Spaniards came from Andalucia, Extremadura and Castilla.	Melendez (1982), Morera (1995)
	The main port of entry of African-derived persons has been Puerto Limón. In 1927 the percentage of these persons was 4% of the country's total population, 94% of them living in the Province of Limón, while the figures for 1950 were respectively 2% and 92%	Melendez and Duncan (1974)
	Detailed data about internal migrations (1883–1950)	Robles (1990)
	Percentage of the urban population: 50	Camargo (1998)
Colombia	The country's population increased from 4 million in 1905 to 27 million in 1980; the contribution of Bogotá's population for these totals also increased from 2% to 15%	Robinson and Gilbert (1985)
	Percentage of the urban population: 73	Camargo (1998)
Cuba	In the period 1902–1919, 450 000 Europeans migrated to Cuba, 97% of them Spaniards. In the same period the country received 208 000 people from other American countries, of which 24% emigrated from Jamaica and 21% from the USA	Niddrie (1985)
	From 1935 to 1954 the migratory flux was reversed, with large number of people leaving the island. This was interrupted during 1955–1959, following the Marxist revolution, but then emigration increased again, due to those who were opposed to Fidel Castro's regime. Between 1960 and 1980 the net migratory flux indicated a negative balance of 709 913 people	Hernández Castellón (1986)
	The total of Cubans who arrived in the USA between 1959 and 1993 is estimated as 862 000. In 1980 emigrants represented 1.3% of the island's population	Chávez (1996)
	Percentage of the urban population: 76	Camargo (1998)

Table 3.7. (*cont.*)

Country or region	Information	Authors
Guiana	Between 1846 and 1917 some 239 000 East Indians entered the country	Robinson *et al.*(1985)
	Percentage of the urban population: 36	Camargo (1998)
México	This is the most populous Spanish-speaking country in the world. Immigration has always been low. The main contribution from Europe was from Spain. Africans came through the East Coast, but also through Acapulco, and the population from the West Coast shows indications of African influence. The degree of non-Indian heritage varies in different regions of the country. In 1970 there were only 191 234 foreigners residing in México, and almost half were from across the border (USA or Guatemala). On the other hand, the rate of emigration is high, especially to the USA. It is estimated that over 16 million Mexicans settled legally in the USA in the last 100 years	Romero Molina *et al.* (1976), Lisker (1981), Fox (1985), Faulhaber and Schwidetzky (1986)
	Interstate migration is important, with metropolitan regions as centers of attraction	Winnie (1965), Fox (1985)
	As many as 17 million people now live in México City	Cerda-Flores *et al.* (1991)
	Percentage of the urban population: 75	Camargo (1998)
Paraguay	Immigration from Europe occurred at low levels, and introduction of people of African ancestry was also in small-scale. Interestingly, in 1811 the dictator J. G. Francia prohibited marriage between 'pure' Spanish people, when generally elsewhere restrictions were made to exogamic unions. The objective was to decrease Spanish power after independence. Until World War I the majority of the immigrants were Spaniards or Portuguese. Afterwards German Mennonites and Japanese were also received, but in 1950, from a population of 1.4 million, only 0.03% were foreign-born. By 1979 300 000 Brazilians had settled spontaneously along the border between the two countries. On the other hand, it is estimated that now around 500 000 Paraguayan political refugees live in Brazil and Argentina	Crossley (1985), Kunter (1987)
	Percentage of the urban population: 53	Camargo (1998)
Peru	Metropolitan Lima (about 5 million inhabitants) is over 10 times larger than the second largest city in the country. Changes in population distribution occurred mainly at the expenses of	Smith (1985)

Table 3.7. (*cont.*)

Country or region	Information	Authors
	the Sierra region, as the figures indicate when comparing percentages of the population in 1876 and 1972 in three contrasting areas: Coast, 24/48; Sierra, 71/42; *Selva*, 5/10	
	Percentage of the urban population: 72	Camargo (1998)
Uruguay	Immigrants, mainly Italian and Spanish, totalled 650 000 between 1836 and 1926, and a quarter of these arrived in 1904–13. In 1860 half of Montevideo's population comprised of foreigners. Presently, half of the country's population lives in Montevideo	Crossley (1985), Sans (1992)
	Extensive data about foreigners, 1840–1980, for Tacuarembó, Melo and Montevideo	Sans *et al.* (1996)
	Detailed information about the arrival of African slaves in Montevideo (1769–1829). In 1829 African-derived individuals represented 15% of the city's population	Caballero (1990)
	Percentage of the urban population: 90	Camargo (1998)
Venezuela	In 1891 Caracas represented 4% of the country's population, while in 1980 this proportion had been raised to 24%	Robinson and Gilbert (1985)
	Detailed information about the growth and composition of Caracas. In 1861 there were 6603 foreigners in this city, mainly Spanish, French, Italians and Germans, constituting 10% of the city's population. By 1950, this value had increased slightly to 12%. In 1950 only 45% of the population had been born there, a value that had increased to 63% by 1990. In 1990 33% of the foreigners were Colombians, 17% Spaniards and 12% Portuguese	Guerra-Cedeño (1996)
	Percentage of the urban population: 93	Camargo (1998)

Table 3.8. *Selected information and bibliographic sources about migratory movements which led to the formation and diversity of the Brazilian population*

Information	Authors
Formation	
Discussion of the relative role that European and modern Asiatic immigration, as compared with indigenous growth, had in the formation of the present population. The most recent estimate places it at 18%. It should be remembered, also, that the immigrants settled mainly in the southeast and southern regions	Levy (1974, 1996), Clevelario (1997)
European and modern Asiatic contributions	
(a) Previous information	Salzano and Freire-Maia (1970), Bergmann (1977) Salzano (1987)
(b) Parental groups	Guaraciaba (1966), Frozi and Mioranza (1975), Lando and Barros (1976), Koch (1980), De Boni (1987, 1990, 1996), Klein (1989), González Martinez (1990), De Boni and Costa (1991), Timm and Timm (1992)
(c) Specific studies about the integration	Klein (1989), Andreazza and Nadalin (1994), Salles (1996)
The African contribution	
(a) Parental groups and adaptation	Goulart (1975), Chiavenato (1986), IBGE (1987), Salles (1988), Rodrigues (1988)
Internal migrations	
(a) Global assessments	Salzano (1971), IBGE (1976), Carvalho (1985), Taschner and Bógus (1986), Martine (1994), Ferreira (1996)
(b) To metropolitan regions	Camarano (1986)

Table 3.8. (*cont.*)

Information	Authors
(c) Rural–urban–rural	Beltrão and Migon (1989)
(d) Within the State of São Paulo	Cunha (1987)
(e) Within and among counties	Martine (1984), Machado and Hakkert (1988)
(f) Within the city of Rio de Janeiro	Smolka (1992)
(g) Percentage of the urban population: 78	Camargo (1998)
Emigrations	
(a) It is possible that the Brazilian population may have decreased by 1.0–2.5 million due to emigration in the past decade	Carvalho (1996)
(b) Estimates of Brazilians in foreign countries: USA, 600 000; Paraguay, 325 000; Japan: 170 000; Europe, 127 000	Patarra (1996)
(c) Study about Brazilian migrants to the USA	Goza (1992)

Table 3.9. *Migration estimates in selected Brazilian communities*

Regions, localities and references	Individual migration	Marital distance: generation			Parent–offspring distance: generation		Exogamy index (%): generation		
		3	2	1	3–2	2–1	3	2	1
North									
Manaus (Santos et al., 1983)	—	—	—	—	—	—	24	9	—
Parintins (Schüler et al. 1982)	116	174	117	35	156	88	13	12	2
Oriximiná (Santos et al., 1987)	255	370	184	49	145	52	12	3	1
Óbidos (Guerreiro et al., 1993)	80	165	102	80	84	138	21	16	7
Santarém (Santos et al., 1996)	379	366	218	74	299	159	9	—	—
Belém (Coimbra et al., 1971; Guerreiro and Chautard-Freire-Maia, 1988)	245	270	192	—	255	—	—	—	—
Castanhal (Ribeiro-dos-Santos et al., 1995)	294	210	221	79	282	203	23	23	9
Northeast									
Migrants (Peceguini and Cabello, 1986)	—	220	81	—	—	—	—	—	—
Lençóis Island (Freire-Maia and Cavalli, 1978)	—	88	—	—	68	—	—	54	—
Natal (Silva et al., 1981)	150	156	128	—	119	—	75	—	—
Aracaju (Conceição et al., 1987)	181	74	—	—	120	—	46	—	—
South									
Curitiba E (Culpi and Salzano, 1984)	385	—	65	—	141	65	—	37	—
Curitiba A (Culpi and Salzano, 1984)	751	—	87	—	348	87	—	39	—
Porto Alegre E (Araújo and Salzano, 1974)	—	274	109	—	—	—	—	—	—
Porto Alegre A (Araújo and Salzano, 1974)	—	258	122	—	—	—	—	—	—

Distances were calculated in kilometers using road maps; in the absence of roads straight lines were used. Generation 1: grandparents; generation 2: parents; generation 3: present. The localities have been ordered in a west–east distribution in the North, and a north–south distribution in the other regions. E, European-derived; A, African-derived subjects.

Definitions: individual migration, the distance between the birthplace of the interviewed individual and the locality studied. Marital distance, the distance that separates the birthplaces of the parents of the individuals studied. Parent–offspring distance, the average distance between the birthplaces of the individual interviewed and those of his or her father and mother. Exogamy index, percentage of marriages between individuals born in different places.

Table 3.10. *Migration estimates in selected Latin American communities*

Locality or region	Period	No. of marriages	Marital distance (km)	Exogamy index (%)	Reference
Carib					
Bimini, Bahamas	Present-generation	191	—	42	Halberstein (1980)
Saint-Barthélemy	1862–1914	297	—	4	Benoist (1966)
	1914–1961	434	—	6	
Nicaragua					
Bluefields	Generation 2	122	135	45	Biondi *et al.* (1993)
	Generation 1	189	157	39	
Costa Rica					
Dota	1918–1962	1068	—	30	Barrantes (1978)
Argentina					
Provincia de Jujuy					
Puna	1987–1992	501	133	26	Alfaro and
Quebrada	1987–1992	513	109	19	Dipierri (1997)
Valle	1987–1992	6972	255	44	
Ramal	1987–1992	3403	195	42	
Provincia del Chaco					
Fortín La Valle	Present generation	36	—	80	Palatnik (1976)
Chile					
Valparaiso	Generations 2+1	723	—	57	Pinto-Cisternas *et al.* (1971*a*)
Valparaiso	1865–1914	50499	—	76	Lazo *et al.* (1978)
Casablanca	1865–1914	3318	—	35	Lazo *et al.* (1978)

Generations 1 and 2 as for Table 3.9.

Table 3.11. *Inbreeding levels, as estimated from church and state legal records, for different periods of time and in 15 Latin American countries*

Country and locality	Period	No. of couples	$F(\times 100\,000)$	Reference
México				
Whole country	1956–1957	28 292	31	Freire-Maia (1968)
Cuba				
Whole country	1956–1957	2277	54	Freire-Maia (1968)
El Salvador				
Whole country	1956–1957	2494	142	Freire-Maia (1968)
Honduras				
Whole country	1956–1957	3759	110	Freire-Maia (1968)
Costa Rica				
Whole country	1954	3833	114	Freire-Maia (1968)
Dota	1918–1962	1068	440	Barrantes (1978)
Panama				
Whole country	1956–1957	350	0	Freire-Maia (1968)
Venezuela				
Whole country	1956–1957	2931	191	Freire-Maia (1968)
Whole country	1967–1979	1517	67	Orioli et al. (1982)
La Sabana	1905–1925	34	0	Castro de Guerra et al. (1990a,b)
	1940–1965	62	328	
Patanemo	Present generation	90	160	
	Past generation	63	0	

Los Teques	1790–1869	979	685	Pinto-Cisternas et al. (1981)
Birongo	1905–1925	34	459	Castro de Guerra et al. (1990b)
	1940–1965	62	792	
Ganga	Present generation	106	260	Castro de Guerra et al. (1993)
	Past generation	50	250	
Colonia Tovar	1843–1977	940	697	Pineda et al. (1985)
Colombia				
Whole country	1956–1957	34470	119	Freire-Maia (1968)
Ecuador				
Whole country	1956–1957	3954	229	Freire-Maia (1968)
Whole country	1967–1979	2261	49	Orioli et al. (1982)
Peru				
Whole country	1956–1957	565	274	Freire-Maia (1968)
Whole country	1967–1979	203	62	Orioli et al. (1982)
Bolivia				
Whole country	1956–1957	4130	28	Freire-Maia (1968)
Brazil				
North	1932–1956	2261	190	Freire-Maia (1957)
Northeast	1932–1956	68797	365	
Center-West	1932–1956	7534	228	
Southeast	1932–1956	121704	191	
South	1932–1956	69776	81	
Whole country	1932–1956	270072	200	
Whole country	1967–1979	2328	54	Orioli et al. (1982)
Whole country	Present generation	16000	182	Sampaio et al. (1986)
Northeast migrants	Present generation	1068	59	Yasuda (1969)[a]

Table 3.11. (cont.)

Country and locality	Period	No. of couples	F(×100 000)	Reference
Northeast migrants	Present generation	1806	35	Peceguini and Cabello (1986)[b]
Lençóis Island	Present generation	76	150	Freire-Maia and Cavalli (1978)
Natal	Present generation	771	351	Silva et al. (1981)
Curitiba	1945–1951	23615	112	Freire-Maia (1957)
	1967	4487	43	Fonseca and Freire-Maia (1970)
	1980	114	21	Bassi (1983)
	1980	629	5	Bassi and Freire-Maia (1985)
Valongo	Average, five generations	20	4774	Souza and Culpi (1992)
Florianópolis	1915–1980	28649	118	Agostini and Meirelles-Nasser (1986)
Porto Alegre	1965–1969	5547	92	Araújo and Salzano (1974)
Argentina				
Whole country	1956–1957	51391	58	Freire-Maia (1968)
	1967–1979	6494	11	Orioli et al. (1982)
	1980–1981	212320	31	Castilla et al. (1991a)
Aicuña	1724–1971	ND[c]	1600	Castilla and Adams (1990)
Uruguay				
Whole country	1956–1957	8822	65	Freire-Maia (1968)
Whole country	1967–1979	921	46	Orioli et al. (1982)
Tacuarembó	1840–1979	11542	130	Sans et al. (1996)
Melo	1840–1979	8453	210	
Montevideo	1840–1979	14578	70	
Chile				
Whole country	1956–1957	28596	74	Freire-Maia (1968)

Whole country	1967–1979	2413	50	Orioli et al. (1982)
Arica	1952–1959	1088	19	Blanco and Chakraborty (1975)
Valle de Elqui (rural)	1900–1975	9938	131	Zuñiga Ide (1980)
Valle de Elqui (urban)	1900–1975	15 245	61	
Caleu	1971	119	604	Blanco and Covarrubias (1971)
Valparaiso (urban)	1888–1967	29 880	64	Villarroel et al. (1980)
Valparaiso, ZI[d]	1917–1966	135 906	45	Lazo et al. (1970)
Valparaiso, ZII[d]	1917–1966	8078	138	
Valparaiso, ZIII[d]	1917–1966	31 961	39	
Valparaiso, ZIV[d]	1917–1966	19 776	58	
Puchuncaví (Valparaiso)	1763–1960	2346	486	Lazo et al. (1973)
Olmué (Valparaiso)	1904–1913	286	654	Quezada and Barrantes (1973)
	1942–1954	374	372	
Santiago	1938–1962	2355	52	Blanco and Chakraborty (1975)
Santiago[e]	1963–1964	684	197	Cruz-Coke (1965a)
Ñuñoa (Santiago)	1850	3825	59	Fernandez et al. (1966)
Ñuñoa (Santiago)	1958–1962	533	14	
Puerto Dominguez	1960–1969	247	392	Blanco and Chakraborty (1975)
Valdivia	1953–1969	3229	22	

[a] A higher value (133) was calculated using a bioassay method. Sample ascertained in 1962–1963.

[b] Sample ascertained in 1969–1970. A higher value (227) was obtained using phenotype assays (Cabello and Krieger, 1991); but on Jacobina, in this same region, Cabello et al. (1997) obtained negative values with this same method.

[c] ND, no data available.

[d] Province of Valparaiso, Zone I (urban), Zone II (dry-agricultural), Zone III (mining-agricultural), and Zone IV (urban-agricultural).

[e] Hospital population.

Table 3.12. *Isonymy studies performed in Latin America*

Population	Period	Methods[a]	Authors
México			
Tzintzuntzan, Paracho	1945–1980	1	Lasker *et al.* (1984)
Monterrey	1990	1,2	Rojas-Alvarado and Garza-Chapa (1994)
Costa Rica			
Escazú	1800–1899	3,4	Madrigal and Ware (1997)
Venezuela			
Colonia Tovar	1900–1977	3,4	Pineda *et al.* (1984)
States of Aragua, Corabobo, Falcon, Lara, Mérida, Sucre	1984	1,5	Rodríguez-Larralde (1989)
Quibor	1866–1875; 1981–1982	1	Pinto-Cisternas *et al.* (1990a)
Central Littoral and north of Barlovento	1780–1967	6	Castro de Guerra *et al.* (1990a)
Birongo, La Sabana	1900–1965	1,4	Castro de Guerra *et al.* (1990b)
			Pinto-Cisternas *et al.* (1990b)
State of Lara	1984	5	Rodríguez-Larralde (1993)
States of Falcón, Mérida, Nueva Esparta, Yaracuy	1984	5–9,11	Rodríguez-Larralde *et al.* (1993)
States Sucre, Táchira	1900–1950	5,8–11,12	Rodríguez-Larralde and Barrai (1997a)
States Anzoátegui, Trujillo	1900–1950	5,7–9,11,12	Rodríguez-Larralde and Barrai (1997b)
Birongo, Ganga, La Sabana, Patanemo	1986–1989	1	Pinto-Cisternas *et al.* (1997)
			Castro de Guerra *et al.* (1999)
Peru			
San José	1908–1958	13	Lasker (1968, 1969)
Cayaltí, Monsefú, Mochumí,	1957–1958	1	Lasker (1978)

San José, Reque	1825–1914	1,14–18	Pettener *et al.* (1998)
Santa Ana Parish		13,19,20	Pettener *et al.* (1997)
(six communities)			
Brazil			
Northeast	1962	21	Azevêdo *et al.* (1969)
Northeast	1969–1970	3,22	Cabello and Krieger (1991)
Argentina			
Puna, Quebrada, Valle, Ramal	1987–1992	3	Alfaro and Dipierri (1997)
Aicuña	1724–1971	ND	Castilla and Adams (1990)
Pocho	1960–1990	3,4,12	Colantonio (1998)

[a] *Methods*: 1. Coefficient of relationship; 2. Coefficient of isonymy; 3. Crow and Mange (1965); 4. Pinto–Cisternas *et al.* (1985); 5. Euclidean distance; 6. Classification in five categories, according with their frequency and distribution; 7. Fisher's (1943) alpha; 8. Measure of microdifferentiation (R_{ST}); 9. Coefficient of consanguinity (ϕ_{ii}); 10. Karlin and McGregor's (1967) n_i; 11. Unbiased isonymy (I_{ii}); 12. Conditional kinship (r_{ij}); 13. Observed and expected frequencies based on random pairing; 14. Shannon index (H); 15. Pielou index (*e*); 16. Crow's (1980) method B; 17. Repeated-pairs counter (RP) and its expected value RP_r; 18. Pairwise similarity index (I_{12}); 19. Within-lineage repeated-pair isonymy (RP_w); 20. Marital isonymy (Im); 21. Adaptation to kinship analysis; 22. Correction of (3) when pedigree information is available.

Table 3.13. *Population numbers (in thousands) in Middle America, over a 55-year period*

Country or territory	Population in 1920	Population in 1975	Degree of increase (×)
Bahamas	55	187	3.4
Barbados	155	263	1.7
Belize	44	147	3.3
Costa Rica	421	2182	5.2
Cuba	2950	9183	3.1
Dominican Republic	1140	5183	4.5
El Salvador	1168	4092	3.5
Guatemala	1450	5976	4.1
Haiti	2124	5956	2.8
Honduras	783	3070	3.9
Jamaica	855	2199	2.6
México	14 500	60 247	4.1
Nicaragua	639	2373	3.7
Panama	429	1650	3.8
Trinidad and Tobago	389	1164	3.0
Netherlands Antilles	55	244	4.4
Guadeloupe	150	399	2.7
Martinique	165	389	2.3
Puerto Rico	1312	3026	2.3
Canal Zone	20	43	2.1
West Indies	280	557	2.0
Other islands	38	97	2.5
Total	29 122	108 627	3.7

Source: Marcílio (1980)

Table 3.14. *Population numbers (in thousands) in South America separated by a 55-year period*

Country or territory	Population in 1920	Population in 1975	Degree of increase (×)
Argentina	8861	26 262	3.0
Bolivia	1918	5272	2.7
Brazil	27 404	107 508	3.9
Chile	3783	10 937	2.9
Colombia	6057	26 397	4.3
Ecuador	1898	7130	3.7
Guiana	295	859	2.9
Paraguay	699	2888	4.1
Peru	4862	15 869	3.3
Surinam	130	461	3.5
Uruguay	1391	3066	2.2
Venezuela	2408	12 735	5.3
French Guiana	26	48	1.8
Total	59 732	219 432	3.7

Source: Marcílio (1980).

Table 3.15. *Behavior in relation to contraception of Brazilian women aged 15–49 years who at the time of the study (1986) had a partner*

Political units	No contraception (%)	Types of contraception (%)		
		Pill	Sterilization	Other
North				
Roraima	27.8	22.2	27.8	22.2
Amapá	25.0	42.8	28.6	3.6
Amazonas	39.1	18.8	37.7	4.4
Pará	38.6	14.8	41.6	5.0
Acre	40.4	23.4	31.9	4.3
Rondônia	37.9	23.6	32.0	6.5
Northeast				
Maranhão	72.4	4.1	22.0	1.5
Piauí	61.5	10.8	24.1	3.6
Ceará	51.1	19.9	20.0	9.0
Rio Grande do Norte	42.5	17.8	29.5	10.2
Paraíba	52.3	16.7	22.8	8.2
Pernambuco	44.2	14.2	35.8	5.8
Alagoas	67.5	9.9	20.9	1.7
Sergipe	51.0	22.7	18.0	8.3
Bahia	47.1	18.9	24.9	9.1
Center-West				
Mato Grosso	57.2	16.8	23.8	2.2
Goiás	31.9	12.8	50.9	4.4
Federal District	24.0	26.6	42.2	7.2
Mato Grosso do Sul	29.8	22.9	43.0	4.3
Southeast				
Minas Gerais	38.7	25.1	25.9	10.3
Espírito Santo	31.5	35.3	29.5	3.7
Rio de Janeiro	27.9	30.0	33.4	8.7
São Paulo	30.6	25.7	30.7	13.0
South				
Paraná	32.7	29.5	31.7	6.1
Santa Catarina	28.7	36.4	22.1	12.8
Rio Grande do Sul	30.0	43.2	15.9	10.9
Brazil	40.2	22.9	29.3	7.6

Source: Oliveira and Simões (1988) based on the Brazilian National Research by Samples of Domiciles.

Table 3.16. *Selected mortality studies among Brazilian populations*

Region	District	Information	Authors
Brazil in general		Mortality in the Brazilian capitals	Freitas-Filho (1956)
		Mortality in the North, Northeast, Center-West, Southeast and South regions	Yunes *et al.* (1976)
		Analysis of the epidemiologic transition. Two distinct patterns were distinguished, characterizing an incomplete transition	Prata (1992)
		Mortality by neonatal tetanus (1979–1987)	Schramm *et al.* (1996)
Northeast	Fortaleza	Detailed analysis of the causes of death, with the calculation of the appropriate indices, 1920–1975	Rouquayrol (1966); Rouquayrol *et al.* (1967); Rouquayrol and Carneiro (1976)
	Natal	$n=221$ early neonatal deaths compared with $n=3367$ controls. Establishment of risk factors. Reduction in these types of death in this population could be accomplished mainly through improved maternal care and prevention of prematurity	Gray *et al.* (1991)
	Recife	Study of 61 neonatal deaths which occurred among 2673 liveborn children	Allain *et al.* (1983)
		$n=1181$ violent deaths. The homicide rate in this city approaches that of El Salvador and Colombia	Lima and Ximenes (1998)
Center-West	Federal District	Deaths in infancy (<15 years), based on statistics of the Secretary of Health. In 1978 it was half that observed in 1963	Pereira and Albuquerque (1983)
Southeast	Minas Gerais	Presentation of Bayesian statistical methods and their application to the state's infant mortality risk in 1994	Assunção *et al.* (1998)
	Duque de Caxias	Violent deaths in the period 1979–1987. The city exists in an area where extreme violence is most prevalent. Main victims are 15 to 39-year-old males	Souza (1993)
	Rio de Janeiro	Same type of data, 1980–1988. The homicide rate generally increased during the period, reaching 45/100 000 in 1988	Minayo and Souza (1993)
		Evolution of neonatal mortality in the state, 1979–1993	Leal and Szwarcwald (1996)

	São Paulo	Maternal mortality, 1993–1996, based on death certificates. High coefficient: 74.3–47.9/100 000	Theme–Filha et al. (1999)
		Discusses the high infant mortality rate of this city (8.6% in 1974), despite being located in the most developed Brazilian state	Laurenti (1978)
	Ribeirão Preto	Relationship between birthweight, social class and infant mortality in a cohort of 9067 children born in eight maternity hospitals	Almeida et al. (1992)
South	Porto Alegre	Relative risks of dying in the first year of life, in 1980, in slums and other areas of the city. The average relative risk of dying was 2.4–3.6 times higher, and as high as 18.2 times for death through septicemia, in slum dwellers	Fischmann and Guimarães (1986)
	Pelotas	Study of a birth cohort of 6011 children, 99% of those born in 1982 in the city. Mortality and health indicators at 12 and 20 months of age. Children born after shorter birth intervals were disadvantaged in relation to the others	Barros et al. (1984); Victora et al. (1985); Huttly et al. (1992)

Table 3.17. *Opportunity for the action of natural selection in several Latin American populations*[a]

Population	Indices of opportunity for selection[b]			Reference
	Im	*If*	*I*	
México				
Whole country	0.49	0.61	1.40	Spuhler (1962)
Cuanalan	1.47	0.28	2.16	Halberstein and Crawford (1975)
San Pablo	1.63	0.31	2.46	Halberstein and Crawford (1973)
Tlaxcala	0.59	0.35	1.14	Halberstein and Crawford (1973)
Village, Oaxaca Valley	0.63	0.69	1.31	Little and Malina (1989)
Belize				
Whole country	0.25	0.64	1.05	Spuhler (1962)
Jamaica				
Whole country	0.21	0.81	1.18	Spuhler (1962)
Barbados				
Whole country	0.26	0.90	1.39	Spuhler (1962)
Trinidad and Tobago				
Whole country	0.13	0.87	1.11	Spuhler (1962)
Guiana				
Whole country	0.14	0.70	0.98	Spuhler (1962)
Brazil				
Whole country	0.50	0.50	1.24	Spuhler (1962)
Argentina				
Fortin Lavalle	0.10	0.17	0.29	Palatnik (1976)
Salsacate	0.08	0.38	0.49	Colantonio and Celton (1996)
Parroquia	0.08	0.42	0.53	Colantonio and Celton (1996)
Chancani	0.08	0.24	0.34	Colantonio and Celton (1996)
Chile				
Arica	0.15	0.45	0.67	Cruz–Coke and Biancani (1966)
Belén	0.33	0.22	0.62	Cruz–Coke and Biancani (1966)
Huallatire	1.38	0.17	1.79	Cruz–Coke and Biancani (1966)
Towns	0.15	0.45	0.67	Crow (1966)
Villages	0.33	0.22	0.62	Crow (1966)
Nomads	1.38	0.17	1.78	Crow (1966)

[a]Only non-Indian communities are listed here, since the latter have been reviewed by Salzano and Callegari–Jacques (1988) and Crawford (1998).

[b]$Im = pd/ps$, where pd are premature deaths and ps is the proportion surviving or $1-pd$. $If = Vf/x_m^2$, where Vf is the variance in offspring number in completed sibships and x_m is the mean number of livebirths per woman at completion of reproductive life.

$I = Im + If/ps =$ Index of Opportunity for Selection; its genetic significance and usefulness are proportional to the genetic component in the phenomena on which it is based (Crow, 1958).

4 *Ecology, nutrition and physiologic adaptation*

Life histories lie at the heart of biology   *Stephen C. Stearns*

Personal and evolutionary destinies

Inevitably and normally, we all are born, develop, mature, age and die. This simple fact, however, has important evolutionary consequences, since survival and reproduction are the key factors which determine individual and species fitnesses. The main life history traits, as indicated by Stearns (1992), are: (a) size at birth; (b) growth pattern; (c) age at maturity; (d) size at maturity; (e) number, size and sex ratio of offspring; (f) age and size-specific reproductive investments; (g) age and size-specific mortality schedules; and (h) length of life. All of them are interrelated, so that disturbance in a given stage of the life cycle may affect the other components in multiple ways.

In humans, culture acts as a buffering system between ourselves and the environment, and the ensuing cultural–biologic interactions often make the interpretation of a given process difficult. The environment has to be partitioned in its physical, biotic and social components; humans' resources, constraints, and specific stressors may lead to short- or long-term responses, at different (individual, populational) levels. These responses can be classified into seven categories: avoid, modify, buffer, distribute, resist, conform, or change. In its most simple characterization, what happens is *energy flow* (Thomas *et al.*, 1979; Thomas, 1998).

As examples, let us consider the extremes of the cycle. Humans are different from most mammals (including nonhuman primates) by the fact that significant quantities of fat are deposited in the fetus *in utero*. Consequently, *Homo sapiens* babies have a fat mass approximately four times that predicted for a mammal of their body size at birth. Kuzawa (1998) suggested that this may be due to the need to store energy, since in infancy the human brain consumes as much as 50–60% of the body's available energy. Storage would also be needed to buffer the disrupted nutrient flow that normally is associated with the transition from placenta to breast, as well as with the weaning process.

103

At the other side of the cycle, the aging process is still poorly under-stood. It is clear that physiologic capacity declines with age, but at different rates. While cell renewal decreases precipitously after age 50, the decline of maximum oxygen consumption starts earlier and is more gradual. Gage (1989) reviewed the state of development of biomathematical models of human mortality, and how they can apply to questions such as the interna-tional variation in mortality patterns. He verified two characteristic pat-terns among contemporary nations: (a) low mid-adult mortality coupled with increasing mortality at the older ages, observed in Latin America; or (b) high mid-adult mortality combined with declining mortality at the older ages, present in Central Africa.

A focus of investigation could also be the reproductive process. Waynforth *et al.* (1998), for example, related life history theory to the timing of first reproduction in Maya and Ache Indian subjects. They found that father absence affects male mating strategy, but that these differences did not translate into reproductive outcomes. Not all predictions were confirmed by the actual data.

Field methods for the study of these questions were reviewed by Weiner and Lourie in 1969, and it seems that a comparable manual with the new techniques now available would be most welcome. An annotated bibliography of topics in human ecology involving aspects as diverse as environmental pollution, rural–urban contrasts, high-altitude adaptation, physical growth, nutrition, menarche and menopause, biocultural interac-tions, aging, secular changes, and physique and sports has been produced by Siniarska and Dickinson (1996). A recent, global evaluation of concepts and methods in body composition was provided in Malina (1999). Less extensive reviews of physical anthropology in México, 1943–1964 (Genovés and Comas, 1964), and 1930–1979 (Villanueva, 1982), and growth studies in Brazil (Marcondes, 1989) are also available, as well as what De Walt (1998) called the political ecology of population increase and malnutrition in southern Honduras. Finally, in the area of growth and development, the treatises by Falkner and Tanner (1986) and Ulijaszek *et al.* (1998) are obligatory references.

Nutrition

General overview (see Potter, 1997)

Latin America has three main food cultures. In areas of Amerindian influence (México through Central America to Colombia, Ecuador, Peru and Bolivia) the traditional staple foods are maize, beans and potatoes. In the tropics of South America and the Caribbean, a large proportion of starchy roots and tubers are consumed. In the temperate areas, however, animal products are more prevalent.

Table 4.1 presents information on the sources of energy derived from two types of foods (cereals and meat) in the countries of the area. We are interested here in main trends only; it should be remembered that considerable heterogeneity may occur within countries. In addition, a series of simplifying assumptions are made in the calculation of dietary energy supplies. The reader is referred to Potter (1997) for details. It is clear, however, that the relative consumption of cereals as compared with meat varies widely among countries and regions. The ratios of per cent cereals to per cent meat are about twice as high in Central America (average: 7.5; range: 3.1–12.6) than in the Caribbean (average: 3.0; range: 0.7–8.8) and South America (average: 2.8; range: 1.3–4.8). Especially high consumption of cereals is observed in Guatemala (63%), El Salvador and Honduras (54%), while high amounts of meat are ingested by Argentinians and Bermudans (25%).

Pulses are a relatively important part of the diet in Central America, Brazil, Cuba and Paraguay. Milk and dairy products represent about 5% of the energy supply in the region. Vegetables and fruits (especially plantains and bananas) are important in the Caribbean, furnishing about 20% of the energy intake there. On the other hand, fat content is not high anywhere (20–25% of dietary energy supply). Consumption of sugar is high in Cuba and Costa Rica. Alcoholic beverages account for about 3% of the energy intake in Latin America as a whole.

Hladik *et al.* (1993) provided an ample review of the tropical forests as sources of food for the people of Latin America and other regions. Many sections of their book deal with the Amazonian region, but there are also reports about México. Of special interest is a contribution by Hladik (1993), with an analysis of sugar perception in forest populations. He argued that in the 'sweet biochemical environment' of the forest there is no reason for a high taste sensitivity to evolve, while this is not true elsewhere. Nishizawa and Uitto (1995), on the other hand, considered different aspects of the management of aquatic and land fauna and flora for food production in Brazilian and Peruvian Amazonia, as well as in the semi-arid Brazilian northeast.

The general ecological conditions of Latin America were reviewed by Toledo and Castillo (1999), while ways of developing apparently barren land for agriculture were given in Paterniani and Malavolta (1999), exemplified by the 'cerrado' area of Central Brazil. Olivo and Ablan (1999) considered the main characteristics of the availability of food energy in Venezuela in the period 1970–1996. Ochoa-Díaz López *et al.* (1999), on the other hand, correlated farmers' socioeconomic conditions with their children's health in La Fraylesca, Chiapas, México. The survey involved 1046 households (5546 individuals), anthropometric measurements, a 24-hour dietary recall, stool tests, and childhood mortality data. They found that in this area the volume of maize production as a classification method proved more valuable than land tenure in predicting health status.

Under- and overnutrition

Malina (1996) stated that while the percentage of chronically undernourished people has decreased from 19% in 1969–71 to 13% in 1988–90, their absolute number has increased from 54 to 59 million. He compared the percentage of low stature (more than two standard deviations from the median) with the amount of energy available in the form of food and the index of human development (see Chapter 3) for 14 Central and South American countries. As expected, these percentages are inversely related to the energy available and the index values.

   In Latin America, about 54% of employed men are engaged in heavy work. Based on data previously obtained by L. A. Arteaga, Spurr (1983) considered the relationships between average daily caloric availability, degree of physical activity, and the prevalence of undernutrition and obesity in five South American countries (Bolivia, Colombia, Chile, Ecuador and Uruguay). People from these countries with lower caloric availability are precisely those performing the hardest work and having the highest prevalence of undernutrition, as well as the lowest frequency of obesity.

   Based on studies in Central American rural populations Frisancho (1978) verified that chronic undernutrition has a disproportionately greater effect on childhood ossification timing than on adolescent epiphysical union. Martorell and González-Cossío (1987), on the other hand, after reviewing data from Latin America and elsewhere, concluded that newborns of well-nourished women showed little, if any, benefit from dietary supplementation of their mothers during pregnancy, while newborns of poorly nourished women show substantial improvements in birth weight as a result of these supplementation programs.

   Protein-energy malnutrition is a heterogeneous condition. Studies in México have shown that severely affected patients are biotin-deficient, a previously unrecognized observation. Genetic errors of metabolism may also lead to problems of nutrient metabolism, and treatments have been devised for the control of these situations (Velázquez, 1997*a,b*).

Specific studies

Table 4.2 summarizes information about 21 studies performed in 10 Latin American countries in different periods of time, to exemplify the kinds of approach that have been undertaken on populations of the region. They indicated that (a) biochemical level profiles and fat patterning, as well as interindividual and intraindividual energy and nutrient intake variability, may be different from those seen in First World countries; (b) there is the possibility of other interethnic variations; (c) no significant differences in the nutritional status of Colombian women or children in female-headed, as opposed to male or dual-headed households occur; and (d) there is little influence of nutrition on maximal oxygen consumption or prevalence of periodontal disease.

The influence of culture on dietary habits cannot be overemphasized. Examples are given by McBryde and Costales (1969) in their study in northwestern Colombia. They list a wide range of beliefs in predominantly African-derived or Amerindian-derived communities. Among the former, for instance, people believe that a mother, by eating eggs, can give her baby umbilical tetanus; sweet things cause a child's spleen to grow and produce diarrhea; 'hot' foods (beef, pork and peccary meat) can cause fever. Among the Chocó and Cuna Indians, some of the beliefs are that if a nursing mother eats papaya, guama, or borojó, it will give the baby colic; and after a woman has given birth she must not eat the meat of animals with teeth or fish with scales – if she does the child's naval will not heal. The authors emphasized that such customs limit the diet and can cause malnutrition to the point of health impairment in the members of these communities.

A curious belief involves the dialectical distinction between 'hot' and 'cold' aspects of internal and external environments, which exist in many tropical and subtropical areas, and in people with different biological and cultural backgrounds. It asserts that health is directly related with the natural equilibrium between these two forces. McCullough and McCullough (1974) investigated the biochemical composition of foods considered 'hot' or 'cold' by the Yucatecans of México, and suggested that changes in their proportion may influence electrolyte and water metabolism. This would explain the equilibrium between these two types of food that are sought by the rural people of this Mexican province.

Physical growth

General overview

Physical growth studies are a traditional area in biological anthropology, and much has been investigated and written about it. Unfortunately, a large proportion of these studies is just descriptive and often repetitive, without specific foci of analysis. Many questions remain unanswered, and seven that we consider to be important are listed in Table 4.3. We still do not know quantitatively the relative roles of genetics and environment in the regulation of the process, which specific stages of it are more subject to such differential influences, whether there are gender-diverse sensitivities to environmental insults, the long-term consequences of deprivation in childhood and adolescence, as well as the factors influencing and the uniformity of secular trends.

The methods available to investigate these questions have improved over the years, and this is especially true in relation to the possibilities of statistical analysis. They now make possible the simultaneous consideration of multiple factors, and the testing of specific hypotheses. One question that has remained unanswered is whether cross-sectional or longitudinal investigations should be favored. Both have advantages and disadvantages when compared one with the other. For instance, in cross-sectional investigations it is possible to recruit for study a much larger fraction of a given population, but the time factor can only be inferred. In longitudinal projects numbers have to be smaller, and the loss of subjects due to various factors is a perennial problem. Financing and many other factors also limit the time available for the investigation, the Fels Research Project being a notable exception (Roche, 1992). Ideally, the study should not be limited to children and adolescents, but also include their families and the community where they live. Again, this would involve the mobilization of resources and personnel that are out of reach for many investigation centers.

For the answer to the first question raised in Table 4.3 an analytical approach would be necessary focusing on specific aspects of the process. Mueller (1983), for instance, concentrated on the genetics of fatness, already a complex issue. He concluded, based in part on studies involving Colombian subjects, that somewhat less than one-third of the variation appears ascribable to genetic causes, such factors probably playing a greater role in the childhood/adolescence periods. Whole genome scans and the discovery of substances which may influence the process (such as leptin, the product of the *ob* gene) may sharpen the approach even more.

Stinson (1985) reviewed studies dealing with the second question in Table 4.3. She verified that the strongest support for the concept that males are less buffered than females to environmental stress is found in investigations of the prenatal period. The postnatal investigations yielded much less consistent results. Also contradictory are the studies on responses to environmental improvements. During the prenatal period, males seem to show a greater response to nutritional supplementation, while postnatal catch-up growth is usually greater in females. Some of the studies from which she based her conclusions were conducted in seven Latin American countries, especially Guatemala.

Spurgeon *et al.* (1978) and Spurgeon and Meredith (1979) concentrated mainly on African and African-derived subjects in the West Indies, North, Central and South America. Age changes and group differences were found for hip width relative to lower limb height, and lower limb height relative to sitting height.

Malina (1978, 1989) dealt with a specific period of growth and maturation, i.e. adolescence. He found that well-off Latin American children mature somewhat earlier and gain less during the adolescent spurt than North American or European children. Median ages at menarche for most Latin American series (12.8–13.4 years) tend to be slightly lower than those reported for girls of northwest European ancestry or North American girls of European ancestry. Other factors considered were the interrelationships among indicators of sexual, skeletal and somatic maturation, as well as secular trends in Mexicans living in rural and urban areas of their country and in the USA. At least among the Japanese the significant gain in stature observed between 1957 and 1977 is due almost entirely to an increase in length of the legs. Macías-Tomei *et al.* (2000) also observed an earlier maturation of Caracas upper-middle class adolescents, as compared with Europeans.

Specific studies

Tables 4.4–4.7 attempt to give an overall picture of the physical growth studies performed in Latin America. Due to the volume of the literature, only selected examples could be listed, from papers available to us at the time of writing.

Forty-three investigations performed on six Central American countries or territories, spanning from 1967 to 1998, are briefly summarized in Table 4.4. Some of the conclusions obtained are: (a) chest circumference responds to mild to moderate malnutrition differently than height and weight; (b) growth dynamics are also different for the cortical, periosteal and medullary metacarpus regions; (c) Guatemalans suffer more from chronic energy undernutrition than from protein undernutrition; and (d) delayed growth in childhood may be a significant factor for delayed age at menarche. The application of Waddington's homeorhesis concept to physical growth is appropriate. According to that author, homeorhesis can be conceived as an equilibrium property characterizing the regulatory pathway of development which is the result of many genes controlling different synthetic processes. These processes interact in a way which assures the delimitation of a pathway exhibiting homeostatic (equilibrium) properties, leading to canalization. A good example is the development of subcutaneous fat. Wolanski (1998), comparing data from Mexican, Polish, USA (including Puerto Rican) and Peruvian populations, established a route with the following stages: preschool loss, prepubertal gain, adolescent loss, stabilization, adult gain, and after maximum fatness, loss with age.

We have previously emphasized the importance of considering the growth process in a holistic way, with the multidisciplinary investigation not only of children and adolescents, but of the whole community. An example of such a study is presented in Table 4.5. Summarized there are findings from the Oaxaca Project, coordinated by Robert M. Malina. A total of 39 publications, dated from 1972 to 1998, are listed in the table, classified under eight items: (a) general assessments; (b) skeletal growth; (c) motor performance; (d) menarche; (e) aging; (f) spatial factors; (g) genetic, mortality, and reproductive factors; and (h) secular factors.

The Oaxaca Valley is located in southern México. The area is predominantly rural, the capital of the same name having around 160 000 inhabitants. Communities consist of Zapotec-speaking or Ladino populations. Nutritional conditions are marginal, with high mortality rates. Some points that can be singled out from the summaries of Table 4.5 are: (a) there is negligible influence of socioeconomic variation in estimated growth velocities; (b) stature and weight are significantly correlated with measures of metacarpal cortical bone; (c) the adaptive significance of small body size in conditions of undernutrition varies between populations and with performance tasks; (d) contrary to Guatemalans (see above), protein rather than energy may be the more limiting nutrient here; (e) urban–rural contrasts are minor; (f) in motor performance, sisters resemble each other more than brothers in the undernourished sample, while brothers are more similar than sisters in the better-nourished sample; and (g) both for stature (comparison including long bones from archeological sites and 1895 records as the baseline) and age at menarche, no evidence for secular trends was observed.

Selected data about physical growth studies in South America are presented in Table 4.6. They refer to populations of six countries, the results being ascertained from 45 bibliographic references. One study suggested that better female canalization could explain the pattern of sexual dimorphism observed in a poor neighborhood, while another asserted that independent of ethnic or cultural backgrounds, there is a redistribution of fat away from the extremities towards the trunk. On the other hand, the study of Brazilian children of Japanese ancestry with high socioeconomic status, whose height was similar to those of Japanese in Japan but lower than those of non-Japanese-derived Brazilians, raised again the issue of universal growth standards (see, for instance, Habicht *et al.*, 1974). It is obviously inappropriate to compare indiscriminately the growth of African pygmies with the USA giants studied by the National Center for Health Statistics, as was proposed by the World Health Organization, and the same can be said in relation to South American Indians (Salzano, 1991).

Growth can also be assessed through dentition. Table 4.7 lists eight studies on this subject performed in four Latin American countries. The timing of deciduous teeth eruption seems to be more closely associated with postnatal weight than with birth weight, but a low correlation with four anthropometric measurements was obtained in one investigation. In such dentition, crown diameters of anterior teeth present more variability than the others, indicating different underlying morphologic fields.

A point that was variously considered in several of the papers listed in Tables 4.4–4.6 was the secular trend that may occur in human populations in relation to stature and age at menarche. In most European populations the trends have been remarkable: 1.0 cm increase in stature per decade and 0.3 year per decade, respectively (Malina *et al.*, 1983a; Bentley, 1999). This trend, however, is not universal, and in given populations may even show reversals.

In terms of the populations considered here, Higman (1979) studied historical height data for African slaves in Trinidad, Guyana and other British Caribbean colonies in the early nineteenth century. Modern Afro-Caribbean adult populations are about 10 cm taller than their parental groups. In the same region, van Wering (1981a) observed a 3.4 cm gain in the period 1954–1974 in Aruba, while Pospísil (1969), comparing data from 1919 and 1969 in Cuba, also found a general increase in height, although he did not quantify the differences. Malina (1985) verified that middle class Mexican children in México City showed greater secular changes in stature (1920s to 1980s) than Mexican American children in the USA. In Brazil, Monteiro *et al.* (1994) reported an increase of 1.4 cm in the years 1952–67 and of 2.4 cm between 1967 and 1982; Kac (1998) observed increases in stature for Navy recruits of 0.1 cm/year in cohorts born from 1940 to 1965; while Kac and Santos (1997) reported gains of 0.3 cm per year in enlisted Navy males born from 1970 to 1977. Finally, Bejarano *et al.* (1996) found an average increase of 4 cm in the Province of Jujuy, Argentina from 1870 to 1960, although with large interregional differences.

In contrast, no such trends were observed in rural Colombia (Himes and Mueller, 1977), or in the Mexican Indian populations of Oaxaca (Malina *et al.*, 1983a) and Yalcoba (Daltabuit and Leatherman, 1998); nor in Puno (Gonzales *et al.*, 1982) or Nuñoa (Leatherman *et al.*, 1995), both in Peru. For age at menarche also, while Rona (1975) reported a reduction from averages of 14.7 to 12.6 years in the period 1887–1970 in Chile, and Farid-Coupal *et al.* (1981) a similar decrease (14.5–12.5 years) between 1935 and 1978 in Venezuela, no such trend was found in Oaxaca over the past 80 years (Malina *et al.*, 1983a).

Bogin and Keep (1999) made an overall analysis of stature trends in Latin America, covering a period of more than 8000 years. Both negative and positive trends were discerned, which they interpreted as being reflections of the socioeconomic and political conditions of the populations sampled.

Physiology

General

To live, humans must capture and convert energy. The quantitative evaluation of energy flow, however, is not an easy task. Ulijaszek (1992) reviewed the methods available, emphasizing recent advances. In the consideration of the difference between mean energy intake and expenditure where both measures were taken from the same subjects he mentioned studies in Guatemala from lactating and nonlactating women, as well as those engaged or not in agricultural work. They showed different negative energy balances, although the data should be considered cautiously, since it is easier to measure energy expenditure than intake.

Shephard (1985), on the other hand, considered the factors related specifically to working capacity. Its evaluation should include aerobic power, body composition, physical performance, and muscular endurance. The results should be interpreted in relation to size, nutrition, fluid balance, lifestyle, family size, and general health. After reviewing different methods and studies he emphasized that the relative contributions of genetics and environment to these variables is still poorly understood. More extended data on two- or three-generation biological and/or cultural relatives are needed.

A series of studies on oxygen consumption and energy expenditures have been conducted by Spurr *et al.* (1992, 1994, 1996, 1997) in children and adults living in Cali, Colombia. Differences between upper and lower socioeconomic classes were small and nonsignificant. In the averaged data of energy expenditure by women at home they observed a gradual increase in the morning, a decline at noon, some increase in early afternoon, and decline in late afternoon and early evening.

A high prevalence (56%) of hypovitaminosis A was observed in children born in public hospitals of Rio de Janeiro, Brazil, by Ramalho *et al.* (1998). The deficiency was independent of other nutritional and anthropometric parameters, such as low birth weight or small for gestational age babies. Ryan (1997), on the other hand, considered the effects of iron deficiency and of its eventual supplementation in infants from México, Chile and Costa Rica. In Chilean children he obtained convincing evidence for the relationship between iron-deficiency anemia and cognitive deficits. In Costa Ricans with severe deficiency, mental and psychomotor development did not improve after a 3-month supplementation.

Flinn and England (1995) examined the relationship between cortisol levels (as a measure of stress) and family environment in a rural village of Dominica. They found that the most important correlate of household composition which affects childhood stress is maternal care, and that there are clear differences in the average of cortisol levels between step- and genetic children living in the same home.

Patterns of activity, comfort, and sleep allocation are strongly determined by environmental factors, although they may also be influenced by inner, biological elements. Aguiar *et al.* (1991) observed clear differences in the patterns observed in an Amazonian island, as compared to those found in the adjacent continental town of Belém, or with previous European studies; while Benedito-Silva *et al.* (1998) investigated the proportion of morningness/eveningness types among 260 adults living in five Brazilian cities. Differences among cities were marked, but did not follow a coherent latitude trend. This is a promising area of study, but which needs methodological progress and standardization of procedures.

Blood pressure

Blood pressure is a much studied characteristic, influenced by many factors (for a comprehensive review see Gerber and Halberstein, 1999). Table 4.8 presents information on a general survey and 12 investigations performed in the Caribbean area and South America. They include questions related to heritability and, among the outside influences, affiliation to a given ethnic group, stress (conformation or not to established social rules), migration and its concomitant, urbanization. It is not yet clear how much of the susceptibility of African-derived subjects to high blood pressure is due to biological or environmental factors. Frisancho *et al.*'s (1999*a*) paper suggests that both are important, since the level of this liability varies from one population to another. The identification of the specific factors that may influence the characteristic is of course important for the correct implementation of preventive measures, as well as for the management of the condition after its establishment.

Physique and motor performance

A total of 11 studies undertaken in four Latin American countries is summarized in Table 4.9, additional investigations developed in Oaxaca, México having been already listed in Table 4.5. The relationship between morphology and physiology is a complex one, factors related to genotype, diet, amount of training, or working environments, being some of the factors that should be considered. Motivation is also of paramount importance in physical achievements. On the other hand, no one presently views constitutional types (somatotypes) as something rigidly controlled by heredity (see, for instance, Lasker, 1999). The prevalent notion involves *plasticity*, so that environmental factors should be carefully considered.

This area has obvious applied importance in the fields of child development, medicine and sport. But a given morphotype cannot guarantee good performances, as exemplified by Venezuelan athletes or Cuban dancers (Table 4.9).

Aging

According to the Book of Genesis in the Old Testament, death was the price of knowledge; more prosaically, death was the price of multicellularity (Raff and Kaufman, 1983; Salzano, 1995). In the course of evolution, diversification beyond that found in unicellular organisms was achieved by cell aggregation. At least two cellular types could be found in the first metazoans: *somatic* and *germinative*. The latter retained several properties of their unicellular predecessors, such as nonadhesiveness and immortality, which are nonexistent in the somatic ones. Presently there is a dual control system for cellular proliferation, based on growth factors and programmed cell death (apoptosis).

The process of aging can be considered either at the cellular or organismal level. At the cell level what occurs is a decline in the immune response (especially in T cells) with aging, as well as a reduction in the proliferation of osteoblasts (perhaps leading to osteoporosis). At the organismal level, it is important to distinguish possible 'intrinsic' aging from age-related pathology. Life table analyses from many human populations reveal that age-specific mortality curves appear to converge to what has been called the 'species-specific life span', estimated as 95 years. This could be a point of reversal, with any of the genotypes which have an effect on survival prior to this age exhibiting the opposite effect after it. Mortality curves would present a triphasic pattern, gradual increase for ages 20–70 years; acceleration from 70 to 90 years, and deceleration afterwards (Gage, 1989; Effros, 1998; Bérubé *et al.*, 1998; Schächter, 1998; Toupance *et al.*, 1998).

Aging studies in Latin American populations are not numerous, probably due to the fact that they are generally young populations. But this situation is changing rapidly due to declining mortality rates, as was indicated in Chapter 3, and more investigations may be anticipated in the next years. Be as it may, Himes and Mueller (1977), studying 634 adults from rural Colombia, estimated a loss in stature due to aging *per se* of 0.12 cm/year for males and 0.03 cm/year for females. In Oaxaca, southern México, Malina *et al.* (1982*a*) observed little difference in body weight between sexes after 40 years of age. Veras and Dutra (1993) considered several characteristics of health and the daily activities of 738 old persons living in three distinct areas of Rio de Janeiro, Brazil. Cytogenetic changes in 60 Brazilian elderly subjects (aged 62–96 years), as compared with 60 controls, were investigated by Mattevi and Salzano (1975*a,b*). As has been observed by other investigators, there was an increase in the level of aneuploidy and of structural abnormalities in the blood of the aged persons. On the other hand, older subjects showed a lower number of cells with satellites and associations in the 13–15 acrocentric chromosomes.

An important physiological change which occurs among women at approximately 50 years of age is menopause, the permanent cessation of menstruation. In evolutionary terms, why might natural selection favor the cessation of reproduction in human females only halfway through the maximum life span of the species? Three explanations had been put forward: (a) the 'grandmother hypothesis'; in primates generally quality is preferred over quantity in number of offspring, postmenopausal females would enhance their fitness by acting as grandmothers, or surrogate care-takers for their children's offspring; (b) menopause would be just a by-product of the hominid increase in life span; and (c) pleiotropy, features having high adaptive value early in the life course are selected for even if they result in reduced fitness later in the life cycle. At present explanations (a) and (c) look more probable than (b) (Pavelka and Fedigan, 1991).

Among the studies related to menopause performed in Latin America we can mention those of Carmenate *et al.* (1997) and Díaz *et al.* (1998), who studied respectively 1000 and 337 women, aged 35–60 years, in the La Habana Province, Cuba. Several biodemographic variables were consider-ed, and in the second sample five anthropometric measurements were obtained. Mean age at menopause was found to be 49.45 years. Obesity was common in pre- and postmenopausal women, but the latter presented a higher prevalence of morbidities. In Brazil Zabaglia *et al.* (1998) found no association between lipid profile variables and bone mineral density in 72 postmenopausal women.

Adaptation to high altitude

Around 140 million people worldwide (36 million in Central and South America) live at high altitude, defined as elevations above 2500 m, and a large number of studies have been performed on them. General reviews considering mostly Andeans can be found in Ruffié *et al.* (1977), Baker (1978), Schull and Rothhammer (1990), Eckhardt and Melton (1992), and Moore *et al.* (1998). Compared with acclimatized newcomers, lifelong residents of the Andes and/or Himalayas have less intrauterine growth retardation, better neonatal oxygenation, and more complete neonatal cardiopulmonary transition, enlarged lung volumes, decreased alveolar–arterial oxygen diffusion gradients, and higher maximal exercise capacity. Tibetans have also other physiological distinctions that are adaptive, probably acquired due to their greater generational length of high-altitude residence. To what extent these specific differences are due to developmen-tal adaptation to hypoxia or are a consequence of genetic adaptation is not clear. But the unusual chest morphology and small body size of the Altiplano people may have a strong genetic base.

Fourteen specific studies performed in four countries are listed in Table 4.10. They indicate that (a) the probability of developing acute mountain sickness is related to the difference between arterial oxygen saturation at low and high altitudes; potassium and sodium levels may also be important; (b) the last trimester of gestation and the first 6 months of life seem to be decisive for the process of high-altitude adaptation; (c) delayed age at adrenarche and menarche is probably a direct effect of the high-altitude environment; and (d) there are definite ethnic influences in the adaptation process, due to genetic–environmental interactions.

Relationship between variables, genetic and reproductive factors

Tables 4.11 and 4.12 list studies that gave emphasis to the relationships between variables (seven studies, five countries) or to genetic and reproductive factors (12 studies, four countries), respectively. To these, investigations related to the Oaxaca Project on these same topics (Table 4.5) should be added. It is clear that all the characteristics considered in growth studies or morphological analyses depend, in a complex way, on the interaction between environmental and genetic factors. Their variability over time should also be taken into account. Undernutrition does not mask genetic effects, and in some cases may even make them more conspicuous. Strategies for survival may differ between populations living at different socioeconomic levels, but some patterns of stabilizing selection may be basically the same in all of them. One important problem that may lead to retarded growth is that of children born of adolescent mothers. This type of environmental problem is one of those that could be prevented through appropriate educational campaigns. The deleterious effects of inbreeding could also be prevented in this way, although its effect in normal children seems to be small.

Table 4.1. *Sources of energy derived from food in Latin America*

Country or region	Sources (%) of total energy		
	Cereals (A)	Meat (B)	Ratio A/B
Central America			
Belize	31	10	3.1
Costa Rica	40	10	4.0
El Salvador	54	5	10.8
Guatemala	63	5	12.6
Honduras	54	5	10.8
México	48	10	4.8
Nicaragua	50	5	10.0
Panama	38	10	3.8
Average			7.5 ± 3.8
Caribbean area			
Antigua and Barbuda	27	20	1.3
Bahamas	25	20	1.2
Barbados	29	15	1.9
Bermuda	17	25	0.7
Cuba	33	10	3.3
Dominican Republic	35	10	3.5
Dominica	27	10	2.7
Grenada	25	10	2.5
Guadeloupe	38	15	2.5
Haiti	38	5	7.6
Jamaica	35	10	3.5
Martinique	33	15	2.2
Netherlands Antilles	35	15	2.3
St Lucia	27	15	1.8
St Kitts and Nevis	21	10	2.1
St Vincent	35	10	3.5
Trinidad and Tobago	44	5	8.8
Average			3.0 ± 2.1
South America			
Argentina	33	25	1.3
Bolivia	46	15	3.1
Brazil	35	15	2.3
Chile	46	15	3.1
Colombia	33	15	2.2
Ecuador	35	10	3.5
French Guiana	31	20	1.5
Guiana	48	10	4.8
Paraguay	31	20	1.5
Peru	46	10	4.6
Surinam	50	15	3.3
Uruguay	35	20	1.7
Venezuela	38	10	3.8
Average			2.8 ± 1.2

Source: Potter (1997).

Table 4.2. *Selected information about nutritional studies in Latin America*

Country	Information	Authors
México	Detailed studies on two occasions (1968 and 1978) of a Zapotec-speaking community of the Oaxaca Valley	Reyes *et al.* (1995)
	Patterns of change in a Maya-speaking Yucatán community and evaluation of children's nutritional status, 1986 and 1996	Daltabuit and Leatherman (1998)
Jamaica	Dietary survey in two localities, comparison with anthropometric and hematologic findings	Miall *et al.* (1967)
	Evaluation of the nutritional status of 452 13–14-year-old girls from nine schools in Kingston. Fat patterns may differ in African-derived, as opposed to European-derived subjects	Walker *et al.* (1996)
Trinidad	Approximately 2000 individuals surveyed. Nutrition does not seem to influence the rate of periodontal disease. Independently of oral hygiene, East Indians presented lower prevalences of these diseases than African-derived subjects	Wertheimer *et al.* (1967)
Guatemala	Diet and serum cholesterol levels among 182 Black Caribs living in Livingston. The diets were markedly low in vitamin A, riboflavin and calcium, but high in fats, probably contributing to the high serum cholesterol levels observed	Scrimshaw *et al.* (1961)
	Nutritional survey in 192 rural groups, in six Central American countries. Adequacy of intake levels, family and food distribution, clinical and biochemical data	Flores (1971)
	The effects of a supplementation program developed in four rural villages was less marked for skeletal maturity than for body size	Martorell *et al.* (1979)
Costa Rica	Comparison, in 68 normal adults aged 17–32 years, of six skinfold and four fasting serum results. Upper trunk fatness may be more relevant than anterior waist fatness to biochemical dysfunction	Bailey *et al.* (1987)
Colombia	Nutritional study of eight communities (473 individuals), information about main dishes as well as taboos, superstitions and beliefs concerning certain foods	McBryde and Costales (1969)
	Data on 60 nutritionally normal and 74 marginally undernourished girls 6–16 years of age and 27 upper and 22 lower socioeconomic class women. Maximal oxygen consumption showed a close relationship (correlation coefficient: 0.81) with lean body mass, independently of nutritional or socioeconomic differences	Spurr *et al.* (1994)

Table 4.2. (*cont.*)

Country	Information	Authors
	No significant differences in the nutritional status of women or children in female-headed, as compared to male- or dual-headed households	Staten *et al.* (1998)
	No differences between pregnant and nonpregnant poor women in Cali	Dufour *et al.* (1999)
Ecuador	Dietary data on 221 residents of a rural highland community indicated low interindividual but high intraindividual energy and nutrient intake variability, compared with developed countries	Berti and Leonard (1998)
Peru	Detailed information from District of Nuñoa, comparing nutritional levels in the 1960s and 1980s. Dietary diversity increased among cooperative and town households, but remained unchanged or decreased among community households	Leatherman (1998)
Brazil	Extensive information about the Center for Nutritional Rehabilitation and Education, a multidisciplinary nucleus which provides assistance to children from São Paulo slums	Sawaya (1997)
	$n = 710$ males plus 473 females who work in a state bank. Males showed better food habits than females	Fonseca *et al.* (1999)
	$n = 428$ males and 613 females, investigated at random in Cotia, Greater São Paulo. Prevalences of undernutrition of 3.9% (males) and 6.2% (females), and of overnutrition of 27.5–34.0% (males) and 25.8–43.6% (females), in different age groups	Martins *et al.* (1999)
	$n = 358$ males and 289 females, 22–59 years old, working in a state bank. Overweight was about three times higher in men than in women	Ell *et al.* (1999)
	Analysis of 209 articles published between 1944 and 1968 in the Arquivos Brasileiros de Nutrição	Vasconcelos (1999)
Chile	Survey on 1906 Chileans found no significant differences in periodontal disease between persons with low or high vitamin levels	Barros and Witkop (1963)

Table 4.3. *Some significant questions in physical growth studies*

1. How much of it is influenced by inheritance, and how much by the environment?
2. Are there differences between males and females in environmental sensitivity?
3. Are there population differences in the growth process? At what specific stages do these differences occur?
4. Is it possible to influence significantly the rates of growth through intervention processes?
5. What is the relationship between the body mass index (weight/stature^2) and fatness? Are there specific fatness patterning types, and does the physiology of fatness differ in children and adults?
6. What are the long-term consequences of deprivation during the growth process?
7. Which factors influence secular trends, and how uniform are they?

Table 4.4. *Selected information about physical growth studies in Central America*

Country or territory	Information	Authors
Caribbean	Selected measurements, several countries, ethnic differences, influence of disease, diet and social factors	Ashcroft (1971)
Aruba	$n = 2559$, 0–14 years, secular growth change between 1954 and 1974	van Wering (1981*a*)
	Same sample as above, comparison with Colombian and Guatemalan series. Also considered: pubertal stages and menarche	van Wering (1981*b*)
Belize	$n = 750$ preschoolers (0–66 months), two administrative districts, one coastal and the other inland. Maya and Garifuna (Black Carib) children are more commonly malnourished	Jenkins (1981)
	$n = 123$, 9–14 years, six growth measures, Mopan Maya. Comparison with socioeconomic and demographic variables	Crooks (1994)
Cuba	$n = 3828$, 6.5–20.5 years, 15 characteristics, sample subdivided by ethnic group but no information on socioeconomic variables	Laska-Mierzejewska (1967)
	$n = 4231$, 6–20 years, height and weight, sample separated by ethnic group. Comparison with a 1919 series and other Latin American samples	Pospísil (1969)
	Laska-Mierzejewska (1967) sample compared to Belgians and Africans, five characteristics	Martínez-Fuentes (1971)
	$n = 249$, longitudinal study, growth channeling	Valiente *et al.* (1987)
	$n = 215$, 2–11 months, body fat, principal components analysis	Diaz *et al.* (1989)
	$n = 505$, 2–11 months, four skinfolds, principal components analysis. Nutritional status is related to fatness, but not to fat distribution	Diaz *et al.* (1994)
	$n = 7286$, 5–20 years, four skinfolds, classified by frame size (as estimated by elbow breadth). The larger the frame size, the greater the amount of subcutaneous fat	Martínez *et al.* (1995)
	$n = 709$, first year of life, five measurements, Santa Clara, mixed-longitudinal. Growth curves correlated with birthweights	Enríquez and Martín (1998)
Guatemala	$n = 1119$, 0–7 years, seven skinfolds. Values lower than those of well-nourished children	Malina *et al.* (1974)
	$n = 5029$, 0–7 years, length and weight. Mild to moderate protein-calorie malnutrition does not affect the relationship between these two variables	Yarbrough *et al.* (1975)
	$n = 710$, 1–7 years, mixed-longitudinal study, severe retardation in metacarpal growth	Himes *et al.* (1975)

Table 4.4. (*cont.*)

Country or territory	Information	Authors
	$n = 1119$, 0–7 years, head and chest circumferences. Chest circumference responds to mild to moderate malnutrition differently than height and weight	Malina *et al.* (1975)
	Same sample as above head circumference/chest circumference ratios. They are not good indicators of malnutrition	Martorell *et al.* (1975)
	$n = 710$, 1–7 years, sexual differences, bone periosteal and medullary diameters, cortical thickness and cortical area	Himes *et al.* (1976a)
	Same sample as above. Different growth dynamics for the cortical, periosteal, and medullary metacarpus	Yarbrough *et al.* (1977)
	$n = 164$, longitudinal, monthly measurements. Pre- and postadolescent children followed a seasonal pattern of growth, but adolescent children did not	Bogin (1978)
	$n = 892$, 7–13 years, mixed longitudinal, height and weight, comparison with previous sample of higher socioeconomic status	Bogin and MacVean (1978)
	$n = 981$, 7–13 years, low socioeconomic status. They suffer more from chronic energy undernutrition than from protein undernutrition	Bogin and MacVean (1981)
	$n = 685$, 7 years, European, Guatemalan, and mixed parentage, height, weight, skeletal age. Greater environmental influence on boys, greater genetic influence on girls	Bogin and MacVean (1982)
	$n = 519$, four cohorts, 1, 3, 5, 7 years, five measurements, sociodemographic characteristics, refugees of an earthquake. General reduction in the measurements, especially on male linear dimensions	Johnston *et al.* (1985)
	Same sample, 10 years later ($n = 271$). Growth status at the first visit was a significant predictor of status after 10 years, but some catch-up growth could be identified	Johnston and MacVean (1995)
	$n = 497$, subjects previously studied in a nutrition survey intervention study were investigated for age at menarche Delayed growth in childhood was a significant factor for delayed age at menarche	Khan *et al.* (1996)
	$n = 533$, measured as children and remeasured as adults. Severely stunted children has significantly greater adult abdominal fatness, especially if they had migrated to urban centers	Schroeder *et al.* (1999)

Table 4.4. (*cont.*)

Country or territory	Information	Authors
Mexico[a]	$n = 583$, longitudinal study (1959–70) starting with age 1 month. Concomitant socioeconomic studies. Comparison with other series (height)	J. Faulhaber (1976)
	Same sample, seasonal variability in growth rate, relationship with temperature, humidity and sun exposure	J. Faulhaber (1982)
	$n = 274$, 7.5–8.5; 9.5–10.5 years, comparison between bone maturation and three anthropometric measurements	M.E.S. Faulhaber (1982)
	$n = 340$, 5–18 years, longitudinal study, upper arm growth velocity	Ramos-Galvan (1982)
	$n = 505$, 7–13 years, six anthropometric measurements, bone maturation, rural–urban comparisons	Villanueva *et al.* (1984)
	$n = 1918$, 7–13 years, standards for school children, comparison with two other series	Casillas and Vargas (1987)
	$n = 900$, relationship of the subcutaneous fat at two sites between newborns and their parents	Jiménez *et al.* (1987)
	$n = 213$, children and adults, prediction of adult height	Ramos-Rodriguez (1987)
	$n = 230$, height, weight and menarche	J. Faulhaber (1987)
	$n = 30$, 1–10 years, with and without food supplementation, height and weight, mathematical models	Schlaepfer (1987)
	Application of the homeorhesis concept (C.H. Waddington) to physical growth in México	Ramos-Rodríguez (1989)
	$n = 742$, 0–19 years, eight measurements, two populations of Yucatan, comparison with two other Mexican series	Dickinson *et al.* (1989)
	$n = 2141$, 5–17 years, six measurements, three populations, Yucatan. Migration to the coast may have had an unfavorable impact on boys	Dickinson *et al.* (1991)
	$n = 3009$, 5–20 years, four samples of different ecology, Heath–Carter somatotype determinations	Murguia *et al.* (1991)
	$n = 1208$, 2–80 years, two fatfolds. Comparison with Peruvians, Puerto Ricans, and Mexicans living in USA	Wolanski (1998)

[a]See also The Oaxaca Project (Table 4.5).

Table 4.5. *The Oaxaca Project*[a]

Subjects	Information	Authors
General assessments	n = 331, 6–14 years, height, weight, arm circumference, measurements made in 1968	Malina *et al.* (1972)
	n = 1410, 6–14 years, four measurements, two urban and two rural communities studied in 1968 and 1972	Malina *et al.* (1981)
	n = 285, 6–13 years, mixed longitudinal, 1978–79	Buschang and Malina (1983)
	n = 1450, 6–17 years, height and weight, six communities studied in 1968, 1971, 1972. Age at menarche (n = 746)	Malina (1983)
	n = 293, 6–13 years, eight measurements, correlation with socioeconomic status evaluated quantitatively. Only boys showed a clear socioeconomic differential. Study made in 1978	Malina *et al.* (1985)
	n = 285, 6–13 years, sitting height and leg length, measurements taken on two occasions (1978, 1979). Differences with USA children due to diminished growth rates of leg length	Buschang *et al.* (1986)
	General evaluation of the project	Malina (1986)
	n = 213, 6–13 years, eight measurements, children measured twice in 1978 and 1979. Negligible influence of socioeconomic variation on estimated growth velocities	Little *et al.* (1988*a*)
Skeletal growth	n = 394, 5–18 years, hand and wrist	Malina *et al.* (1976)
	n = 346, 6–14 years, stature and weight are significantly correlated with measures of metacarpal cortical bone	Himes *et al.* (1976*b*)
	n = 324, 6–10 years, four metacarpal dimensions, sex differences	Himes and Malina (1977)
	n = 354, 6–13 years, skeletal ages derived from the Tanner-Whitehouse original method and its revised version. The second yields, on average, consistently lower ages.	Malina and Little (1981)
Motor performance	n = 364, 6–15 years, right and left grip strength and three other measurements. No superior functional efficiency associated with reduced body size	Malina and Buschang (1985)
	n = 126 boys, 9–15 years, compared with 95 Mexican Americans. Effect of body composition on physical performance. Different relationships between fatness and performance in the two series	Malina and Little (1985)
	Same sample reported in Malina and Buschang (1985). Comparison with a Papuan series. The adaptive significance of small body size in conditions of undernutrition varies between populations and with performance tasks	Malina *et al.* (1987)

Table 4.5. (*cont.*)

Subjects	Information	Authors
	$n = 320$, 9–14 years, four skinfolds used to predict body density, calculation of fat-free and fat mass. Comparison with Mexican Americans. Protein rather than energy may be the more limiting nutrient in the Oaxaca sample	Malina *et al.* (1991)
	$n = 373$, 6–14 years, comparison with African- and European-derived Americans. The body mass index has a different meaning for the strength and motor fitness of individuals from the three series	Malina *et al.* (1998)
Menarche	$n = 315$, 9–16 years, comparison with five other Mexican samples. Timing of menarche in Oaxaca (14.27 years) is similar to that for rural Mayans in Guatemala	Malina *et al.* (1977)
	$n = 418$ for the Valley of Oaxaca, $n = 328$ for rural communities outside the Valley. Median age considering the two groups (that do not differ between themselves): 14.40 years	Malina *et al.* (1980*a*)
Aging	$n = 229$, 20–82 years, five measurements. Body weight shows little sex difference after 40 years of age. Grip strength per unit body weight is similar to that of better-off subjects	Malina *et al.* (1982*a*)
Spatial factors	$n = 1410$, 6–14 years, two urban colonias, two rural Ladino, and two rural Zapotec-speaking communities. Urban–rural contrasts are minor	Malina (1982)
	$n = 336$, 20–23 years, anthropometric dimensions. About half of the individuals studied in 1968 had migrated to México City in 1978. No significant differences in growth status (1968 data) of sedentes and migrants were found	Malina *et al.* (1982*b*)
	Significant child growth differences between children born to native parents ($n = 287$) and those of native–immigrant matings ($n = 38$) were found in fatness, body proportions, and size	Little *et al.* (1988*b*)
Genetic, mortality and reproductive factors	Infant and childhood mortality (birth to 14 years) and growth status of 143 school children (5–14 years) are considered. About 59% of all deaths occur in children under 15 years of age. This high mortality (especially at younger ages) suggests poor living conditions, which was confirmed by the growth studies	Malina and Himes (1978)

Table 4.5. (*cont.*)

Subjects	Information	Authors
	Mortality statistics (1945–70) indicated that mortality of children 1–4 years of age due to gastrointestinal disorders occurred predominantly during the rainy season	Malina and Himes (1977*a*)
	Live births during the same period (1945–70) showed a nonrandom distribution, which may reflect cultural patterns associated with the annual agricultural cycle	Malina and Himes (1977*b*)
	Fertility, offspring survival and offspring prereproductive mortality were regressed on stature and weight of 205 multiparous adults (ages 20–72 years). The exceptionally short stature of Zapotec Indians in this community is probably due to poor environmental conditions and not to genetic selection or adaptation	Little *et al.* (1989)
	Four craniofacial dimensions in 322 children from families with an average inbreeding coefficient of 0.01 and 36 children from families with an inbreeding coefficient of zero were compared. A significant difference was found in bizygomatic diameter, while biparietal diameter and fronto-occipital length followed the same trend	Little *et al.* (1991)
	198 sibling pairs, 6–13 years of age were compared for seven measurements, taking into consideration socioeconomic status (SES) also (lower or higher SES). The correlation between them vary according to sex and SES	Little *et al.* (1986)
	Same sample as above, comparisons on motor performance. Sibling correlations for the same tasks tend to be higher in well-nourished children. However, sisters resemble each other more than brothers in the undernourished sample, while brothers are more similar than sisters in the better-nourished sample	Malina *et al.* (1986)
	Same sample as above. Partial correlations of the log-transformed data resulted in a slight overall decrease in sibling correlations for all pairings. This suggests a genetic–environmental interaction and may explain the generally higher sibling correlations observed in non-European as compared to European-derived populations	Little *et al.* (1987)

Table 4.5. (*cont.*)

Subjects	Information	Authors
	Same sample as above. Correlations negative or close to zero were observed for annual growth increments	Little *et al.* (1990)
Secular factors	$n = 111$ males, 18–86 years, stature, compared with three other series, the oldest dating back to 1895. No evidence for a secular trend	Himes and Malina (1975)
	Schoolchildren, 6–14 years of age, measured in 1968 ($n = 341$) and 1978 ($n = 353$) showed little changes in five measurements	Malina *et al.* (1980*b*)
	Both for stature (comparisons including long bones excavated in various archeologic sites in the Valley of Oaxaca) and age at menarche (adults, $n = 220$, 14.53 years; schoolgirls, $n = 101$, 14.70 years), no evidence for secular trends was obtained	Malina *et al.* (1983*a*)

[a]Data about this project were also referred to in Table 4.2 of this chapter (Reyes *et al.*, 1995), Chapter 3 (Little and Malina, 1989) and will be mentioned in Chapter 5 (Malina *et al.* 1983*a*) and Chapter 6 (Buschang and Malina, 1980).

Table 4.6. *Selected information about physical growth studies in South America*

Country	Information	Authors
Argentina	Comparison between La Plata ($n = 303$) and San Antonio de los Cobres ($n = 162$), 5–15 years, 62 (!) measurements. General information about the investigation, with few details	Ringuelet (1978)
	$n = 6494$, 8–13 years, mean age at menarche 12.53 years	Lejarraga *et al.* (1980)
	$n = 765$, 6–14 years, seven measurements. Better female canalization could explain the sexual dimorphism observed in a poor neighborhood	Pucciarelli *et al.* (1993)
	$n = 321$, 6–13 years, four measurements, rural community. The children tend to be fatty and overweight, while their muscle mass and height are proportionally low, but with values within the reference	Bolzan *et al.* (1999)
Brazil	Physical growth studies have a long history in Brazil. Earlier investigations have been summarized in the reference works indicated in the next column	Castro Faria (1952), Pourchet (1957/1958), Salzano and Freire-Maia (1970)
	$n = 558$, 7–8 years, height and weight, sample divided by socioeconomic status, no significant differences from the results of three other series	Carneiro *et al.* (1970)
	$n = 9258$, 0–12 years, height and weight, influence of socioeconomic status and foreign ancestry	Levy (1975)
	$n = 5737$ newborn children, birthweights, influence of ethnic group, socioeconomic level, mother's height and weight, parity, age of father, and parents' marital distance	Araújo and Salzano (1975a)
	$n = 2891$ newborns, nine measurements. Multiple regression analyses between the variables, socioeconomic status	Arena (1976a)
	$n = 2444$, 4–16 years, four measurements, relationship with ethnic classification. The mean upper-segment/lower-segment decreases as the proportion of African-derived genes increases	Oliveira and Azevêdo (1977)
	$n = 84$, 12–60 months, advantages of using arm circumference in the evaluation of malnutrition	Gewehr *et al.* (1984)
	$n = 433$ newborn children, significant association between their weight and the level of placental alkaline phosphatase	Silva *et al.* (1985)
	$n = 169$, 0–3 years, neural and psychomotor development related to socioeconomic status	David *et al.* (1985)
	$n = 550$, 6–12 years, height and weight, suggestions of undernutrition in relation to weight, but not to height	Pereira *et al.* (1985)

Table 4.6. (*cont.*)

Country	Information	Authors
	$n = 1013$, less than 5 years, height and weight, São Paulo, prevalence of undernutrition as high as 26%; no progress in relation to a survey undertaken 10 years ago	Monteiro *et al.* (1986*a,b*)
	$n = 3368$, 10–19 years, biacromial and ilium crest diameters are significantly influenced by menarche	Hegg and Bonjardim (1988)
	$n = 437$, 7–11 years, 11 measurements, sample stratified in two socioeconomic levels	Anjos *et al.* (1989)
	$n = 145$, 8 years, low socioeconomic status, 10 measurements, five physical fitness, and three psychological tests. Information about food intake. Reasonable growth and motor performance, but poor results in the psychological tests	Rocha-Ferreira *et al.* (1991)
	Special volume of Cadernos de Saúde Pública, with 11 main contributions, considering different aspects of the physical growth and nutrition of Brazilian children	Santos and Anjos (1993)
	$n = 352$, 6–26 years, self and physician assessments of sexual maturation. Concordance between the two varied from 60% to 71%, depending on the body region considered	Matsudo and Matsudo (1994)
	National surveys, cohorts born in 1952, 1967, and 1982. Height gains of 1 cm per decade in the 1952–67 period, and 2.4 cm per decade in the 1967–82 period	Monteiro *et al.* (1994)
	$n = 375$, 6–8 years, classified for height in below 2 Z-scores and above 1 Z-score. Classroom performance of the former was lower than those of the latter	Lei *et al.* (1995)
	$n = 2248$, zero to 60 months of age, height and weight. As much as 20% of undernutrition (all forms)	Marins *et al.* (1995)
	Two cohort studies of mothers and children (1982 and 1993), maternal–child health, all hospital deliveries, Pelotas, Rio Grande do Sul. Although most health care indicators improved over the decade, health services are still biased towards those who least need them	Barros and Victora (1996)
	$n = 124$, 7–10 years, Japanese ancestry, high socioeconomic status, measured for height. The medians are similar to those of Japanese in Japan, but lower than those of non-Japanese-derived Brazilians. The authors raise doubts about the hypothesis of universal uniformity of growth potential asserted by the World Health Organization	Kac and Santos (1996)

Table 4.6. (*cont.*)

Country	Information	Authors
	n = 57 033, 18–19 years. Increase in height according to the year of birth (1970–77)	Kac and Santos (1997)
	n = 3269, 18–19 years, height according to year of birth (1940–65). Increase of 0.1 cm/year for the country as a whole	Kac (1998)
	n = 12 644 mother–children pairs, inverse relationship between the level of stunting and maternal education, income, and housing	Engstrom and Anjos (1999)
	Case–control investigation of 165 pairs of children classified respectively as one Z-score below and one Z-score above the norm. Preschool short stature was associated with several socioeconomic factors and with parents' stature	Guimarães *et al.* (1999)
Chile	n = 354 females, 10–17 years, height. Children with 2–4 foreign grandparents generally taller than those with less foreign ancestry	Rona and Pierret (1973)
	Same sample as above, age at menarche of 12.6 years. Menarche ensued significantly earlier in those with more subscapular or subcoastal adipose tissue	Rona and Pereira (1974)
	Three cross-sectional studies performed in 1887 (n = 3557), 1940 (n = 990) and 1971 (n = 354) indicated a trend towards lower ages at menarche (14.73–12.64 years)	Rona (1975)
	n = 313, 10–19 years, Mapuche Indians + neo-Chileans, height. The higher sexual dimorphism observed in Chile as compared with France was attributed to differential contributions of European sex chromosomes to the present-day population, due to asymmetrical European–Amerindian crossing according to sex	Valenzuela (1975)
Colombia	Children, 7–10 years (n = 728) and adults 20–60 + years (n = 718). Skinfolds obtained at four sites, principal components analysis. Three components emerged, fatness (70–80% of the variance), trunk-extremity (10–15%) and upper–lower body (10–15%) patterns. The fatter tended to be more patterned in both age groups. Sibs, parent–offspring, and spouses were all correlated in fatness	Mueller and Reid (1979)
	n = 810 for children 6–15 years and n = 715 for adults, 25–55 + years. Skinfolds at four sites, as above. Comparison with 606 Black and 598 White adolescents. Independently of ethnic or cultural backgrounds, there is a redistribution of fat away from the extremities towards the trunk	Mueller (1982)

Table 4.6. (*cont.*)

Country	Information	Authors
Peru	$n = 969$, cross-sectional, 7–20 years, five measurements, in Puno (3800 meters). No secular increase in height (1945–80), but earlier sexual maturation in the present generation	Gonzalez *et al.* (1982)
	$n = 1466$, cross-sectional, 2.0–60.0 + years and $n = 404$, mixed-longitudinal, 3–22 years, measured in 1983–84, 14 measurements, compared with similar data obtained in 1964–66. Differences in growth timing or velocity are not reflected in adulthood. No secular trends are discerned	Leatherman *et al.* (1995)
Venezuela	$n = 955$, 9–20 years, age at menarche 12.68 years. Time trend towards earlier ages observed when comparison is made with other series, beginning in 1935	Farid-Coupal *et al.* (1981)
	Comparative analysis of five Venezuelan series (height and weight) studied from 1939 to 1974, among themselves, and with English and Cuban data	Arechabaleta (1982)
	$n = 944$, 2–69 years, metacarpophalangeal pattern profiles. Variances of Venezuelan males tended to be larger than those of Americans, especially under 9 years of age	Arias-Cazorla and Rodríguez-Larralde (1987)
	$n = 28\,752$ urban and $n = 10\,557$ rural subjects, 0–18 years, height and weight, measured in 1981 and 1986. Urban children were taller and heavier than their rural counterparts	Lopez-Blanco *et al.* (1992)
	$n = 214$, mixed longitudinal, entry in the cohort at ages 4, 8 and 12. Mean follow-up interval 4.1 years. Age at peak height velocity in boys and girls occurred 0.4 and 0.6 years earlier than in European children	Lopez-Blanco *et al.* (1995)

Table 4.7. *Selected information about studies on growth and dentition in Latin America*

Country	Information	Authors
México	$n = 510$, 10–16 years, mixed-longitudinal. Median ages of eruption are similar to those obtained in European-derived populations. Comparison with two other Mexican, rural series. Low correlation with four anthropometric measurements	J. Faulhaber (1989)
	$n = 1041$, 7–11 years, two socioeconomic levels, no marked differences in eruption time between the two groups	Mejia-Sanchez and Rosales-Lopes (1989)
Dominican Republic	$n = 100$ Mulattoes, 3–5 years, crown diameters of deciduous teeth. Differences between sexes and in the variability of different teeth, anterior teeth being more variable	Garcia-Godoy *et al.* (1985)
Guatemala	$n = 273$, 0–36 months. The timing of deciduous teeth eruption seems more closely associated with postnatal weight than with birth weight	Delgado *et al.* (1975)
	$n = 1277$, 0–9 years, mixed longitudinal, teeth emergence. Guatemalans exhibited emergence times intermediary between the Japanese (fast) and Bangladeshi and Javanese (low)	Holman and Jones (1998)
Brazil	$n = 490$ males, mixed ancestry, 18–24 years. The frequency of missing third molars is around 8% per quadrant, and absence of the four third molars appears in 2%. These values are of the same order of magnitude as those obtained in European or European-derived populations	Crispim *et al.* (1972)
	$n = 302$ European-derived and $n = 904$ African-derived individuals, 6, 9, and 12 years. Differences in teeth eruption between the two ethnic groups were not large, but the African-derived subjects were more precocious at the beginning of the process. In later years these dissimilarities disappeared	Melo e Freitas and Salzano (1975)
	$n = 2162$, African-derived, low socioeconomic index, 5–13 years. Teeth eruption more precocious than in European-derived subjects, independent of socioeconomic status	Rummler *et al.* (1985)

Table 4.8. *Selected blood pressure studies performed in Latin American populations*

Country or region	Information	Authors
Carib		
General	Review of data from the whole region; comparison with Caribbean Americans and British Afro-Caribbeans	Halberstein (1999)
St Vincent Island	Detailed investigation, involving social and biologic variables. No evidence of intergenerational transmission for the susceptibility to high blood pressures	Hutchinson and Crawford (1981); Hutchinson (1986); Hutchinson and Byard (1987)
La Désirade Island	Blood pressures and prevalence of hypertension showed a significant positive correlation with degree of African ancestry	Darlu *et al.* (1990)
Brazil		
Marajó Island	$n = 46$, two communities. Low values compared with urban groups	Silva *et al.* (1995)
Northeast	$n = 1068$ families. Genetic heritability is 0.41 for systolic pressure in children, 0.14 for systolic pressure in adults, and 0.34 for diastolic pressure in both generations	Krieger *et al.* (1980)
Rio de Janeiro	$n = 3282$, age above 20. The best cut-off points for waist: hip girth ratio as a predictor of arterial hypertension was 0.95 for men and 0.80 for women	Pereira *et al.* (1999)
Ribeirão Preto	$n = 208$. Investigation of the influence of African- or European-derived ancestry in association with an index called 'cultural consonance in lifestyle', which was found to significantly influence blood pressure	Dressler *et al.* (1999)
Rio Grande do Sul	$n = 4565$, 20–74 years, 2050 households. The degree of genetic determination was investigated in 557 families. Heritability for systolic blood pressure was estimated as 0.22, and for diastolic blood pressure in 0.40	Robinson *et al.* (1991)
Porto Alegre and Itapuã	$n = 470$, over 20 years of age. No urban–rural differences	Martins *et al.* (1982)
Pelotas	$n = 1657$, 20–69 years, of whom 328 were hypertensive. Continued care by the same physician was the only factor significantly associated with appropriate control of the symptoms	Piccini and Victora (1997)

Table 4.8. (*cont.*)

Country or region	Information	Authors
Bolivia		
Chicaloma	n = 159, African- and European-derived subjects. The prevalence of hypertension was higher (6–7%) in descendants of Africans than in descendants of Europeans, but 3 times lower than that found among USA persons of African ancestry	Frisancho *et al.* (1999*a*)
Chile		
Caleu	n = 642, semi-isolated community. Fifty-five per cent of the subjects could be grouped in a single, large pedigree. Correlation analyses between relatives	Cruz-Coke and Covarrubias (1962, 1963)
Easter Island	n = 179. The fraction (n = 50) which periodically migrates to the continent, returning afterwards, shows an increase in blood pressure variance. In this fraction, the blood group MN locus seems to influence blood pressure	Cruz-Coke *et al.* (1964*a,b*)

Table 4.9. *Selected information about studies on physique and motor performance in Latin America*

Country	Information	Authors
México[a]	Two groups of males ($n = 171$ and $n = 94$) and one of females ($n = 72$). Differences between the two male groups probably due to training. Poor results of the women may be due to lack of a sport tradition among them	Vargas and Casillas (1977)
	$n = 84$ female and 108 male swimmers from five Latin American countries. Comparison with three nonathlete Mexican series for four anthropometric characteristics, and bone age determinations. Specialized characteristics related to sport performance, and general precocity in relation to the growth expected for given chronological ages were observed	Peña *et al.* (1984)
	$n = 85$ females, rural community, state of Michoacán, dynamometer measurements. Positive correlations with height and weight, lower strength than those observed in Otomi subjects	d'Aloja (1987)
	$n = 236$, miners, Oaxaca. Pulmonary function considered in relation to six other anthropometric variables. Comparison between those working inside and outside of the mine did not disclose any harmful effects among the former	d'Aloja (1989)
	$n = 65$ swimmers 8–17 years, vital capacity and body composition, two evaluations at a 6-month interval. Discussion about training effects	Cardenas Barahona and Peña Reyes (1989)
	$n = 84$ females grouped according to skinfolds, somatotyped by the Parnell and Heath–Carter classifications. The latter gives inconsistent results in the relationship between adiposity and muscularity	Villanueva (1989)
Cuba	$n = 330$ baseball players, 17 anthropometric measurements, Heath–Carter somatotyping. No differences in somatotype indices in relation to the diverse players' positions	Alvarez *et al.* (1989)
	$n = 59$ ballet dancers, Cuba's National Ballet, 44 variables considered, including Heath–Carter somatotyping. Cluster analysis by individual does not clearly separate the most gifted dancers from the others	Martinez *et al.* (1989*a*)
	$n = 58$ females, eight skinfold sites studied; and $n = 50$ females and 31 males, 12 sites examined, all with ages around 18 years. Cluster and principal components analysis identify a general fatness component, plus others related to trunk–extremity and cephalocaudal distributions	Martinez *et al.* (1989*b*)

Table 4.9. (*cont.*)

Country	Information	Authors
Venezuela	$n = 114$ highly competent athletes, eight specialities, Heath–Carter somatotyping. Comparison with Olympic athletes does not disclose significant differences. The low performance of Venezuelan athletes in the Olympic games is related to other factors, not to their physiques	Méndez de Pérez (1982)
	$n = 67$ athletes, both sexes, three specialities, national teams, Heath–Carter somatotyping. Differences among specialities	Menéndez (1984)
Brazil	$n = 131$ children, 8–9 years, maximal mechanical aerobic and anaerobic power output on a cycle-ergometer considered in relation to seven anthropometric variables. Despite their underprivileged environments, they maintain a level of performance proportional to their smaller body size	Anjos *et al.* (1992)

[a]See also the Oaxaca Project (Table 4.5).

Table 4.10. *Selected studies about high altitude adaptation in South America*

Country	Information	Authors
Colombia	The symptoms of mountain sickness can be better controlled by increasing potassium and decreasing sodium intake	Waterlow and Bunjé (1966)
Ecuador	$n = 116$ at high and $n = 101$ at low altitudes. The first 6 months of life seem to be decisive for the height differences between the two samples	Leonard *et al.* (1995)
Peru	$n = 27$ at high and $n = 45$ at low altitudes, newborns, 14 anthropometric and bone measurements. They are significantly reduced at high altitude; critical period: last trimester of gestation	Haas *et al.* (1977)
	$n = 2099$ at high and $n = 2676$ at low altitudes, birthweights linked to death records. The high altitude population has a lower mean birthweight and a lower optimal birthweight	Beall (1981)
	Microenvironment of Nuñoa babies, 0–12 months, $n = 12$	Daltabuit and Tronick (1987)
	$n = 113$ at high and $n = 169$ at low altitudes. The delayed age at adrenarche (period in childhood characterized by a significant increase in serum adrenal androgen levels) at high altitude is not due to nutritional status	Gonzales *et al.* (1994)
	$n = 488$ at high and $n = 712$ at low altitudes. A later age at menarche was observed at high altitude after controlling for socioeconomic status and the Benn index (weight/height$^{2.15}$)	Gonzales and Villena (1996)
	$n = 17$, 23–30 years, soccer players. Thirty-five per cent developed acute mountain sickness within 6 hours after arrival by air from Lima to Cusco (3400 m). The lower the difference between arterial oxygen saturation at sea level and at arrival, the higher the probability of developing acute mountain sickness	Gonzales *et al.* (1998)
Bolivia	$n = 323$, 8–14 years, European descent, middle and upper socioeconomic class, living in La Paz, five measurements. Their height was essentially similar to those of Guatemalan children of high socioeconomic status, the effect of high altitude being small.	Stinson (1982)
	$n = 446$, 10–19 years, six measurements, living in La Paz. The process by which large chests are achieved may be different in La Paz and Nuñoa	Greksa *et al.* (1984)
	$n = 175$ at high and $n = 71$ at low altitudes, newborns, nine measurements. Hypoxia may exert a direct effect on human fetal growth	Ballew and Haas (1986)

Table 4.10. (*cont.*)

Country	Information	Authors
	$n = 268$, 13–49 years, body composition, exercise test, studied in La Paz. Age at arrival to high altitude is inversely related to maximum oxygen consumption. Approximately 20–25% of the variability in aerobic capacity at high altitude may be due to genetic factors	Frisancho *et al.* (1995)
	$n = 357$, 13–49 years, vital capacity and residual lung volume, evaluated in La Paz. Vital capacity is much influenced by environmental factors, but the attainment of an enlarged residual volume is related to both developmental acclimatization and genetic factors	Frisancho *et al.* (1997)
	$n = 108$ at high and $n = 120$ at low altitudes, 10–12 years, high and low socioeconomic status, six measurements, dietary and daily physical activities data recorded. Socioeconomic status was more important than altitude in the growth of these children	Post *et al.* (1997)
	$n = 61$ at high and $n = 54$ at low altitudes, 11–35 years, resting ventilation is higher in highlanders, but 40–80% lower than that of Tibetans	Frisancho *et al.* (1999*b*)

Table 4.11. *Relationships between variables*

Country	Information	Authors
México	$n = 216$ females, 32 years or older, 17 socioeconomic, demographic, and environmental properties of their families plus migrant status were reduced by principal components analysis to five factors which were regressed in relation to 19 anthropometric and 21 reproductive traits, age adjusted. In this Yucatecan sample larger families were a strategy to cope with poverty and uncertainty in employment and income	Dickinson (1994)
	$n = 657$ girls and 315 mothers. Age at menarche related to several factors. Correlation daughter/mother ($n = 205$): 0.19	Wolanski *et al.* (1994)
Colombia	$n = 1572$ women, 18–44 years, three different socioeconomic groups. Greater tendency for upper body fat distribution in the lower socioeconomic groups. These differences diminished with age	Dufour *et al.* (1994)
Ecuador	$n = 114$, dietary adequacy, water quality, influence on linear growth. Retardation is explained by the synergistic effect of mild to moderate malnutrition and disease	Berti *et al.* (1998)
Peru	$n = 59$, 7–8 years, born to adolescent mothers plus 73 controls. The first showed delayed visuomotor maturation and a low performance at school	Silvestre *et al.* (1997)
Brazil	In the state of São Paulo the prevalence of undernutrition decreased from 31% (1984–85) to 14% (1995–96). But the incidence of low birthweight babies remained constant (10%) during this period, probably due to the high frequency of Cesarean deliveries (47%) and to the increase of adolescent mothers (1970s: 1.4%; 1990s: 7%)	Monteiro (1997)
	$n = 125$ children, 7–8 years, below $-2Z$-scores of NCHS/WHO standards and $n = 139$, same age, above $-1Z$-scores. Significant factors for the growth retardation (evaluated by odds ratios) were family income, parents' level of schooling, and housing conditions	Lei *et al.* (1997)

Table 4.12. *Genetic and reproductive factors*

Country	Information	Authors
Guatemala	$n = 1480$ comparisons brother–brother, $n = 1299$ comparisons, brother–sister, and $n = 1314$ comparisons, sister–sister for the number of ossification centers, ages 1–7. Correlation coefficients (0.39–0.44) similar (with the exception of the sister–sister comparison) to those found in USA. The hypothesis that sibling correlations should be lower in malnourished children was not confirmed	Martorell *et al.* (1978)
St Vincent and Grenadines	$n = 329$ adults age adjusted. Significant intergenerational transmission was found for height and upper arm circumference	Hutchinson and Byard (1987)
Colombia	$n = 704$, ages below 29 to higher than 65 years, malnourished. The number of surviving offspring (fertility) shows a positive association with measurements of soft tissue, but bony measurements tend to have a curvilinear relationship to fertility. The latter is the same pattern of stabilizing selection observed in well-nourished, industrial societies	Mueller (1979)
Brazil	$n = 3465$ children, no significant inbreeding effect on height and weight	Krieger (1969)
	$n = 1068$ families. Sib correlations for height and weight decrease with absolute age difference. Parent–child correlations increase with age of child, with greater resemblance to the mother than to the father. Genotype effects are estimated as amounting to 40–44% (height) and 36–42% (weight)	Rao *et al.* (1975)
	$n = 534$, 6–15 years, 63 anthropometric variables. Inbreeding effects are always in the direction of reduced values	Meirelles-Nasser (1983)
	$n = 1067$ (same set studied by Rao *et al.*, 1975). There is considerable temporal variation in family resemblance over time for height, weight and Quetelet index (weight/height^2)	Province and Rao (1985)
	$n = 63$ couples, 242 normal children, quantitative orbital traits. Heritabilities of 0.34–0.51 were obtained, depending of the characteristic considered; the first principal component summarizing the variability of the four measurements yielded a heritability value of 0.42	Raposo-do-Amaral *et al.* (1989)
	$n = 80$ families, 146 young adults, seven face measurements. Mid-parent–offspring significant heritabilities were found for bizygomatic (0.65), bigonial (0.91) and tragion-prosthion (0.85) diameters	Cabral-Alexandre and Meirelles-Nasser (1995)

Table 4.12. (*cont.*)

Country	Information	Authors
	$n = 442$ nuclear families, 1500 individuals, interphalangial mobility, heritability estimate of 0.67	Moura *et al.* (1996)
	$n = 1158$ twin pairs and 12 609 singletons, birthweights. In all cases female fetal growth was slightly but consistently lower than that for males. Fetal genetic effects were observed only from children born of undernourished women	Beiguelman *et al.* (1998*a*,*b*)

5   *Morphology*

> ... human aesthetic sense is based on general principles of perception
> that have been important during the evolution of biological signs
>
> *Magnus Enquist and Anthony Arak*

Significance

The dictionary definition of morphology asserts that it is the branch of
biology which deals with the form and structure of animals (including our
species) and plants. That human populations are morphologically much
diversified, both within and between groups, has been recognized since
early times. The question was, how could we systematize and quantify
these differences? Comas (1966) reviewed all the most important earlier
attempts at a classification. By the end of the nineteenth century a series of
instruments designed to perform measurements in living subjects and on
bones were developed, and the corresponding anthropometric characteristics defined. This, however, was not done without discussion, different sets
of measurements being suggested in France, Germany and England. At
several international congresses attempts were made to reach a consensus,
but this was, however, only partially achieved in the early part of this
century.

As the instruments and machines to investigate variability progressed, it
was realized that assessments could be made at different levels of analysis
(molecular, biochemical, cytologic, organismal), and that shape, color and
size are achieved through a complicated process. Static evaluations (based,
for instance, on the type concept, an abstract construct that would sum-
marize features of different individuals or populations) were abandoned,
and dynamic approaches followed.

In terms of anthropometry, two components, *size* and *shape*, are generally considered, and there are statistical methods that can be used to
separate these components (e.g., Spielman, 1973). Generally, similarity or
differences in shape are considered more important than sheer size.

Several factors may influence the degree of morphologic variability
present within and between populations. The most basic relates to its
heritability, or degree of genetic determination. If the trait is much
influenced by geographically diversified environmental factors, high

143

interpopulation variability may occur. This could or could not be true if the trait's determination is mainly genetic.

Assortative mating, on the other hand, is undoubtedly important, and may involve factors such as the availability of potential mates (influenced by population size and reproductive cultural rules) and the concept of beauty. The latter includes body symmetry and underlying cues of health and reproductive potential (see, for instance, Enquist and Arak, 1994; Tovée *et al.*, 1999; Waynforth, 1999). Deviations from body symmetry may also indicate the degree of disturbing influences that may occur in a given person, through his or her ontogenetic development. An example of a study involving sociocultural and morphological characteristics is that performed by Azevêdo *et al.* (1986). They found for 982 couples a general tendency towards homogamy for education, religion and skin colour in the polymorphic and multicultural population of Salvador, Bahia, Brazil.

Anthropometrics

Interpopulation comparisons

One of the first objectives for the development of anthropometric techniques was the investigation of interpopulation comparisons. As the studies progressed, it was soon verified that some characteristics are more influenced by genetic factors than others, in which the environment plays a more important role (see Heritability below). For instance, height has a strong genetic component, although it is also influenced by environmental factors, while weight is mainly an indicator of life (especially nutrition) quality. Given these premises, what is the distribution of these characteristics among Latin American populations?

Table 5.1 gives an answer to this question, in relation to 1960 young (average 23 years), healthy males residing in 15 Latin American countries, and uniformly measured in Panama. Values for height range from 1.63 m (El Salvador) to 1.70 m (Brazil), the mean being 1.66 m ± 0.05. Equivalent numbers for weight are 61.5 kg (Colombia) and 68.0 kg (Venezuela); 63.8 ± 8.2 kg. The differences are statistically nonsignificant, probably expressing standard selection for what should be considered a good, healthy soldier.

When females are considered (Table 5.2; 21 samples, 8890 subjects) the range in height more than doubles (since the difference between extreme observed mean values is 7 cm for males and 16 cm for females). In this case the main interpopulation differences may probably be attributed to different degrees of Amerindian ancestry, since the lowest values were found in samples where this type of heritage is high.

Within-country differences should not be disregarded. Table 5.3 presents data on height, weight and upper arm circumference on 20 to 24-year-old males and females living in seven Brazilian regions in 1975. There are differences of respectively 5 cm and 5 kg in males, and of 4 cm and 4 kg in females, between southeast and northeast inhabitants belonging to this age class interval. The lower values obtained in the latter may be due to poorer living conditions, although, as above, they also have higher percentages of Amerindian genes.

Other surveys include the analyses performed by Montemayor (1961, 1984, 1987) and by Faulhaber and Schwidetzky (1986) in México; by Sauvain-Dugerdil (1987, 1989) in Central America; by Schwidetzky (1987) in the Caribbean Islands; and by Rothhammer (1987) in Chile. They have considered, however, mostly Indian populations.

Specific studies performed in Brazil can exemplify the types of approaches that can be used in these investigations. Gurjão and Macedo (1963) considered, in 290 subjects subdivided by ethnic group in Rio de Janeiro, 15 anthropometric measurements and 19 anthroposcopic characteristics; Eveleth (1972) investigated 198 northeasteners (their gene pool is a mixture of European-derived, African-derived, and Amerindian genes) in relation to 11 measurements and nine anthroposcopic traits; and Pollitzer *et al.* (1982) studied 391 individuals living in Jacobina, Bahia for nine measurements. All of them considered the question of interethnic variation. In principle, intermixed populations should present intermediate values between those of the parental groups, but this was not uniformly true for the traits considered in these populations. The question has intrinsic value and should be pursued further in other groups.

Migration effects

The human species is mobile, and the effects of these displacements have been considered in previous chapters of this book. Its influence on somatic characteristics can be examined under two categories. Studies can be made on the unmixed migrants and their offspring in the native and new environments; or the results of mixed migrant–sedente marriages can be considered. In relation to the first group of investigations, it is important to try to sort out preselection (the fact that migrants do not constitute a random sample of the parental population) from environmental effects.

Studies concerning these points were performed, for instance, by Lasker

(1960), who observed that the variances of some body dimensions were higher in the adult offspring of one or two immigrant Peruvian parents than in the adult offspring of natives. Lasker and Evans (1961) compared subjects born in México and who had always lived there to those born in México but had spent various periods of time in the USA, as well as with those born in the USA of Mexican parents who subsequently returned home. They found evidence of both preselection and of environmental effects influencing the traits under consideration.

In Brazil Saldanha *et al.* (1960) and Beiguelman (1962, 1963*a*) investigated respectively Dutch and Japanese migrants. In the first case the averages of nine measurements were found to be similar to those observed in the Netherlands, with stature and cephalic index showing values closer to those of northern Dutch populations. The unmixed Japanese descendants of both sexes presented an increase in 5 of 12 measurements.

Hulse (1979) found that Cubans of Spanish descent resembled upper class Spaniards from Andalucia; while Livshits and Kobyliansky (1984) and Kobyliansky and Goldstein (1987), besides verifying also a definite tendency for larger size in migrant parents and their children as compared with sedentes, obtained indications of changes in the interactions between the average values of the 20 morphological traits considered with their genetic and environmental variances.

It is clear, due to the subtleties of the above-indicated interactions, that no general answer can be given to the problem of migration effects. However, well-designed investigations could throw light, in specific cases, on the plasticity of morphological traits, and the limits set by heredity to their range of variation.

Heritability

Two studies performed in Latin America can be mentioned to exemplify the types of approach that can be used in relation to this topic. Da Rocha *et al.* (1972) considered 16 anthropometric measurements in 48 monozygous and 51 dizygous Brazilian twin pairs. The heritability estimates (h^2) yielded significant values in both sexes for 12 traits, the exceptions being mandibular and minimum frontal breadth, calf and chest circumferences. The largest heritabilities (0.74–0.92) were observed for stature, head breadth, and lip thickness.

Kobyliansky (1984), on the other hand, examined 20 traits in 305 Mexican families. For most of them high h^2 values (0.60–0.90) were obtained. Principal components and segregation analyses indicated that the traits conform to Mendelian inheritance. Nine of the characteristics had been also studied by da Rocha *et al.* (1972). The heritability estimates were higher in da Rocha *et al*'s. (1972) study in four instances, intermediate in three, and lower in two.

Comparison with other traits

Studies by Lasker and Kaplan (1974) in México and Peru, and of Go *et al.* (1977) in Brazil did not disclose significant associations between anthropometric characteristics and respectively isonymy, or a series of hematologic and behavioral attributes. In Chile Acuña *et al.* (1988) obtained good agreement (correlation coefficient, r: 0.74) between the geographic distances separating seven localities in the Elqui Valley and P.C. Mahalonobis' generalized distance statistics (a multivariate method which reduces anthropometric or other types of quantitative measurements in a single number, summarizing the variability generated, in this case, by 14 anthropometric traits). A similar analysis, comparing spatial distribution with M. Nei's genetic distance yielded an r of 0.55 (genetic distances are also multivariate constructs which summarized, in this case, variability in 12 genetic systems).

The comparisons made by Basu *et al.* (1976) on data from five Haitian populations, which involved geographic, genetic (seven systems), and morphological (14 anthropometric measurements, finger and palm dermatoglyphics) distances did not yield such clear-cut results. No clear association between the geographic distances which separated these places and the distances based in dermatoglyphics were found; and the estimates of the relative contribution of Africans, Europeans, and Amerindians to the populations' gene pool were different when diverse sets of characteristics were employed.

Rothhammer *et al.* (1982), in Chile, investigated how hand size would influence dermatoglyphic features. They studied 105 adults and 108 children. The correlations obtained were generally small (maximum 0.31), but they were able to conclude that children with square hands exhibited higher main line indices, a–b ridge counts, and more open atd angles. Adults with broader hands showed more arches. Taller individuals with larger hands presented higher a–b ridge counts and leaner subjects with long, narrow hands, closer atd angles.

Assortative mating

Men and women have different mating strategies, that involve complex psychological and socio-cultural factors, besides morphology. A review of such strategies, together with specific hypotheses, is given in Buss (1994). The same author (Buss, 1989) analyzed cross-cultural data about mate preferences that included samples from Brazil, Colombia and Venezuela. Waynforth (1999) verified in rural Belize that, as expected, facially more attractive men spent more time in mating effort than their less well-endowed companions. This question of physical attractiveness has been investigated by Jones (1995, 1996) in five populations, three of them from South America (Bahia, Brazil; Ache, Paraguay; Hiwi, Venezuela). He concluded that females with neotenous facial proportions (a combination of large eyes, small noses, and full lips) are favoured by males, regardless of age. But the degree of desirability of other body characteristics varied among the groups considered.

The subject of assortative matings in relation to physical characteristics in humans has been thoroughly revised and discussed in a classical paper by Spuhler (1968). Its genetic consequences in relation to multifactorial characteristics with significant heritability include, *inter alia*, the following: an increase of the genic (or genotypic) variance; the emergence of correlations (genetic associations) among the genes from different loci having similar effects on the phenotype under assortment; and several secondary side-effects on the variance value within and between families. Also, as Crow (1986) pointed out, another consequence of positive assortative matings is that, since they increase the genic variance without much changing the other variance components, they tend to increase the heritability of the trait under assortment. It should also be stressed that under positive assortment several correlation effects arise secondarily; the classical example is the correlation between arm length that is detected when spouses mate assortatively in relation to height. This happens simply due to the correlation which exists between the two characteristics. All these aspects should be taken into account in the evolutionary analyses of this phenomenon (see, for instance, Buss, 1985).

Unfortunately, the number of such studies, however, is small in Latin America. Malina *et al.* (1983*b*) considered assortative mating in relation to age and several anthropometric characteristics in a sample of 68–70 husband-wife pairs in a community of the Oaxaca Valley, México, finding significant correlations only for age (r: 0.96), stature (r: 0.35) and grip strength (r: 0.29).

Trachtenberg *et al.* (1985) studied two groups of couples, 98 in which the husband was a military serviceman and 63 of Jewish ancestry, living in Porto Alegre, Brazil. The characteristics studied involved age at marriage, age at time of survey, 15 anthropometric variables, hair and eye color. Proper adjustments of the data were made taking into consideration time of cohabitation, aging and secular effects, and skewness in the distributions. Correlation coefficients between spouses were generally positive, the highest, being r: 0.53 for stature among the Jews. Size and factors related to body build are the most readily detectable agents in the assortative mating occurring in this population, but in the military sample relative lengths of leg and trunk, rather than stature itself, were important in conditioning mate choice.

Studies in Jacobina, northeastern Brazil, were performed by Pollitzer *et al.* (1990). In a total of 159 couples studied for eight anthropometric measurements they found correlation coefficients varying from −0.52 (face width) to 0.38 (nose length). After applying a factor analysis they concluded that assortative mating in the population emphasized ethnicity over body size features.

Asymmetry and selection

Fluctuating asymmetry (deviation from perfect symmetry in bilateral physical traits that do not display any directional tendency) in relation to eight anthropometric measurements were correlated with life histories in 56 subjects living in rural Belize by Waynforth (1998). He found that low fluctuating asymmetry successfully predicted lower morbidity and more offspring in this sample. A similar investigation was performed by Trivers *et al.* (1999) in Jamaica. Ten paired morphometric traits were considered in 286 children. Fluctuating asymmetries of the legs tended to be related and were less than half as great as those of the arms and ears. Boys showed significantly lower asymmetries than girls, and age was negatively correlated with asymmetry, whereas body size was positively correlated.

Studies by Lasker and Thomas (1976, 1978) in 480 Mexican adults involved 25 anthropometric variables and their relationships with four measures of reproductive fitness: fertility, fecundity, fertility of the parents, and number of surviving siblings. They found that individuals of both sexes with greater head length were more highly fit. Further analyses indicated less variability in the cephalic index of more fertile persons.

Another Mexican sample of 305 families was investigated by Goldstein and Kobyliansky (1984). The average number of children tended to be greater in the modal (approximating the mean) type of 10 parental traits, but directional in 10 others. The differences, however, were not statistically significant, although they exceeded 1% in the actual average measurements.

Two other studies also performed in México should be mentioned. Mueller *et al.* (1981) used principal components analysis to establish the smallest set of anthropometrics which would be least intercorrelated, yet explained major portions of their variation. The investigations were performed in Uruapan, and the anthropometric data thus elaborated were related to three measures of Darwinian fitness: number of surviving children, number of living siblings, and marital status. No consistent pattern, however, was obtained. Little *et al.* (1989), in data from the Oaxaca Project, regressed fertility, offspring survival, and offspring prereproductive mortality on stature and weight of 205 multiparous adults, obtaining no evidences that their short stature could be due to genetic selection.

Skin color

This characteristic has always played a prominent role in interethnic classifications. However, until the early 1950s measurements were made entirely on a subjective basis or with the help of some form of visual matching with color standards. The development of portable reflectance spectrophotometers, which measure, on a chosen wavelength, the amount of light reflected from a surface as compared with a pure white standard, furnished an objective quantitative way of measuring this variability.

An investigation using one of these spectrophotometers was performed as early as 1954 in a Mexican Mestizo population by Lasker. Others followed: for instance, in 1967 Harrison *et al.* studied 484 European-derived and 185 African-derived subjects from Porto Alegre, Brazil. On the basis of three different scales of measurements, comparing the results with autochthonous African samples, they concluded that the 'Blacks' of this city presented from 40% to 50% European skin color genes, in good agreement with independent evaluations obtained using protein genetic markers. The distribution of light reflectance indexes observed in these African-derived subjects suggested that only a few genes would be responsible for the skin color differences between Africans and Europeans, a finding that is now being confirmed by molecular studies.

A review of the studies performed up to that date was undertaken by Harrison (1971). More recent investigations were those of: (a) Rife (1972) in 113 schoolchildren from Aruba, Netherlands Antilles. Those from

Noord were significantly darker than those from three other places on the island; (b) Frisancho *et al.* (1981) in 209 Peruvian Mestizos. They found a high degree of positive assortative mating for this characteristic, and estimated as 55% its additive genetic component; (c) Byard and Lees (1982) in 308 Garifuna (Black Caribs) and 175 Creoles from Belize. A significant degree of intra- and interpopulation variability was observed; and (d) Greksa (1998 *a*,*b*), $n = 257$ in Santa Cruz and $n = 106$ in La Paz, Bolivia, all of European ancestry. A slightly higher degree of vascularity-induced darkening in highlanders was found in relation to lowlanders, maybe reflecting the higher hemoglobin concentrations that are typical of high-land populations.

Another characteristic related to skin is the so-called Mongolian spot, a hyperpigmented area that occurs in the sacrococcygeal region of new-born babies. Sans *et al.* (1986) found that 42% of 226 newborns from Montevideo, Uruguay presented this characteristic, its frequency varying according to the presence or absence of non-European admixture, subjectively estimated. Similar results (Sans *et al.*, 1991) were obtained in Tacuarembó, in northeastern Uruguay (frequency of 43% in 118 newborn children).

Dermatoglyphics

General overview

Skin markings on hands and feet are one characteristic that has been extensively studied in physical anthropological studies. The reasons for this preference is that: (a) the trait can be easily investigated (fingerprints, particularly, are available for thousands of individuals in police files); (b) the patterns of skin ridging are determined early in prenatal life and are not disturbed thereafter by environmental factors; and (c) for some features, such as finger total ridge counts, the mode of inheritance is established as multifactorial.

Against these favorable conditions, there are several drawbacks. The most important are the interobserver differences in the demarcation of critical points. This has to do not only with the quality of the prints obtained, but also with the interpretation of the figures. Coope and Roberts (1971) registered low agreement (sometimes as low as 64%) in the classification of palm main lines by two subjects, and this approach was extended to include almost all dermatoglyphic features, independently scored by four individuals, by Penhalber *et al.* (1994). They concluded that

ridge counts are the measurements which involve larger observation errors; but no discrepancies were noted in the identification and scoring of
digital and palmar dermatoglyphic patterns, types of palmar transverse
flexion crease, and end-points of palmar main lines (*contra* Coope and
Roberts, 1971), nor in the classification of all types of digital patterns.

Other complications exist, such as the correlation of patterns between
fingers, bimanual, sex and interethnic differences. Therefore, interpopulation variability in these characteristics should be considered with caution.

Coope and Roberts (1971) reviewed most of the earlier reports on
studies in Latin American populations. Additional reviews concerning
México (Faulhaber and Schwidetzky, 1986) and Brazil (Salzano, 1987) are
available. Table 5.4 lists studies in 10 Latin American countries not
covered, or only partially covered, in these reviews. As can be seen, sample
sizes and types of characteristics considered vary widely. Demarchi and
Marcellino (1996) departed from the classical univariate comparisons by
subjecting 22 quantitative digital variables (ulnar and radial ridge count in
each of the 10 fingers, plus the total number of loops and whorls by
individual) studied in four localities to a thorough statistical evaluation.
They found an association between reproductive isolation and ridge
counts; and an agreement between the two first canonical variables with
the geographic distribution and the demographic relationships of the
populations studied.

Genetic determination and other influences

The relative role of heredity on dermatoglyphics has been variously considered over the years. In relation to Latin American populations, studies
on the heritability of fingerprints (Beiguelman and Pinto, 1971; Peña *et al.*,
1973; Kolski and Cordido, 1974; Kolski and Salvat, 1977) and palm prints
(Beiguelman, 1971; Pinto and Beiguelman, 1971; Peña *et al.*, 1973;
Callegari-Jacques *et al.*, 1977) were performed in Brazilian and Uruguayan
twins. Generally, the studies indicate: (a) more marked hereditary influences in quantitative rather than qualitative traits; (b) despite the fact that
digital ulnar and radial ridge counts seem to display different population
distributions, correlation coefficients in twins do not show major differences between them or between digital ulnar and radial counts, and total
ridge count; (c) the genetic effect in the distribution of patterns is highest in
the interdigital III and lowest in the interdigital IV regions, the hypothenar
and thenar showing intermediate values. Genetic factors probably play a
relatively minor role in the pattern distribution of interdigital II; (d) the

degree of transverseness of the palm main lines and types of flexion creases, although influenced by chromosomal aberrations, do not conform with a multifactorial mode of inheritance.

Interethnic differences occur in dermatoglyphic traits (see, for instance, the papers listed in Table 5.4), but their exact nature has not been completely elucidated. Salzano and Benevides (1974) did not find a clear gradient in fingerprint characteristics, in either sex, when African-derived subjects from Porto Alegre, Brazil were classified according to amount of African ancestry. Standard deviations and coefficients of variation, as well as the asymmetry in total ridge count were all very similar in the different subgroups. Almeida-Melo and Azevêdo (1977), on the other hand, observed a clear effect of the degree of African ancestry in admixed individuals from Bahia, Brazil in the patterns of interdigital area IV.

Several studies indicated that the digital ridge counts tend to form consistent sets of correlations. The highest correlation values are found between pairs of homologous fingers, but there exist also significant correlations between groups of nonhomologous fingers. Pospisil (1984) and Meier (1995) have considered these questions respectively in populations in Cuba and Colombia. The first found that the results may differ in samples with different degrees of African ancestry. Meier (1995) applied a factor analysis to the data of 880 Mestizos, and from the 20 variables consisting of the ulnar and radial digit counts extracted four factors. He first identified a radial-ulnar dimension, followed by components which identified I, II–III, IV–V subsets. Similar results were obtained by some, but not all, researchers who investigated this process. Be that as it may, analysis such as this one may induce specific hypotheses in relation to the ridges' ontogenetic development that may ultimately be tested by actual fetal observations.

Association studies

Dermatoglyphics are good indicators of the normality of the intrauterine environment, and since the classical observation that they showed specific changes in carriers of Down syndrome, many investigations tried to verify the association of them with a wide variety of pathologic conditions, involving or not chromosome changes. Some of the studies performed in the 1970s and 1980s in Latin America and related to this topic are listed in Table 5.5. Otto *et al.* (1989) verified that all four of the methods which had been established for the dermatoglyphic identification of Down syndrome led to a certain degree (5–14%) of misclassification, and that their overall efficiency was not much different. The other studies yielded mainly inconclusive results, and with the development of more efficient biochemical and/or cytogenetic methods this approach has been abandoned.

Cephalometrics and oral traits

García *et al.* (1984) studied three cephalometric angles of 65 12-year-old children from México City, and observed that they showed more protrusive faces than European-derived subjects. Witkop and Barros (1963), on the other hand, investigated the presence of 14 oral anomalies in 1906 Chilean soldiers. Other studies included the frequency of Carabelli's cusp and shovel-shaped incisors (Pinto-Cisternas and Figueroa, 1968; Devoto, 1969; Sans *et al.*, 1991), as well as taurodontism (Torres Carmona and Márquez Monter, 1984) in Chile, Argentina, Uruguay and México. The degree of genetic determination of five dental characteristics (including shovel-shaped incisors) were evaluated by Blanco *et al.* (1973) in Chile. Problems of definition, however, make the comparison between the frequencies obtained and those of other groups difficult.

Hair and other characteristics

Hair diameter was studied by Kolski and Nunes de Langguth (1971) in 22 monozygotic and 21 dizygotic Uruguayan pairs of twins. They found that a difference of 4.73 microns would indicate dizygosity with 95% of probability. Hairs in a specific region of the body (in the midphalanges) have been extensively studied in many regions of the world. In Latin America there are studies in Yucatan, México (Giles *et al.*, 1968), Recife, Salvador and Holambra, Brazil (Saldanha *et al.*, 1960; Matznetter, 1977) and Santiago, Chile (Covarrubias, 1965). There is a variable amount of sexual dimorphism, in these different samples, and the frequency of hairlessness increase, from low values among European-derived subjects, to higher prevalences among those with significant African or Amerindian ancestry.

Other morphological characteristics (such as ear lobe attachment, finger relative lengths, hair whorls) have been investigated in Latin America but they are not numerous, and difficulties in their precise delimitation make interpopulation comparisons difficult.

Bones and autopsy material

Mexican anthropology has a long tradition of study of skeletal material. Most investigations, however, deal with pre-Colombian remains, and their evaluation is therefore outside the scope of the present book. Some selected examples of research performed in more recent material are: (a) Comas (1959) studied four series of tibiae from the Valley of México and found a clear sexual dimorphism, with a tendency towards greater platicnemia (lateral flattening) among prehistoric as compared with modern series; (b) Genovés and Messmacher (1959) found in 101 crania of known age at death that age determinations based on cranial sutures are subjected to a large amount of error; (c) Vargas *et al.* (1973) applied discriminant functions for the diagnosis of sex in 216 contemporary femurs and obtained an error of 6% only; and (d) Gallardo Velazquez and Pimienta Merlín (1997), based on material from 147 living subjects, tried to estimate their stature using metacarpal lengths obtained radiographically. The highest correlation coefficient (0.73) was obtained for the 2nd left metacarpal among females.

In Brazil Vellini Ferreira (1967/68) obtained statistically significant differences in cranial length and width between European- and African-derived crania ($n = 281$); while Oda *et al.* (1977) studied 271 adult mandibles, also of these two derivations, and of the two sexes. The characteristic investigated was spina mentalis, a median elevation on the internal surface of the mandible, at the region of the incisive teeth. They observed variations in tubercle number and their position; total absence, found in 10% of the cases, occurred mainly in European-derived toothless individuals.

Variations in gross anatomy, obtained from cadaver-dissected material, have been extensively investigated in Brazil by J. Pereira Ramalho and coworkers. Details are too extensive to be covered here, and can be found in the four volumes of Arquivos de Anatomia e Antropologia edited by Ramalho (1975, 1977, 1978, 1979/80).

Morphology at the cellular level

When the chromosomes of a cell are stained two distinct regions can be observed, the *euchromatin* and the *heterochromatin*. A series of experiments and observations showed that cell functions are commanded by genes located on euchromatin. Heterochromatin can be also classified as *facultative* (that related to chromosome X inactivation, fundamental for the dosage compensation phenomenon which equalizes the physiology of males and females) or *constitutive* (present in certain chromosome regions, and possibly involved with regulatory or protective properties).

Heterochromatin is variable in the regions near the centromeres (where the chromosomes attach to the spindle in mitosis or meiosis) of human chromosomes 1, 9 and 16, as well as in the distal portion of chromosome Y's long arm. Several studies have been performed in Brazil and Colombia to more closely investigate the nature of this variability. In relation to Y this variation has a more general evolutionary interest. Since primitive vertebrates have no sex chromosome dimorphism, the question as to why the extreme size differences between X and Y developed has been much considered. The classical explanation has been given by Muller 85 years ago (Muller, 1914). In the absence of exchange between X and Y (the first step needed for the diversification of the two chromosomes) recessive changes would accumulate, since they would be protected from the action of natural selection by their normal counterpart in X. This would lead to an 'erosion' of Y. Fisher (1935) argued that in panmixia (random mating) the elimination of lethals would be of the same order of magnitude in males and females (in the latter case this would occur through homozygosis). Later studies, however, have demonstrated that the greater the inbreeding coefficient, the greater Muller's effect (Frota-Pessoa ad Aratangy, 1968). Therefore, although the effect should be small, populations with different, long histories of inbreeding could show different degrees of Y chromosome reduction.

Investigation of this problem has been hampered by the fact that chromosomes may contract differentially within and between metaphase plates. Material processed at different times could also be affected by diverse factors; and the method used to visualize and register the heterochromatic C-bands would also be important. These questions were considered by Erdtmann *et al.* (1982), who developed a reliable densitometric method of study.

Another basic question concerned the heritability of these structures. This problem had been considered by Viégas and Salzano (1978) at the predensitometric time, and later the chromosomes of the same set of 16 monozygotic and 16 dizygotic twins were investigated by densitometry (Pedrosa *et al.*, 1983). In relation to chromosomes 1, 9 and 16, concordance was not absolute among monozygotics, and this could be due either to artifactual problems or to unequal mitotic crossing-over. For the Y chromosome the studies by Cavalli *et al.* (1984) and Agostini *et al.* (1989) indicated high heritability between the Ys of fathers and sons, the main variability being restricted to the heterochromatin.

How much interpopulation variability occurs in the heterochromatin of these four chromosomes? Data on that are listed in Table 5.6. The series considered involved European-, African-, and Asian-derived subjects, as well as Amerindians, investigated using two different methods. The differences are generally small, the overall impression obtained being that the total amount of heterochromatin of these chromosomes remains constant across ethnic or geographic boundaries. A control therefore should exist, of functional significance, restricting the variation of this highly repetitive DNA. How the control is exercised is presently a matter of pure conjecture. It is possible that these segments undergo ectopic (nonhomologous) pairing at interphase. Unequal crossing-over may lead to interchromosomal differences, while autosome–Y translocations may restore eventual losses caused by the accumulation of deleterious genes in Y. Such saltatory phenomena have already been suggested for the evolution of repeated DNA sequences in general (Britten and Kohne, 1968).

Table 5.1. *Height and weight of military personnel by country of residence measured in the Canal Zone (1965–1970)*

| Country | Average | | Number studied |
	Height (cm)	Weight (kg)	
Dominican Republic	168	64.9	84
Guatemala	164	63.3	62
Honduras	165	62.2	91
El Salvador	163	62.9	47
Nicaragua	166	63.3	84
Panama	168	65.8	131
Venezuela	167	68.0	139
Colombia	166	61.5	207
Ecuador	164	61.7	210
Peru	165	63.5	161
Brazil	170	65.4	94
Bolivia	165	62.9	135
Paraguay	168	67.2	61
Uruguay	169	67.6	43
Chile	168	63.3	411

Source: Dobbins and Kindick (1972).

Table 5.2. *Height of adult women from 16 Latin American countries or places*

Country or place	Time of study	Average height (cm)	Number studied
México	1948–67	149	440
Haiti	1956–58	159	1079
Puerto Rico (rural)	1963–65	152	126
(urban)	1962	160	78
Jamaica (Chinese)	1963–64	156	26
(African-derived)	1959–64	159	420
(African/European)	1963–64	161	68
(European-derived)	1963–64	162	62
Guatemala (European-derived)	1965	159	23
Tortue island	1961	157	40
Colombia	1960	151	184
Venezuela	1965	153	765
Guiana (African-derived)	1960	153	80
Surinam (Javanese)	1960	151	48
Ecuador	1959	146	561
Peru	1960–65	148	535
Brazil (northeasteners)	1963	150	1593
Bolivia	1962–67	150	410
Uruguay	1962	155	1936
Chile	1960	152	398
Easter island	1951	161	18

Source: Meredith (1971).

Table 5.3. *Three anthropometric traits studied in 20 to 24-year-old males and females of seven Brazilian regions during 1975*

Sex of subjects and region	No. studied	Height (cm)	Weight (kg)	Upper arm circumference (cm)
Males				
1	1199	170	60.5	27
2	1334	170	61.9	27
3	1761	170	62.6	27
4	1328	169	59.6	27
5	2793	165	58.0	26
6	520	169	60.0	27
7	1003	167	60.3	27
Females				
1	1114	158	52.3	25
2	1293	158	53.4	25
3	1649	158	54.5	26
4	1353	157	51.4	25
5	2774	154	50.2	24
6	625	157	52.1	25
7	1114	156	51.1	25

Regions: 1. Rio de Janeiro; 2. São Paulo; 3. Paraná, Santa Catarina and Rio Grande do Sul (South); 4. Minas Gerais and Espirito Santo (part of Southeast); 5. Maranhão, Piauí, Ceará, Rio Grande do Norte, Paraíba, Pernambuco, Alagoas, Sergipe and Bahia (northeast); 6. Brasília (Federal District); 7. Rondônia, Acre, Amazonas, Roraima, Pará, Amapá, Mato Grosso and Goiás (north and center-west).
Source: Salzano (1987), based on data of the Brazilian Institute of Geography and Statistics.

Table 5.4. Overview of the dermatoglyphic information available on Latin American populations (descriptive results)[a]

Population	Sex	No. studied	Fingerprints Qual.	Fingerprints Quant.	Palm prints Qual.	Palm prints Quant.	Reference
México							
Puebla-Tlaxcala	M	627			x		Serrano (1975)
	F	250			x		
Sierra de Nayarit	M	74	x		x		Serrano (1982)
Guatemala							
Livingston	M	98	x	x			De Stefano and Calicchia (1986)
	F	48	x	x			
Nicaragua							
Bluefields	M	79	x	x			De Stefano and Calicchia (1986)
	F	71	x	x			
Cuba							
Five provinces	M	3000	x	x			Pospisil (1977)
	F	3000	x	x			
Not given	M	202			x	x	Pospisil (1980)
	F	131			x	x	
Haiti							
Five villages	M	448	x				Basu et al. (1976)
	F	508	x				
French Antilles							
St Barthélemy	M	77	x	x	x	x	Benoist and Dansereau (1972)
	F	72	x	x	x	x	
Netherlands Antilles							
Aruba and Curaçao	M	154	x	x	x		Rife (1972)
	F	98	x	x	x		

	Sex	n					Reference
Venezuela							
Caracas	M	98	x	x	x	x	Larrauri and Rodriguez-Larralde (1984)
	F	92	x	x	x	x	
Brazil							
Curiau	M	58	x	x	x	x	Guerreiro and Schneider (1981)
	F	48	x	x		x	
Humaitá	M	190	x	x	x	x	De Lucca (1975b)
	F	189	x	x	x		
Recife	M	215	x	x	x	x	Matznetter (1978)
	F	186	x	x	x	x	
Salvador	M	232	x	x	x		Matznetter (1978)
	F	169	x	x	x		
Rio de Janeiro	M	92	x	x	x		Matznetter (1978)
	F	107	x	x	x		
São Paulo	M + F	200	x	x	x		Abramowicz (1969)
São Paulo	M	106	x		x		Toledo et al. (1969)
	F	100	x		x		
São Paulo	M	300	x	x	x	x	Penhalber et al. (1994)
	F	300	x	x	x	x	
Porto Alegre	M + F	4800	x	x	x		Kanter (1963)
Argentina							
San Antonio de los Cobres	M	47	x	x	x	x	Marcellino et al. (1980)
	F	23	x	x	x	x	
Four localities, Prov. Córdoba	M	161		x	x		Demarchi and Marcellino (1996)
	F	216		x	x		
Uruguay							
Montevideo	M	200	x				Kolski et al. (1965)
	F	200	x				
Tacuarembó	M + F	1270	x				Sans et al. (1991)

[a] Only studies not considered in Coope and Roberts (1971) were listed here. Qual., qualitative; Quant., quantitative.

Table 5.5. *Selected association studies performed in Latin America between dermatoglyphics and diseases*

Condition	Number of		Results	Reference
	Affected	Controls		
Schizophrenia	97	110	Lower atd average angles in male schizophrenics	Rothhammer *et al.* (1971)
	54	54	Higher atd angles. Differences in the frequencies of digital patterns	Gonçalves and Beiguelman (1974)
Color blindness	123	100	Increase in t line inversions	Arévalo *et al.* (1974)
	55	60	No significant differences	Marques and Marques (1977)
Prader–Willi syndrome	8	100	Differences on the terminations of main lines A and T	Batista *et al.* (1985)
Ullrich–Turner syndrome	108	161	Higher A'-d ridge counts	Otto and Otto (1980); Ferreira *et al.* (1985)
Infertility	52	52	Overall differences in the male digital patterns	Gonçalves and Gonçalves (1985)
Rett syndrome	12	144	Higher atd angle, differences in the frequencies of digital patterns	Martinho *et al.* (1989)
Down syndrome	101	100	Comparison of four methods used for the dermatoglyphic diagnosis of the syndrome. Their overall efficiency was found to be very similar	Otto *et al.* (1989)
Juvenile epilepsy	136	136	Possible epigenetic connection between the embryonic regions I–III and the central nervous system functions	Mattos-Fiore and Saldanha (1996)

The study by Rothhammer *et al.* was made in Chile; all the others were performed in Brazil.

Table 5.6. *Chromosome normal heterochromatin variation in Latin America*

Population	Method	No. studied	Chromosome 1	9	16	Y	Reference
Italian-derived							
Bogotá	1	20	—	—	—	0.86	Monsalve et al. (1980)
São Paulo	1	20	—	—	—	0.88	
Japanese-derived							
Bogotá	1	20	—	—	—	1.00	
São Paulo	1	20	—	—	—	0.93	
Xingu Indians	1	20	—	—	—	0.90	
Caingang Indians 1	1	20	—	—	—	0.91	Ribeiro et al. (1982)
Italian-derived, Curitiba	1	25	—	—	—	0.90	
Japanese-derived, Curitiba	1	25	—	—	—	0.96	
Caingang Indians 2	1	25	—	—	—	0.85	
European-derived, Porto Alegre	2	T40 / M21	2.67	2.22	1.62	0.94	Erdtmann et al. (1981a,b)
Amazonian Indians	2	T278 / M132	2.80	2.38	1.72	0.85	
Caingang Indians 3	2	T116 / M51	2.88	2.39	1.70	0.97	
European-derived, Porto Alegre	2	T38 / M20	1.13	0.94	0.73	0.94	Zanenga et al. (1984)
African-derived, Porto Alegre	2	T38 / M17	1.01	0.79	0.58	0.87	
European-derived, Curitiba	2	T60 / M30	1.17	0.94	0.71	1.19	Cavalli et al. (1985)
Japanese-derived, Curitiba	2	T60 / M30	1.14	0.90	0.70	1.34	
African-derived							
Devotional surnames	2	30	—	—	—	0.95	Barbosa et al. (1997)
Non-devotional surnames	2	30	—	—	—	1.06	

Methods: 1. Ratios of Y/F chromosome sizes, obtained from photographic negative projections. 2. Densitometric profiles. In Erdtmann et al. (1981a) the lengths of the two homologous autosome heterochromatic regions have been added, while in the other reports averages between the two were obtained.

Observations: Caingang Indians 1: Nonoai and Ligeiro reservations, Rio Grande do Sul; Caingang Indians 2: Ivaí and Tamarana reservations, Paraná; Caingang Indians 3: Ligeiro, Cacique Doble, and Nonoai reservations, Rio Grande do Sul. T, total number of individuals studied; M, males only.

6 Health and disease

Why, in a body as wonderfully structured as ours, are there thousands
of failures and weaknesses which make us so vulnerable to disease?
Randolph M. Nesse and George C. Williams

Historical aspects

To a large extent we can only infer what the health conditions were in Latin
America in prehistoric and historical times. However, much can be learned
through the careful examination of osseous or mummified remains, and
written documents. Vargas (1990) gave information about epidemics that
occurred in México during the Conquest and afterwards. He cited studies
of E. Malvido, in which she listed 16 epidemics in the sixteenth, 27 in the
seventeenth, and 17 in the eighteenth centuries; there are indications that
they involved smallpox, cholera, parotiditis, typhus, and measles. Mansilla
and Pijoan (1995) described findings in the remains of a 2-year-old child
who lived in the seventeenth or eighteenth century in what is now México
City, which strongly suggested a case of congenital syphilis. They also
reviewed other evidences for treponemal infection in remains from other
places in México. Márquez-Morfín (1998) mentioned, for the first half of
the nineteenth century in México City, epidemics of influenza (1804),
smallpox (1825) and cholera (1833). She described in detail the mortality
which occurred in 1813 from a typhus epidemic.

'Survival guides', prepared for British colonists planning to travel to the
Carib region, were listed by Halberstein (1997). They included the treatises
of G. Trapham, written in 1679 in relation to Jamaica, of W. Hillary in
1766 concerning Barbados, and of R. Moseley in 1787, for the West Indies
in general. The history of goiter, on the other hand, was considered by
Greenwald (1957, 1969) for several South American countries. He maintained that the disease was absent in the Inca Empire, its first recorded
appearance occurring in Bolivia in 1638. Equivalent information was
obtained for Peru, also in 1638, Argentina in 1783, Paraguay in 1790,
Brazil in 1810, and Chile in 1814. The problems of high altitude in the
Andes were examined by the Jesuit priest José de Acosta as early as 1590,
and Antonio de la Calancha, in 1639, noted the differences in reproductive

165

success between the resident highland Amerindians and the immigrant lowland Spaniards (Cruz Coke, 1997).

The first physician to arrive in Brazil, Master João, came with Pedro Alvares Cabral in 1500. But the first treatise on Brazilian diseases was published only 148 years afterwards, written by Willem Piso (1611–78). His *De Medicina brasiliensi* appeared within the book *Historia Naturalis Brasiliae*, written by him with George Marcgraf (1610–44) and published in Amsterdam. Books in Portuguese, written by physicians living in Brazil, were published even later, in Lisbon: one about smallpox by Simão Pinheiro Morão, which appeared in 1683; another by João Ferreira da Rosa, probably the first in the world to deal exclusively with yellow fever (1694); and a third by Miguel Dias Pimenta, about gangrenous ulceration of dysentheric rectitis (1707). This information was given by Santos (1979), who also mentioned epidemics of cholera (1855–56), which caused 200 000 deaths, yellow fever, smallpox, and bubonic plague (1899). He also attributed to José Francisco Xavier Sigaud the most complete book about Brazilian diseases published in the past, *Du Climat et des Maladies du Brésil, ou Statistique Médical de cet Empire*, published in Paris by Masson in 1844. Confalonieri *et al.* (1981), on the other hand, described the presence of *Trichuris trichiura* (whipworm) eggs in a mummy from an adult male who lived in Minas Gerais in the eighteenth or beginning of the nineteenth century.

All in all, the health situation in Latin America in the past centuries was far from satisfactory. The different pathologies brought from Europe and Africa merged with those already present among the Amerindians. The number of physicians was small, and their training was generally deficient; while therapeutics were more debilitating than stimulating. This led a famous Brazilian physician, Miguel da Silva Pereira, to compare the country to an 'immense hospital' in the beginning of the twentieth century.

General concepts, prevalences and health facilities

Limitations

If it is difficult to evaluate the health conditions of whole nations or regions that obtained in the past, nor is it easy to establish the present state of affairs. This is due to: (a) the concepts of health and disease vary between different cultures; (b) the number and competence of professionals and institutions concerned with the diagnosis and treatment of illnesses differ widely among countries or regions; and (c) the efficiency of the notification programs and of the transfer of information thus obtained to central storage data centers is also variable. It is with these limitations in mind that we should examine the information given below.

Cultural beliefs

Almost everywhere, official medicine occurs in parallel with an empirical set of rules and beliefs that involve both superstition and the utilization of herbs and other natural substances. As was asserted by Du Toit (1997), the Caribbean offers one of the most complex ethnologic reservoirs of folk beliefs, derived from African, Amerindian, and other sources. In Haiti the voodoo ritual pervades the daily life of most of the population, involving family altars and several events through the year. The nature of a patient's problem is frequently established by divination. But these systems of diagnosis and treatment of supernatural or natural ailments differ markedly in diverse parts of the region.

Another aspect to be considered is culture-specific illnesses. A syndrome frequently seen in people from Spanish-speaking regions is related to 'nerves', and is frequently expressed as 'ataques de nervios'. Symptoms may vary from auditory hallucinations and sleeplessness to falling to the floor and even attempted suicide. Other folk illnesses may be physical, and several related to reproduction which occur in Haiti are mentioned by Du Toit (1997). As part of the so-called 'Gran Expedición Humana' Centeno *et al.* (1996) described the concepts of health and disease prevailing in 19 rural Colombian communities of mainly Amerindian or African derivation.

An interesting analysis was performed by Romney (1999) concerning the cultural beliefs of 24 urban Guatemalan women about whether each of 27 diseases were contagious or not. No age differences were found, but women who had more children showed more competence in the discrimination. The amount of agreement among them about which conditions would require 'hot' or 'cold' remedies (see Chapter 4 for an explanation of these categories) was much less marked than in the case of contagion. The type of statistical method used by this researcher (cultural consensus analysis) was also of value in the investigation of 'empacho', a belief related to gastrointestinal disorders. Studies performed in Guatemala, México, Puerto Rico, and a hispanic community in Texas, USA, suggested a common origin for this concept.

General prevalences

Health surveys can be performed on a macro scale or, in a more intensive way, on specific communities. General surveys about environmental (Fleming *et al.*, 1997), infectious (De Santis, 1997), and mental (Hickling, 1997) diseases in the Caribbean have been recently reviewed. Arends *et al.* (1978), on the other hand, performed an in-depth medical-genetic investigation in Tapipa, a Venezuelan African-derived community. They emphasized that many biochemical alterations and subclinical states may occur in 'apparently healthy' populations, exemplified by a prevalence of intestinal parasitosis of 90%, low hematocrits (suggesting anemia), increased gamma globulin levels, and a 20% positive frequency when using Venereal Disease Research Laboratories (VDRL) tests, although only a few individuals had a history of yaws.

Two collective studies, edited by Monteiro (1995) and Minayo (1999) considered Brazilian health problems in a wide context, which involved questions already discussed in Chapters 3 and 4, namely changing demographic patterns, under- and overnutrition, and causes of mortality. Later, they considered the main public health problems of the country, especially infectious diseases, social violence, cardiovascular illnesses, cancer and occupational diseases. Minayo (1999) aptly summarized some of the findings. Despite the negative social indices observed in the 1980s, patterns of nutrition improved; and advances occurred in the areas of housing, sanitation, infrastructure and access to educational and health services. Life expectancy increased, while the incidence of infectious diseases decreased, as did rates of infant and maternal death. On the other hand, the rates of violent deaths and of AIDS increased. Monteiro (1995) stressed the eradication of old and severe problems – measles and poliomyelitis – and significant progress in the control of infant tuberculosis, tetanus, smallpox and other diseases preventable by vaccination; there was also a marked reduction in the fraction of the population exposed to Chagas' disease, and favourable perspectives for the control of the severe forms of schistosomiasis. On the negative side, however, adult tuberculosis was not sufficiently controlled, leprosy is still highly prevalent (second highest prevalence in the world!), morbidity due to malaria increased in the Amazon region, there was an intensification of the leishmaniasis urbanization process, and the reintroduction of cholera and dengue.

Also in Brazil, the investigation of underprivileged or especially vulnerable groups shows a worrisome picture. Prevalence rates of referred morbidity in five areas of the state of Bahia, in the northeast, varied from 19% to 40%, substantially higher than those obtained in official censuses

(Carvalho *et al.*, 1988). In the semi-arid region of the same state, a survey performed among 745 children of 1–72 months of age obtained a frequency of 22% for anemia and 6% for severe anemia, respectively (Assis *et al.*, 1997). In a panning mine located in the state of Pará, in the north, the situation was much worse. Among 223 individuals tested, the prevalences of intestinal parasites was 96%, of anemia 66%, of hepatitis B virus infection 85%, of malaria 35%, of syphilis 42%, and of symptoms compatible with chronic mercury intoxication 10% (Santos *et al.*, 1995). In a southern community widely recognized as being heavily polluted (Cubatão, state of São Paulo) a study conducted between 1980 and 1983 ($n = 1450$) indicated that half of the workers presented some type of morbidity, the frequency of work accidents being 2.7 times higher than that observed in a nearby, also industrialized town (Faria, 1988).

WHO targets of health for all

Even remembering that there is considerable underreporting in official statistics, a rough idea about the health conditions in Central and South America can be obtained through the data presented in Table 6.1. It reproduces WHO information related to six important infectious diseases. Three groups of countries are distinguished, taking into consideration whether they have or have not reached the three WHO targets of health for all for the year 2000. Independent of this classification, however, some points are clear: (a) malaria, tuberculosis and leprosy still are important public health problems in Latin America, especially malaria, with over one million people affected; (b) if we remember that South America has a population number only twice as large as Central America, then the prevalences observed indicate a much worse condition in South than in Central America. For instance, in the first group of nations there are 41 times more cases of leprosy reported in the South (42 270 versus 1029). A large difference is also observed in relation to tuberculosis (170 935 versus 32 500; 5.2 times more); and (c) Haiti stands alone as the only Latin American country in which none of the three WHO targets have been attained. If a comparison is made between this nation and Dominican Republic, which has a population similar to that of Haiti, it is seen that although no cases of AIDS were notified there, its prevalence of malaria is much (12.8 times) higher (23 140 versus 1808 affected).

Cancer

Data concerning six different types of cancer (prevalence and mortality) are given in relation to selected Latin American registries and countries in Table 6.2. Confirming the assertion that males are the more vulnerable sex, with the exception of colorectal cancer, prevalences and mortality are higher among men. In terms of world prevalences, rates of colorectal and breast cancer are low. Variability in frequencies between countries is high especially for esophageal, lung and stomach cancer. In relation to the last, for instance, there is a fourfold difference between Cuba and Costa Rica. As for mortality rates, while those from Costa Rica and Chile show about the same order of magnitude (with the exception of esophageal cancer), those from México are noticeably lower, perhaps indicating substantial underreporting of cancer on death certificates in that country.

More specific studies are those of Prolla *et al.* (1985) and Prolla and Dietz (1985) on mortality of several types of cancer, and mortality and prevalence of breast cancer in Rio Grande do Sul, Brazil; a special issue of the journal Cadernos de Saúde Pública about the epidemiology of stomach cancer in Brazil (Koifman, 1997); and the epidemiology of cutaneous melanoma in Martinique (Garsaud *et al.*, 1998). See Carcinogenesis below for additional information.

Genetic burden

Since infectious diseases are highly prevalent in Latin America, it could be asked whether genetic conditions should also be considered in terms of public health services. Table 6.3 presents selected information about the frequencies of genetic conditions in hospitals and other institutions of the area. It can be verified that these frequencies are far from negligible. In addition, Carnevale *et al.* (1985) and Pinto *et al.* (1996) pointed out that patients with genetic or partly genetic conditions are admitted to hospital earlier, have more and longer admissions, a higher number of surgeries, and are more likely to die than those with nongenetic diseases.

The centralized Chilean system of vital statistics and health services detected a high prevalence there of rare inherited illnesses such as chondrocalcinosis, Creutzfeld–Jakob disease, and achromatopsia, as well as, in Amerindians, of intrahepatic cholestasis of pregnancy, mydriatic response and supernumerary nipples (Cruz-Coke and Moreno, 1994). Strictly following a WHO protocol Chouza *et al.* (1996) also found, among 4468 individuals from two Uruguayan populations, a relatively high prevalence (2.01 per thousand) of Parkinson's disease.

The health system

There are about 15 000 hospitals (that is, institutions with a minimum of five beds which the patients can occupy for at least 24 hours), and around one million beds in Latin American countries. Of these, in 1994, 44% of the total were administered by the state. But this proportion, as well as the number of beds per 1000 inhabitants, differ markedly between countries. For instance, in Guatemala the rate is 1.1, while in Uruguay it is 4.8 beds per thousand. Two to three million people, distributed over 300 labor categories, work in these hospitals (Novaes, 1997). Details about the market for the different types of work in Brazil were presented by Medici *et al.* (1992).

The use of the health system in general (including other types of assistance besides hospitals) was considered in the IBGE (1984) publication for Brazil. As expected, there are marked differences between regions. The focus of the study was on mother–child care, a subject that was also dealt with by Ortiz (1990) in relation to a northeastern Brazilian state, Ceará.

In relation specifically to medical genetics services, the publication edited by Penchaszadeh and Beiguelman (1997) is particularly useful, since it reviews the situation in 13 countries and in Puerto Rico. In addition, it presents information about the health conditions of these populations.

Infectious diseases

The world is witnessing the re-emergence of infectious diseases previously considered to be under control, and Latin America is no exception. This is due to the fact that their patterns are conditioned by a complex interrelationship between the infectious agent, its eventual vector, and its host. The parasite–host coevolutionary process can be viewed as an arms race, in which genetic adaptive changes in one are eventually matched by alterations in the other. This answers, at least in part, the question asked in the quotation at the beginning of this chapter.

It is clear, therefore, that only an ecologic approach can lead to the understanding of changes in this pattern of coexistence (Levin *et al.*, 1999). Most obvious evaluations involve macroeconomics (Cabello, 1991), but more sophisticated analyses can be conducted examining ecoepidemiologic aspects of a given region. This was done, for instance, in relation to Brazilian Amazonia, in a series of 10 papers which considered protozoa, arbovirus, rotavirus, hantavirus and yellow fever (Travassos, 1992). In another special volume of the Brazilian journal *Ciência e Cultura* different aspects of the genetics of human–parasite interactions have been considered (Krieger and Feitosa, 1999; Petzl-Erler, 1999; Villa, 1999).

Table 6.4 gives information about 82 specific studies, conducted in eight Latin American countries, related to 17 infectious conditions. They range from simple epidemiologic investigations to complex molecular studies involving diagnosis, characterization of parasites, vectors and hosts, as well as the production of new vaccines. In most of the conditions listed, the public health situation is far from being under control. On the contrary, prevalences are increasing for the vast majority. A notable exception is the AIDS epidemic, which shows signs of decline since 1996 in São Paulo. The median survival time from diagnosis of patients with this condition is, however, terribly low (14 months) in another Brazilian city (Belo Horizonte). AIDS and tuberculosis, two infections that are now epidemiologically interrelated, show prevalence patterns in Brazil that are intermediate between those seen in Africa and Western developed countries.

The great killer, malaria, remains a huge problem in many Latin American countries. In Brazil, the old objective of eradication was changed to those of mortality prevention and morbidity reduction. Progress towards these targets has been partially achieved, but the development of a really effective vaccine remains elusive.

The molecular investigation of parasites, vectors and hosts is providing much information about the biology of the infectious agents and its reflection in the symptomatology of the diseases. Especially noteworthy are the investigations involving HPV-16, *Trypanosoma*, *Leishmania*, and *Schistosoma*. They include differential tissue distribution of diverse *T. cruzi* clones, evidence for the transfer of nuclear to mitochondrial DNA in *Schistosoma mansoni*, and a pandemic spread of human papilloma virus (HPV)-16 in past centuries.

Another approach is that labeled genetic epidemiology. Through complex statistical techniques it is possible to decide about alternatives related to the genetic factors that may predispose humans to susceptibility or resistance to different infectious illnesses. Studies involving such investigations in Chagas' disease, leishmaniasis, leprosy, and schistosomiasis are listed in Table 6.4.

In a small sample ($n = 43$), Pereira *et al.* (1971) concluded that immunoglobulin G (IgG), IgA and IgM levels among apparently healthy Brazilians were similar to those found in First World countries. On the other hand, an extensive survey of Latin American military recruits furnished detailed information about antibody levels to arboviruses, polioviruses, measles, mumps, *Entamoeba histolytica*, *Schistosoma mansoni*, *Trypanosoma cruzi*, *Echinococcus granulosus*, and *Leishmania tropica* in Brazil (Florey *et al.*, 1967; Niederman *et al.*, 1967; Cuadrado and Kagan, 1967); arboviruses, polioviruses, respiratory viruses, tetanus and trepanomatosis in Colombia (Evans *et al.*, 1969); and to influenza viruses in Argentina, Brazil and Colombia (Cuadrado and Davenport, 1970). More recent studies were concerned with the prevalence of antibodies to nine infectious diseases in Brazilian Amazonian communities (Ferraroni *et al.*, 1983); and to human T cell leukemia virus (HTLV)-I in blood donors, elderly people, and patients with hematopoietic diseases in the French West Indies (Schaffar-Deshayes *et al.*, 1984).

Other immunologic conditions

Prediction and treatment of hemolytic disease of the newborn brought about the first type of situation in which geneticists worked with immunologists and clinicians in the treatment of a given disease. Due to prophylatic measures this condition, caused by maternal–fetal incompatibility, was considerably reduced. The situation in Brazil, prior to the introduction of intervention measures (supply of immunoglobulin D in Rh negative mothers who had already given birth to an Rh positive baby) is summarized in Lacaz *et al.* (1951) and Oliveira (1983).

Another problem relates to the development of human leukocyte antigens (HLA) antibodies during pregnancy. Studies in Cuba were performed by Gómez Arbesú *et al.* (1984) in 295 women, searching for the factors that could enhance or decrease the rates of such antibodies. The levels of these antibodies were also investigated by Trachtenberg *et al.* (1988a) in 481 healthy females from different ethnic affiliations, as well as 49 affected by leprosy or systemic lupus erythematosus, living in Porto Alegre, Brazil. The main factor determining the presence of these antibodies is recency of pregnancy, independent of health status or ethnic affiliation. Such antibodies are useful for HLA typing and matching, a procedure that is important for tissue transplantation survival (Trachtenberg *et al.*, 1993).

Mendelian diseases: nonmolecular approaches

Mendelian diseases (those due to single-factor genes) have been extensively studied in Latin America. Examples of such investigations are given in Table 6.5, which gives information about studies performed on 73 pathologic entities plus three groups of conditions, by authors working in six Latin American countries. Special mention should be given to the work of N. Freire-Maia and Pinheiro (1984*a,b*), which provided a detailed classification of all cases of ectodermal dysplasias described up to that time. The delineation of these nosologic groups has both academic and applied interest, providing precious guidelines for genetic counseling.

The long-term project of study of the hemophilias and von Willebrand disease (developed by Israel Roisenberg and coworkers), as well as that related to inborn errors of metabolism (Roberto Giugliani and colleagues) should also be emphasized. In relation to the latter, the report by Veláz-quez *et al.* (1996) of a possible focus of high prevalence of phenylketonuria in Jalisco, México, is of interest. Jardim *et al.* (1994) also suggested that the frequency of tetrahydropterin deficiency may be higher than expected in southern Brazil.

Huntington's disease research in Venezuela involved a 7-year prospective investigation of 593 persons from an extended family living in the state of Zulia. Of these, 128 had symptoms of the disease. Paternal or maternal inheritance did not affect the rate of disease progression. No discrete age of onset was observed; instead, a prolonged period of time during which symptoms unfolded was noted. These studies (reported by Penney *et al.*, 1990) served for detailed linkage analyses, to be mentioned in the next section of this chapter.

Mendelian diseases: molecular approaches

The importance of performing molecular studies not only in First World countries, but everywhere, is indicated by the investigations summarized in Table 6.6. Information about 37 pathologic conditions, studied in patients from 14 Latin American countries, is given there. It is clear that the patterns of mutations found differ from those observed in other populations, as well as the genotype/phenotype correlations. Moreover, a series of novel variants were found, which could have been missed if these studies had not been performed. Of special evolutionary interest is the evidence for founder effects in conditions as diversified as Bloom syndrome, Creutzfeld–Jakob disease, Hermansky–Pudlak syndrome, hypercholesterolemia, propionic aciduria, and vitamin D-dependent rickets type II. Founder effects are those that can lead to genetic heterogeneity among populations due to the fact that the parental populations which originated them are not a random sample of the founding groups. For instance, individuals from a given geographic region or ethnic subdivision could be differentially represented among the migrants. The identification of these effects may give precious information about past events.

Studies concentrating in cystic fibrosis are listed in Table 6.7. This is one of the most common autosome recessive diseases in European or European-derived individuals. It is characterized by chronic suppurative lung disease, pancreatic failure, high sweat electrolyte concentrations, and reduced life expectancy. To date, more than 850 mutations have been identified as causative factors for this condition. The two most common in Latin America (508 and 542) are those also most common elsewhere in the world. Their worldwide frequencies are 67% (508) and 3% (542) (Cystic Fibrosis Consortium 1992, unpublished data), although there are some significant departures from these values in certain regions. The numbers found for Latin America, in general, are those expected from populations derived from southern Europe (50% and 6%, respectively; Nunes *et al.*, 1991). But, as can be seen from Table 6.7, significant variability may also be found in these frequencies (508: 27–66%; 542: 0–11%). Sample sizes are not large and may account for some of this variation. Other factors, however, may be important; for instance, the disease seems to be less frequent in Africans or Amerindians, and therefore individuals with these ethnic backgrounds are less likely to have it. The investigation of haplo-types associated with the mutations (as was done by Collazo *et al.*, 1995 and Raskin *et al.*, 1997*a,b*) is also important. Different genetic back-grounds associated with the same mutation may condition diverse clinical susceptibility.

The muscular dystrophies are an intriguing group of pathologic conditions that have been extensively studied from the clinical, biochemical, genetic and molecular points of view. In Brazil, the group presently coordinated by Mayana Zatz and Maria Rita Passos-Bueno has been performing important investigations in these diseases for almost three decades now. Beginning in 1989, the studies were extended to the molecular level, and a brief synopsis of their most recent findings is presented in Table 6.8. They include the three main nosologic subdivisions (Duchenne–Becker, X-linked; limb-girdle, autosome recessive; facioscapulohumeral, auto-some dominant) and deal with varied aspects of these illnesses. Their genetic basis is highly complex, the X-linked gene being classified as 'mammoth' due to its gigantic size (2.3 million base pairs, 79 exons). Those responsible for the limb-girdle form occur in at least seven different chromosomes (2, 4, 5, 9, 13, 15 and 17), while the facioscapulohumeral subtype seems to be due to deletions of 3.3. kb (kilobases) repeated units at chromosome 4, in areas that appear to be transcriptionally inactive. Their pathologic effect may be due to changes in a proximally located gene or to modifications in regulatory factors. A myriad of problems appear when attempts are made to relate genetic changes to gene action and phenotypic manifestations at the cellular, organ and individual levels. But significant advances have been made, and much more is expected for the near future as the pace of investigations in this area is remarkably fast.

The multiplicity of molecular markers now available resulted in a tremendous boom in linkage studies. The objective is to map all the genetic conditions in our species to specific chromosome sites. Studies conducted in Latin American, or with Latin American subjects, with this aim are not numerous, but a list of those that came to our attention is given in Table 6.9. It should be stressed that the results included families from outside Latin America as well; but if at least one kindred was living in this area at the time of the study and was included, the whole investigation was chosen and listed. A total of 14 pathologic conditions are indicated in the table, which includes complex categories such as bipolar mood disorder, a debilitating syndrome characterized by episodes of mania and depression, as well as simple entities such as cherubism, in which there is loss of bone, replaced by fibrous tissue, restricted to the jaws (this gives to carriers a 'cherubic' look, like that in Renaissance portrayals of angels). Subjects and studies came from five Latin American countries, and they resulted in the localization of the responsible genes in 11 chromosomes (1, 2, 4, 5, 11, 13, 15, 17, 18, 21 and 22). A variety of techniques was employed to reach these assignments, such as: (a) simulation models; (b) genome-wide screening; (c) standard linkage (family) analysis; (d) homozygosity analysis; (e) linkage disequilibrium (association between markers and the trait under consideration); and (f) construction of new markers from specified genetic regions which were isolated and reproduced in YAC (yeast artificial chromosomes) and physically mapped. These results have obvious importance for genetic counseling and prenatal diagnosis.

Cytogenetics

Presently, a large number of people work in cytogenetics laboratories throughout Latin America. They are, however, mainly concerned with routine diagnoses related to genetic counseling, prenatal tests, or infertility clinical evaluations. Cytogenetic investigations in this geographic area started soon after the classical study of H.J. Tjio and A. Levan in 1956, which made possible the detailed observation of the chromosomes of our species. But most of these researches have now historical interest only, due to the development of more powerful cytogenetic and molecular techniques. Be as it may, a review concerning these earlier efforts, with emphasis on Brazilian studies, was performed by Mattevi and Salzano in 1982. To set the perspective, a recent review by Llerena and Cabral de Almeida

(1998) on cytogenetic and molecular aspects of mental retardation can be consulted. Table 6.10 gives information about research groups, their leaders, and selected information about studies performed in seven Latin American countries.

A problem that interestingly combines cytogenetic and population genetics questions is that related to the incidence and prevalence of Down syndrome. As is well established, trisomy 21 carriers occur mainly through meiotic nondisjunction due to advanced maternal age. Therefore, maternal age at reproduction, and the number of children born from mothers at risk, are important parameters for the estimation of incidence numbers (see, for instance, Beiguelman *et al.*, 1996). Xavier *et al.* (1978) and Lima *et al.* (1996) evaluated this problem in the Brazilian populations of Ribeirão Preto and Porto Alegre, finding incidences of respectively 1.7 and 2.2 per thousand, while Martello *et al.* (1984) performed a whole country analysis taking into consideration the 1970 Brazilian census. These last authors observed three sharply defined demographic patterns in this regard, including the north-northeast, southeast-south and center-west regions of the country, that would be associated with respectively high, medium and low incidences of affected individuals. Castilla *et al.* (1998), on the other hand, examined the first-year survival of 360 liveborn children with the syndrome living in five South American countries (Argentina, Brazil, Chile, Paraguay, and Uruguay). Overall mean survival was 0.74, but those with congenital heart defect had a significantly diminished (0.66) survival probability. As expected, death rates for Down syndrome without cardiac defects (21%) were significantly higher than that reported for developed countries (7% to 16%).

A combination of genetic and molecular studies was used by Pereira *et al.* (1996), Fridman *et al.* (1998), and Schwartzman *et al.* (1999) in Brazil to investigate respectively: (a) clinical aspects of an Xp22.3 deletion; (b) an uniparental disomy due to a translocation 15q15q; and (c) the presence of Rett syndrome in a boy with a 47, XXY karyotype. This last case is consistent with the hypothesis that two X chromosomes are required for the manifestation of Rett syndrome. Similar approaches were developed by Farah *et al.* (1991), Pereira *et al.* (1991), Barbosa *et al.* (1995) and Ramos *et al.* (1996) in the investigation of Brazilian cases of disturbed sex determination.

A study of the Fragile X syndrome (FMR-1) has been continuing in São Paulo, Brazil for more than 10 years now. Representative publications are those of Navajas and Vianna-Morgante (1989), Mingroni-Netto *et al.* (1990, 1997), and Vianna-Morgante *et al.* (1999). They considered different aspects of this condition, such as age and mental status, prevalence among mentally retarded subjects, correlation between cytogenetic, phenotypic, and molecular data, and transmission and segregation patterns of the FMR-1 mutation. As is well known, the syndrome (the most frequent inherited disease causing mental retardation) is due to an unstable expansion of a CGG trinucleotide repeat in the 5' untranslated region of the gene. The risk of affected offspring increases with the number of repeats a given carrier possesses. But no indication of segregation distortion (preferential transmission of alleles with larger repeat numbers), as was observed in other diseases caused by trinucleotide repeat expansion, was found in these investigations.

Multifactorial conditions

About 5–10% of babies present some kind of congenital defect, which could be morphologic (malformations, deformities) or functional (mental retardation, blindness, deafness). They are due to a series of genetic or environmental causes, generally both sets of characteristics being influential. The need to standardize observations on these conditions led to a WHO study, developed between 1961 and 1964 in 24 centers of 16 nations, concerned with the occurrence and type of congenital malformations found in stillborn and liveborn infants. This investigation included six centers in Brazil, Chile, Colombia and México, furnishing important baseline data on 117 307 newborn babies (Stevenson *et al.*, 1966*a,b*).

Shortly afterwards Eduardo E. Castilla devised the idea of developing such a program among Latin American countries. The study started in 17 Argentinian maternity units in July 1967, and progressively included more and more countries. The Estudio Colaborativo Latinoamericano de Malformaciones Congénitas (ECLAMC) was thus established and continued to thrive. In 1995 the project involved 183 hospitals in 30 regions of 12 Latin American countries. The amount of information obtained over these 32 years is enormous. Annual meetings are held by the researchers involved in the investigation, where problems and significant advances are noted and discussed; a document is issued at each of these meetings. A complete list of the publications and theses which resulted from these efforts, for the period 1967–92, is given in Prado (1992). Other overviews are presented in Castilla and Villalobos (1977), Castilla and Orioli (1983), and Castilla *et al.* (1991*b*). A South American geographic atlas covering the incidence of 38 selected malformations was produced by Castilla *et al.* (1995), and a manual of primary prevention of congenital defects by Castilla *et al.* (1996*a*). Other selected recent publications involved patterns in multimalformed babies and the question of the relationship between sirenomelia and the VACTERL (V, vertebral; A, anal; C, cardiovascular; TE, tracheoesophageal; R, renal; L, limb defects) association (Schüler and Salzano, 1994); an epidemiologic analysis of risk factors in the last condition (Rittler *et al*., 1997); hand and foot postaxial polydactyly, including a segregation distortion in the offspring of African-derived fathers (Orioli, 1995; Castilla *et al*., 1997*a*); and natural family planning and unintended pregnancies as risk factors for malformations (Castilla *et al*., 1997*b*; Gadow *et al*., 1998).

A total of 19 studies performed in five countries independently of the ECLAMC is listed in Table 6.11. Schmidt *et al*. (1981) were able to document that the prevalence of severe malformations can be significantly underreported at birth. Follow-up at 3 years of age in children ascertained at birth as normal indicated three times more defects than in the original survey. Cardiopathies were responsible for 26% of the deaths in the first months of life due to congenital malformations in Costa Rica; they were also detected in 18% of the series of autopsies performed by Mattos *et al*. (1987) in Porto Alegre, and were responsible for 13% of these deaths.

Facial clefts are important and relatively frequent congenital malformations, and have been extensively studied in many parts of the world. Brazilian investigations involved corrective procedures (Oliveira, 1954; Zuiani *et al*., 1998); association with fetal loss (Menegotto and Salzano, 1990); epidemiology and familial recurrence (Menegotto and Salzano, 1991*a,b*); and oblique facial clefts and relationship to blepharophimosis, ptosis, and epicanthus inversus syndrome (Richieri-Costa and Gorlin, 1994; Kokitsu-Nakata and Richieri-Costa, 1998). Recently the group headed by Maria Rita Passos-Bueno started a search for candidate genes, results of which are still unpublished. In Chile Hernan Palomino and his group are also investigating important aspects of these conditions, recent examples being the relationship to Amerindian ancestry (Palomino *et al*., 1997*a*) and complex segregation analysis (Palomino *et al*., 1997*b*, Blanco *et al*., 1998).

Animal experiments indicate that chronic dietary iodine deficiency can have a marked effect on neurologic maturation and therefore on the behavior of affected individuals. Goiter, enlargement of the thyroid gland, occurs widely in the world (WHO estimates are of some 200 million affected individuals), especially in localized areas, due to such iodine deficiency. It is not clear, however, why in endemic areas some develop goiter and others are not affected, or what are the relationships between isolated and endemic goiter.

Between 1963 and 1968 an extensive investigation on endemic goiter was

performed, under the sponsorship of the Pan American Health Organization, in eight Latin American countries (Argentina, Brazil, Chile, Colombia, Ecuador, Paraguay, Peru and México). The results have been published in Stanbury (1969). In his book Stanbury emphasized that there
were many questions still requiring investigation: for instance, the success
of prophylactic programs for cretinism or endemic deafmutism, and
whether nonaffected persons may be mildly delayed in terms of growth and
development.

Additional studies have been performed by Greene (1973, 1974) in two
Ecuadorian communities and by D.V. and A. Freire-Maia in an area of
Central Brazil (D.V. Freire-Maia, 1981; D.V. Freire-Maia and A. Freire-
Maia, 1981*a,b*, 1982; A. Freire-Maia *et al.*, 1982). In Ecuador there were
indications that people sensitive to the taste of phenylthiocarbamide
limited their ingestion of naturally occurring goitrogens, and therefore
enjoyed an advantage in relation to those who were insensitive, or less
sensitive. In Brazil, segregation analysis led to exclusion of single locus
determination of resistance or susceptibility, four factors however stand
out as the most important in this case: paternal phenotype, maternal
phenotype, age and sex.

The International Atherosclerosis Project had as its major objective the
comparison of the frequency and severity of atherosclerotic lesions in
autopsied persons from selected populations with different environmental
and genetic backgrounds. A total of 23 207 sets of coronary arteries and
aortas from autopsied cadavers, 10 to 69 years of age, were studied. Some
of them were derived from nine Latin American countries (Brazil, Chile,
Colombia, Costa Rica, Guatemala, Jamaica, México, Peru, Venezuela), as
well as from Puerto Rico. The results showed that, in general, atherosclerosis in autopsied people is closely associated with coronary heart
disease mortality rates. In addition, there were indications that the differences in atherosclerosis among populations begin relatively early in life,
and it was found that different arterial segments develop similar degrees of
atherosclerosis (Tejada *et al.*, 1968).

Twins are classical subjects of study for the investigation of the nature–
nurture dichotomy, and they have been investigated for the assessment of
the degree of genetic influence in relation to a series of biochemical traits
by Colletto *et al.* (1981, 1983) and Rapaport *et al.* (1991). Eventual genetic
effects on the Apgar score (a score that evaluates the biologic condition of
newborn babies) were also researched in twins by Franchi-Pinto *et al.*
(1999).

Other investigations involved pedigree analysis of febrile convulsions
(Raffin *et al.*, 1980), mental retardation (Braga *et al.*, 1977; N. Freire-Maia

et al., 1977), immunoglobulin levels (Hatagima *et al.*, 1999) and nonsyndromic deafness (Braga *et al.*, 1999) in Brazil; etiology of blindness in Colombia (Zarama de Martinez, 1992); prevalence of Parkinson's disease in Uruguay (Chouza *et al.*, 1996); and complex segregation analysis of diabetes mellitus in México (Zavala *et al.*, 1979).

Inbreeding studies were extensively pursued in the 1950s and 1960s; in Brazil, the investigations especially of Newton Freire-Maia and coworkers should be emphasized. Most of the early data are adequately covered in Salzano and N. Freire-Maia (1970) and N. Freire-Maia (1975). More recent works are those of Azevêdo *et al.* (1980*a*), Castilla and Parreiras (1982), and Kaku and N. Freire-Maia (1992). It is quite clear that inbreeding tends to increase precocious mortality and general morbidity, but its effect is variously modulated by environmental and socioeconomic factors, so that detailed analyses are needed in the consideration of specific situations.

There is much resistance in accepting eventual genetic effects on human behavior. A voluminous literature exists, however, showing that inherited factors cannot be ignored. The fear that these results may lead to discrimination is unwarranted. Facts of science should be separated from any eventual misapplication, and the latter strongly opposed. Some of the studies involving normal behavior in Latin America are those of Llop *et al.* (1975), who found no relationship between degree of Amerindian parentage and intellectual performance in Chile; and those of school achievement, intelligence and personality, investigated in twins (Telles da Silva *et al.*, 1975; Salzano and Rao, 1976; Go *et al.*, 1977), as well as language disabilities, studied in families (Borges-Osório and Salzano, 1985, 1986, 1987), in Brazil. As for deviations from the norm, a lack of association between homosexuality and inbreeding was found by Fragoso *et al.* (1994); Bau and Salzano (1995) observed changes in personality type, besides the two polar ones observed among alcoholics; while Flores *et al.* (1998) investigated factors which may predispose to incest (extreme violence in the family environment being one of them) and examined the offspring of some incest cases. All these studies were performed in Brazilian populations.

Mutagenesis, teratogenesis, carcinogenesis

There is a clear relationship between mutagenesis, teratogenesis and carcinogenesis. A change in the genetic material (mutation) can lead to serious morphologic changes, therefore becoming teratogenic (from the Greek *teras* or *teratos*, meaning 'wonders' or 'monsters'); and it may also be the triggering event in the chain process that leads to cancer (therefore being carcinogenic, a term also derived from the Greek, in this case *karkinos* or cancer). These subjects are being actively investigated in Latin America, with the functioning of a Latin American, a Mexican, and a Brazilian Association of Mutagenesis, Carcinogenesis and Environmental Teratogenesis (in relation to the Brazilian Association, the presentations given at its third meeting were published in the *Revista Brasileira de Genética*, **20**, **3**, (Suppl.).

A series of agents can increase the mutation rates, and therefore enhance the probability for the development of these events. They are: (a) temperature, (b) viruses or transposons, (c) other genes, (d) radiation, and (e) chemical substances.

That high-energy radiation can induce mutations was proved by Herman J. Muller in the 1920s, and for this he received the Nobel Prize in 1946. In relation to Latin America (and, more specifically, Brazil), N. Freire-Maia *et al.* (1965) found, however, no clear evidence of genetic damage in the offspring of physicians working with ionizing radiation (indicating that appropriate protection measures had been implemented). The whole subject has been discussed by Pavan and Brito da Cunha (1968); N. Freire-Maia (1972), A. Freire-Maia and N. Freire-Maia (1982), and A. Freire-Maia (1984), while two symposia sponsored by the Brazilian Academy of Sciences and held in Rio de Janeiro dealt, the first, with questions of radiobiology and photobiology (Tyrrell, 1978) and the second with results on areas of high natural radioactivity (Cullen and Penna Franca, 1977).

Areas of high natural radioactivity occur in the USA (Illinois), India and Brazil. Similarities and dissimilarities in relation to the studies performed there and their results were discussed in Cullen and Penna Franca (1977), and the Brazilian investigations were extensively considered in a series of nine papers by Ademar Freire-Maia and coworkers (A. Freire-Maia, 1969, 1971*b*, 1974*d,e,f*, 1975*c*; A. Freire-Maia and D.V. Freire-Maia, 1967; A. Freire-Maia and Krieger, 1975, 1978). No detectable effect of natural radiation on the sex ratio at birth, occurrence of congenital anomalies, pregnancy terminations, stillbirths, livebirths, and postinfant mortality in the children, nor in the fecundity and fertility of the couples was observed. Other causes of mortality and morbidity are relatively so important, that the effect of the low-level natural radiation becomes negligible.

In September 1987 a radiation accident occurred in Goiânia, Brazil, in which a radiotherapy [137]Cesium source was removed from its housing and damaged, leading to external exposure and/or internal contamination of a large number of people. A detailed account of the accident was given in Candotti *et al.* (1988). During the period of 30 September to 22 December 1987, the unit established by the Brazilian National Commission of Nuclear Energy in the city's Olympic Stadium monitored 112 800 people. Of these, 249 were identified as having been contaminated, and 49 of them were interned in local hospitals. Four deaths occurred, and one patient had a forearm amputated. Cytogenetic analyses of 110 exposed or potentially exposed individuals were performed by Ramalho *et al.* (1988) and Ramalho and Nascimento (1991). Dose estimates for 21 subjects exceeded 1 Gy (100 rads), and for eight people they exceeded 4 Gy. Follow-up of the 10 most exposed patients indicated that the average half-life disappearance of lymphocytes containing dicentric and centric chromosome rings was 130 days, which is shorter than the usually accepted value of 3 years that had been reported in the literature up to that time.

The National System of Information about Teratogenic Agents (SIAT) was set up in Porto Alegre in 1990, and in two other major Brazilian cities (São Paulo and Rio de Janeiro) in 1992. Their integrated work is furnishing important prospective data about agents that may lead to congenital malformations in our species (Clavijo *et al.*, 1992; Schüler *et al.*, 1993). Results of another type have been obtained by Nazer and Valenzuela (1973) in Chile (effects of contraceptives); as well as by Sousa *et al.* (1990; delta-aminolevulinate dehydrase levels and their possible significance in lead poisoning), and by Fontoura-da-Silva and Chautard-Freire-Maia (1996; butyrylcholinesterase variants and mild pesticide poisoning) in Brazil.

Thalidomide is the prototype of a drug whose use was introduced for medical reasons, and found to be teratogenic. Schmidt and Salzano (1980, 1983) studied the records of 204 affected persons who were pursuing legal action for damages against the pharmaceutical industry and the Brazilian government. Ninety-three of them showed malformation patterns clearly indicative of the drug's action, and have been studied in detail. An interesting feature, whose investigation was only possible once the victims became adults, was the occurrence of three cases of gynecomasty in patients having major upper limb defects. A detailed review of the syndrome and its occurrence in Brazil was provided by Saldanha (1994). After the drug's removal from the general pharmaceutical market in 1965, it continued to be used on a restricted basis for the treatment of leprosy, and in recent years its indications were extended to a wide variety of medical conditions. By a case-reference approach Castilla *et al.* (1996b) were able to identify 34 thalidomide embryopathy cases born in South America after 1965, in areas endemic for leprosy. They emphasized that phocomelia alone (aplasia or hypoplasia of the limbs) is neither specific nor sufficient to serve as a sentinel for the occurrence of the syndrome. According to Oliveira *et al.* (1999), 61 Brazilian second generation victims had been identified by 1994; they discussed the factors that should be considered for the regulation and rational use of the drug in Brazil.

Patients with upper gastrointestinal ulceration may be treated with misoprostol, a synthetic prostaglandin E analog, but the drug is not recommended for pregnant women because it may stimulate uterine contractions and cause vaginal bleeding and miscarriage. In Brazil, where abortion is not a legal procedure, misoprostol is being misused as an abortifacient. Since in many cases the desired pregnancy termination does not occur, concerns were raised about fetal safety. A prospective study carried out by the SIAT program mentioned above led to the conclusion that misoprostol would not be a potent teratogenic agent in pregnancy (Schüler *et al.*, 1999). However, medical histories from mothers who had children affected by Möbius syndrome, as well as paired comparisons between infants with this syndrome and those with neural tube defects, strongly suggested that attempted abortion with misoprostol is associated with an increased risk of Möbius syndrome in infants (Gonzalez *et al.*, 1993; Pastuszak *et al.*, 1998).

Carcinogenic investigations can involve the appropriate causal agent (such as the molecular diagnosis for human papilloma virus suggested by Guimarães *et al.*, 1992 in Brazil), or studies in the patients themselves, as those performed in México by Fialkow *et al.* (1969), and Lisker *et al.* (1973, 1982*a,b*) or those carried out in Brazil by Ayres *et al.* (1966, 1967) or Ribeiro *et al.* (1990). The prevalence of homozygosity for the deleted alleles of glutathione S-transferase mu (GSTM1) and theta (GSTT1) among distinct Brazilian ethnic groups may be relevant to an eventual predisposition to environmental carcinogenesis (Arruda *et al.*, 1998*b*); in this connection, several studies in which agents related to environmental and occupational cancer were searched for in Latin American populations were reported in Koifman (1998).

In São Paulo, Brazil, a specific effort is being undertaken, generously funded by its state foundation, with the objective of generating one million gene sequences from material obtained from the most common tumours found in the country. This multicenter project is already well under way, with the number of sequences obtained in October 1999 reaching 45 000. An additional important development is that the expressed sequence tags, which signal the places where protein synthesis occurs, are being obtained from the center of the genes and not from its periphery, due to a special technology developed by the group (Anonymous, 1999).

Some genetic conditions may determine an increase in susceptibility to chromosome breaks, which in turn may enhance the probability of carcinogenesis. This happens with familial (but not sporadic) adenomatous polyposis patients, as reported in a study developed in Brazil by Sales *et al.* (1999). The test was made in relation to bleomycin sensitivity; the drug induced significantly more chromatid breaks in cultures from familial patients as compared with controls or sporadic cases.

Many studies are being conducted in Latin America concerning potential mutagenic or antimutagenic effects of drugs and other agents (examples are the investigations that are being performed by Catarina S. Takahashi in Brazil, or Máximo E. Drets in Uruguay). Since in these cases the emphasis is not on peculiarities of the individuals or groups tested, but in the establishment of generalities, these studies will not be covered here.

Association studies

Attempts to relate genetic markers to other (normal or abnormal) conditions has a long tradition, early studies dating back to the 1930s. These investigations have been criticized because generally no clear connection could be visualized between the traits and markers considered, although unexpected side-effects of a given gene expression are always a possibility. With the development of methods that could assess the variability of our major histocompatibility complex (MHC) (and especially of its main regions, called HLA, or human leukocyte antigens) these studies gained a new impetus, since in these cases a specific relationship could be invoked between the pathologic state and human's immunological defenses.

Such approaches have been pursued in Latin America, as is documented in Tables 6.12 and 6.13. Forty-six studies in eight countries related to 29 conditions are listed in Table 6.12. They involved a wide range of systems at the protein and DNA levels, including some considered heterogeneous (cause of death) or complex (African ancestry) entities. Be that as it may, over half of them (59%) claimed positive associations. In some cases the relationship may be indirect, like that of the dopamine D4 receptor gene (involved in neurotransmission) with a given personality trait (harm avoidance) rather than with alcoholism *per se*. As for the HLA studies, information about 65 studies conducted in eight countries involving 34 conditions is listed in Table 6.13. Almost all of them reported positive associations. As with the example above, some of these associations may be with factors that are linked to the studied marker, and not with it directly. Detailed examination of the molecule involved may sharpen the observed association. For instance, specific residues at the DRβ1 third hypervariable region may be the determining agents for the susceptibility of Mexicans to rheumatoid arthritis (Debaz *et al.*, 1998).

Genetic screening in populations and selected families, genetic counseling

Although individually pathologic genetic conditions are generally rare, in the aggregate they constitute a considerable burden, both to the community or to specific families or persons. It is only natural, therefore, that steps had been taken either to prevent or to deal with such a burden. This can be done in several ways: (a) by population screening at the prenatal or neonatal periods, establishing measures that would prevent the birth of affected individuals, or treatment regimes that would correct the genetic defect; or (b) by genetic counseling of adults at the familial or individual levels, explaining the risk of development of a given inherited condition or of its recurrence.

Knowledge is power; therefore, any enterprise which aims at dealing with these subjects should be guided by strict ethics rules. An extensive discussion about possible choices of action was given by Wertz and Fletcher (1989), who considered this problem in a cross-cultural way. Medical geneticists from 19 countries (which included a Latin American nation, Brazil) were asked for information on the situation in their countries regarding legislation on genetics and its implementation, cultural attitudes, sources of challenge and support, national expenditures in medical genetics, as well as what their conduct would be in relation to specific cases. Salzano and Pena (1989) gave information about the Brazilian situation and summarized the opinions of 32 medical geneticists. Although a dominant pattern was observed in the answers, significant interpersonal differences were also found. Both similarities and dissimilarities were encountered when Brazilians were compared with their colleagues from the other 18 countries, indicating the complexities that surround these problems.

A follow-up and extended survey, now including 37 nations but coordinated by the same persons (D.C. Wertz and J.C. Fletcher), was conducted between 1992 and 1995, and the complete results are due for publication next year. Partial results concerning Brazil (Salzano and Schüler, 1998) and México (Carnevale *et al.*, 1998; Lisker *et al.*, 1998) are already available. Earlier discussions about these matters were held in different Latin American congresses of genetics. The situation in relation to ethics and genetics as viewed from Argentina, Brazil, Chile and México was considered in the 1979 meeting (Hunziker *et al.*, 1980), again in 1981, including Venezuela (Cruz-Coke and Brncic, 1982), and also in 1983, with information about, in addition to these countries, Colombia and Cuba (Lemoine, 1984). The right of the people involved in risk situations for hereditary diseases to receive unbiased information was always emphasized.

Where can these individuals receive this kind of information? A satellite meeting held on the occasion of the Ninth International Congress of Human Genetics, which took place in Rio de Janeiro in August 1996, surveyed the facilities available in Argentina, Brazil, Colombia, Costa Rica, Cuba, Dominican Republic, Ecuador, México, Paraguay, Peru, Puerto Rico, Uruguay and Venezuela, and produced a document with specific suggestions concerning improvements that should be sought in such services (Penchaszadeh and Beiguelman, 1997). Details about the functioning of these genetic counseling units in Brazil (Geiger *et al.*, 1985, 1987; Pina-Neto and Petean, 1999) and Argentina (Giorgiutti, 1989) are also available.

Cuba is the only Latin American nation in which whole-country studies of pregnant mothers and their newborns are performed to assess risks for neural tube defects, sickle cell anemia and phenylketonuria, followed by adequate information about the results and the offering of elective abortion of severely affected fetuses (see, for instance, Heredero-Baute, 1984; Körner *et al.*, 1986).

In Brazil, a continent-size country, such a comprehensive public health survey is at present not feasible. However, a number of regional voluntary programs are under way or have been held concerning Tay–Sachs disease among Porto Alegre Jews (Buchalter *et al.*, 1983), neonatal amino acid disorders and hypothyroidism in the southern state of Rio Grande do Sul (Camargo Neto *et al.*, 1991; Jardim *et al.*, 1992*b,c*), congenital adrenal hyperplasia due to 21-hydroxylase deficiency in the same population (Pang and Clark, 1993), and hemoglobin disorders in São Paulo (Paiva e Silva and Ramalho, 1997; Ramalho *et al.*, 1999). Surveys in México first included infants born in México City and the neighboring states of México and Tlaxcala, investigated for hypothyroidism and phenylketonuria (Velázquez *et al.*, 1994), and afterwards a whole-country screen for hypothyroidism (Vela *et al.*, 1999).

Several aspects of prenatal studies in Brazilian mothers at risk have been reported in Gollop *et al.* (1986, 1987, 1990); Costanzi *et al.* (1987); Gollop and Eigier (1987); Franchi-Pinto *et al.* (1994); and Santos *et al.* (1998).

A problem that was exhaustively investigated almost everywhere is that related to heterozygote detection in families in which autosome or X-linked recessive alleles are segregating, to provide risk assessment to carriers. With the advent of the DNA era such evaluations are becoming easier to perform, and in given cases a diagnosis can be made with certainty. This, however, depends on a series of factors, including gene size and heterogeneity of the mutations that may lead to the clinical condition. Examples of studies using both protein and DNA approaches performed in Brazil (Levisky *et al.*, 1983; Falcão-Conceição *et al.*, 1983, 1988; Gonçalves-Pimentel *et al.*, 1988; Roisenberg, 1995), and México (Arenas *et al.*, 1996; Alvarez-Leal *et al.*, 1997) can be examined to obtain details about the intricacies of such evaluations.

Table 6.1. *Targets of health for all and prevalences of six infectious diseases in Latin America*

Countries	Number of notified cases, 1995–96					
	Leprosy	AIDS	Tuberculosis	Malaria	Measles	Neonatal tetanus
First group						
Bahamas	—	374	59	—	0	0
Barbados	—	130	3	—	0	0
Belize	—	38	53	9413	0	1
Costa Rica	15	192	162	—	24	—
Cuba	262	94	1579	20	0	0
Dominican Republic	229	367	6006	1808	0	0
El Salvador	—	417	1686	3362	1	5
Guatemala	—	831	3496	24178	1	12
Honduras	—	797	4176	59446	4	4
Jamaica	—	527	121	10	4	0
México	523	4216	10852	7316	180	60
Nicaragua	—	25	3003	69444	0	1
Panama	—	243	1099	730	0	0
Trinidad and Tobago	—	412	205	35	0	0
Total, Central America	1029	8663	32500	175762	214	83
Argentina	565	2067	13397	1065	59	3
Brazil	39792	16469	87254	565727	580	51
Chile	—	323	4038	—	0	0
Colombia	709	1042	9702	49669	160	27
Ecuador	115	67	6327	18128	42	34
Paraguay	401	50	2148	898	13	8
Peru	90	998	41739	192629	105	45
Surinam	64	2	53	6606	0	1
Uruguay	—	156	701	—	2	0

Table 6.1. (cont.)

Countries	Number of notified cases, 1995–96					
	Leprosy	AIDS	Tuberculosis	Malaria	Measles	Neonatal tetanus
Venezuela	534	634	5576	16371	89	12
Total, South America	42270	21808	170935	851093	1050	181
Grand total	43299	30471	203435	1026855	1264	264
Second group						
Antigua and Barbuda	—	13	5	—	0	1
Dominica	—	14	10	—	0	0
Grenada	—	18	—	1	0	0
St Lucia	—	14	—	—	0	0
St Christopher and Nevis	—	6	3	—	0	0
St Vincent and Grenadines	—	19	—	—	0	0
Total, Central America	—	112	18	1	0	1
Bolivia	32	28	10194	46911	7	14
Guiana	—	144	314	59311	0	0
Total, South America	32	172	10508	106222	7	14
Grand Total	32	284	10526	106223	7	15
Third group						
Haiti	72	0	6632	23140	1	—

First group: countries which attained the three WHO targets of health for all by the year 2000, namely: life expectancy at birth higher than 60 years; mortality rate below age 5 years lower than 70 per 1000 liveborn babies; and infant mortality rate below 50 per 1000 liveborn children. Second group: countries which attained one or two of the targets. Third group: countries which did not attain any of the targets.
Source: WHO (1998).

Table 6.2. *Cancer prevalence and mortality in selected registries and countries of Latin America*[a]

Country	Region or population	Esophagus		Lung		Stomach		Colon/Rectum		Breast	Prostate
		Men	Women	Men	Women	Men	Women	Men	Women	Women	Men
Prevalences											
Martinique	All	13.7	2.5	11.0	3.0	24.9	10.6	9.2	8.8	28.2	48.2
Cuba	All	5.2	1.7	44.3	15.7	9.8	5.0	13.7	14.6	35.0	27.3
Costa Rica	All	3.8	1.2	12.7	4.7	46.9	21.3	9.1	9.6	26.7	23.7
Colombia	Cali	3.6	1.6	24.6	9.8	36.3	19.9	8.0	9.2	34.8	26.1
Brazil	Goiânia	9.8	2.6	26.0	11.6	28.2	14.9	13.4	12.7	40.5	29.0
Ecuador	Quito	4.4	0.5	8.3	3.9	29.5	22.7	8.7	8.4	26.2	23.0
Paraguay	Asunción	11.2	1.7	18.2	3.6	14.4	5.8	7.3	10.6	36.3	22.0
Peru	Trujillo	1.0	0.6	9.5	4.2	28.9	26.4	6.0	9.0	28.3	19.9
Mortality											
México	All	1.9	0.8	14.9	5.8	9.8	7.8	2.2	2.3	7.1	8.9
Costa Rica	All	4.1	1.6	15.7	5.9	46.0	22.5	4.3	4.2	12.3	14.4
Chile	All	8.3	3.7	21.8	5.9	35.3	14.7	4.2	4.6	12.7	12.4

[a] Rates are per 100 000, adjusted to the World Standard Population. The mortality figures in the ninth and tenth columns refer to colon cancer only.

Source: Potter (1997).

Table 6.3. *Selected studies about the prevalence of genetic diseases in four Latin American countries*

Country and population	Type of ascertainment	Findings	References
México, México City	Pediatric hospital	$n = 2945$ admissions, 4.3% genetic, 33.5% partly genetic conditions	Carnevale *et al.* (1985)
Colombia, Bogotá	Pediatric necropsies	$n = 1463$, 5.7% genetic, 21% partly genetic conditions	Bernal *et al.* (1983)
Brazil, Porto Alegre	High risk ward, pediatric hospital	$n = 849$, 12.7% genetic or partly genetic conditions	L. Pinto *et al.* (1996)
Chile, whole country	Centralized National Health Services System	Prevalences of about 50 single gene defects, frequencies similar to those of non-Chilean populations	Cruz-Coke and Silva (1969); Cruz-Coke *et al.* (1972); Cruz-Coke and Valenzuela (1973, 1975); Cruz-Coke and Moreno (1994)

Table 6.4. *Selected information about infectious diseases in Latin America*

Condition	Country	Findings	References
Acquired immunodeficiency syndrome (AIDS)	Brazil	Pattern of the prevalence intermediate between those seen in Africa and in Western developed nations	Anonymous (1989)
		Incidence, period 1980–92, metropolitan regions and Roraima	Bastos *et al.* (1993)
		Median survival following diagnosis 14.3 months, Belo Horizonte	Acurcio *et al.* (1998)
		Decline of the epidemic in São Paulo, starting 1996	Waldvogel and Morais (1998)

Disease	Country	Description	References
Chagas' disease	Argentina	32% of positive serology ($n = 298$) in Empedrado Department, Corrientes	Bar et al. (1997)
	Brazil	Molecular methods for the identification of strains and species of Trypanosoma, and of the analysis of their relationships	Macedo et al. (1992a, 1993); Vallejo et al. (1993, 1994); Oliveira et al. (1998); Gomes et al. (1998)
		Immunoglobulin (IgA, IgG, IgM) levels in a Chagasic population ($n = 300$) Differential tissue distribution of diverse T. cruzi clones	Barbosa (1993) Vago et al. (1996); Macedo and Pena (1998); Andrade et al. (1999)
		Heritability of T. cruzi infection: 0.56 ($n = 525$, 146 pedigrees) 86 T. cruzi field stocks studied in relation to DNA sequences of the mini-exon and 245 α ribosomal ribonucleic acid (rRNA) genes could be separated in two lineages, one associated with human isolates and the other with the sylvatic cycle of the parasite Comparative diagnosis, using polymerase chain reaction (PCR) and other methods ($n = 113$)	Williams-Blangero et al. (1997) Fernandes et al. (1998) Gomes et al. (1999)
Dengue	Brazil	116 900 notified cases in the first 12 weeks of 1998. Persistent epidemy in Belém since October 1996	Machado (1998)
Diarrhea and respiratory infections	Brazil	Survey in Olinda and Recife ($n = 5436$ children under 5 years of age). High prevalences	Vázquez et al. (1999)
Echinococcosis	Uruguay	Sonographic evidence of asymptomatic Echinococcus granulosus in the liver of 156 of 9515 persons in the Department of Florida	Carmona et al. (1998)
Filariasis	Brazil	Prevalences of microfilaria in Recife, 1981–91 showing increasing rates (1.7–3.5%) Prevalence in Recife, 1992, 6%! Belém is the other city in which filariasis is a public health problem	Albuquerque (1993) Maciel et al. (1999)
Hepatitis	Argentina	Genomic characterization of type C (HCV). Genotype prevalences similar to those of Venezuela, but different from those of Brazil	Quarleri et al. (1998)
	Brazil	Prevalence of the Australia antigen (HbsAg) in healthy persons, leprosy and leukemic patients, two cities, southern Brazil HbsAg prevalences in two cities, northeastern Brazil HBV prevalences in seven urban and 17 Indian communities ($n = 4593$)	Salzano and Blumberg (1970) Conceição et al. (1979) Ribeiro-dos-Santos (1993)

Table 6.4. (cont.)

Condition	Country	Findings	References
		2.7% of HCV positivity in 2557 asymptomatic blood donors from Rio de Janeiro	Vanderborght et al. (1993)
		$n = 740$ individuals above 9 years tested in Mato Grosso, central Brazil. General prevalence: 1.2%	Souto et al. (1997)
		Hepatitis G (HGV) prevalence tested in 200 asymptomatic blood donors from São Paulo using reverse transcriptase – PCR (RT-PCR); 9% were positive	Bassit et al. (1998)
Human papilloma virus type 16 (HPV 16)	Brazil	12 genomic variants observed in 25 cervical biopsy specimens, comparison with samples from Singapore, Tanzania and Germany	Ho et al. (1991); Chan et al. (1992)
Human T-lymphotropic virus types I and II (HTLV-I, HTLV-II)	Brazil	Prevalence of 1.35% (both types) in 51 135 blood donors from Belo Horizonte. Guidelines for counseling	Passos et al. (1998)
	Colombia	Association between tropical spastic paraparesis and HTLV-I infection	Arango et al. (1988)
	Jamaica	$n = 47$ native-born Jamaicans with tropical spastic paraparesis reinforce the association with HTLV-I infection	Rodgers-Johnson et al. (1988); Ceroni et al. (1988)
		Infection by HTLV-I found in 11 of 13 polymyositis patients	Morgan et al. (1989)
Intestinal parasites	Brazil	$n = 1081$, 1–5 years, 97% with parasites, correlation with red cell and serum protein levels	Bruno et al. (1981/1982)
		No differences, in a group of 164 children of Bauru, regarding nutritional conditions in those with or without parasites	Costa et al. (1985)
		Prevalence of 57% among 1190 children 0–15 years in Maringá	Teodoro et al. (1988)
		Prevalence of 54% in 1381 preschool children living in Rio de Janeiro slums	Costa-Macedo et al. (1998)
		$n = 491$, Duque de Caxias, increase of Ascaris lumbricoides prevalences with age	Costa-Macedo et al. (1999)
		Prevalence of 88%, Novo Airão, little variation with age	Boia et al. (1999)

Disease	Location		Reference
Leishmaniasis	Colombia	$n = 1016$ boys from Cali, 63% infected, leading to growth problems	Wilson et al. (1999)
	Venezuela	$n = 360$ children, Aracaju, general prevalence 42%	Tsuyuoka et al. (1999)
	Brazil	$n = 890$, 7 months–75 years, general prevalence 64%	Chacin-Bonilla et al. (1998)
		DNA fingerprinting and random amplified polymorphic DNA (RAPD) used to identify species and strains of Leishmania and to investigate their genetic relationships	Macedo et al. (1990; 1992b); Gomes et al. (1995)
		RAPD polymorphisms are limited in two species of Leishmania, but pronounced in their intermediate host Biomphalaria glabrata	Simpson et al. (1995)
		$n = 502$ individuals from 94 families, segregation analysis. Evidence for a major genetic mechanism acting on infection	Feitosa et al. (1999)
	México	$n = 75$ isolates from localized cutaneous lesions, characterization by isoenzyme markers, two species (L. mexicana and L. braziliensis) identified	Canto-Lara et al. (1998)
	Venezuela	Three rural communities, NE, prevalence of the disease increased with altitude	Jorquera et al. (1998)
Leprosy	Brazil	Review of genetic factors that may influence the therapeutics of the condition	Beiguelman (1975)
		$n = 544$ families, 2925 individuals, segregation analysis of the Mitsuda reaction, which indicates susceptibility/resistance to the disease. Suggestion of the influence of a major gene	Feitosa et al. (1996)
Malaria	Brazil	Information about the cases treated in Manaus, 1974–77, more than half from workers involved in the opening of two roads along the forest	Ferraroni and Hayes (1979)
		Epidemiology and clinical aspects of patients living in Humaitá, Amazonas	Meira et al. (1980, 1984a)
		Prevalence along pioneer highways in the Amazonian Region	Hayes and Ferraroni (1981)
		Relatively short persistence of anti-Pfs2400 repeat peptide antibodies, that could be used for a transmission-blocking vaccine	Marrelli et al. (1997)
		Prevalence and levels of IgG antibodies to the heat-shock protein Pf72/Hsp70-1, a candidate for a multivalent vaccine	Alexandre et al. (1997)
		Drop of the prevalence of autochthonous cases in the upper Paraguay Basin (15% in 1991, 1.6% in 1996)	Matsumoto et al. (1998)
		Plasmodium falciparum may have three different mechanisms for resistance to chloroquine, mefloquine and quinine respectively	Zalis et al. (1998)
		No protective effect of SPf66 vaccine ($n = 572$)	Urdaneta et al. (1998)

Table 6.4. (*cont.*)

Condition	Country	Findings	References
		The old objective of eradication was changed to mortality prevention and morbidity reduction. This led, from 1992 to 1996, to a 21% reduction in the mortality rate in the Amazon	Gusmão (1998)
		The 19 kDa C-terminal region of the *Plasmodium vivax* merozoite surface protein 1 (PvMSP1$_{19}$) is highly immunogenic in individuals exposed to this parasite	Soares *et al.* (1999)
	Guiana	General epidemiologic information	Giglioli (1968)
Onchocerciasis	Colombia	First case in 1965. Study of a focus on the Colombia–Ecuador border ($n = 655$). *Simulium exiguum* was found infected with *Onchocerca volvulus*	Corredor *et al.* (1998)
Schistosomiasis	Brazil	Twenty pedigrees (269 individuals) suggest a major codominant gene controlling susceptibility/resistance to *Schistosoma mansoni* infection	Abel *et al.* (1991)
		Identification and mapping of *S. mansoni* genes	Dias Neto *et al.* (1993); Franco *et al.* (1995a,b); Tanaka *et al.* (1995)
		Evidence for the transfer of nuclear to mitochondrial DNA in *S. mansoni*	Pena *et al.* (1995)
		Epidemiology of the disease in Ravena, Minas Gerais	Coura-Filho *et al.* (1995)
		Evidence for a major gene controlling interleukin-5 production in subjects infected by *S. mansoni*. No relationship with the gene identified by Abel *et al.* (1991)	Rodrigues *et al.* (1996)
		Prevalence of the disease in 15 counties of the Metropolitan Region of Belo Horizonte, Minas Gerais	Coura-Filho (1997)
		Study in 17 counties, 1424 localities, and a population of 485 200 inhabitants, undertaken for 14 years, indicated an area of high endemicity in Pernambuco's southeast	Carvalho *et al.* (1998)
Streptococcus infections	Brazil	Specific agglutination of group C streptococci	Ottensooser *et al.* (1974)
Tuberculosis	Brazil	RFLP comparison between *Mycobacterium tuberculosis* isolates from patients with and without AIDS. Higher strain clustering among the AIDS patients	Ivens-de-Araujo *et al.* (1998)
		Time-series mortality in São Paulo, 1900–97	Antunes and Waldmann (1999)

Table 6.5. *Studies on Mendelian diseases in Latin America: nonmolecular approaches*

Condition	Country number	Information	References
Acheiropoidia	1	Extensive clinical, population, and genetic investigations	A. Freire-Maia (1970; 1974a,b,c; 1975a,b; 1981); A. Freire-Maia and Chakraborty (1975); A. Freire-Maia et al. (1975a,b,c; 1978); Morton and Barbosa (1981); Fett-Conte and Richieri-Costa (1990)
Achondroplasia plus Down syndrome	1	Second report of this very rare association (probability of random association: one in 25 million!)	Carakushansky et al. (1998)
Amino acidopathies and congenital hypothyroidism	1	Pilot neonatal screening, around 28 500 children tested	Jardim et al. (1992b,c); Wajner et al. (1994)
Aplasias and hypoplasias	1	Report of a family with four cases of peromelia, AR	N. Freire-Maia et al. (1959)
Atrichias and hypotrichoses	1	Short review of nine forms and description of a family with universal atrichia	Pinheiro and N. Freire-Maia (1985)
Brachydactyly	1,5	Clinical studies of the A2 type	N. Freire-Maia et al. (1980); Lisker et al. (1983)
Branchio-oto-renal syndrome	1	Report of a family	Casadei et al. (1998)
C3 deficiency, complement system	1	The deficiency causes susceptibility to infections, AR	Grumach et al. (1988)
Cerebrofaciothoracic syndrome	1	One family, AR	Guion-Almeida et al. (1996)
Cherubism	1	Report of a Brazilian family, AD, afterwards studied for molecular markers (see Table 6.6)	Salzano and Ebling (1966)
Cleydocranial dysostosis	1	Clinical review of the condition, AD	Laredo-Filho et al. (1987)
Coagulation Factors V and VIII	1	Report of an isolated Factor V and a combined Factor V/VIII deficiencies	Fischer et al. (1984, 1988)
Coffin–Siris syndrome	1	Report of one family, AR	Ferreira et al. (1995)
Color blindness	2	Detailed examination of the condition	Cruz-Coke (1970)

Table 6.5. (cont.)

Condition	Country number	Information	References
Congenital adrenal hyperplasia	1	Form due to 21-hydroxylase deficiency. Screening in 13 countries, including Brazil	Pang and Clark (1993)
Congenital deafness	6	Detailed clinical and genealogic information in an inbred community	Arias (1974)
Congenital fascial dystrophy	1	Review of the literature and description of a family, AR	Matiotti et al. (1992)
Cryptophthalmos	1	Description of two families, one with a pair of concordant monozygous twins, AR	Azevêdo et al. (1973)
Cystinuria and urolithiasis	1	Laboratory, population, and clinical studies	Giugliani and Ferrari (1980a,b; 1981; 1987; Giugliani et al. (1985a, 1986, 1987, 1989a, 1990a)
Ectodermal dysplasias	1	Thorough review and classification based on clinical and genetic characteristics	Settineri et al. (1976); N. Freire-Maia and Pinheiro (1984a,b); Pinheiro and N. Freire-Maia (1994)
Ectrodactily	1	One family, AR	A. Freire-Maia (1971a)
EEC syndrome	1	Reports on several patients, clinical and genetic review	Salzano and Schmidt (1988); Rodini and Richieri-Costa (1990a)
Ehlers–Danlos syndrome	5	In one family, the syndrome occurs together with a plasma thromboplastin component deficiency	Lisker et al. (1960)
Ellis van Creveld syndrome	1	Detailed clinical–genetic review and study of large families	Pilotto (1978); Silva et al. (1980)
Epidermolysis bullosa	1	Association with the Pelger–Huet anomaly	Cat et al. (1967)
Galactosemia	1	Description of four cases	Luft et al. (1990)
GAPO syndrome	1	Report of one family, AR	Gagliardi et al. (1984)
Geleophysic dysplasia	1	Report of one family, AR	Boy et al. (1998a,b)
Gilbert syndrome	1	Test of caloric restriction and mode of inheritance	Magalhães et al. (1980)
Glaucoma	1	Juvenile, autosome recessive form	Beiguelman and Prado (1963)
GM1 gangliosidosis	1	Clinical and laboratory studies in several families	Giugliani et al. (1985b); Barbosa-Coutinho et al. (1987); Folberg et al. (1992)

Disease/condition	No.	Description	References
Growth hormone deficiency	1	Study of an extensive genealogy in a semi-isolated population	Souza (1997)
Hemophilias A, B	1,2	Detailed clinical, laboratory and genetic studies	Roisenberg (1964); Roisenberg and Morton (1971); Marinho et al. (1972/1973); Cruz-Coke and Rivera (1980); Alexandre and Roisenberg (1985); Quezada Díaz et al. (1988); Ferrari (1988)
Homocystinuria	1	Improved specific laboratory test	Wannmacher et al. (1982a)
Huntington disease	6	7 years follow-up of a very large family (n = 593)	Penney et al. (1990)
Ichthyosis	1	X-linked, due to steroid sulfatase deficiency	Romiti et al. (1996)
Krabbe disease	1	Discussion of the diagnosis, AR	Jardim et al. (1992a)
Lysosomal diseases	3	n = 450, high-risk patients, flow-chart for diagnosis	Villasante et al. (1990)
Malpuech syndrome	1	Report of one family, AR	Guion-Almeida (1995)
Mandibulofacial dysostosis	1	New (Bauru) type	Marçano and Richieri-Costa (1998)
Maple syrup urine disease	1	Treatment and poor outcome of one case	Jardim et al. (1995)
Marden–Walker-like syndrome	1	Report of one family, AR	van den Ende et al. (1992)
Metabolic disorders	1	Screening, especially on high-risk patients	Wannmacher et al. (1982b); Wajner et al. (1986); Giugliani et al. (1989b, 1991); Barth et al. (1990); Giugliani and Coelho (1997); Pereira et al. (1997); Coelho et al. (1997)
Mitochondrial acetoacetyl-coenzyme A deficiency	1	Case report, the second in Latin America	Wajner et al. (1992)
Mucopolysaccharidoses	1	Diagnosis, clinical aspects and treatment	Wannmacher et al. (1984); Giugliani et al. (1990b); Leistner and Giugliani (1998)
Multiple cartilaginous exostoses	1	Report of a family, AD	L. Freire-Maia and A. Freire-Maia (1981)
Myotonic dystrophy	4	The associated hypogonadism had different bases in four patients	Febres et al. (1975)
Neurofibromatosis	1	11 patients studied from the dermatologic point of view	Saddy et al. (1985)
New syndromes	1	Reports on three possible new syndromes, AR, XR	Aguilar et al. (1978); Santos et al. (1991); Paes-Alves et al. (1991)
Niemann–Pick disease	1	Relationship to sea-blue histiocytosis, cholesterol metabolism	Viana et al. (1990, 1992); Coelho et al. (1995); Scalco et al. (1999)

Table 6.5. (*cont.*)

Condition	Country number	Information	References
Oculomaxillofacial dysostosis	1	Four patients with oblique facial clefts, suggesting a unique dysmorphogenetic process	Richieri-Costa and Gorlin (1994)
Osseous distrophy	1	Rare condition, resembling both Camurati–Engelmann disease and hyperostosis corticalis generalisata familiaris	Ilha and Salzano (1961)
Osteochondrodysplasias	1	Clinical, genetic, and anatomopathologic findings in 17 patients with the lethal form	Galera et al. (1998)
Osteochondromatosis	1	Two families, clinical and radiologic findings, AD	L. Freire-Maia (1961)
Osteopetrosis	1	Review of dominant cases and frequency in a Brazilian state	Salzano (1961)
Phenylketonuria	1,5	Diagnosis, heterozygote detection, and apparent higher frequency in Jalisco, México	Esperon (1978); Blau et al. (1989); Camargo-Neto et al. (1993); Velázquez et al. (1996); Silva et al. (1997)
Poland–Moebius syndrome	1	Association between the two conditions and family recurrence suggest autosomal dominant inheritance	Larrandaburu et al. (1999)
Progeria	1	One pair of concordant-affected monozygotic twins is reported	Viégas et al. (1974)
Ramon syndrome	1	Neoplasms were caused by a fibromatous process similar to that found in the gingivae of these patients	Pina-Neto et al. (1998)
Rapp–Hodgkin syndrome	1	Varied expressivity in a family with 11 affected patients, AD	Rodini et al. (1990)
Retinoblastoma	3	Clinical data and results of treatment of 235 cases	Gaitan-Yanguas (1978)
Ring-shaped skin creases syndrome	1	Report of a four-generation family, AD	Guion-Almeida et al. (1998a)
Rotor's syndrome	1	Report of an extensive investigation including 72 relatives in six generations	Pereira-Lima et al. (1966)
Say syndrome	1	Two reports, one involving discordant monozygotic twins	Ashton-Prolla and Félix (1997); Guion-Almeida et al. (1998b)

Sensorineural hearing loss	1	The patients also show cataracts. New AR syndrome	De Vitto et al. (1997)
Sialic acid storage disease	1	Method of detection and report of one case	Folberg et al. (1990); Utagawa et al. (1999)
Spherocytosis	1	Associated with polydactyly, large family (117 related members, five generations), AD	Roisenberg et al. (1962)
Spondyloepimetaphyseal dysplasia	1	Associated with joint laxity, AR	Pina-Neto et al. (1996)
Tetrahydrobiopterin deficiency	1	$n = 5200$ patients suspected of having some inborn error of metabolism, 30 cases detected in 28 sibships	Jardim et al. (1994)
Tibial hemimelia-polysyndactyly-triphalangeal thumbs syndrome	1	Three-generation family, variable expressivity, AD	Richieri-Costa et al. (1990)
von Willebrand disease	1	Prevalence of severe form, production of reagents for diagnosis	Fischer et al. (1989); Meissner et al. (1992); Fischer et al. (1996)
Waardenburg I syndrome	1	Extensive studies in large kindreds	Silva et al. (1990, 1993); Silva (1991); Silva and Batista (1994)
Whistling face syndrome	1	Report of a family, AR	Alves and Azevêdo (1977)
Wilson's disease	1	Hypodensity on cerebral white matter disclosed by tomography	Jardim et al. (1991)
Zellweger syndrome	1	Case report, AR	Carvalho et al. (1990)
Zlotogora–Ogur syndrome	1	Three affected brothers, AR	Rodini and Richieri-Costa (1990b)

Key for country numbers: 1, Brazil; 2, Chile; 3, Colombia; 4, Cuba; 5, México; 6, Venezuela. AR, autosomal recessive; AD, autosomal dominant; XR, X-linked recessive.

Table 6.6. *Studies on Mendelian diseases in Latin America: Molecular approaches*

Condition	Country number	Information	References
General	9	Overview of molecular diagnosis of three groups of conditions	Barrera *et al.* (1991)
Antithrombin deficiency type I	2	Novel splice site mutation, intron 1, nucleotide position + 5	Arnaldi *et al.* (1999)
Apolipoprotein AI	1	Differential methylation in liver and leucocyte	Dewey and Vidal-Rioja (1993)
Arylsulfatase A pseudodeficiency	2	$n = 171$ healthy subjects, frequency of the defective allele 8%	Pedron *et al.* (1999)
Ataxia-telangiectasia	5	Four mutations in 41 patients account for 86–93% of the cases	Telatar *et al.* (1998*a,b*)
Bipolar disorder and schizophrenia	2	$n = 47$ for bipolar disorder and $n = 39$ for schizophrenia. No differences among them or controls in the 5-HTT serotonin transporter gene frequencies	Mendes de Oliveira *et al.* (1998)
Bloom syndrome	1,2,7,8,9	Ashkenazic *blm*^{Ash} mutation in non-Jews of USA, México and El Salvador suggests a founder effect	Ellis *et al.* (1998)
Craniosynostoses	2	$n = 50$. Mutations in the *FGFR2* gene account for 93% of the syndromic cases. General review	Passos-Bueno *et al.* (1998, 1999*a*) Misquiatti *et al.* (1998)
Creutzfeldt–Jakob disease	3	Prevalence of the 200K variant especially high, perhaps due to a mutation which occurred in Spain	Lee *et al.* (1999)
Fabry disease	2	Novel mutation (30 del G), carrier detection	Ashton-Prolla *et al.* (1999)
Fragile X syndrome	2	$n = 256$ mentally retarded boys, 2% had the full mutation	Haddad *et al.* (1999)
Galactosemia	2	Three new disease-causing mutations in three patients from two families	Sommer *et al.* (1995)
Gaucher disease	1,3,10	$n = 31$, 13 different mutations account for 93% of the cases	Cormand *et al.* (1998)
GM1-gangliosidosis	2,13	20 patients, six novel and two previously described mutations account for 90% of the cases	Silva *et al.* (1999)
Growth hormone deficiency	2	$n = 23$, three with deletions of the *GH* gene and severe deficiency	Arnhold *et al.* (1998)
Hemophilias A, B	2,9,11	Patterns of mutations	Figueiredo *et al.* (1992*a*, 1994*a*); Figueiredo (1993); Heit *et al.* (1998)
Hermansky–Pudlak syndrome	2,12	A 16-bp (base pair) frameshift duplication is nearly ubiquitous among Puerto Rican patients, indicating a founder effect	Hazelwood *et al.* (1997); Oh *et al.* (1998)
Huntington disease	9	$n = 83$ patients and $n = 96$ controls, (CAG)n repeat distribution. The juvenile onset is associated with expansions greater than 49 repeats	Alonso *et al.* (1997)

Disease	Country	Comments	References
21-Hydroxylase deficiency	2	Detailed mutation analysis in over 200 patients	De-Araujo et al. (1996); Bachega et al. (1999); Paulino et al. (1999)
Hypercholesterolemia	2	High frequency (9/31 families) of the Lebanese mutation	Figueiredo et al. (1992b); Alberto et al. (1999)
Laron syndrome	7	Mutation creating a new splicing site which leads to an 8 amino acid deletion in the receptor hormone	Berg et al. (1992)
Machado–Joseph disease	2	Large number of families studied. Expanded CAG repeats are strongly correlated with age at onset, but other modifying factors must exist	Maciel et al. (1995); Lopez-Cendes et al. (1997)
Myotonic dystrophy	2	41 families with 235 patients were studied. The size of the CTG repeat was markedly larger in skeletal muscle as compared with lymphocytes	Passos-Bueno et al. (1995); Zatz et al. (1995a)
Phenylketonuria	2,5,14	Extensive surveys, with the observation of novel mutations	Santos et al. (1996); Pérez et al. (1996); De Lucca et al. (1998)
Polycystic kidney disease	1	New mutation at exon 44 of the PKD1 gene	Iglesias et al. (1999)
Porphyria cutanea tarda	1	Four missense mutations, a microinsertion, a deletion, and a novel exonic splicing defect observed in 10 families	Mendez et al. (1998)
Prader–Willi and Angelman syndromes	2	Five patients characterized, but no mutation in 256 mentally retarded boys	Vercesi et al. (1999)
Propionic acidemia	2,3,7	Nine patients with mutation profile similar to that found in Spanish subjects	Rodriguez-Pombo et al. (1998)
Protein S deficiency	2	Novel nonsense mutation in exon 2 of the PS gene	Pugliese et al. (1999)
Pseudohermaphroditism (male)	2	Five patients with female assignment and behaviour. In three of them single transitions occurred in different codons	Cabral et al. (1998)
Retinoblastoma	2	Five families and six isolated cases studied by PCR-Xba1-RFLP	Costanzi et al. (1993)
Spinal muscular atrophy	2	Deletions of exons 7 and/or of the SMN gene were found in 69% of 74 families	Kim et al. (1999)
Sensorineural deafness	6	Three of the four mtDNA A1555G mutation investigated probably arrived in Cuba from the Canary Islands	Torroni et al. (1999)
Vitamin D-dependent rickets type II	4	More than 200 affected with a possibly distinct form of receptor-positive resistance to vitamin D	Giraldo et al. (1995)
Waardenburg syndrome type I	2	Identification of the mutation in the HuP2 gene	Baldwin et al. (1992)

Key for country numbers: 1, Argentina; 2, Brazil; 3, Chile; 4, Colombia; 5, Costa Rica; 6, Cuba; 7, Ecuador; 8, El Salvador; 9, México; 10, Paraguay; 11, Peru; 12, Puerto Rico (USA); 13, Uruguay; 14, Venezuela.

Table 6.7. Selected studies about the prevalence of mutations among cystic fibrosis patients in Latin America

Population	No. individuals studied	No. of sites searched	% of identified mutations	Frequency (%)			References and notes
				508	542	Others	
Argentina							
Buenos Aires	79	1	61	61	0	—	1
Córdoba	19	2	66	66		—	2
Brazil							
São Paulo 1	24	1	31	31	—	—	3
São Paulo 2	58	1	52	52	—	—	4
Minas Gerais	31	1	53	53	—	—	4
Paraná	17	1	44	44	—	—	4
Santa Catarina	24	1	27	27	—	—	4
Rio Grande do Sul 1	60	1	49	49	—	—	4
Espírito Santo	12	6	54	42	4	8	5
Rio Grande do Sul 2	61	1	51	51	—	—	6
South and Southeast	124	4	17	—	11	6	7
Rio de Janeiro	44	8	34	31	2	1	8
Cuba							
Not indicated	72	1	52	52	—	—	9
México							
Three localities	40	16	56	45	5	6	10
Not indicated	76	1	7	—	7	—	11
Eight countries	24	15	66	48	2	16	12

References and notes: 1. Luna *et al.* (1996); 2. Saleh *et al.* (1996); 3. Martins *et al.* (1993); 4. Raskin *et al.* (1993). Slightly different figures were obtained by Raskin *et al.* (1997*a,b*) who investigated also the KM-19/XV-2C haplotype distribution in *508* and non-*508* bearing chromosomes (*n* = 133 individuals). They found that 88% of the *508* alleles are linked to haplotype B (KM-19 +/XV-2C−). Other analyses involved linkage disequilibrium among populations, and the specific search for mutations R1162X and 2183AA → G (Pereira *et al.*, 1999); 5. Rabbi-Bortolini *et al.* (1998). The other sites considered (number of chromosomes found with mutations in parentheses) were: *551*(0), *553*(1), *1282*(1), and *1303*(0); 6. Maróstica *et al.* (1998). No association between the genetic status in relation to the *508* mutation and pulmonary status was found; 7. Raskin *et al.* (1999). Studies restricted to 247 non-*508* alleles. 8. Cabello *et al.* (1999). The other sites considered (number of chromosomes found with mutations in parentheses) were: *551*(0), *1282*(0), *1303*(0), *1507*(0) and *1717*(0); 9. Collazo *et al.* (1995). A total of 71% of 21 *508* alleles was linked to KM-19/XV-2C haplotype **B**; 10. Villalobos-Torres *et al.* (1997). The other mutations detected were (numbers in parentheses): *549*(1), *621*(1), *1303*(1) and *3849*(2); 11. Villareal *et al.* (1996); 12. Arzimanoglou *et al.* (1995). The other sites considered (number of chromosomes found with mutations in parentheses) were: *1717*(0), *334*(3), *507*(0), *51D*(0), *551S*(0), *553*(1), *560*(0), *621*(0), *1162*(0), *1282*(0), *1303*(3), *1717*(0), *3849*(0).

Table 6.8. *Muscular dystrophy studies performed in Brazil and México*[a]

Clinical condition	Results	References
Duchenne/Becker	General review of early studies	Zatz *et al.* (1993)
(The most common form, with severe –	Multiplex reactions as an initial screening method of diagnosis	Falcão-Conceição *et al.* (1992)
Duchenne – or milder – Becker – clinical	Carrier detection by haplotype microsatellite analysis	Falcão-Conceição *et al.* (1994)
severity. Due to a 2.3 million base pairs gene	Deletion analysis in 50 patients from 41 families	Alho *et al.* (1995)
with 79 exons located on the short arm of	Utrophin (dystrophin-related protein) is increased in patients, but its	Vainzof *et al.* (1995)
chromosome X)	quantity shows no correlation with clinical severity	
	Deletion analysis and new multiplex assay to amplify exons 44–52	Coral-Vázquez *et al.* (1997)
	Highly skewed X-chromosome pattern inactivation does not predict	Sumita *et al.* (1998)
	muscular weakness in carriers	
	Paternal inheritance or different mutations in maternally related patients	Zatz *et al.* (1998*a*)
	occur in 3% of familial cases	

Table 6.8. (cont.)

Clinical endition	Results	References
Limb-girdle (onset at the 2nd or 3rd decade, without facial involvement and slow progression. Autosomal recessive)	General review of early studies. Of four patients with the same γ-sarcoglican mutation, three had a severe Duchenne-like disease, but one showed much milder symptoms	Zatz et al. (1993) McNally et al. (1996)
	First missense mutation in the δ-sarcoglican gene associated with a severe phenotype	Moreira et al. (1998)
	140 patients (40 families) studied, distributed by seven different subgroups (LGMD2A to LGMD2G). All LGMD2E and LGMD2F patients had a severe condition	Passos-Bueno et al. (1996b)
	General survey of the calpainopathies (LGM2A) which included two mutations identified in Brazilian patients	Richard et al. (1999)
	Two patients (one of the LGMD2C, and the other of the LGMD2F subgroups) reinforce the view of a closer relationship between the α, β and δ-sarcoglycans (SG) as compared to γ-SG, the latter being associated with dystrophin	Vainzof et al. (1999)
Facioscapulohumeral (facial and shoulder girdle muscles are variably affected. Autosomal dominant, gene at 4q 35)	General review of early studies	Zatz et al. (1993)
	34 Brazilian families studied at the clinical and molecular levels. New mutations and somatic mosaicism are not rare, and anticipation occurs in multigenerational families	Zatz et al. (1995b)
	The gene affects males more severely and more frequently than females	Zatz et al. (1998b)

[a]The studies by Coral-Vázquez et al. (1997) were performed in México, but all the others were done in Brazil. For a recent review of the Brazilian studies in these conditions and in the craniofacial disorders see Passos-Bueno (1999).

Table 6.9. *Linkage studies using molecular markers performed in Latin America*

Condition	Country number	Chromosome location	References
Acheiropodia	2	7q36	Escamilla *et al.* (2000)
Ataxia-telangiectasia	3	11q22-23, between S384 and S535	Uhrhammer *et al.* (1995)
Bipolar disease	3	18q22-23; 18pter	Escamilla *et al.* (1996, 1999); Freimer *et al.* (1996); McInnes *et al.* (1996)
Calcium pyrophosphate dihydrate deposition disease	1	5p15, between S416 and S2114	Andrew *et al.* (1999)
Cherubism	2	4p16, telomeric to S1582	Tiziani *et al.* (1999)
Ectodermal dysplasia (Margarita Island form)	5	11q23, between S4171 and S4460	Suzuki *et al.* (1998)
Kmobloch syndrome	2	21q22.3, between S171 and S1446	Sertié *et al.* (1996)
Limb-girdle muscular dystrophy			
(LGMD2A)	2	15q15.1-q.15.3, between S222 and S779	Allamand *et al.* (1995)
(LGM2B)	2	2p12-p16	
(LGM2C)	2	13q12	
(LGM2D)	2	17q12-q21.33	
(LGM2E)	2	4q12	
(LGM2F)	2	5q33-34	Passos-Bueno *et al.* (1996a,b)
Mitochondrial neurogastrointestinal encephalomyopathy syndrome	4	22q13.32-qter, distal to S1161	Hirano *et al.* (1998)
Multiple endocrine neoplasia type 1	1	11q13	Guadagna *et al.* (1998)
Nephronophthisis	5	3q22	Omran *et al.* (2000)
Sclerosteosis	2	17q12-q21, between S927 and S791	Balemans *et al.* (1999)
Proximal spinal muscular atrophy	2	5q11.2-13.3	Whittle *et al.* (1993)
van der Woude syndrome	2	1q32-41, modified by a gene at 17p11.2-11.1	Sertié *et al.* (1999)

Key for country numbers: 1, Argentina; 2, Brazil; 3, Costa Rica; 4, Puerto Rico (USA); 5, Venezuela.

Table 6.10. *Information about human cytogenetic studies performed in Latin America*

Country	Some of the key workers (in alphabetical order)	Selected references
Argentina	S. Brieux de Salum, E.C. Gadow, N. Magnelli, T. Matayoshi	Barreiro (1997)
Brazil	A.M. Vianna-Morgante, B. Erdtmann, C. Casartelli, C. Bottura, C.P. Koiffmann, E.H. Tajara da Silva, G. Carakushansky, G.A. Paskulin, H.M.M. Mendez, H.R.S. Nazareth, I.J. Cavalli, I. Ferrari, J.C. Cabral de Almeida, J.A.D. Andrade, J.C. Llerena Jr., M.I.S.A. Melaragno, M.S. Mattevi, M.A.C. Smith, M. Varella-Garcia, O. Frota-Pessoa, R.C. Mingroni-Netto, S.R. Rogatto, W. Pinto Jr., W. Beçak	Mallmann *et al.* (1970); Suñe *et al.* (1970); Erdtmann *et al.* (1971, 1975) Mattevi *et al.* (1971, 1981, 1982); Xavier *et al.* (1978); Tajara *et al.* (1982*a,b*); Mattevi and Salzano (1982); Mendez *et al.* (1982); Martello *et al.* (1984); Cavalcanti *et al.* (1988); Navajas and Vianna-Morgante (1989); Cabral de Almeida *et al.* (1989*a,b*); Sampaio *et al.* (1989); Mingrone-Netto *et al.* (1990, 1997); Farah *et al.* (1991); Pereira *et al.* (1991, 1996); Barreto and Salzano (1992); Barbosa *et al.* (1995); Moreira *et al.* (1995); Mantovani *et al.* (1995); Lima *et al.* (1996); Ramos *et al.* (1996); Fernandez *et al.* (1997); Acosta *et al.* (1998); Boy *et al.* (1998*a*); Castilla *et al.* (1998); Fridman *et al.* (1998); Llerena and Cabral de Almeida (1998); Pagni and Melaragno (1998); Ribeiro *et al.* (1998); Schwartzman *et al.* (1999); Carvalho *et al.* (1999)
Chile	R. Youlton, S. Castillo-Taucher	Youlton (1971); Castillo-Taucher (1997)
Ecuador	C. Paz y Miño	Paz y Miño (1991, 1997)
México	A. Carnevale, F. Salamanca, H. Rivera, J.M. Cantú, O. Mutchinick, R. Lisker, S. Armendares	Zavala *et al.* (1972); Lisker *et al.* (1973, 1978, 1982*a*); Aguilar *et al.* (1981); Dominguez *et al.* (1999); Mutchinick *et al.* (1999)
Peru	J. Descailleaux	Orrillo *et al.* (1976); Perez de Gianella (1997)
Uruguay	J.H. Cardoso, G.A. Folle, M.E. Drets	Drets (1997)

Table 6.11. *Selected studies on congenital malformations performed in Latin America outside the ECLAMC[a] project*

Country	Information	References
Argentina	Prevalence of anencephaly in the Province of Jujuy	Dipierri and Ocampo (1985)
Brazil	Recurrence risks of bone aplasias and hypoplasias of the extremities	N. Freire-Maia and A. Freire-Maia (1967)
	Epidemiology of congenital malformations in Porto Alegre	Araújo and Salzano (1975*b*)
	Epidemiology of Legg Calvé Perthes disease	Alves *et al.* (1976)
	Prevalence of congenital malformations and associated variables in Campinas	Arena (1976*b*, 1977)
	Follow-up at 3 years of age of children ascertained at birth in Porto Alegre	Schmidt *et al.* (1981)
	Frequency of malformations in 731 autopsies of children aged 0 to 14 years	Mattos *et al.* (1987)
	Monozygotic twins discordant for the Brachmann–de Lange syndrome	Carakushansky *et al.* (1996)
	Study of 41 craniofrontonasal syndrome patients	Saavedra *et al.* (1996)
	Clinical data on two Costello syndrome patients	Pratesi *et al.* (1998)
Colombia	Incidence of congenital malformations and anthropometry in newborn babies of Popayan	Silva (1983)
Costa Rica	Mortality due to congenital malformations in the whole country and separated by provinces, 1970–77	Barrantes (1980)
México	Prevalence of brachymesophalangia-V in Oaxaca	Buschang and Malina (1980)

[a]ECLAMC, Estudio Colaborativo Latinoamericano de Malformaciones Congénitas.

Table 6.12. *Selected studies on associations between given conditions and several genetic markers in Latin America*

Condition	Genetic marker(s)	Main result	Country number	Reference
Alcoholism	Color blindness	Positive association	3	1
	ADH2, ALDH2	No relationship	6	2
	Dopamine D4 receptor gene	No relationship with alcoholism, but positive for harm avoidance	2	3
Alzheimer disease	5-HTTLPR	Positive association	2	4
Anemia (iron deficiency)	Ahaptoglobinemia	No relationship	2	5
Body weight	ABO blood groups	Selection may act very early in pregnancy on females	2	6
	Paraoxonase gene family	Homozygosity for *A148/A148* is associated with lower birth weight and, eventually, coronary heart disease	7	6
Bone mineral density	*ApaI, BsmI, TaqI* sites at the vitamin D receptor gene	No association	6	7
Cardiovascular diseases	ABO blood groups, haptoglobin (Hp) levels, Apo AI-CIII-AIV cluster, Apo B	Significant associations with ABO types, Hp levels and the *SstI*S2* allele	2	8
Cataracts (senile)	Lactose absorption	No association	6	9
Chagas' disease	African ancestry	Increased susceptibility in African-derived subjects	2	10
Congenital malformations	ABO, Rh blood groups, Albumin, Haptoglobin,Transferrin	No association	6	11
Death (cause of)	Total liver protein concentration	Positive association	2	12
Duodenal ulcer	ABO blood groups	Significant association	6	13
Filariasis	ABO, MN, Rh blood groups, ABH Secretion, Hemoglobin, Haptoglobin	Significant association with ABO and Secretor phenotypes	2	14
Irritable bowel syndrome	Lactose absorption	No association	6	15
Joint hypermobility	African ancestry	Negative association	2	16
Leishmaniasis	ABO blood groups	Positive association	2	17

Disease	Genetic marker	Association		Ref.
Leprosy	ABO, MN, Rh blood groups	No association	2	18
	Phenylthiourea (PTC)	Positive association	2	19
	Haptoglobin	No association	2	20
	Atypical pseudocholinesterase	Positive association	6	21
	Isoniazid acetylation	Association with resistance of *Mycobacterium leprae* to diaminodiphenyl sulfone	2	22
	Glucose-6-phosphate deficiency	No association	2	23
	Hemoglobin types	No association	2	24
	Thalassemia	No association	2	25
Lung cancer	CYP1A1, CYP2E1	CYP1A1 exon 7 polymorphism showed different distributions in patients and controls	2	26
Malaria	ABO, Rh, Hemoglobin, Albumin, Ceruloplasmin, Haptoglobin, Transferrin	The combined frequency in the ABO blood groups of B + AB as compared with O + A individuals was significantly higher among subjects with malaria	2	27
	Hemoglobin A2	Level significantly higher in affected persons	8	28
Mental diseases	ABO blood groups	No association	1	29
Neuropathy (optic)	Mitochondrial DNA haplotypes	No association	5	30
Renal diseases	Phenylthiourea (PTC)	Positive association between proteinuria and the *T* allele	3	31
Retinoblastoma	Esterase D (ESD)	ESD levels significantly lowered in patients	2	32
Rheumatic diseases	ABO, MN, Rh, ABH secretion, Haptoglobin, Amylase	No association	2	33
	Alpha 1-antitrypsin levels	No association	2	34
Schistosomiasis	Haptoglobin levels	Higher frequency of anhaptoglobinemics among the patients	2	35
	African ancestry	Greater resistance associated with degree of African ancestry	2	36
	13 genetic systems	Difference in the Glyoxalase 1 system between patients of the intestinal and hepatosplenic forms	2	37
Stomach cancer	ABO, Rh blood groups	Positive association with ABO	2	38
Systemic lupus erythematosus	ABO blood groups, ABH Secretion	Positive association with ABO	2	39
Tuberculosis	ABO blood groups	Conflicting results	2,4	40
	Phenylthiourea	Conflicting results	2	41
	Hemoglobin types	Positive association	2	42
	Isoniazid acetylation	No association	2	43

Table 6.12. (*cont.*)

Condition	Genetic marker(s)	Main result	Country number	Reference
Thyroid diseases	Phenylthiourea	Positive association	2	44
Thyphoid fever	ABO, MN, Rh	Positive association with Rh	3	45

References: 1. Cruz-Coke (1965b); Cruz-Coke et al. (1971); 2. Lisker et al. (1995); 3. Roman et al. (1999); Bau et al. (1999); 4. Oliveira and Zatz (1999); 5. Azevêdo et al. (1971); 6. Kelso et al. (1992); Busch et al. (1999); 7. Jaramillo-Rangel et al. (1999); 8. Robinson and Roisenberg (1975, 1980); Oliveira (1977); Colônia and Roisenberg (1979); Barretto et al. (1987); Bydlowski et al. (1996); 9. Lisker et al. (1988a); 10. Nunesmaia and Azevêdo (1973); Azevêdo et al. (1979); 11. Lisker et al. (1982b); 12. Lima et al. (1981); 13. Lisker et al. (1964); 14. Ayres et al. (1976); 15. Lisker et al. (1989); 16. Santos and Azevêdo (1981); Azevêdo and Santos (1982); 17. Cabello et al. (1995); 18. Beiguelman (1963b, 1964a); Salzano (1967); 19. Beiguelman (1964b); 20. Schwantes et al. (1967); 21. Navarrete et al. (1979); 22. Costa et al. (1993); 23. Beiguelman et al. (1968); 24. Cézar et al. (1974); 25. Ramalho et al. (1983); 26. Sugimura et al. (1995); 27. Santos et al. (1983); 28. Arends (1967); 29. Solá (1931); 30. Torroni et al. (1995); 31. Cruz-Coke and Iglesias (1963); 32. Costanzi et al. (1989a,b); 33. Robinson et al. (1984); 34. Asensio et al. (1986); 35. Nunesmaia et al. (1975); 36. Bina et al. (1978); Azevêdo (1984); 37. Weimer et al. (1991); 38. Ribeiro (1963); 39. Ottensooser et al. (1975); 40. Soriano Lleras (1954); Saldanha (1956b); Mourão and Salzano (1978); 42. Ramalho and Beiguelman (1977); 43. Beiguelman et al. (1977); Massud et al. (1978); 44. Azevêdo et al. (1965); Mendez de Araújo et al. (1972); 45. Cruz-Coke et al. (1968).

Key for country numbers: 1, Argentina; 2, Brazil; 3, Chile; 4, Colombia; 5, Cuba; 6, México; 7, Trinidad and Tobago; 8, Venezuela.

Table 6.13. *Selected studies on associations between given conditions and human leukocyte antigen (HLA) (or other major histocompatibility complex, MHC) markers in Latin America*

Condition	Main positive associations	Country number	References
Actinic prurigo	B40, Cw3	4	Bernal et al. (1988)
Alport syndrome	DR2	2	Donadi et al. (1998)
Ankylosing spondylitis	B27	7	Arellano et al. (1984)
Asthma	DQ2	2	Gerbase-de-Lima (1997)
Atopy	A7	7	Escobar-Gutiérrez et al. (1973)
C2 deficiency	A25, B18, DR2	2	Araújo et al. (1997)
Celiac disease	DR3, DR7, DQ2, others	1	Satz et al. (1989); Palavecino et al. (1990); Herrera et al. (1994)
Cervical carcinoma	DQA1*03011	5	Ferrera et al. (1999)
Chagas' disease	Conflicting results	2,3,8	Llop et al. (1988); Dalalio (1994); Fernandez-Mestre et al. (1998); Deghaide et al. (1998)
Chromoblastomycosis	A29	2	Tsuneto et al. (1989)
Diabetes mellitus	Several	2,3,6	Eizirik et al. (1987); Valette et al. (1988); Pérez-Bravo et al. (1996); Marques et al. (1998)
Geographic tongue	B13, Cw6	2	Gonzaga et al. (1996)
Glaucoma	Segregation distortion of HLA haplotypes	2	Gerbase-de-Lima et al. (1992a)
Glomerulosclerosis (focal segmental)	DR4	2	Gerbase-de-Lima et al. (1998a)
Graves disease	DR3, DQA1*0501	2	Kraemer (1998)
Hemophilia	DR4 and antifactor VIII formation	8	Simonney et al. (1985)
Hepatitis	B8, DR3, DR4 (autoimmune); DR1 (cryptogenic)	2	Gerbase-de-Lima et al. (1992b)
Hypertension (essential)	Segregation distortion of HLA haplotypes, DR4	2	Gerbase-de-Lima et al. (1989, 1992c)
IgA-deficiency	B8, B14, DR1, DR3, DR7	2	Gerbase-de-Lima et al. (1998b)

Table 6.13. (*cont.*)

Condition	Main positive associations	Country number	References
Leishmaniasis	DQ3, TNFA, TNFB	2,8	Barbier et al. (1987); Petzl-Erler et al. (1991); Lara et al. (1991); Cabrera et al. (1995); Blackwell et al. (1997)
Leprosy	DR2, C4B*Q0	2,7	Escobar-Gutiérrez et al. (1973); De Messias et al. (1993); Visentainer et al. (1997)
Malaria	A9, DR4	2,4	Meira et al. (1984b,c); Restrepo et al. (1988)
Multiple sclerosis	DR2, DQ1, B7	2	Gerbase-de-Lima (1997)
Neurocysticercosis	Conflicting results	2,7	Del Brutto et al. (1991); Bompeixe et al. (1999)
Paracoccidioidomycosis	Conflicting results for HLA, association with C4B*Q0 and C4A*Q0	2	Lacerda et al. (1988); Goldani et al. (1991); De Messias et al. (1991); Visentainer et al. (1993); Rebelatto (1996)
Pemphigus foliaceus	DR1, DR4	2	Petzl-Erler and Santamaria (1989); Moraes et al. (1991, 1997)
Psoriasis	Cw6, B13, B17	2	Gonzaga et al. (1996)
Rheumatic diseases	DR53, DR7, DR1, DRB1*0404, DRB1*0401, DRB1*1001	2	Monplaisir et al. (1986a); Guilherme et al. (1991); Gerbase-de-Lima et al. (1994); Gerbase-de-Lima (1997); Debaz et al. (1998)
Schistosomiasis	DQB1*0201	2	Secor et al. (1996); Chiarella et al. (1998)
Silicosis	No association	2	Gerbase-de-Lima (1997)
Spondyloarthropathies (seronegative)	B27 is not a good diagnostic test	7	Pérez-Rojas et al. (1984)
Systemic lupus erythematosis	B8, B15, B53, DR3, DR4	2,6	Balthazar et al. (1985); Monplaisir et al. (1988); Silva and Donadi (1996)
Urinary tract infection	No association with DQ alpha	2	Albarus et al. (1997)
Vogt–Koyanagi–Harada disease	DR4 (especially DRB1*0405), DR1, DQ4	2,7	Goldberg et al. (1998a); Arellanes-García et al. (1998)

Key for country numbers: 1, Argentina; 2, Brazil; 3, Chile; 4, Colombia; 5, Honduras; 6, Martinique; 7, México; 8, Venezuela.

7 Hemoglobin types and hemoglobinopathies

Hemoglobin is probably the best analyzed genetic system in humans
Friedrich Vogel and Arno G. Motulsky

Hemoglobin disorders, which include sickle cell disorders and the thalassemias, are the commonest of human inherited diseases
World Health Organization

A paradigmatic subject

Human hemoglobin is formed by four peptide (globin) chains. The formula which describes its molecule is, in adults, $\alpha_2\beta_2$, indicating that the four chains comprise two identical pairs. These chains are differentially produced during ontogeny, so that they are diverse at the embryonic, fetal, and adult stages. Interest in the molecule arose due to its changes in a series of common clinical conditions which are caused by its structural variants (hemoglobinopathies) or defects of synthesis (thalassemias).

Table 7.1 presents some of the key events that led to a better understanding of the ways in which hemoglobin can vary and the resulting consequences. Investigation of these conditions was first limited to cytologic methods; later details of the protein sequences was obtained, culminating with studies at the DNA level. Presently several hundred structural and regulatory variants are known, due to changes in almost any step which leads from DNA to protein.

Hemoglobin is a paradigmatic subject for genetic research for two particular reasons. First, the difference in electrophoretic mobility between hemoglobin (Hb) A and Hb S (the one responsible for the sickling phenomenon) led Pauling *et al.* (1949) to the concept of molecular disease, that is, to pathologic conditions caused by specific changes in the protein molecule; today a wide array of diseases can be explained in this way. Second, the hemoglobin diseases are still the only undisputed examples of balancing selection in humans. Their peculiar world distribution suggested a possible selective advantage of heterozygotes in relation to malaria. This was first suggested for the thalassemias by J.B.S. Haldane, one of the founders of the synthetic theory of evolution, in 1949. But it was A.C.

Allison, during the 1950s, who gathered sufficient evidence to demonstrate unequivocally the interaction between sickle cell anemia and malaria.

Speaking of predecessors, it is interesting to know that a Brazilian physician, Jessé Accioly, proposed in the same year (1947), and independently from James V. Neel, that the sickle cell trait would represent the heterozygous state of a gene that, in homozygosis, would condition sickle cell anemia (in modern nomenclature $Hb_\beta*A/Hb_\beta*S$ and $Hb_\beta*S/Hb_\beta*S$, respectively; see Azevêdo, 1973). The necessary evidence for this view, however, was only presented two years later, by Neel (1949) and Beet (1949).

The relationship between the hemoglobin genes and malaria conditioned that the hemoglobinopathies became particularly common in African peoples and their descendants who had migrated to the Americas (the first description of sickle cell anemia was made in a West Indian subject), while the thalassemias spread mainly among Asiatics. But the joint occurrence of both conditions is not an uncommon finding, especially in the intermixed Latin American populations.

Hemoglobin types

The three most common alleles

As was indicated previously, normal adult hemoglobin is formed by the association of two protein chains present in duplicate which are called alpha and beta. Normally the sixth amino acid (they are the protein units) in the β chain is a glutamic acid. Changes in the genetic region (locus) which codifies this amino acid in this specific region, however, can lead to the substitution of glutamic acid for valine. This is the molecular change responsible for the formation of sickle cell hemoglobin (Hb S) instead of Hb A (the normal component of human hemoglobin). In this same position another modification may occur, changing glutamic acid for lysine. In this case the variant hemoglobin is called Hb C. These are the two most common forms that occur, besides Hb A, in African people or in persons derived from them.

The distribution of the alleles (alternative forms of a gene) which condition these three types of hemoglobin was extensively investigated in Latin America. At the beginning, taking advantage of the sickling phenomenon, that can be easily observed *in vitro* in anoxic conditions, only the prevalence of HbS was studied. The disadvantage of these cytologic methods is that generally it is not easy to separate homozygotes ($Hb_\beta*S/Hb_\beta*S$) from heterozygotes ($Hb_\beta*A/Hb_\beta*S$). The development of electrophoretic techniques (which separate molecules according to their electric charges) led to a new level of analysis, making possible the identification not only of Hb S, but of all other forms that would be different from Hb A (including Hb C) in relation to this characteristic.

One of us (FMS) has been interested in the distribution of hemoglobin types in Latin America (and more specifically Brazil) for a long time. Periodic reviews of this distribution in that country can be found in Salzano (1965, 1979, 1985, 1986); Salzano and Tondo (1982); and Bortolini and Salzano (1999). Important workers in this area of study in Brazil include Cassio Bottura, Marco A. Zago and Fernando F. Costa (see, for instance, Zago and Costa, 1985; Zago *et al.*, 1999); J. Targino de Araújo (Targino de Araújo, 1971); Paulo C. Naoum (Naoum *et al.*, 1984; 1986); and Antonio S. Ramalho (Teixeira and Ramalho, 1994). The last two have produced textbooks on hemoglobin and hemoglobinopathies (Ramalho, 1986*a*; Naoum, 1987, 1997) and are also involved with programs of detection of hemoglobin and thalassemia genes in the general population, as well as with the prevention of the associated disorders.

As for other countries in Latin America, mention should be made of Rubén Lisker (Lisker *et al.*, 1990); and Guillermo Ruiz Reyes (Ruiz Reyes, 1983) in México; Bruno Colombo and Gisela Martínez in Cuba (Colombo and Martínez, 1981, 1985); Graham R. Serjeant in Jamaica (Serjeant, 1974; Serjeant *et al.*, 1986); H. Fabritius *et al.* (1980) in Guadeloupe; N. Monplaisir *et al.* (1981, 1985) in Martinique; German F. Sáenz (Sáenz, 1986; Sáenz *et al.*, 1993) in Costa Rica; and especially Tulio Arends in Venezuela (Arends, 1961*a*, 1966, 1971*a,b*, 1984; Arends *et al.*, 1971, 1982, 1990).

More than half a million people had been studied in Latin America in relation to the three most common alleles of the Hb_β locus (not included in this total are the earlier studies, which involved the sickling tests only). A summary of the findings is provided in Tables 7.2 to 7.5. The findings in people of mainly African descent from Middle America (continental or Caribbean islands) and South America are listed in Tables 7.2 and 7.3. The highest average frequencies of $Hb_\beta*S$ were observed in Surinam (8.2%), Belize (7.9%) and in the Bahamas (7.4%). The prevalences of $Hb_\beta*C$ were generally much lower, with an extraordinarily high value of 10.1 in Esmeraldas, Ecuador, as well as another of 3.3% in Curaçao. In contrast, and as expected, the prevalences of any one of these two alleles generally do not reach 1% in populations classified as mainly of European descent (Table 7.4). In surveys in which there was no classification in relation to parental ancestry, or the subjects were classified as Mestizo, also as expected, the values were intermediate (Table 7.5). Populations especially well studied in relation to these markers were those of Cuba (where systematic screening for Hb types is performed at birth; no less than 257 514 determinations done in unclassified subjects); Jamaica (102 538 individuals of African ancestry) and southeast Brazil (53 276 persons unclassified in relation to parental continental groups).

Using the statistics developed by Nei (1987) it is possible to evaluate whether the genetic variability in relation to Hb_β is different in the several geographic regions and in the people of diverse ethnic background listed in Tables 7.2–7.5. The average gene diversity naturally follows the sequence European-derived (0.01), unclassified or Mestizo (0.03) and African-derived (0.12), almost all of it being due to its intrapopulational fraction, and with no marked differences between regions, within the classification made by parental ancestry.

Rare variants

Besides the three common alleles mentioned in the previous section, several hundred have been described and characterized around the world. Variants that occur in the alpha chain in Latin American populations are listed in Table 7.6. We have been able to locate 31 of them, distributed over 13 countries. The changes occur all over the molecule, from position 15 (J-Oxford) to 142 (Constant Spring). Three of them (J-Cubujuqui, Suresnes and J-Camagüey) are due to changes in position 141. The most widespread type is G-Philadelphia, identified in seven countries from continental Middle America, the Caribbean islands, and South America. The second most widespread is J-Broussais, found in five countries.

The list of the β chain variants is given in Table 7.7. They are more numerous (46, distributed over 18 countries), and this is probably due to the fact that while the β chains are synthesized after birth only, the α chains are produced at both intrauterine and extrauterine stages of life. Therefore, any change in the latter is potentially more damaging than in the former.

The three most widespread variants are HbE (which is polymorphic, that is, its allele occurs in frequencies higher than 1%, in Asiatic populations), Korle-Bu and D-Punjab, all three being detected in eight countries of the region. J-Baltimore was observed in six nations, while N-Baltimore and K-Woolwich were found in five. Among the variants listed in Table 7.7 Hb Porto Alegre is especially interesting because due to the change that occurred in the molecule it tends to polymerize *in vitro* (heterozygous blood forming octamers instead of the normal tetramers, and homozygous blood dodecamers). This tendency does not occur *in vivo*, the mutation therefore being classified as 'silent'. Another interesting finding is that of Hb Costa Rica, since it represents the first instance of a somatic mutation in this system (somatic mutations are those that occur outside the germ tissue, and therefore are not inherited).

Other protein chains are synthesized by the hemoglobin genetic regions or loci. The δ chain is produced in small amounts throughout postnatal life and is possibly the product of a gene which had lost its function, and therefore is almost inactive. Its variants are generally more difficult to identify, and only two were detected in Latin American populations, the relatively common A'2 (B2) and A2-Babinga (Table 7.8). Variants of γ chains, which are produced in fetal life only, are also indicated there. Gamma chains are of two types, synthesized by different loci; the only difference between them, however, occurs at position 136, where either an alanine ($^a\gamma$) or glycine ($^g\gamma$) can be found. As is indicated in Table 7.8, with one exception, all γ variants were identified in Jamaica, where extensive screening of newborns had been performed. The other types of variants listed in the table involved either fusion or deletion events, both very rare.

The thalassemias

Introduction

Several conditions show genetically determined diminished or absent synthesis of the hemoglobin chains, and they are collectively named as thalassemias, a word derived from *thalassa*, the Greek term for the Mediterranean Sea. This is due to the fact that these conditions are relatively common in the Mediterranean area, but it is now known that they are widespread all over the world.

The thalassemias are classified in accordance with the chain that is absent or reduced. The α- and β-thalassemias are the most common; those that affect the δ and γ chains do not lead to clinical conditions and have not been much studied. Some thalassemias are caused by hybrid or fusion genes, and a vast array of other genetic mechanisms can also lead to them. The World Health Organization has estimated that as many as 7% of the world's total population are carriers of α- or β-thalassaemias (Vogel and Motulsky, 1997).

Before the development of molecular techniques, diagnoses of the thalassemia syndromes were difficult. For the α-thalassemias, the presence of Hb Bart's (a tetramer of γ globin chains) in newborns was used (see, for instance, Pedrollo *et al.*, 1990), but studies at the DNA level indicated that just this determination would be insufficient for the establishment of reliable population frequencies. As for the β-thalassemias, the hetero-zygotes are usually characterized by microcytic and hypochromic red blood cells. Hb A_2 and Hb F levels are also usually raised in some, but not all types. Therefore, frequencies based on these criteria give only approximate values. The determination of the globin chain syntheses established a new level of analysis; but in the two next sections we will limit our review to data at the DNA level in which the exact genetic change has been established.

Alpha-thalassemias

Most α-thalassemias are caused by gene deletions. Since there are in the genome duplicated copies of the gene that conditions the α chain, pairing at meiosis sometimes is not perfect, and eventual crossing-overs can orig-inate regions without one of the two copies of the gene, as well as, complementarily, triplicates of it. Depending on the position in which this crossing-over takes place, considering the polarity of the DNA molecule, it is called either a 'leftward' or 'rightward' event. The rightward single α 3.7-kb deletion is the most common cause of α-thalassemia in Africa and the Mediterranean region, while in Asia both types of deletions can be found (Vogel and Motulsky, 1997). The leftward deletion involves a 4.2-kb region.

Table 7.9 lists six studies in five Latin American countries, involving 1641 subjects, in which the frequency of the 3.7-kb deletion has been investigated. The numbers vary from 10% to 22%, differences existing whether the variant was studied in individuals with or without other abnormal hemoglobin types. These differences may be due to ascertain-ment problems, and due to this fact their interpretation is difficult.

Six family studies performed in México, Cuba, and Brazil are sum-marized in Table 7.10. Hemoglobin H disease is a condition in which there is deletion of three of the four Hb α alleles. Severe anemia occurs, with the production of Hb H, a β chain tetramer. As can be seen in the table, the genetic basis for this disease in patients from the three indicated nations can be different. The association between α chain variants and deletions may be due to some kind of adaptive interaction.

Beta-thalassemias

Unlike the α-thalassemias, most β-thalassemias are not caused by gene deletions. The heterogeneity of the genetic changes is enormous, leading to nonfunctional mRNA (nonsense, frameshift, initiator codon mutants), defects in the processing of RNA, and changes both in the intervening sequences (IVS) or in the codon (CD) regions (Vogel and Motulsky, 1997).

Twenty-eight different types of β-thalassemia mutations observed in series from México, Cuba, Guadeloupe, Brazil and Argentina are listed in Table 7.11. Four of the most common are CD39 C → T, IVS1.110 G → A, − 29 A → G and IVS 2.1 G → A. The initiation codon A → G change may be indigenous to México. Losekoot *et al.* (1990) also found a novel frameshift mutation [FSC 47 (+ A)] in a Surinam patient, and Muniz *et al.* (2000) another [IVS 1.108 (T → C)] in Cuba. An unusual combination of genetic changes concentrated in a single person was reported by Costa *et al.* (1992). They found the IVS.1.10 G → T and CD39 C → T β-thalassemia mutations in association with the $-\alpha^{3.7\text{-kb}}$ deletion and Hb Hasharon (α 47 Asp → His) in this patient. Sonati *et al.* (1998) also studied an African-derived girl with a mild clinical expression of S-β-thalassemia, in which the IVS.1.6 T → C mutation (frequently referred as the 'Portuguese' β-thalassemia) was found associated with the CD2 C → T polymorphism and with HbS (an African marker). The patient provides, therefore, a good example as how variants which presumably have originated in different continents can come together in admixed populations.

Beta S-globin haplotype diversity

The β locus is located in chromosome 11, in a region that has suffered a series of duplications over time. The DNA molecule is composed of a polynucleotide chain, with an alternating series of sugar and phosphate residues. The 5' position of one pentose ring is connected to the 3' position of the next pentose region via a phosphate group. The terminal nucleotide at one end of the chain has a free 5' group, while the terminal nucleotide at the other end has a free 3' group. It is conventional to write nucleic acid sequences in the 5'–3' direction. The specific region considered, of about 65 kb, presents a series of coding regions separated by introns responsible respectively for the formation of embryonic (ε), fetal ($^{G}\gamma$, $^{A}\gamma$), and adult (δ, β) hemoglobins. Using a series of restriction enzymes, which identify specific portions of DNA, it is possible to verify the constitution of the adjacent regions of the beta locus. A given array of markers is denominated a haplotype.

Studies performed in the 1980s indicated independent origins of the sickle-cell mutation in different regions of Africa and Asia, since they were associated with different haplotypes, and these haplotypes were named after the regions where they had been first identified. Table 7.12 presents the distribution of the five most common β-globin haplotypes in several world populations. Not being autochthonous to the New World, the sickle-cell mutation was introduced into America mainly by gene flow from Africans, during slavery times, and it could give an indication of places of origin of these forced migrations.

The Cameroon haplotype, restricted to the central-western region of Africa, was identified only in Guadeloupe, Venezuela and Surinam, at low frequency (3%). There is a marked difference in the relative frequencies of the Bantu and Benin haplotypes, if we consider Central America, Venezuela and Surinam versus Brazil. While in the former group of nations Benin is the most frequent (with the exception of México), the opposite is true in Brazil, where Bantu is the most prevalent. As for the Senegal haplotype, it shows a somewhat irregular distribution in Latin America, with higher prevalences in Cuba, Guadeloupe, Venezuela and northern Brazil.

Figure 7.1 presents a graphical view of the data of Table 7.12. There is a clear separation between the Arab-Asian and the African populations situated at the extreme western part of that continent. At the tip of the other branch five southern African populations appear, while the other groups (including Latin Americans) occupy an intermediate position. This is, of course, what would be expected for the highly admixed populations of the American continent. Since these haplotypes are generally obtained from sickle-cell patients, problems of ascertainment, however, may introduce some uncertainty in these distributions.

Table 7.13 gives information about several levels of diversity considering these haplotypes. First intra- and interpopulation variation is considered, in Africa, Europe and Latin America. As expected, the highest level of total diversity (0.689) is found in Africa, but this value is not much different from that found in Latin America (0.597); both are almost twice as high as the European value, but the sample of Europeans is limited to three countries only. Also of note is the much lower interpopulation variability observed in Latin America (14% versus 53% or 62%). Brazil alone shows levels of diversity equivalent to those of Latin America as a whole, and there are variations within Africa in regions where different haplotypes prevail.

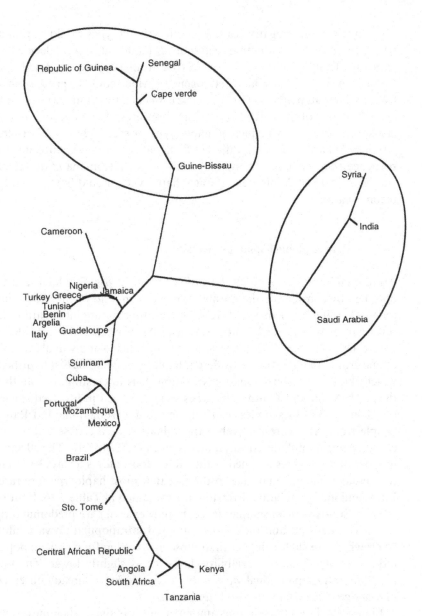

Figure 7.1. Unrooted tree based on a specific type of genetic distance (D_A; Nei *et al.*, 1983). The tree was obtained using the neighbor-joining method (Saitou and Nei, 1987), and compares the relationships among populations of different ethnic background taking into consideration the β S-globin haplotypes.

In almost all investigations a few 'atypical' haplotypes would be found, which received little attention. Zago *et al.* (2000), however, decided to study in detail 40 chromosomes derived from subjects of Brazil and Cameroon. They verified high heterogeneity, which could be grouped in at least 14 different patterns. They could be explained by (a) single nucleotide changes in one of the polymorphic restriction sites; (b) simple and double crossovers between two typical β^S haplotypes, or more frequently between a typical β^S haplotype and a different β^A haplotype; and (c) nonreciprocal sequence transfer (conversion). In a similar study Romana *et al.* (2000) verified that 15 of 20 different atypical haplotypes should have arisen by recombination.

Beta A-globin haplotype diversity

Additional information about the variability of the β-globin locus can be obtained by considering the normal sequences, instead of those carrying the sickle cell gene. Table 7.14 furnishes the appropriate information. In this analysis we included South American Amerindians, not considered before because the *Hb*S allele is absent in umixed individuals of this ethnic group. Not less than 16 different haplotypes could be distinguished, which show wide interpopulation variation. It is interesting to note that the Arab-Asian and Bantu haplotypes were observed in low frequencies (1.7% and 4.3%) in samples of β^A chromosomes from India and Bantu people from Africa, respectively. An evaluation of the observed versus expected intrapopulational variability is given in Table 7.15. The effective number of haplotypes (defined in the table's footnote) is always less than the actual number found, due to the fact that most haplotypes are rare. Intra- and interpopulation diversities are evaluated in Table 7.16. Four of the five Neo-American samples (exception: Mexicans) are predominantly of African descent, and they show total and intrapopulation variability levels which are slightly higher than those observed in four African populations. Interpopulation variability, however, is slightly lower. The low total and intrapopulational variability of Amerindians, already observed in other genetic systems, is also found here.

Figure 7.2 shows the relations among groups obtained when these data are submitted to a genetic distance analysis. A clear separation is obtained between Asians, Amerindians and Pacific Islanders (lower left of the figure), Europeans (upper left) and Africans (right), the four mixed Latin American populations occurring, as expected, between the last two sets. A curious exception is the Xavante Indians' position within the European

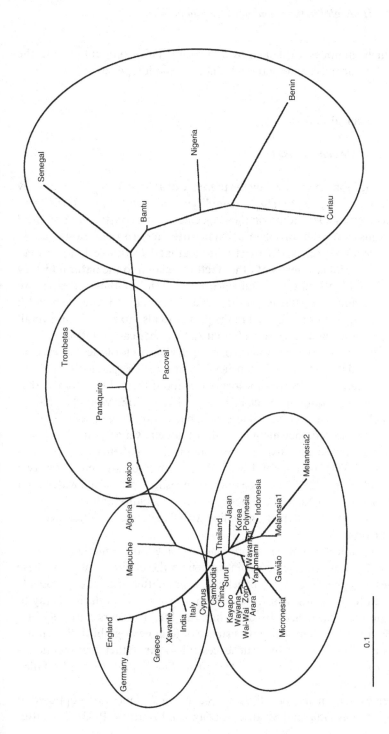

0.1

Figure 7.2. Same type of analysis described in Figure 7.1, using now the β A-globin haplotypes.

set, which cannot be explained by admixture. This is possibly due to the unusual occurrence, among these Indians, of haplotype 16.

Clinical studies

Sickle cell diseases

These conditions are defined by the presence of at least a $Hb_\beta*S$ gene. They include, therefore, $Hb_\beta*S/Hb_\beta*S$ (simplifying, SS) $Hb_\beta*S/Hb_\beta*C$ (SC) and S/Thalassemics. Depending on these specific genetic constitutions, clinical severity may vary. The problem is that in earlier reports these entities could not be adequately separated, due to the fact that detection techniques were still deficient. Since a series of other factors (related to the natural history of the individual) may also influence the severity of these diseases, we opted to include at least some of the earlier reports, to indicate the breadth of the investigations performed in the past, and also to provide an overall picture of these patients in the different Latin American countries.

The classic book, which described in depth all aspects of these diseases in Jamaica, had its first edition published 26 years ago (Serjeant, 1974). But it is still a magnificent source of information about them. Data about other selected investigations are given in Tables 7.17 and 7.18. More than half of them include studies performed in Brazil, but 12 other countries or regions are also represented; they include results obtained over 65 years (1934–99). It is impossible to establish the exact number of patients considered in Serjeant's (1974) monumental work, but the numbers given in the two tables add up to 2828. The scope of these investigations varied widely, but they generally emphasized the hematologic, morbidity and mortality aspects, with only a small fraction dealing with the diseases' interference with reproduction.

In accordance with observations made outside Latin America, it is clear that the level of fetal hemoglobin influences the clinical course of these diseases, a higher level being beneficial. As a matter of fact, one possible approach to the therapy of these conditions would be to devise a way to maintain the fetal locus's function in adult life, as occurs naturally in a specific condition (hereditary persistence of fetal hemoglobin). But efforts in this direction have proved unsuccessful so far. Other therapeutic regimes (for instance, hydroxyurea treatment, Saleh *et al.*, 1997) also failed to provide conclusive results.

Splenic function may be different in sickle cell anemia (SS) as opposed to Hb S-β^0-thalassemia and SC disease (Zago and Bottura, 1983). The latter is in general clinically more uniform, perhaps because its pathogenesis depends mainly on the hemoglobin concentration (Marmitt *et al.*, 1986). On the other hand, recent molecular studies indicated that the genetic constitution at this level is important. Thus, the Central African Republic (or Bantu) haplotype may be associated with more severe sickle cell anemia (Figueiredo *et al.*, 1996); and persons with type 2 Hb S-β^+-thalassemia (Hb A levels between 20% and 30%) have milder hematologic manifestations of the disease compared with those with Hb S-β^0-thalassemia (negligible production of Hb A) or Hb S-β^+-thalassemia type 1 (5–15% of Hb A) (Romana *et al.*, 1996).

People with sickle cell diseases show delayed sexual maturation in both sexes (Zago *et al.*, 1992*b*), and women with sickle cell anemia have reproductive problems. Hutz and Salzano (1983*b*) reported that among 117 SS women living in Rio de Janeiro and who were at reproductive age, only 21 have had at least one child alive. Their average offspring (1.7 ± 0.7 children) indicated a degree of subfertility equivalent to those observed outside Latin America. This problem is probably related to the high fetal loss (48%) observed among them, mainly due to spontaneous abortions (31%).

Thalassemias

Thalassemia major, a disease in which there is a complete or almost complete absence of β chains (β^0), also called Cooley anemia, is a condition much more severe than sickle cell anemia, and often leads to death in adolescence or earlier. Deletion of the four α alleles, on the other hand, is a lethal entity clinically diagnosed as hydrops fetalis. But there are a whole array of genetic changes in the globin loci that may lead to milder conditions, and the problem with studies involving the clinical picture of the thalassemias is that until recently, as was already mentioned, their characterization had to rely on imperfect methods. In the older studies this classification depended on hematologic and clinical findings only. With the development of techniques that could quantify the level of α or β chain syntheses the situation improved, but it was only with the molecular methods that a complete specification of the type of thalassemia found could be achieved.

Be as it may, a list of selected 13 clinical studies on these conditions, almost all performed in Brazil, is given in Table 7.19. They involved 14 investigations (912 patients) performed between 1961 and 1992, but only two included molecular techniques. Zago *et al.* (1982) verified that the association of Hb Hasharon (an α chain mutant) with β°-thalassemia improved the clinical picture for a patient. This could be due either to the fact that α^{Hasharon} may be removed from the cytoplasm at a faster rate than the normal α, and reducing the α chain excess would contribute to a milder clinical expression; or to an eventual association of the Hasharon mutation with an α-thalassemia defect (it is well known that the simultaneous presence of α-thalassemia ameliorates the clinical severity of homozygous β-thalassemia).

Table 7.1. *Main events in the study of haemoglobin*

Year	Discovery	Main contributors
1910	First description of 'peculiar elongated and sickle-shaped red cells' in a severely anemic West Indian student	J.B. Herrick
1917	Production of sickling by incubation of susceptible cells in a sealed chamber	V.E. Emmel
1925	Description of the thalassemia syndrome	T.B. Cooley, P. Lee
1927	The sickling effect was due to falling oxygen concentration	E.V. Hahn, E.B. Gillespie
1948	Metabisulphite method for the presence of sickle cells	G.A. Daland, W.B. Castle
1949	Determination of the genetic basis of sickle cell anemia	J.V. Neel, E.A. Beet
	Different electrophoretic mobility between Hb A and Hb S. Concept of molecular disease	L. Pauling, H.A. Itano, S.J. Singer, I.C. Wells
1954	The sickle cell trait would protect carriers against malaria	A.C. Allison
1956	Hb A and Hb S differed among themselves in a single amino acid substitution	V.M. Ingram
1961/1963	Determination of the amino acid sequence	G. Braunitzer and colleagues, W. Konigsberg, R.J. Hill, J. Goldstein
1978	Description of the first DNA polymorphism adjacent to the β-globin locus	H.W. Kan, A.M. Dozy

Source: Serjeant (1974); Vogel and Motulsky (1997).

230   *Hemoglobin types and hemoglobinopathies*

Table 7.2. *Frequencies (%) of the three most common alleles of the* Hb$_\beta$ *locus in continental Middle America and the Caribbean islands: populations classified as mainly of African descent*

Country or region	No. individuals studied	Alleles (%)			References
		A	S	C	
Continental Middle America					
Belize	2024	91.4	7.9	0.7	1,2
Guatemala	432	95.5	4.3	0.2	2,3
Honduras	1431	93.5	6.0	0.5	2
Nicaragua	173	93.9	5.2	0.9	4
Costa Rica	1107	93.1	5.4	1.5	5–7
Panama	777	91.4	6.9	1.7	8
Caribbean Islands					
Bahamas	492	91.0	7.4	1.6	9
Cuba	5713	94.8	4.4	0.8	10–12
Dominican Republic	798	95.9	3.5	0.6	6
Haiti	844	93.4	5.8	0.8	13,14
Jamaica	102 538	92.6	5.5	1.9	15–18
Puerto Rico	602	96.2	2.9	0.9	19
St Thomas	1769	93.3	4.7	2.0	6
Guadeloupe	26 135	93.5	5.1	1.4	20,21
Martinique	4959	94.2	4.0	1.8	22,23
Dominica	1000	92.9	6.5	0.6	24
St Lucia	825	91.1	7.0	1.9	20
St Vincent	748	94.3	4.4	1.3	20
Barbados	912	94.2	3.5	2.3	20
Curaçao	2292	92.3	4.4	3.3	25,26
Trinidad and Tobago	751	94.5	4.0	1.5	20,27

References: 1. Firshein (1961); 2. Crawford (1983); 3. Crawford *et al.* (1981); 4. Biondi *et al.* (1988); 5. Sáenz *et al.* (1971); 6. Sáenz (1988); 7. Madrigal (1989); 8. Ferrell *et al.* (1978); 9. Halberstein *et al.* (1981); 10. Hidalgo *et al.* (1974); 11. Colombo and Martínez (1985); 12. Cabrera Llano *et al.* (1992); 13. Arends (1971b); 14. Basu *et al.* (1976); 15. Went (1957); 16. Went and MacIver (1958); 17. Miall *et al.* (1967); 18. Serjeant *et al.* (1986); 19. Suarez *et al.* (1959); 20. Arends (1971b); 21. Fabritius *et al.* (1980); 22. Collet (1958); 23. Montestruc *et al.* (1959); 24. Sáenz *et al.* (1993); 25. Jonxis (1959); 26. van der Sar (1959); 27. Henry (1963).

Table 7.3. *Frequencies (%) of the three most common alleles of the* Hb$_\beta$
locus in South American populations classified as mainly of African descent

Country or region	No. individuals studied	Alleles (%)			References
		A	*S*	*C*	
Outside Brazil					
Venezuela	1233	95.5	3.7	0.8	1–8
Colombia	1536	93.8	4.4	1.8	9,10
Ecuador	223	84.7	5.2	10.1	11
Surinam	2159	89.6	8.2	2.2	12–14
Brazil					
North	1165	96.2	3.0	0.8	7,15–18
Northeast	4118	97.2	2.2	0.6	19–21
Center-West	ND	ND	ND	ND	
Southeast	1881	95.3	3.9	0.8	22–24
South	4574	96.5	2.9	0.6	7,21,24

References: 1. Arends and Layrisse (1956); 2. Arends (1961*a,b*); 3. Arends (1971*a*); 4. Arends *et al.* (1978); 5. Arends (1984); 6. Castro de Guerra *et al.* (1993); 7. Bortolini *et al.* (1992); 8. Castro de Guerra *et al.* (1996); 9. Restrepo (1971); 10. Monsalve *et al.* (1987); 11. Martínez-Labarga *et al.* (1999); 12. Liachowitz *et al.* (1958); 13. Jonxis (1959); 14. Pik *et al.* (1965); 15. Costa and Borborema (1974); 16. Schneider *et al.* (1987); 17. Santos *et al.* (1987); 18. Guerreiro *et al.* (1999); 19. Sampaio (1987); 20. Conceição *et al.* (1987); 21. Bortolini *et al.* (1997*b*); 22. Ramalho *et al.* (1976); 23. Martins *et al.* (1987); 24. Salzano (1986). ND, no data.

Table 7.4. *Frequencies (%) of the three most common alleles of the* Hb_β
*locus in Latin American populations classified as mainly of European
descent*

Country or region	No. individuals studied	Alleles (%)			References
		A	*S*	*C*	
Outside Brazil					
Costa Rica	6911	99.5	0.4	0.1	1
Cuba	8091	99.7	0.3	0.0	2–4
Puerto Rico	1487	99.97	0.03	0.0	5
Venezuela	99	100.0	0.0	0.0	6
Brazil					
North	268	99.8	0.2	0.0	7–9
Northeast	2214	98.7	1.1	0.2	9–11
Center-West	ND	ND	ND	ND	
Southeast	1178	99.5	0.2	0.3	9,12,13
South	1984	99.5	0.3	0.2	9,12

References: 1. Sáenz (1988); 2. Hidalgo *et al.* (1974); 3. Colombo and Martínez (1985); 4.
Cabrera Llano *et al.* (1992); 5. Suarez *et al.* (1959); 6. Arends (1984); 7. Costa and
Borborema (1974); 8. Santos *et al.* (1987); 9. Salzano (1986); 10. Sampaio (1987); 11.
Conceição *et al.* (1987); 12. Naoum *et al.* (1986); 13. Teixeira and Ramalho (1994).
ND, no data.

Table 7.5. *Frequencies (%) of the three most common alleles of the* Hb$_\beta$ *locus in Latin American populations unclassified in relation to ethnic groups or admixed (mestizo)*

Country or region	No. individuals studied	Alleles (%)			References
		A	*S*	*C*	
Middle America					
México	21 148	99.4	0.5	0.1	1–7
El Salvador	600	99.0	1.0	—	8
Costa Rica	5793	97.7	2.1	0.2	9,10
Cuba	257 514	96.8	2.8	0.4	11–15
South America					
Venezuela	23 685	98.3	1.4	0.3	16–24
Colombia	3227	98.6	1.1	0.3	25–26
Bolivia	378	100.0	0.0	0.0	27
Chile	673	99.6	0.2	0.2	27
Brazil: North	4700	98.6	1.2	0.2	28–32
Brazil: Northeast	16 042	97.2	2.2	0.6	28,32–36
Brazil: Center-West	3322	98.1	1.6	0.3	28
Brazil: Southeast	53 276	99.0	0.8	0.2	28,32,37
Brazil: South	2136	99.0	0.9	0.1	28,32
Uruguay	76	100.0	0.0	0.0	38

References: 1. Lisker *et al.* (1969); 2. Lisker (1971); 3. Grunbaum *et al.* (1980); 4. Franco-Gamboa *et al.* (1981); 5. Ruiz Reyes (1983); 6. Tiburcio *et al.* (1978); 7. Lisker *et al.* (1988*b*); 8. Bloch and Rivera (1969); 9. Elizondo and Zomer (1970); 10. Sáenz (1988); 11. Hidalgo *et al.* (1974); 12. Heredero *et al.* (1978); 13. Martínez and Cañizares (1982); 14. Granda Ibarra and Hernández Fernández (1984); Granda Ibarra *et al.* (1984); 15. Colombo and Martínez (1985); 16. Arends and Layrisse (1956); 17. Layrisse *et al.* (1958); 18. Arends (1961*b*); 19. Nuñez-Montiel *et al.* (1962); 20. Guevara *et al.* (1963); 21. De Piñango and Arends (1965); 22. Arends (1971*a*); 23. Arends (1984); 24. Arends *et al.* (1990); 25. Restrepo (1965); 26. Restrepo (1971); 27. Arends (1971*b*); 28. Naoum *et al.* (1986); 29. Guerreiro *et al.* (1993); 30. Ribeiro-dos-Santos *et al.* (1995); 31. Santos *et al.* (1996); 32. Salzano (1986); 33. Conceição *et al.* (1987); 34. Weimer *et al.* (1991); 35. Lima *et al.* (1996); 36. Pantaleão *et al.* (1998); 37. Teixeira and Ramalho (1994); 38. Sans *et al.* (1995). ND, no data.

Table 7.6. *Alpha-hemoglobin variants observed in Latin America*

Variant	Substitution	Abnormal characteristics	Country	References
J-Oxford	15 Gly → Asp	—	Brazil	1,2
I	16 Lys → Glu	—	Brazil, Costa Rica	1,3
Handsworth	18 Gly → Arg	—	Jamaica	1,4
Le Lamentin	20 His → Gln	—	Martinique	5
J-Medelin	22 Gly → Asp	—	Colombia	1,6
Chad	23 Glu → Lys	—	Surinam	1
Campinas	26 Ala → Val	—	Brazil	27
Fort Worth	27 Glu → Gly	—	Jamaica	1,4
Spanish Town	27 Glu → Val	—	Jamaica, Martinique	1,4,5
Prato	31 Arg → Ser	—	Venezuela	1,7
Fort de France	45 His → Arg	—	Martinique	5
Hasharon	47 Asp → Hys	Unstable	Brazil, Colombia	1,5,8–10
J-Rovigo	53 Ala → Asp	Unstable	Brazil	1,2,11–13
Shimonoseki	54 Gln → Arg	—	Jamaica	1,4
J-México	54 Gln → Glu	—	Colombia, México	1,6,14,15
M-Boston	58 His → Tyr	↓O$_2$ affinity; Fe^{+3}	Brazil	1,2
G-Philadelphia	68 Asn → Lys		Brazil, Colombia, Costa Rica, Cuba, Jamaica, Martinique, Panama	1,2,4,5,16,17
J-Habana	71 Ala → Glu	—	Cuba	1,16
Daneshgah-Tehran	72 Hys → Arg	—	Argentina	18
Q-Iran	75 Asp → His	—	Panama	1,17
Aztec	76 Met → Thr	—	México	1
Stanleyville II	78 Asn → Lys	—	Brazil	19,20
M-Iwate	87 Hys → Tyr	↓O$_2$ affinity; Fe^{+3}	Colombia	1,14,21
J-Broussais	90 Lys → Asn	—	Dominica, Guadeloupe, Jamaica, Martinique, Venezuela	1,5,7,17,22,23
Georgia	95 Pro → Leu	—	Martinique	5

Chiapas	114 Pro → Arg	—	México	1,15,24,25
Tarrant	126 Asp → Asn	↑O₂ affinity	México	1,15,24
J-Cubujuqui	141 Arg → Ser	↑O₂ affinity	Costa Rica	1
Suresnes	141 Arg → His	↑O₂ affinity	Costa Rica	1
J-Camagüey	141 Arg → Gly	↑O₂ affinity	Cuba	1,16,26
Constant Spring	142 Term. → Gln(+31 amino acids)	—	Brazil	2

References: 1. Sáenz et al. (1993); 2. Naoum et al. (1984); 3. Naoum (1997); 4. Serjeant et al. (1986); 5. Monplaisir et al. (1985); 6. Arends (1966); 7. Arends (1984); 8. Restrepo (1971); 9. Zago et al. (1982); 10. Costa et al. (1992); 11. Targino de Araüjo (1977); 12. Targino de Araüjo et al. (1980); 13. Silva et al. (1992); 14. Arends (1971b); 15. Ruiz Reyes (1983); 16. Colombo and Martinez (1985); 17. Sáenz (1988); 18. Weinstein et al. (1985); 19. Costa et al. (1987); 20. Costa et al. (1991a); 21. Arends (1971a); 22. Fabritius et al. (1980); 23. Arends et al. (1982); 24. Franco-Gamboa et al. (1981); 25. Ibarra et al. (1981); 26. Martínez et al. (1978); 27. Wenning et al. (2000).

Table 7.7. *Beta-hemoglobin variants observed in Latin America*

Variant	Substitution	Abnormal characteristics	Country	References
Deer Lodge	2 His → Arg	↑O₂ affinity	Venezuela	1-4
HbS-Antilles	6 Glu → Val + 23 Val → Ileu	Sickling	Martinique	4,5
G-San José	7 Glu → Gly	Slightly unstable	México	4,6
Siriraj	7 Glu → Lys	—	Martinique	7
Rio Grande	8 His → Thr	—	México	4
Porto Alegre	9 Ser → Cys	↑O₂ affinity, tendency to polymerize	Argentina, Brazil, Cuba, Venezuela	4,8-14,41,42
J-Baltimore	16 Gly → Asp	—	Brazil, Colombia, Guadeloupe, Martinique, México, Trinidad	4,6,7,16-18
Alamo	19 Asn → Asp	—	Cuba, Venezuela	4,12,19
D-Iran	22 Glu → Gln	—	Guadeloupe, Jamaica	4,20

Table 7.7. (cont.)

Variant	Substitution	Abnormal characteristics	Country	References
E	26 Glu → Lys	—	Brazil, Costa Rica, Cuba, Guadeloupe, Jamaica, Martinique, Mexico, Surinam	4,6,7,12,18,20,21
Knossos	27 Ala → Ser	Affects synthesis	Martinique	4,7,18,22
Genova	28 Leu → Pro	Unstable	Cuba	4,12,23
Chile	28 Leu → Met	Unstable	Chile	43
Rio Claro	34 Val → Met	—	Brazil	44
Bucuresti	42 Phe → Leu	Unstable; ↓ O_2 affinity	Cuba	4,12
Willamette	51 Pro → Arg	Unstable; ↓ O_2 affinity	Cuba, Venezuela	4,12
Osu Christiansborg	52 Asp → Asn	—	Jamaica	4,20
Ocho Rios	52 Asp → Ala	—	Jamaica	4,20
Dhofar	58 Pro → Arg	—	Jamaica	4,20
Zürich	63 His → Arg	Unstable; ↓ O_2 affinity	Brazil	4,24
I-Toulouse	66 Lys → Glu	Unstable; Fe^{+3}	Cuba	4
Rambam	69 Gly → Asp	—	Argentina	45
Korle-Bu	73 Asp → Asn	↓ O_2 affinity	Colombia, Costa Rica, Cuba, Guadeloupe, Jamaica, Martinique, Panama, Venezuela	4,7,12,17,18,20, 25
J-Chicago	76 Ala → Asp	—	Venezuela	4
Costa Rica	77 His → Arg	—	Costa Rica	26
Buenos Aires	85 Phe → Ser	Unstable; ↑ O_2 affinity	Argentina	4
Santa Ana	88 Leu → Pro	Unstable	Brazil, Cuba	4,12,14
Roseau-Pointe a Pitre	90 Glu → Gly	Slightly unstable ↓ O_2 affinity	Guadeloupe	27
Caribbean	91 Leu → Arg	Slightly unstable ↓ O_2 affinity	Jamaica	4,20,28
Saint-Etienne	92 His → Gln	—	Argentina	46
N-Baltimore	95 Lys → Glu	—	Brazil, Cuba, El Salvador, Guadeloupe, Martinique	4,7,12,17,18,29, 30
J-Cordoba	95 Lys → Met	↓ O_2 affinity	Argentina	4,31

Köln	98 Val → Met	Unstable	Brazil	15,32
Mainz	98 Val → Glu	Unstable	Brazil	33
Camperdown	104 Arg → Ser	Unstable	Brazil	15,34
New York	113 Val → Glu	Slightly unstable $\downarrow$ O_2 affinity	Costa Rica	4
Fannin-Lubbock	119 Gly → Asp	Slightly unstable	México	4,6
Riyadh	120 Lys → Asn	—	México	4,6,35
D-Punjab	121 Glu → Gln	$\downarrow O_2$ affinity	Argentina, Brazil, Cuba, Guadeloupe, Jamaica, Martinique, Mexico, Venezuela	1,3,4,6,7,12,16, 20,36–38,47
O-Arab	121 Glu → Lys	—	Jamaica, Martinique, Puerto Rico	4,7,18,20
Hofu	126 Val → Glu	—	Peru, Venezuela	1,3,4
J-Guantanamo	128 Ala → Asp	Unstable	Chile, Cuba	4,12,39,40
K-Woolwich	132 Lys → Gln	Unstable	Brazil, Dominica, Guadeloupe, Jamaica, Martinique	4,7,15,17,18,20
North Shore-Caracas	134 Val → Glu	Unstable	Venezuela	1,3,4
Hope	136 Gly → Asp	Unstable $\downarrow$ O_2 affinity	Cuba, Martinique	4,7,12

References: 1. Arends *et al.* (1982); 2. Arends *et al.* (1984); 3. Arends (1984); 4. Sáenz *et al.* (1993); 5. Monplaisir *et al.* (1986b); 6. Ruiz Reyes (1983); 7. Monplaisir *et al.* (1985); 8. Bonaventura and Riggs (1967); 9. Seid-Akhavan *et al.* (1973); 10. Peñalver and Miani (1974); 11. Martínez *et al.* (1977a); 12. Colombo and Martínez (1985); 13. Tondo *et al.* (1963); 14. Gonçalves *et al.* (1994a); 15. Sonati *et al.* (1996); 16. Arends (1966); 17. Fabritius *et al.* (1980); 18. Sáenz (1988); 19. Arends *et al.* (1987); 20. Serjeant *et al.* (1986); 21. Targino de Araújo *et al.* (1982); 22. Galacteros *et al.* (1984); 23. Martínez *et al.* (1983); 24. Miranda *et al.* (1994a); 25. Restrepo (1971); 26. Rodriguez Romero *et al.* (1996); 27. Merault *et al.* (1985); 28. Rabb and Serjeant (1982); 29. Arends (1971b); 30. Naoum (1997); 31. Bardakdjian *et al.* (1988); 32. Miranda *et al.* (1997); 33. Passos *et al.* (1998); 34. Miranda *et al.* (1996); 35. Franco-Gamboa *et al.* (1981); 36. Naoum *et al.* (1976); 37. Naoum *et al.* (1988); 38. Zago and Costa (1988); 39. Martínez *et al.* (1977b); 40. Sciarrata *et al.* (1990); 41. Silva *et al.* (1981); 42. Rosa *et al.* (1984); 43. Hojas-Bernal *et al.* (1999); 44. Grignoli *et al.* (1999); 45. Plaseska-Karanfilska *et al.* (2000); 46. Weinstein *et al.* (2000); 47. Perea *et al.* (1999b).

Table 7.8. *Other structural hemoglobin variants observed in Latin America*

Variant	Substitution	Country	References
A2-Babinga	δ 136 Gly → Asp	Panama	1,2
A'2 (B2)	δ 16 Gly → Arg	Brazil, Costa Rica, Cuba, Jamaica, Panama, Venezuela	2–5
F-Texas-I	$^{A}\gamma$ 5 Glu → Lys	Jamaica	2,8
F-Jamaica	$^{A}\gamma$ 61 Lys → Glu	Jamaica	9
F-Victoria Jubilee	$^{A}\gamma$ 80 Asp → Tyr	Jamaica	2,8
F-Hull	$^{A}\gamma$ 121 Glu → Lys	Jamaica	2,8
F-Kingston	$^{A}\gamma$ 55 Met → Arg	Jamaica	2,8
HPFH	$^{A}\gamma$ 195 (C → G)	Brazil	13
F-Port Royal	$^{A}\gamma$ 125 Glu → Ala	Jamaica	2,8
Lepore-Baltimore	δβ (D1-F2)	Brazil	12
Lepore-Boston	δβ (F3-G18)	Argentina, Cuba, Jamaica	2,6–8
P-Nilotic	δβ (B4-D1)	México	2
Hb Niteroi	β 42–44 or 43–45 Phe, Glu, Ser deletion	Brazil	10
HPFH (type 2)	ψβdeletion, 105 kb	Brazil	11

References: 1. Saénz (1988); 2. Sáenz *et al.* (1993); 3. Arends *et al.* (1982); 4. Arends (1984); 5. Sonati *et al.* (1996); 6. Fabritius *et al.* (1980); 7. Colombo and Martinez (1985); 8. Serjeant *et al.* (1986); 9. Ahern *et al.* (1970); 10. Praxedes and Lehmann (1972); 11. Gonçalves *et al.* (1995); 12. Miranda *et al.* (1994b); 13. Bordin *et al.* (1998).

Table 7.9. *Types of alpha-thalassemia observed in four Latin American populations*[a]

Country and non-α-thalassemic genotypes	No. individuals studied	Frequency (%) of α-thalassemic genotypes				−α frequency (%)	Reference
		αα/αα	αα/−α	−α/−α	Other		
Cuba							
AA	60	77	23	0	0	12	1
SS	208	66	31	3	0	18	1
SC	34	79	21	0	0	10	1
Sβ^Thal	31	74	26	0	0	13	1
AG^Phila	5	0	5	0	0	—	1
Jamaica							
AA	205	66	30	3	1	18	2
SS	268	63	34	3	0	20	2
SC	143	67	31	1	1	17	2
Sβ^+Thal	32	78	22	0	0	11	2
Sβ^0Thal	12	58	42	0	0	21	2
Guadeloupe							
SS	269	62	34	1	3	19	3
SC	138	66	31	1	2	17	3
Venezuela	148	62	33	5	0	22	4
Brazil							
SS	41	78	22	0	0	11	5
AA	45	73	18	2	7	11	6
AS	2	0	2	0	0	—	6

[a]The deletions are of the − 3.7 kb type.

References: 1. Martínez *et al.* (1990); 2. Serjeant *et al.* (1986); 3. Kéclard *et al.* (1997); 4. Arends *et al.* (2000); 5. Costa *et al.* (1989); 6. Sonati *et al.* (1991).

Table 7.10. *Family studies involving different α-thalassemia conditions in Latin Americans*

Country and conditions	Findings	References
México		
Hb H disease	Two patients, with the $-\alpha^{3.7}/--^{SEA}$ and $\alpha\alpha^{Hph}/--^{FIL}$ conditions	Ibarra et al. (1995)
Cuba		
Hb H disease	Nine patients, African/Chinese ancestry, all $-\alpha^{3.7}/--^{SEA}$	Martinez et al. (1986)
Brazil		
Alphathalassaemia/mental retardation	One patient, breakpoint for the 16p13.3 deletion near the region identified by probe EKMDA2	Wilkie et al. (1990)
HbH disease	One patient showing the $-\alpha^{3.7}/--^{MED}$ constitution	Sonati et al. (1992a)
Association with other variants	One homozygote and three heterozygotes for Hb Stanleyville II ($\alpha78$ Asn → Lys), always associated with the $-\alpha^{3.7}$ deletion	Costa et al. (1991a)
	One patient showing compound heterozygosity for β-thalassemia (IVSI.110 (G → T)/CD39(C → T), plus Hb Hasharon ($\alpha47$ Asp → His) and the $-\alpha^{3.7}$ deletion	Costa et al. (1992)

Table 7.11. *Beta-thalassemia mutations observed in five Latin American populations*

Mutations	Thal. type	Surveys				
		México[a] (n = 20)	Cuba[b] (n = 76)	Guadeloupe[c] (n = 132)	Brazil[d] (n = 68)	Argentina[e] (n = 81)
Initiation codon A → G, TC	0	1	1	—	—	—
CD 11 -T	0	—	1	—	—	—
CD 14/15 (+ G)	0	—	—	—	—	—
CD 15 (G→A)	0	—	1	—	—	—
CD 24 T → A	+	—	—	4	—	—
CD 39 C → T	0	5	25	—	45	40
CD 41/42 (-TTCT)	0	—	5	—	—	—
CD 121 (G→T)	0	—	2	—	—	—
FSC 6 -A	0	—	—	—	—	1
IVS 1.1 G → A	0	2	4	—	4	8
IVS 1.1 (G → A) (CD30)	/	—	1	—	—	—
IVS 1.2 T → C	0	—	—	5	—	—
IVS 1.5 G → A	+	2	1	12	—	—
IVS 1.5 G → C	+	—	—	15	—	5
IVS 1.6 T → C	+	—	3	—	5	5
IVS 1.108 T → C	/	—	1	—	—	—
IVS 1.110 G → A	+	3	7	19	14	19
IVS 2.1 G → A	0	—	7	—	—	3
IVS 2.745 C → G	+	—	—	—	—	2
IVS 2.849 A → G	0	—	1	7	—	—
-28 A → C	0	2	—	—	—	—
-29 A → G	+	—	11	51	—	2
-87 C → T	0	1	—	—	—	—
-88 C → T	+	—	2	4	—	—
Poly A T → C	+	—	3	9	—	—
Spanish δβ°	0	3	—	—	—	—
Hb Lepore	+	—	—	3	—	1
Hb E	+	—	—	3	—	—

[a]Economou et al. (1991); Ibarra et al. (1995); Villalobos-Arámbula et al. (1997). The list given is of the 1997 paper, but some alleles previously reported are not indicated there: FSC6 − A (2), IVS 1.5 G → C (2), IVS 2.745 C → G (1), 619 bp deletion (2); [b]Muniz et al. (2000); [c]Romana et al. (1996); [d]Martins et al. (1993); [e]Roldán et al. (1997).

CD, codon; FSC, frameshift; IVS, Intervening sequences or introns.

Table 7.12. *β^S haplotype distribution in various Afro-American, African and other human populations*

Region	Place[a]	No. of chromosomes	Haplotypes (%)						References
			Arab–Asian	Bantu	Benin	Senegal	Cameroon	Others[b]	
Europe	Greece	96	0	0	95	0	0	5	1,2
	Italy	131	0	0	100	0	0	0	1,3
	Portugal	46	0	46	30	24	0	0	4,5
Asia	India	111	86	1	0	4	0	9	6,7
	Goa	2	100	0	0	0	0	0	5
Near East	Saudi Arabia	72	62	0	24	0	0	14	6
	Turkey	214	1	0	96	0	0	3	1
	Syria	13	100	0	0	0	0	0	1
Africa North	Argelia	20	0	0	100	0	0	0	8
	Tunisia	115	0	0	95	0	0	5	1
West	Senegal	46	0	0	14	86	0	0	8
	Guine-Bissau	4	0	0	25	50	0	25	5
	Republic of Guinea	47	0	0	4	96	0	0	9
Central-West	Benin	20	0	0	100	0	0	0	8
	Cameroon	80	0	0	52	0	16	32	10
	Nigeria	657	0	1	92	1	4	2	1,6
Bantu-speaking	Angola	60	0	94	6	0	0	0	1,5
	Mozambique	4	0	50	50	0	0	0	5
	Kenya	111	0	98	2	0	0	0	1
	South Africa	23	0	87	0	0	0	13	11
	Tanzania	41	0	100	0	0	0	0	1
	Central African Republic	28	0	86	7	0	0	7	8

	Location	n							References
Islands	Cape Verde	11	0	0	18	82	0	0	5
	Sto. Tomé and Principe	15	0	73	27	0	0	0	5
America Central	México[c]	63	3	52	0	0	0	15	12
	Jamaica	338	0	11	75	2	0	12	13,14
	Cuba[d]	198	0	41	51	8	—	—	15
	Guadeloupe	226	0	11	73	8	3	5	16
	Venezuela[c]	272	0	30	51	12	3	4	24
	Surinam	77	0	30	53	3	3	11	1
South	Brazil								
	Amapá	6	0	50	17	33	0	0	17
	Pará	74	0	66	26	8	0	0	17,18,19
	Pernambuco	114	0	81	14	0	0	5	20
	Bahia	42	0	55	45	0	0	0	21
	São Paulo	209	0	65	34	1	0	0	21,22
	Rio Grande do Sul	49	0	80	18	2	0	0	23
	Brazil total	494	0	61	34	3	0	2	

[a]When more than one series was available in a given country, weighted averages were obtained, except for Brazil, where the data were also grouped by state.

[b]Including rare types wich have arisen by mutation, recombination, or gene conversion (18).

[c]Mestizos.

[d]The authors excluded from the final computation the atypical haplotypes (Cameroon and others), and it is impossible to evaluate the frequency of the Cameroon haplotype.

References: 1. Öner et al. (1992); 2. Boussiou et al. (1991); 3. Schilirò et al. (1990); 4. Monteiro et al. (1989); 5. Lavinha et al. (1992); 6. Kulozik et al. (1986); 7. Labie et al. (1989); 8. Pagnier et al. (1984); 9. Sow et al. (1995); 10. Lapouméroulie et al. (1992); 11. Ramsay and Jenkins (1987); 12. Peñaloza et al. (1995); 13. Antonarakis et al. (1984); 14. Wainscoat et al. (1983); 15. Muniz et al. (1995); 16. Kéclard et al. (1996); 17. Pante-de-Souza et al. (1999); 18. Pante-de-Souza et al. (1998); 19. Figueiredo et al. (1994b); 20. França et al. (1998); 21. Zago et al. (1992a); 22. Gonçalves et al. (1994b); 23. Wagner et al. (1996); 24. Arends et al. (2000).

Table 7.13. *β^S haplotype diversity considering seven African and Afro-American human groups*[a]

Groups	No. of populations	Absolute haplotype diversity			Relative interpopulational haplotype diversity (%)[b]
		Total	Intrapopulational	Interpopulational	
I	13	0.689	0.251	0.471	62
Ia	5	0.189	0.170	0.023	12
Ib	3	0.321	0.251	0.105	30
Ic	2	0.078	0.076	0.002	3
Id	4	0.357	0.309	0.047	13
II	6	0.597	0.526	0.085	14
IIa	6	0.491	0.451	0.048	10
III	3	0.382	0.214	0.255	53

[a]The values were obtained from the data furnished in Table 7.12, classifying the populations into eight groups:
I. All Sub-Saharan countries: Cape Verde, Central African Republic, Senegal, Guine-Bissau, Republic of Guinea, Mozambique[c], Angola, Nigeria, Benin, Cameroon, Kenya, Tanzania, Sto. Tomé, Príncipe.
Ia. Countries from Sub-Sahara where the Bantu β^S haplotype distribution is majority: Tanzania, Sto. Tomé and Príncipe, Kenya, Central African Republic, Angola.
Ib. Countries from Sub-Sahara where the Benin β^S haplotype distribution is majority: Nigeria, Cameroon, Benin.
Ic. Countries from Sub-Sahara where the Benin β^S haplotype distribution is majority, but excluding Cameroon: Nigeria, Benin.
Id. Countries from Sub-Sahara where the Senegal β^S haplotype distribution is majority: Senegal, Guine-Bissau, Republic of Guinea, Cape Verde.
II. Latin American countries in general: Brazil, Surinam, México, Cuba, Guadeloupe, Jamaica.
IIa. Brazilian states only: Amapá, Pará, Bahia, Pernambuco, São Paulo, Rio Grande do Sul.
III. Europeans: Greece, Italy, Portugal.
[b]Coefficient of gene differentiation (Nei, 1987).
[c]Mozambique was not included in the subgroups because it could not be unambiguously placed in any of them.

Table 7.14. β^A haplotype distribution in several human populations

Place and ethnic group	No. studied	Haplotypes (%)[a]																Reference
		1	2	3	4	5	6	7	8	9	10	11	12	13	14	15	16	
Europeans																		
England	37		43.8			40.0	13.5		2.7									1
Germany	16		43.8	6.2		31.3	6.2		12.5									1
Greece	64		46.7	1.6	1.6	14.1	25.0	6.3			1.6						3.1	1
Italy	169		66.9		0.6	21.3	11.2											1
Cyprus	120		70.0		2.5	20.0	7.5											1
Asians																		
India	234	0.4	55.2		0.4	24.4	14.5	1.3			1.7	0.4			1.3	0.4		1
Cambodia	65		80.0		3.1	13.8	3.1											1
Thailand	131		85.5		2.3	6.9	4.6					0.7						1
China	158		83.0		0.6	6.3	7.6					2.5						1
Korea	60	1.7	85.0			10.0	1.7									1.6		1
Japan	56		83.9			5.4	3.6						5.3				1.8	1
Africans																		
Algeria	70	1.4	54.3	7.1	1.4	12.9	8.6		2.9		1.4	8.6	1.4					1
Bantu	47		12.8	40.4	10.6	10.6	14.9							2.1	4.3		4.3	1
Benin	30		3.3	56.7	13.3		6.7							10.0	3.3	6.7		1
Nigeria	35		8.6	60.0	17.1	11.4											2.9	1
Senegal	14		7.1	28.6	14.3	50.0												2
Afro-Americans																		
Panaquire	96	5.2	35.4	20.9	12.6	7.3	3.1		4.2	1.0		1.0	1.0	1.0			7.3	3
Curiau	42		7.4	45.5	9.9							15.0	7.4	7.4			7.4	4
Pacoval	50	8.3	31.3	25.0	2.1	10.4	4.2				2.1	6.2	2.6	8.3	2.1			4
Trombetas	40	7.7	38.3	30.8	5.1		2.6						2.6	10.3			2.6	4

Table 7.14. (cont.)

Place and ethnic group	No. studied	Haplotypes (%)[a]																Reference
		1	2	3	4	5	6	7	8	9	10	11	12	13	14	15	16	
Mestizo-Americans																		
México	97	4.3	50.7	3.2	6.5	16.2	2.2	1.0	1.0	4.3		2.1	3.2	2.1			3.2	5
Amerindians																		
Mapuche	86	5.8	57.0	2.3	2.3	3.5	26.7	1.2								1.2		6
Xavante	60		60.0			15.0	18.3	5.0									1.7	7
Zoró	60		93.0				7.0											7
Gavião	58	1.7	88.0	1.7			3.4	5.2										7
Suruí	44		81.8			4.5	11.4											7
Wai-Wai	56		87.5				12.5											7
Yanomami	34	2.9	91.2				5.9											1
Kayapo	44		79.5				20.5											1
Wayana-Apalai	34		82.4				14.7			2.9								1
Wayampi	30		86.7			3.3	6.7	3.3										1
Arara	30		83.4		3.3		10.0											1
Pacific Islanders																		
Indonesia	60		90.0		3.3	1.7			3.3								1.7	1
Melanesia[b]	266	0.4	73.3	1.5	4.5	9.4	3.0		4.9				1.5	1.1			0.4	1
Melanesia[c]	266	3.0	66.3			17.9			6.5	3.9		1.8				0.3	0.3	1
Polynesia	339	0.6	90.8		1.2	4.4	0.9		0.9	0.3				0.3			0.6	1
Micronesia	206		93.7					1.9	1.0	1.5		1.5		0.5				1

[a]Haplotype 3 is defined as Benin, 5 as Senegal, 6 as Cameroon, 10 as Arab-Asian, and 16 as Bantu in studies with β^S chromosomes.
[b]Highlanders.
[c]Islanders.

References: 1. Guerreiro *et al.* (1994); 2. Long *et al.* (1990); 3. Castro de Guerra *et al.* (1997); 4. Pante-de-Souza *et al.* (1999); 5. Villalobos-Arámbula *et al.* (1997);

Table 7.15. *Genetic diversity considering β^A haplotype distributions*

Places and peoples	Genetic diversity		
	Intrapopulational haplotype diversity[a]	Observed no. of different haplotypes	Effective no. of haplotypes[b]
Europeans			
England	0.647	4	2.8
Germany	0.733	5	3.7
Greece	0.705	8	3.4
Italy	0.497	4	2.0
Cyprus	0.468	4	1.9
Asians			
India	0.598	8	2.5
Cambodia	0.344	4	1.5
Thailand	0.263	5	1.4
China	0.303	5	1.4
Korea	0.271	5	1.4
Japan	0.294	5	1.4
Africans			
Algeria	0.677	10	3.1
Kung San	0.739	6	3.8
Bantu	0.788	8	4.7
Benin	0.662	7	2.9
Nigeria	0.607	7	2.5
Senegal	0.692	4	3.2
Afro-Americans			
Panaquire	0.807	12	5.2
Curiau[c]	0.759	7	4.1
Pacoval[c]	0.816	10	5.4
Trombetas[c]	0.747	8	3.9
Mestizo-Americans			
Mexicans	0.712	13	3.5
Amerindians			
Mapuche	0.605	7	2.5
Xavante	0.591	4	2.4
Zoró	0.132	2	1.2
Gavião	0.225	5	1.3
Suruí	0.323	3	1.5
Wai-Wai	0.223	2	1.3
Yanomami	0.169	3	1.2
Kayapo	0.333	3	1.5
Wayana-Apalai	0.308	3	1.4
Wayampi	0.250	4	1.3
Arara	0.304	3	1.4

Table 7.15. (*cont.*)

	Genetic diversity		
Places and peoples	Intrapopulational haplotype diversity[a]	Observed no. of different haplotypes	Effective no. of haplotypes[b]
Pacific Islanders			
Indonesia	0.190	5	1.2
Melanesia[d]	0.450	10	1.8
Melanesia[e]	0.523	8	2.1
Polynesia	0.174	8	1.2
Micronesia	0.122	6	1.1

[a]Expected heterozygosity (Nei, 1987).
[b]Calculated according to Long *et al.* (1990). When all haplotypes in a system are equally frequent, the effective number of haplotypes is maximized, being equal to the observed number of haplotypes; otherwise, the former will be smaller than the latter. The formula used was ne = 1/1-GSI, where GSI is the Gini–Simpson index of diversity. GSI is the probability that an individual will be heterozygous in a randomly mating population.
[c]Two, two, and one rare haplotypes were not considered in the analysis of Curiau, Pacoval, and Trombetas, respectively.
[d]Highlanders.
[e]Islanders.

Table 7.16. *β[A] haplotype diversity analysis using the 16 usual haplotypes*

Groups[a]	No. of populations	Absolute haplotype diversity			Relative interpopulational haplotype diversity (%)[b]
		Total	Intrapopulational	Interpopulational	
Africans	4	0.668	0.609	0.079	11
Europeans	5	0.623	0.594	0.036	6
Amerindians	11	0.328	0.308	0.022	7
Neo-Americans	5	0.809	0.757	0.065	8
Neo-Brazilians	3	0.797	0.761	0.054	7

[a]Africans included: Bantu, Benin, Senegal and Nigeria; Europeans: Cyprus, Italy, Germany, England, Greece; Amerindians: Arara, Wayampi, Wayana-Apalai, Kayapo, Yanomami, Wai-Wai, Suruí, Gavião, Zoró, Xavante, Mapuche; Neo-Americans: Panaquire, Mexicans, Curiau, Pacoval, Trombetas; Neo-Brazilians: Curiau, Pacoval, Trombetas.
[b]Coefficient of gene differentiation (Nei, 1987).

Table 7.17. *Selected sickle cell disease studies performed in Latin America, outside Brazil*

Country or region	No. of patients	Type of study			Main findings	Reference
		Hemat.	Clin.	Reprod.		
México	4	Yes	Yes	No	One AS, two SS and one S/Thalassemia subjects	1
Cuba	15	Yes	Yes	No	General description	2
Dominican Republic	6	Yes	Yes	No	Clinical picture of subjects simultaneously affected by sickle cell disease and tuberculosis	3
Jamaica	17	Yes	Yes	Yes	Maternal and fetal hazards in SS, SC, and S/Thalassemia conditions	4
	Large	Yes	Yes	Yes	Classical analysis on the clinical symptoms in these conditions	5
Puerto Rico	1	Yes	Yes	No	Simultaneous occurrence of sickle cell disease and spherocytosis	6
	14	Yes	No	No	Delayed half-life of erythrocytes in SS, SC, and CC individuals	7
Guadeloupe	67	Yes	No	No	SC subjects, some genetic factors may modulate their clinical picture	8
	43	Yes	No	No	Patients with HbS/β^+-thalassemia type 2 have milder clinical picture compared with those with HbS/β^0-thalassemia or HBS/β^+-thalassemia type I	9
Curaçao	1	Yes	Yes	Yes	Detailed study of a pregnant S/S woman	10
	34	Yes	Yes	Yes	Detailed discussion of the clinical picture of SS, SC, and CC subjects	11
	66	Yes	Yes	No	Discussion of the clinical findings in SS, SC and S/Thalassemia individuals	12
	9	Yes	Yes	No	Hydroxyurea therapy in sickle cell anemia with good results in young, but not in older patients	13
Trinidad	5	Yes	Yes	No	Clinical picture of the patients	14
Venezuela	22	Yes	Yes	No	Clinical picture of the patients	15
	22	Yes	Yes	No	Clinical study of the patients	16
Colombia	22	Yes	Yes	No	Interaction between thalassemia and respectively HbS and HbC	17
Argentina	115	Yes	No	No	Revision of cases studied along 30 years	18
Chile	2	Yes	Yes	No	Study of a sickle cell/thalassemia family	19

References: 1. Lisker *et al.* (1962); 2. Chediak *et al.* (1939); 3. Herrera Cabral (1950); 4. Anderson *et al.* (1960); 5. Serjeant (1974); 6. De Torregrosa *et al.* (1956); 7. Suarez *et al.* (1959); 8. Lee *et al.* (1998); 9. Romana *et al.* (1996); 10. van der Sar (1943); 11. van der Sar (1959); 12. van Zanen (1962); 13. Saleh *et al.* (1997); 14. Henry (1963); 15. Barnola *et al.* (1953); 16. Torrealba (1956); 17. Echavarria and Molina (1971); 18. Abreu and Peñalver (1992); 19. Rona *et al.* (1973).

Table 7.18. *Selected sickle cell disease studies performed in Brazil*

No. of patients	Type of study			Main findings	Reference
	Haemat.	Clin.	Reprod.		
4	Yes	Yes	No	General description of the patients	1
8	Yes	Yes	No	General description of the patients	2
141	Yes	Yes	No	Review of cases studied between 1958 and 1969	3
9	No	Yes	No	Pulmonary gas transfer defect in sickle cell disease may be due to right to left shunting and also ventilation/perfusion disturbances	4
22	Yes	Yes	No	Review of cases referred for study	5
40	Yes	Yes	No	Suggestion of a high incidence of S/β°-thalassemia in certain Brazilian populations	6
156	Yes	Yes	No	Sickle cell anemia and S/β°-thalassemia showed similar clinical severity	7
20	Yes	Yes	No	Splenic function may be different in sickle cell anemia as opposed to S/β°-thalassemia and SC disease	8
409	Yes	Yes	No	Low frequencies of some clinical signs, sex difference in relation to the types of crises	9
209	No	No	Yes	High rate of fetal losses (48%)	10
354	Yes	Yes	No	Relationships between Hb F level, morbidity and mortality	11
156	Yes	Yes	No	Higher rate of splenomegaly among S/β°-thalassemia as compared with sickle cell anemia	12
12	No	Yes	No	Prevalence of leg ulcers in relation to controls	13
17	Yes	No	No	Similar β-chain deficiency in reticulocytes and bone marrow cells of S/β°-thalassemia patients	14
32	Yes	Yes	No	Hb SC disease symptomatology did not differ from those of other series, perhaps because the pathogenesis depends mainly of the hemoglobin concentration	15
10	Yes	No	No	Increased circulating immune complexes in S/β°-thalassemia subjects	16
95	Yes	No	No	Low percentages of $^{G}\gamma$ chains in both sickle cell anemia and S/β°-thalassemias	17
89	Yes	Yes	No	The level of Hb F is a useful indicator of clinical complications in sickle cell anemia	18
27	No	Yes	No	Sickle cell anemia subjects with leg ulceration showed decreased ^{99m}Tc clearance in calf blood flow	19
54	Yes	No	No	No interaction between glucose-6-phosphate dehydrogenase deficiency and sickle cell diseases	20

Table 7.18. (*cont.*)

No. of patients	Type of study			Main findings	Reference
	Haemat.	Clin.	Reprod.		
125	No	Yes	Yes	Growth and sexual maturation are delayed in persons with sickle cell diseases	21
47	Yes	No	No	Increased (19%) prevalence of hepatitis C antibody in patients with sickle cell diseases	22
25	Yes	Yes	No	The Central African Republic haplotype may be associated with more severe sickle cell anemia	23
106	Yes	No	No	Renal dysfunction was observed and quantitated both in SS and AS individuals	24
73	Yes	Yes	No	Three polymorphisms with alleles that may increase the risk of hypercoagulability were studied, but their clinical impact seems to be small among sickle cell disease patients	25
63	Yes	No	No	The spontaneous formation of circulating burst-forming unit-erythroid in sickle cell diseases may be due to an expanded erythropoiesis secondary to the hemolysis which occur in these individuals	26

References: 1. Serra de Castro (1934); 2. Frimm (1947); 3. Marinho (1970); 4. Lemle *et al.* (1976); 5. Silva (1979); 6. Zago *et al.* (1980); 7. Zago *et al.* (1983*a*); 8. Zago and Bottura (1983); 9. Hutz and Salzano (1983*a*); 10. Hutz and Salzano (1983*b*); 11. Hutz *et al.* (1983); 12. Zago *et al.* (1983*c*); 13. Ramalho *et al.* (1985*a*); 14. Costa and Zago (1986); 15. Marmitt *et al.* (1986); 16. Donadi *et al.* (1989); 17. Zago *et al.* (1989); 18. Bordin *et al.* (1998); 19. Saad and Zago (1991); 20. Saad and Costa (1992); 21. Zago *et al.* (1992*b*); 22. Arruda *et al.* (1993); 23. Figueiredo *et al.* (1996); 24. Sesso *et al.* (1998); 25. Andrade *et al.* (1998); 26. Perlingeiro *et al.* (1999).

Table 7.19. *Selected clinical studies in thalassemia from two Latin American countries*[a]

No. of patients	Diagnostic criteria			Main findings	Reference
	Haemat.	Chain synthesis	Molecular		
77	Yes	No	No	High heterogeneity	1
17	Yes	Yes	No	Diversity similar to that observed in areas of high or moderate incidence of the disease, for instance, Italy	2
1	Yes	Yes	No	Hb Hasharon heterozygosity may lead to a less severe expression of homozygous β°-thalassemia	3
2	Yes	Yes	No	β^+-thalassemia homozygotes with low Hb F levels	4
3	Yes	Yes	No	Hb H disease in three families. Unbalanced globin chain synthesis in the patients	5
24	Yes	Yes	No	Review of the β-thalassemia cases	6
2	Yes	Yes	No	Association of silent and high Hb A_2 β-thalassemias	7
424	Yes	No	No	Contribution of thalassemia minor carriers to cases of anemia in a population	8
195	Yes	No	No	Higher (16%) incidence of carriers of Hb H intraerythrocyte aggregates in subjects with anemia of unknown cause as compared with controls (4%)	9
62	Yes	No	No	Classification of cases studied in the period 1969–85	10
37	Yes	Yes	No	γ^G level for the β-thalassemia homozygotes (59%) was significantly lower than that of the controls (71%)	11
21	Yes	No	No	No significant differences in plasma folic acid and vitamin B_{12} levels between β-thalassemia heterozygotes and controls	12
1	Yes	No	Yes	Clinical features of a patient with an α-thalassemia/mental retardation syndrome	13
1	No	No	Yes	The patient is homozygote for the T $\rightarrow$ C substitution at position 6 of the first intervening sequence (IVS 1.6) of the β-globin gene (Portuguese type)	14
46	Yes	No	Yes	Red cell indices present considerable overlap between patients and controls	15

[a]With the exception of study no. 1, performed in Jamaica, all the others involved Brazilian patients.

References: 1. Went and MacIver (1961); 2. Zago et al. (1981); 3. Zago et al. (1982); 4. Zago et al. (1983b); 5. Zago et al. (1984); 6. Zago and Costa (1985); 7. Zago et al. (1985); 8. Ramalho (1985b, 1986b); 9. Castilho (1987); 10. Targino de Araújo et al. (1987); 11. Zago (1989); 12. Silva and Varella-Garcia (1989); 13. Wilkie et al. (1990); 14. Costa et al. (1991b); 15. Sonati et al. (1992b).

8   Normal genetic variation at the protein, glycoconjugate and DNA levels

Change is a characteristic of all systems and all aspects of all systems
Richard Levins and Richard Lewontin

A world of differences

The human species is highly variable, and Latin Americans are no exception. In fact, due to the merging of groups that previously had genetically differentiated in other continents, this variability is accentuated. The study of normal variation among humans has a respectable past, going back to the eighteenth century, with scholars such as Carl von Linné (1707–78), Georges L. Leclerc de Buffon (1707–88), and Johann F. Blumenbach (1752–1840). At the beginning the traits considered were mainly morphologic. It was only with the seminal discovery in 1900, by Karl Landsteiner (1868–1943), of the ABO blood groups, that the door was open for the discovery of simple-inherited traits that could provide more rigorous evaluations of such differences, separating the effects of genes from those of their environments.

In terms of methodology, three distinct key developments should be noted. The first was based on immunologic techniques, leading to the discovery of some two dozen blood groups. Their specificity is determined by the chemical structure of the antigen determinants located on proteins and glycoconjugates (glycolipids and glycoproteins). In the mid-1950s, electrophoretic methods (based on the differences in the electric charges of the proteins) were devised; the investigator to be mentioned here is Oliver Smithies. Finally, with the spectacular developments in molecular biology in the 1970s and 1980s, we are now considering the variability present in the genetic material *per se* (deoxyribonucleic acid, or DNA), and not only its primary product (proteins).

255

The Latin American microcosm

The first population study on blood groups in Latin America that we could locate was that of Montenegro (1925) in Brazil. In Argentina the work of Mazza *et al.* (1928) should be remembered. A large number of investigations followed; the blood group and protein polymorphisms (common variants) had been reviewed at the world level at regular intervals, the most recent compilations being those of Mourant (1954), Mourant *et al.* (1976), Tills *et al.* (1983), and Roychoudhury and Nei (1988). Specifically for Latin America the reviews of Salzano (1965, 1971*b*, 1987), Salzano and Freire-Maia (1970), and Lisker (1981) should be consulted (not listed here are treatises dealing with Amerindian populations only).

Due to Brazil's continental size and the amount of information available for its population we decided in some cases to separate our analysis, including first all the other countries, and then Brazil separately. Also, as was done in previous chapters of this book, we will examine the populations taking into consideration their genetic ancestry and historical processes of interchange. In the next chapter this rationale will be considered and appropriately discussed.

Protein and blood group genetic variability in Latin America (except Brazil)

Information available

Tables 8.1 to 8.3 list the protein and blood group genetic systems for which data exist considering respectively European-derived, African-derived and Mestizo (or unclassified in relation to ethnic group) populations. The hemoglobin data were discussed in the previous chapter. European-derived subjects with few indications of admixture have been less studied than the two other sets. There is information available for 21 genetic systems, distributed over three Caribbean and six South American countries or regions. Uruguay and Venezuela are the countries with the highest numbers of systems investigated (nine and eight, respectively). When African-derived subjects are considered the number of genetic systems available for comparison (48) more than doubles, and there are data for six continental Central American, 10 Caribbean, and six South American countries or regions. Guatemala (with 20), Venezuela (20), and Ecuador (18) are those best studied in terms of numbers of genetic systems. The amount of information concerning Mestizo subjects is similar (46 systems, investigated in six continental Central American, nine Caribbean, and 11

South American countries or regions). By far the country best studied in number of systems (36) is México, followed by Uruguay (20) and Colombia (18).

Genetic diversity analysis

It is clear, as was indicated above, that the information available outside Brazil is too uneven for evaluations using a uniform set of variables. The analysis of genetic diversity considering the three ethnic categories, presented in Tables 8.4 to 8.6, had to be made, therefore, considering different sets of genetic information. For the European-derived populations (Table 8.4) the intraethnic gene diversity observed in Venezuela (0.301), Cuba (0.418) and Uruguay (0.502), which involved numbers and types of genetic systems not too dissimilar, are probably the best estimates. As for the African-derived groups (Table 8.5), the most representative samples in Continental Central America and the Caribbean Islands are those of Panamá (0.179, nine systems) and Guatemala (0.270, 20 systems). The high value found for Jamaica (0.721) is due to the fact that most of the systems considered were from the HLA loci, that due to their special nature (connection with the immune response) may be atypical. In South America the most representative values would be those found in Ecuador (0.251, 19 systems), and Venezuela (0.393, 24 systems). Corresponding figures for the Mestizo category are shown in Table 8.6. They are, for Continental Central America and the Caribbean region, those of Haiti (0.214, 16 systems) and Guadeloupe (0.619, seven systems); and for South America those of Chile (0.340, 12 systems) and Uruguay (0.409, 20 systems).

Formally, considering the standard deviations calculated for these figures, significant differences occur between them. Since, however, the genetic systems considered were different, this is not surprising. More interesting is the finding that in estimates based in more than two systems, and leaving aside those in which HLA is in a majority, the values are distributed in a continuous way, from 0.179 (African-derived subjects, Panama) to 0.619 (Mestizos, Guadeloupe), independently of ethnic categories.

Interethnic differences within countries or regions could be evaluated in 15 instances only. The results are displayed in Figure 8.1. The highest value (4% only) was observed in the comparison between Mestizo and African-derived subjects from Honduras. Other estimates, especially those in the Caribbean islands, are much lower. Theoretically African-derived and European-derived people should present the highest interethnic differences, but this was the case in Cuba only.

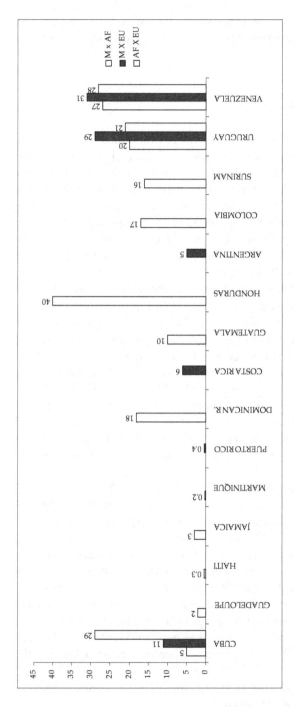

Figure 8.1. Relative proportion (×1000) of the genetic diversity in Latin American countries due to interethnic differences. M, mixed; AF, African-derived; EU, European-derived.

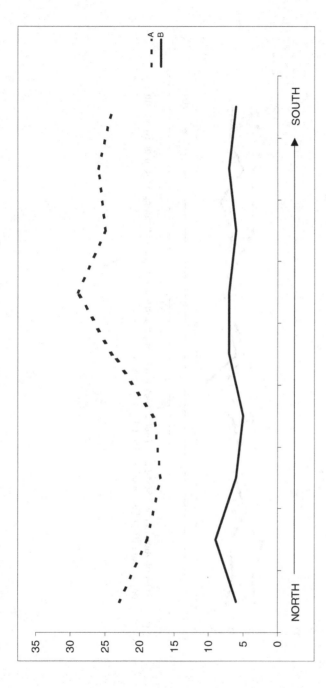

Figure 8.2. Distribution of the *ABO*A* and *ABO*B* alleles (%) along a north–south gradient among European-derived Latin American populations. For this analysis Brazil was included together with the other Latin American countries. The geographic points used for that country were the centers of its five regions (north, northeast, center-west, southeast, south).

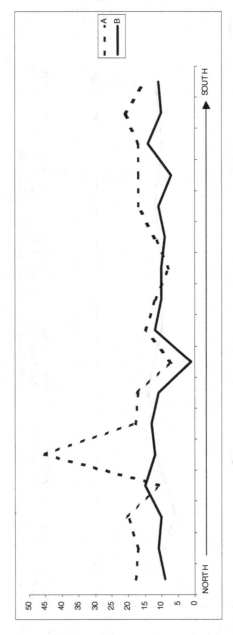

Figure 8.3. Distribution of the *ABO*A* and *ABO*B* alleles (%) along a north–south gradient among African-derived Latin American populations (Brazil included; see legend to Figure 8.2).

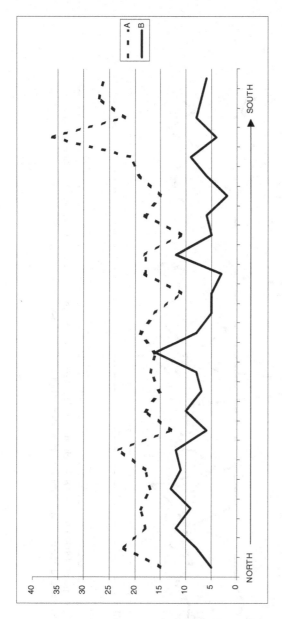

Figure 8.4. Distribution of the *ABO*A* and *ABO*B* alleles (%) along a north–south gradient among Mestizo Latin American populations (Brazil included; see legend to Figure 8.2).

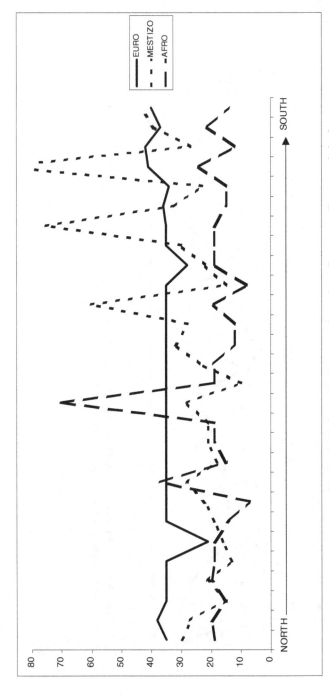

Figure 8.5. Ratio of *ABO*A* to *ABO*B* (biochemical index × 10) along a north–south gradient among European-derived, African-derived and Mestizo Latin American populations (Brazil included; see legend to Figure 8.2).

The ABO blood group system is the most studied in Latin America, as elsewhere in the world. There is information about it for 23 countries/ territories when Mestizos are considered, for 13 when African-derived individuals are examined, and for five when European-derived persons apparently free of admixture are evaluated. Figures 8.2–8.4 present the frequencies of the $ABO*A$ and $ABO*B$ alleles when they are plotted in a north–south distribution. As can be seen, no clear geographic trend is discernible. The $ABO*A$ to $ABO*B$ ratio is also generally uniform throughout the countries/regions, its average values being 1.9 for African-derived, 2.7 for Mestizo, and 3.5 for European-derived individuals (Figure 8.5). Considering this locus only, the total diversity does not vary much among ethnic categories (0.423 for African-derived, 0.424 for Mestizo, and 0.445 for European-derived subjects), but the percentage of it that is due to intercountry/region diversity is somewhat higher (3.7%) among the African-derived, as compared with the values obtained for the Mestizo (1.7%) and European-derived (0.8%) segments. As a comparison, values for the HLA-A and HLA-B loci were considered for 10 populations identified as Mestizo (México, Guadeloupe, El Salvador, Panamá, Puerto Rico, Colombia, Uruguay, Venezuela, Ecuador, Brazil/North) five identified as African-derived (Guatemala, Jamaica, Colombia, Venezuela, Brazil/ South) and three identified as European-derived (Uruguay, Brazil/South, Brazil/Southeast). Although the total diversity in these loci were much higher (African-derived, 0.922; Mestizo, 0.924; European-derived, 0.919) the intercountry/region numbers remained low (2.9%, 3.4%, and 0.9%, respectively).

Brazil

Information available

Table 8.7 presents the information available on protein and blood group genetic systems for Brazil, separately by geographic region and ethnic classification. With the exception of the Center-West region, for which few genetic data are available, the information is more uniform than that available for the rest of Latin America. The best-studied region is the Southeast, with totals of 20–23 genetic systems available for its three ethnic subgroups.

Genetic diversities

Estimates of intraethnic genetic diversities obtained for the different Brazilian regions and their ethnic subgroups are displayed in Table 8.8. Leaving aside the two Center-West values, based on small numbers of systems, they are distributed in an uniform way from 0.193 (European-derived sample, North) to 0.475 (same ethnic category, Southeast). As in the remainder of Latin America, the values show a distribution that is independent of ethnic categorization.

Since the data at hand are more uniform, we considered the variability found in Brazil in more detail. Table 8.9 gives the values for the total genetic diversity and its fraction that is due to interethnic differences separately by system and geographic region. Interethnic differences are small, in three of the evaluations considering all loci amounting to 1% only. The most variable systems were haptoglobin (HP), and the RH, MN and ABO blood groups.

Similar evaluations can be made using a more restricted (12) but uniform set of systems, and excluding the Center-West region. The results are given in Tables 8.10 to 8.12. The estimates now become, as expected, much more uniform, the amount of interregion and interethnic differences varying around 1% only. Despite the fact that they are small, however, if these sets of data are subjected to a genetic distance and tree analyses, the distribution showed well-defined ethnic clusters, the European- and African-derived samples occupying extreme positions and the mixed ones occurring midway (Figure 8.6).

DNA markers

Information available

Tables 8.13–8.15 list the DNA systems for which there is information in Latin America. The amount of data is much more restricted than those available for the protein markers, although the number of loci (places where the genetic units are located) is higher. In addition, the unevenness among countries is most marked. For the European-derived populations (Table 8.13), although some information exists for Costa Rica, Colombia and Venezuela, the bulk of the results were obtained in Argentina and Brazil, with the latter presenting about 6 times more systems investigated than the former. A total of 70 loci was investigated in this ethnic group.

The data available for the African-derived populations (Table 8.14) is

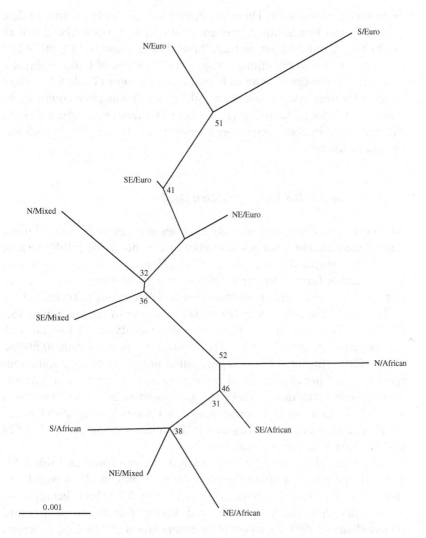

Figure 8.6. Unrooted neighbor-joining genetic distance tree showing the relationships of populations from several Brazilian geographic regions differentially classified in relation to ethnic group. The D_A genetic distance (Nei *et al.*, 1983) was used, and the neighbor-joining method is that described by Saitou and Nei (1987). The tree reliability (indicated by numbers) was tested using Felsenstein's (1985) bootstrap test with 2000 replications (Hedges, 1992). The first letters indicate the geographic region, and those after the virgule the ethnic classification.

more evenly distributed. There is information available for five Middle American and five South American countries. But again, the Brazilian results are far more numerous than those of other nations. In total 74 loci were investigated in this ethnic category for the whole of Latin America.

As for the Mestizo or ethnically unclassified groups (Table 8.15), there are data for four Middle American and five South American countries. In this category the Argentinian results are more numerous, exceeding the Brazilian data in about one order of magnitude. In this category 65 loci have been studied.

A nuclear DNA polymorphism: D1S80

As an example of the gene diversity analyses that can be performed using Latin American and other populations we have chosen the D1S80 system, for which a reasonable amount of information exists around the world. This system is formed by the repetition of a 16-base pair core sequence situated at the distal end of the short arm of chromosome 1. Technically, it is classified in the category of the variable number of tandem repeat loci (VNTRs). They are characterized by high diversity of alleles and high heterozygosities. Specifically the D1S80 marker has more than 30 alleles. Large alleles (higher than 41 repeat units) are many times population-specific, occurring in just one or a few groups. For instance, in African-derived South Americans, allele *42 was observed in Cametá (northern Brazil) only, and *45 in Curiepe (northern Venezuela) only (Silva *et al.*, 1999). The most frequent alleles vary in the different ethnic groups but *24 and *18 have a very widespread distribution.

The results of the gene diversity analysis are presented in Table 8.16. Total diversity was of about the same order of magnitude in Neo-Latin Americans (0.877) and sub-Saharan Africans (0.872), both being about 10% higher than the values obtained among Europeans (0.797) and Amerindians (0.790). In terms of interpopulation differences, however, they present a value (3%) which is intermediate between those of Europeans (1%) and Amerindians (12%). This is reflected in the parameter *NmFM*, a standardized value that relates *N*, the effective size of a population (that is, the gene pool that is available for genetic change and recombination) with *m*, the amount of gene flow that occurs between this population and the outside world. Since the exchange between groups leads to uniformity, there is an inverse relationship (observed in Table 8.16) between interpopulation variability and gene flow.

Mitochondrial DNA

Data related to the variability of this organelle in non-Indian Latin American populations are available for seven Latin American countries (Table 8.17). But they differ greatly in scope and methodology. Therefore, strict comparisons are difficult, and we preferred to concentrate in specific aspects of the African-derived populations, as indicated below.

The question asked was, in which way the variability found in this organelle among African-derived Latin Americans compares with that found in the three putative Latin American founder stocks? Sixty, 84, 62 and 58 variable sites were observed in the first hypervariable segment (HVS-I) of, respectively, African-derived Latin Americans, Africans, Europeans and Amerindians (references in Table 8.18). Their overall interpopulation variability in this region was, in this same order, 16%, 8%, 5%, and 31%. There is more homogeneity among Europeans and Africans, the highest value occurring among Amerindians. As expected, the African-derived Latin Americans present an intermediate value.

Table 8.18 and Figure 8.7 show the data related to the 10 most variable HVS-I sites. As is clearly observed, they are not the same in the four ethnic categories. This is probably due to variability in the sensitivity to molecular changes among these regions, coupled with different population histories of the groups studied.

Figure 8.8 presents the unrooted neighbor-joining genetic distance tree obtained when these data are subjected to this analysis. A separation between African or African-derived, European, and Amerindian populations is clearly shown, as well as the restricted (European) and much more pronounced (Amerindian) interpopulation differences. The African-derived Latin Americans of Porto Alegre, Brazil and the Black Caribs of Belize cluster more clearly with the African groups than the Brazilian Salvador or Paredão. The latter show peculiarities, disclosed by its long branch, due to its history of isolation, probably coupled with bottleneck effects.

Y chromosome

Although the first population studies related to this chromosome reported low variability levels, more recent investigations disclosed variable sites appropriate for population comparisons. Many of them are located in the Y nonrecombining region, constituting an appropriate counterpart of the

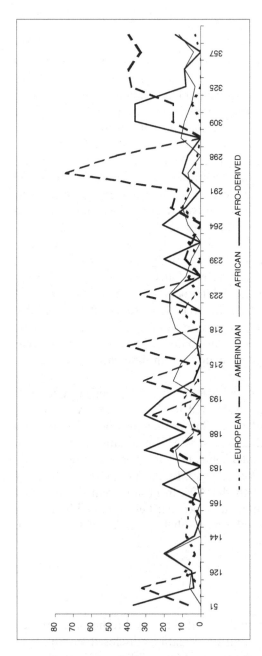

Figure 8.7. Interpopulation single-site diversity (%) considering the 10 most variable specific HVS-I mtDNA sites in Africans, Europeans, Native Americans, and African-derived Latin Americans.

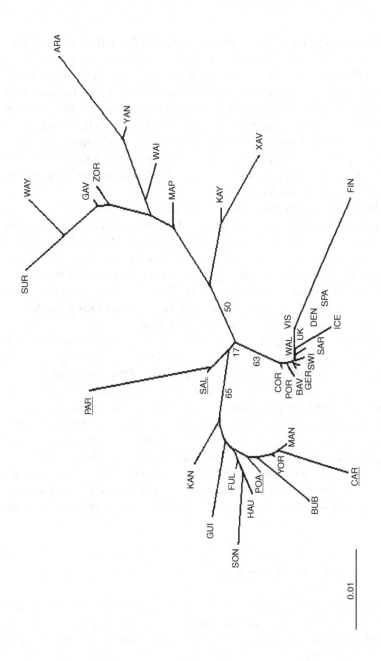

Figure 8.8. Unrooted neighbor-joining genetic distance tree showing the relationships obtained among African and African-derived, European, and Amerindian groups considering the 10 most variable sites of the HVS-I mtDNA. Numbers represent bootstrap values, and the code for the populations considered is given in the footnote to Table 8.17.

mtDNA system. The latter is inherited strictly from the maternal side, while the Y is transmitted exclusively by the father–son route.

The DYS19 locus is one of the most studied in population comparisons. Located in the short arm of the Y chromosome, it is classified as a microsatellite or short tandem repeat (STR) marker. Variability occurs in relation to a GATA repeat unit, and up to the present 10 alleles (varying in length between 174 and 210 bp) have been found. Allele distributions are generally unimodal. In Africans living in the south of the Sahara the *194* allele is the most common. In Europeans *190* and in Native Americans *186* are the most prevalent.

Table 8.19 presents the gene diversity analysis performed considering Latin Americans and representatives of their putative ancestors. The highest total diversity (0.758) was observed among African-derived Latin Americans, and the lowest (0.389) among Amerindians. But the interpopulation fraction of this variability is twice as high in Amerindians (33%) than in African-derived Latin Americans (16%). The estimated male gene flow per generation (N_{mM}) is 35 times higher between European-derived Latin Americans, as compared with Amerindians.

Table 8.19 also presents data related to an analysis of molecular variance (AMOVA) performed in relation to the DYS19 results. The numbers obtained are presented in colums 7 and 8 of the Table. The parameter ϕ_{ST} is obtained as the ratio of the estimated variance component due to differences among the populations over the estimated absolute total variation (Michalakis and Excoffier, 1996). The results roughly parallel those obtained using Nei's (1986) G_{ST}. AMOVA also permits a hierarchical analysis considering the Native American, European, African, and Neo-Latin American categories. It indicated that only 12% of the variance could be attributed to differences among these categories, 4% to differences among populations/within category, and 84% to within-population variability.

Protein–DNA comparisons

Generally, the same population was not investigated using both protein and DNA markers. The data obtained by our group among African-derived South American urban and rural communities, and displayed in Table 8.20, are an exception. Estimates of intrapopulation variability for the 6–11 protein loci considered are much lower than those obtained using eight VNTR/STR loci. However, this is not reflected in the interpopulation variation, which shows similar values for the two sets. Results with mtDNA sequences of the HVS-I region, however, present a very high (21%) interpopulation variability for three Afro-Brazilian populations.

Variation in two key molecules

Introduction

The fate of a given genetic change will depend on the region of the molecule in which it will be expressed and the eventual repercussions of this change at the phenotype and population levels. Besides hemoglobin, considered in the previous chapter, two other molecules attracted the interest of Latin American scientists, glucose-6-phosphate dehydrogenase (G6PD) and albumin. They will now be considered in turn.

Glucose-6-phosphate dehydrogenase

The G6PD monomer is formed by 515 amino acids, giving a molecular weight of 59 kDa. The active form of this enzyme functions as a dimer and contains lightly bound NADP (nicotinamide adenine dinucleotide phosphate, important in the oxireduction processes essential for life). Variants have been classified by the World Health Organization into four classes, depending whether they condition: (1) nonspherocytic hemolytic anemia; (2) severe enzyme deficiency; (3) moderate deficiency; and (4) no deficiency. Complete deficiency of the enzyme seems to be lethal.

The G6PD gene was independently cloned and sequenced in 1986 by two groups of investigators. It contains 13 exons (regions in the DNA which codify for proteins) and is over 20 kb in length. Around 100 mutants have been characterized so far, and those which have been observed in Latin America are listed in Table 8.21.

A total of 19 variants have been adequately investigated at the molecular level in seven countries or regions of Latin America, due to a large extent to the efforts of Dr Ernest Beutler and his group. They are amply and evenly distributed along the molecule, from position 14 to 488. About half of them condition nonspherocytic hemolytic anemia (understandably, since it is easier to identify traits which lead to severe clinical symptoms). More systematic studies of these variants have been conducted in Brazil (they included around 10 300 healthy subjects, and 48% of the mutants were found in this country), México and Costa Rica. Their ethnic distribution is quite remarkable: up to now no definite Amerindian variant has been identified in this molecule (although seven of the carriers in whom they have been found were classified as Mestizos). European derivation was established for about half of them.

Albumin

Human serum albumin consists of a single polypeptide chain and is synthesized in and exported to the liver as nonglycosylated protein. Albumin is the most important transport and deposition protein in the circulation. Of the nearly 4000 possible effective point mutations in the 585 codons of the albumin gene, about 800 would lead to a change in electrophoretic mobility and could therefore be discovered in standard population surveys. To date, however, only about 50 different alloalbumins have been detected and characterized at the molecular level, and shown to cluster in three regions of the polypeptide chain: the propeptide, the N-terminus, and the C-terminal regions of subdomains IIB and IIIB.

Variants are generally rare, but may occur more frequently, especially in Amerindians. Table 8.22 lists nine types observed in four Latin American countries. These variants show quite distinct geographic distributions. For instance, Yanomama 2 occurs in 14% of 3504 individuals of this tribe of Venezuela and Brazil, but was never found anywhere else. Bilirubin-binding studies of this variant showed that heterozygotes could not be distinguished from those who had normal (A) albumin only. Homozygotes Yanomama 2/Yanomama 2, however, presented a lower (75%) binding ability (Lorey *et al.*, 1984). It is possible, therefore, that this reduced adaptation may provide a check to the spread of this allele.

Another relatively common variant is albumin México, found in polymorphic frequencies in Amerindians from México and USA and also in one of 2941 individuals from the Brazilian city of Belém. The widest geographic distribution is that of albumin Porto Alegre 2; besides occurring in southern (Porto Alegre), northeastern (Fortaleza), and northern (Coari, Manaus) Brazil, this specific mutation (501 Glu → Lys) had already been observed in Adana (Turkey), Lambadi (India), Birmingham and Glasgow (UK) and Vancouver (Canada). In the last three cases, it was identified in unrelated families who had emigrated from the Punjab, Kashmir, and northern India, respectively. This variant shows a decreased (56–67%) binding to total nonesterified fatty acids (Nielsen *et al.*, 1997). No advantage in relation to this parameter could therefore explain its wide geographic distribution.

Table 8.1. *Information on protein and blood group genetic systems
available for European-derived Latin American populations (except Brazil;
the hemoglobin information was provided in the previous chapter)*

Region and country	Genetic systems
Caribbean Islands	
Cuba	ABO, KELL, MN, RH (1 serum), RH (5 sera), PGD, PGM1, PGM2, TF
Puerto Rico	ABO, RH (3 sera), G6PD
St Barthélemy	HLA-A, HLA-B, HLA-C
South America	
Argentina	RH (1 serum)
Colombia	HP (iso)[a]
Ecuador	ABO
Peru	GC, TF
Uruguay	ABO, DI, FY, KELL, RH (1 serum), RH (5 sera), HLA-A, HLA-B, HLA-C
Venezuela	RH (1 serum), AK, ESD, PGM (iso)[a], AL, CP, HP, TF

Sources: Mourant *et al.* (1976); Tills *et al.* (1983); Roychoudhury and Nei (1988), and
Appendix 1. Key to the symbols in Appendix 2.
[a]iso, Isoelectric focusing.

Table 8.2. *Information on protein and blood group genetic systems available for African-derived Latin American populations (except Brazil; the hemoglobin information was provided in the previous chapter)*

Region and country	Genetic systems
Continental	
Central America	
Belize	RH (unknown number of sera), MN, DI, FY, KELL, KIDD, LE, LU, MNSs, P, PGM1 (iso), ALB, CP, GC (iso), HP, TF
Costa Rica	G6PD
Guatemala	ABO, DI, KELL, MN, MNSs, RH (3 sera), ACP, AK, ESD, PGD, PGM1, PGM1 (iso), PGM2, ALB, CP, GC (iso), HP, TF, HLA-A, HLA-B
Honduras	ABO, FY, KELL, MN, RH (3 sera), HP, TF
Nicaragua	ACP, ADA, DI, ESD, GPT, LDH, PEPA, PEPB, PEPC, PGM1, PGM2, G6PD, GPI, SODA, UMPK
Panama	ACP, ESD, PGP, PGM1, PGM2, HP, TF
Caribbean Islands	
Cuba	ABO, KELL, MN, RH (1 serum)
Dominica	ABO, KELL, MN
Guadeloupe	ABO, HP, TF
Haiti	ABO, CP, HLA-A, HLA-B
Jamaica	ABO, RH (3 sera), G6PD, HLA-A, HLA-B, HLA-C, HLA-DR, HLA-DQ
Martinique	ABO
Puerto Rico	G6PD, HP
Dominican Republic	ABO, RH (4 sera)
St Vincent	PGM1 (iso), GC (iso)
Trinidad and Tobago	G6PD
South America	
Colombia	ABO, RH (1 serum), HP, HLA-A, HLA-B, HLA-DR, HLA DQ
Ecuador	ACP, ADA, AK, CA2, ESD, GLOI, G6PD, PGD, PGM1, PGM1 (iso), PGM2, F13A, F13B, ORM1, AHSG, C6, C7, APOC2
French Guiana	HP
Surinam	ABO, FY, KELL, MN, RH (1 serum), RH (4 sera), GPX1, G6PD, GC, HP, TF
Uruguay	ABO, RH (5 sera), ACP, AK, ESD, G6PD, PGD, PGM1, PGM2, TF
Venezuela	ABO, DI, FY, KELL, MN, RH (1 serum), RH (3 sera), RH (5 sera), ESD, G6PD, PGD, PGM1, ALB, HP, TF, HLA-A, HLA-B, HLA-C, HLA-DQ, HLA-DR

Sources: Mourant *et al.* (1976); Tills *et al.* (1983); Roychoudhury and Nei (1988), and Appendix 1. Key to the symbols in Appendix 2. iso, Isoelectric focusing.

Table 8.3. *Information on protein and blood group genetic systems available for Mestizo Latin American populations (except Brazil; the hemoglobin information was provided in the previous chapter)*

Region and country	Genetic systems
Continental Central America	
Costa Rica	ABO, G6PD
El Salvador	ABO, G6PD, HLA-A, HLA-B, HLA-C, HLA-DR
Guatemala	ABO
Honduras	FY
México	ABO, DI, FY, BF, KELL, LE, MN, MNSs, P, RH (1 serum), RH (3 sera), RH (4 sera), RH (5 sera), ACP, ADA, AK, ESD, GALT, GLOI, G6PD, PEPA, PGD, PGI, PGP, PGM1, PGM1 (iso), PGM4, ALB, HP, TF, GC, HLA-A, HLA-B, HLA-C, HLA-DR, LAC-DEF
Panama	HLA-A, HLA-B, HLA-C
Caribbean Islands	
Dutch Antilles	ABO, MN, G6PD
Cuba	ABO, KELL, MN, RH (1 serum), RH (5 sera), G6PD, GLOI, NADH-MR, HP, TF, GM
Guadeloupe	ABO, LE, RH (3 sera), GM, HLA-A, HLA-B
Haiti	ABO, DI, FY, KELL, LE, MN, ACP, AK, PGD, PGM1, PGM2, CP, HP, TF, GC
Jamaica	ABO, G6PD, GC
Martinique	ABO, HLA-A, HLA-B
Puerto Rico	ABO, Goᵃ, RH (1 serum), HP, TF, HLA-A, HLA-B
Dominican Republic	ABO, MN, RH (1 serum)
Trinidad and Tobago	ABO, RH (1 serum)
South America	
Argentina	ABO, KELL, LE, MN, RH (1 serum), RH (4 sera), HLA-A, HLA-B, HLA-C, HLA-DR, HLA-DQ, LAC-DEF
Bolivia	G6PD, LAC-DEF
Chile	ABO, FY, KELL, KIDD, MN, RH (1 serum), RH (unknown number of sera), ACP, CHE1, ESD, G6PD, PGM1, PGM2, HP
Colombia	ABO, DI, FY, KELL, MN, RH (1 serum), C3, C4, BF, GLOI, G6PD, GPT, HP (iso), HLA-A, HLA-B, HLA-C, HLA-DR, HLA-DQ
Ecuador	ABO, HLA-A, HLA-B, HLA-C, HLA-D, HLA-DR, HLA-DQ
French Guiana	ABO
Paraguay	ABO, RH (5 sera), AHSG, ORM1
Peru	ABO, KELL, MN, RH (1 serum), RH (3 sera)
Surinam	ABO, MN
Uruguay	ABO, DI, FY, KELL, MN, RH (1 serum), RH (5 sera), ACP, AK, ESD, G6PD, PGD, PGM1, PGM2, HP, HLA-A, HLA-B, HLA-C, HLA-DR, HLA-DQ
Venezuela	ACAH-C5, PI, G6PD, α2M, GC, HP, TF, GM, HLA-A, HLA-B, HLA-C, HLA-DQ, HLA-DR

Sources: Mourant *et al.* (1976); Tills *et al.* (1983); Roychoudhury and Nei (1988), and Appendix 1. Key to the symbols in Appendix 2.

Table 8.4. *Genetic diversity observed using different genetic systems within European-derived Latin American populations (except Brazil)*

Region and country	Intraethnic gene diversity[a]	Genetic systems[b]	References
Central America			
Costa Rica	0.010 ± 0.001	HB	Sáenz (1988)
Caribbean Islands			
Cuba	0.418 ± 0.112 (0.006, HB – 0.688, RH (5))	ABO, MN, RH (1), RH (5), HB	Colombo and Martínez (1985); Hidalgo et al. (1995)
Puerto Rico	0.595 ± 0.157 (0.439, ABO – 0.751, RH (3))	ABO, RH (3)	Torregrosa (1945/1946) Mourant et al. (1976)
St Barthélemy	0.766 ± 0.045 (0.729, HLA-C – 0.883, HLA-B)	HLA-A, HLA-B, HLA-C	Fumeron et al. (1981)
South America			
Argentina	0.457 ± 0.000	RH (1)	Etcheverry (1949)
Colombia	0.600 ± 0.001	HP (iso)	Mastana et al. (1994)
Ecuador	0.369 ± 0.027	ABO	Hoffstetter (1949)
Peru	0.107 ± 0.035 (0.000, TF – 0.107, GC)	GC, TF	Mourant et al. (1976)
Uruguay	0.502 ± 0.111 (0.147, RH (1) – 0.933, HLA-B)	ABO, DI, FY, KELL, RH (1), RH (5), HLA-A, HLA-B, HLA-C	Alvarez et al. (1993); Sans et al. (1993)
Venezuela	0.301 ± 0.080 (0.041, ALB – 0.554, PGM1 (iso))	RH (1), AK, ACP, ESD, PGM1 (iso), ALB, TF, HP	Mourant et al. (1976); Weimer et al. (1999)

[a] Average and range.
[b] Numbers in parentheses indicate the number of sera used in the Rh studies.

Table 8.5. *Genetic diversity observed using different genetic systems within African-derived Latin American populations (except Brazil)*

Region and country	Intraethnic gene diversity[a]	Genetic systems[b]	References
Central America			
Belize	0.222 ± 0.076 (0.012, KELL – 0.502, HP)	DI, FY, KELL, MN, RH (1), CP, HP, TF	Sauvain-Dugerdil (1989)
Guatemala	0.270 ± 0.066 (0.010, ALB – 0.866, HLA-A)	ABO, DI, KELL, MN, RH (3), ACP, HB, ADA, AK, ESD, PGD, PGM1, PGM1 (iso), PGM2, ALB, CP, HP, TF, HLA-A, HLA-B	Scrimshaw *et al.* (1961); Tejada *et al.* (1965); Crawford *et al.* (1981); Sáenz (1988); Santiago-Delpin (1991)
Honduras	0.366 ± 0.107 (0.124, HB – 0.699, RH (3))	ABO, FY, HB, KELL, MN, RH (3)	Mourant *et al.* (1976)
Nicaragua	0.222 ± 0.052 (0.012, PGM2 – 0.503, HP)	HB, ACP, ADA, ESD, GLOI, PGM1, PGM2, PGD, G6PD, CP, HP	Sauvain Dugerdil (1989); Biondi *et al.* (1988)
Panama	0.179 ± 0.050 (0.109, TF – 0.451, HP)	HB, ACP, ESD, PGD, PGM1, PGM2, ALB, HP, TF	Ferrell *et al.* (1978); Sáenz *et al.* (1993)
Caribbean Islands			
Cuba	0.433 ± 0.134 (0.087, HB – 0.735, RH (5))	ABO, MN, RH (5), HB	Hidalgo and Heredero (1985); Hidalgo *et al.* (1995); Sáenz *et al.* (1993)
Dominica	0.184 ± 0.086 (0.029, KELL – 0.430, MN)	ABO, KELL, MN, HB	Mourant *et al.* (1976); Sáenz *et al.* (1993)
Guadeloupe	0.339 ± 0.144 (0.051, TF – 0.485, HP)	ABO, HP, TF	Languillon (1951); Sauvain-Dugerdil (1989)
Haiti	0.336 ± 0.106 (0.229, CP – 0.422, ABO)	ABO, CP	Mourant *et al.* (1976)
Jamaica	0.721 ± 0.105 (0.132, HB – 0.730, HLA-B)	RH (3), HB, HLA-A, HLA-B, HLA-C, HLA-DR, HLA-DQ	Mourant *et al.* (1976); Went (1957); Went and MacIver (1958); Blank *et al.* (1995)
Martinique	0.439 ± 0.014	ABO	Languillon (1951)

Table 8.5. (*cont.*)

Region and country	Intraethnic gene diversity[a]	Genetic systems[b]	References
Puerto Rico	0.075 ± 0.010	HB	Suarez et al. (1959)
Dominican Republic	0.390 ± 0.164 (0.079, HB − 0.630, RH (4))	ABO, RH (4), HB	Sáenz et al. (1993); Alvarez (1951); Mourant et al. (1976)
South America			
Colombia	0.599 ± 0.109 (0.159, HB − 0.928, HLA-B)	ABO, RH (1), HB, HP, HLA-A, HLA-B, HLA-DR, HLA-DQ	Restrepo and Gutierrez (1968); Restrepo (1971); Monsalve et al. (1987); Blank et al. (1995); Trachtenberg et al. (1996)
Ecuador	0.251 ± 0.049 (0.006, PGM2 − 0.507, AHSG)	HB, ACP, ADA, AK, CA2, ESD, G6PD, GLOI, PGD, PGM1, PGM1 (iso), PGM2, AHSG, F13A, F13B, C6, C7, ORM, APOC2	Martinez-Labarga et al. (1999)
Surinam	0.306 ± 0.062 (0.035, KELL, DUFFY − 0.647, RH (4))	ABO, FY, KELL, MN, RH (1), RH (4), HB, G6PD, GC, HP, TF	Kahn (1936); Butts (1955); Liachowitz et al. (1958); Jonxis (1959); Nijenhuis and Gemser-Runia (1965); Peetoom et al. (1965); Smink and Prins (1965); Mourant et al. (1976); Sáenz et al. (1993)
Uruguay	0.262 ± 0.077 (0.114, ESD − 0.741, RH (5))	ABO, RH (5), ACP, AK, ESD, G6PD, PGD, PGM1, PGM2, HP, TF	Sans et al. (2000)
Venezuela	0.393 ± 0.065 (0.010, PGD − 0.929, HLA-B)	ABO, DI, FY, KELL, MN, RH (1), RH (3), RH (5), HB, ACP, AK, ESD, G6PD, PGD, PGM1, ALB, CP, HP, TF, HLA-A, HLA-B, HLA-C, HLA-DR, HLA-DQ	Tejada et al. (1965); Mourant et al. (1976); Arends et al. (1978); Arends (1984); Castro de Guerra (1991); Bortolini et al. (1992); Castro de Guerra et al. (1996); Bortolini et al. (1998)

[a] Average and range. Minimum and maximum values in parentheses.
[b] Numbers in parentheses indicate the number of sera used in the Rh studies.

Table 8.6. *Genetic diversity observed using different genetic systems within Mestizo or ethnically unclassified Latin American populations (except Brazil)*

Region and country	Intraethnic gene diversity[a]	Genetic systems[b]	References
Central America			
Costa Rica	0.240 ± 0.191 (0.049, HB − 0.431, ABO)	ABO, HB	Echandi (1953); Sáenz (1988); Madrigal (1989)
El Salvador	0.557 ± 0.146 (0.016, HB − 0.910, HLA-B)	ABO, HB, HLA-A, HLA-B, HLA-C, HLA-DR	Cabrera (1950); Fleischhacker (1959); Sáenz (1986)
Guatemala	0.319 ± 0.010	ABO	Santiago-Delpin (1991)
Honduras	0.357 ± 0.020	FY	Spencer *et al.* (1978)
México	0.399 ± 0.054 (0.006, HB, ALB − 0.937, HLA-B)	ABO, DI, FY, KELL, LE, MN, MNSs, HB, ACP, ADA, AK, ESD, G6PD, PGD, PGP, PGM1, PGM1 (iso), ALB, GC, HP, HLA-A, HLA-B, HLA-C, HLA-DR, TF, RH (1), RH (3), RH (4), RH (5)	Salazar-Mallen and Portilla (1944); Salazar-Mallen and Arteaga (1951); Lisker *et al.* (1967, 1986); Gorodezky *et al.* (1972); Mourant *et al.* (1976); Garza-Chapa *et al.* (1978); Tiburcio *et al.* (1978); Grunbaum *et al.* (1980); Ruiz Reyes (1983); Tills *et al.* (1983); Roychoudhury and Nei (1988); Santiago-Delpin (1991); Cerda-Flores *et al.* (1994); Weckmann *et al.* (1997)
Panama	0.845 ± 0.047 (0.753, HLA-C − 0.899, HLA-B)	HLA-A, HLA-B, HLA-C	Santiago-Delpin (1991)
Caribbean Islands			
Dutch Antilles	0.455 ± 0.044 (0.411, ABO − 0.499, MN)	ABO, MN	Rife (1972)
Cuba	0.412 ± 0.106 (0.031, KELL − 0.751, RH (5))	ABO, KELL, MN RH (1), RH (5), HB, GM	Soto Pradera *et al.* (1955) Colombo and Martinez (1985); Hidalgo *et al.* (1995)
Guadeloupe	0.619 ± 0.100 (0.123, HB − 0.880, HLA-B)	ABO, LE, RH (3), HB, GM, HLA-A, HLA-B	Montestruc and Ragusin (1944); Languillon (1951); Fabritius *et al.* (1980); Darlu *et al.* (1990)

Table 8.6. (*cont.*)

Region and country	Intraethnic gene diversity[a]	Genetic systems[b]	References
Haiti	0.214 ± 0.047 (0.011, AK − 0.500, MN)	ABO, DI, FY, KELL, LE, MN, ACP, AK, PGD, PGM1, PGM2, HB, CP, HP, TF, GC	Basu *et al.* (1976); Mourant *et al.* (1976); Roychoudhury and Nei (1988)
Jamaica	0.245 ± 0.111 (0.116, GC − 0.467, ABO)	ABO, HB, GC	Snyder (1929); Serjeant *et al.* (1986); Roychoudhury and Nei (1988)
Martinique	0.444 ± 0.009	ABO	Montestruc and Ragusin (1944)
Puerto Rico	0.483 ± 0.157 (0.035, TF − 0.915, HLA-B)	ABO, HB, HP, TF, HLA-A, HLA-B	Van der Sar (1959); Tills *et al.* (1983); Santiago-Delpin (1991)
Dominican Republic	0.511 ± 0.076 (0.402, RH (1) − 0.733, RH (3))	ABO, MN, RH (1), RH (3)	Alvarez (1951); Mourant *et al.* (1976)
Trinidad-Tobago	0.269 ± 0.215 (0.056, RH − 0.482, ABO)	ABO, RH (1)	Arneaud and Young (1955)
South America			
Argentina	0.399 ± 0.109 (0.093, KELL − 0.602, ABO)	ABO, RH (1), KELL, MN	Rahm (1931); Gargiulo *et al.* (1939); Palazzo and Tenconi (1939); Carneiro (1945); Lébron and Bonatti (1949); Pezzi *et al.* (1951); Sepich (1965); Chalar *et al.* (1966); Palatnik (1966); Milone *et al.* (1966); Mourant *et al.* (1976)
Chile	0.340 ± 0.058 (0.019, KELL, PGM2 − 0.642, RH)	ABO, FY, KELL, KIDD, MN, RH (undetermined number of sera), ACP, ESD, PGM1, PGM2, HP	Rahm (1931); Schepeler and von Dessauer (1945); Sandoval (1961); Sandoval and Hidalgo (1961); Alvial and Henckel (1963); Cruz-Coke *et al.* (1967); Pinto-Cisternas *et al.* (1971*b*); Henckel (1977); Acuña *et al.* (1988); Harb *et al.* (1998)
Colombia	0.467 ± 0.105 (0.043, KELL − 0.942, HLA-B)	ABO, DI, FY, KELL, MN, RH (1), HB, HLA-A, HLA-B, HLA-C, HLA-DR, HLA-DQ	Paez Perez and Freudenthal (1944); Soriano Lleras (1954); Restrepo *et al.* (1965); Fajardo and Lavalle (1966); Mourant *et al.* (1976); Caraballo *et al.* (1992)

Ecuador	0.695 ± 0.084 (0.275, ABO − 0.905, HLA-B)	ABO, HLA-A, HLA-B, HLA-C, HLA-D, HLA-DR, HLA-DQ	Hoffstetter (1949); Santiago-Delpin (1991)
French Guiana	0.420 ± 0.017	ABO	Floch and Lajudie (1948)
Paraguay	0.483 ± 0.195 (0.289, ABO − 0.676, RH (5))	ABO, RH (5)	Mourant et al. (1976)
Surinam	0.476 ± 0.010 (0.466, ABO − 0.485, MN)	ABO, MN	Collier et al. (1952); Nijenhuis and Gemser-Runia (1965)
Uruguay	0.409 ± 0.071 (0.009, PGM2 − 0.939, HLA-B)	ABO, DI, FY, KELL, MN, RH (1), RH (5), ACP, AK, ESD, G6PD, PGD, PGM1, PGM2, HP, HLA-A, HLA-B, HLA-C, HLA-DR, HLA-DQ	Mazzela (1949); Gispert et al. (1950); Soto Pradera et al. (1955); Baraibar (1961); Mourant et al. (1976); Tills et al. (1983); Santiago-Delpin (1991); Sans et al. (1995)
Venezuela	0.520 ± 0.101 (0.029, TF, HB − 0.703, HLA-B)	HB, G6PD, HP, GC, TF, GM, HLA-A, HLA-B, HLA-C, HLA-DR, HLA-DQ	Gallango (1967); Mourant et al. (1976); Boada and Yates (1979); Arends (1984); Santiago-Delpin (1991)

[a]Average and range. Minimum and maximum values in parentheses.
[b]Numbers in parentheses indicate the number of sera used in the Rh studies.

Table 8.7. *Information on protein and blood group genetic systems available for populations of different Brazilian regions*

North
European-derived: ABO, MN, RH, HB, ACP, AK, CHE1, ESD, G6PD, PGD, PGM1, PGM2, ALB, CP, HP, TF
African-derived: ABO, MN, RH, HB, ACP, AK, CA2, CHE1, CHE2, ESD, G6PD, GLOI, PGD, PGM1, PGM2, CP, HP,TF, GM
Mixed: ABO, DI, FY, MN, RH, HB, ACP, AK, CA2, CHE1, CHE2, ESD, G6PD, GLOI, PGD, PGM1, PGM2, ALB, CP, HP, TF, HLA-A, HLA-B, HLA-C, HLA-DR

Northeast
European-derived: ABO, MN, RH, HB, ACP, ABH, AK, ALADH, ALPP, CHE1, ESD, G6PD, PGD, PGM1, PGM2, PL, CP, GC, ALB, HP, TF
African-derived: ABO, RH, HB, ACP, ABH, ACONs, ALDH, ALPP, AK, CHE1, CHE2, ADH1, ADH2, ADH3, ESD, G6PD, GLOI, PGM1, PGM2, PGM3, PL, PGD, ALB, CP, HP, TF
Mixed: ABO, DI, MN, RH, HB, ABH, ACP, ADH1, ADH2, ESD, G6PD, PGD, PGM1, PGM2, CP, ALB, HP, TF

Center-West
European-derived: ABO, RH, HB, HLA-A,HLA-B
African-derived: ABO, RH, HB
Mixed: ABO, RH

Southeast
European-derived: ABO, FY, KELL, LE, MN, MNSs, P, RH, HB, ABH, ACP, ADH2, ADH3, ALDH2, CA2, CHE1, ESD, HP, GM, HLA-A, HLA-B, LAC-DEF, BAND 3 PROTEIN
African-derived: ABO, FY, LE, MN, MNSs, RH, HB, ACP, ABH, ADH2, ADH3, ALDH2, CA2, ESD, G6PD, HP, GM, HLA-A, LAC-DEF, BAND 3 PROTEIN
Mixed: ABO, KELL, MN, P, RH, HB, ABH, ACP, ADA, CA2, ESD, G6PD, GLOI, PGM1, PGD, TF, HLA-A, HLAB, HLA-DR, LAC-DEF

South
European-derived: ABO, MN, FY, RH, HB, ACP, AK, CHE1, CHE2, ESD, G6PD, PI, PGD, PGM1, PGM2, PGM3, PGM4, ALB, CP, GC, HP, TF, HLA-A, HLA-B, HLA-C, HLA-DQ, HLA-DR
African-derived: ABO, MN, RH, HB, ACP, AK, CHE1, CHE2, ESD, G6PD, GLOI, PI, PGD, PGM1, PGM2, PGM3, PGM4, ALB, CP, GC, HP, TF, HLA-A, HLA-B, HLA-C, HLA-DQ, HLA-DR
Mixed: ABO, MN, RH, GC, HLA-A

Sources: Mourant *et al.* (1976); Tills *et al.* (1983); Roychoudhury and Nei (1988), and Appendix 3. Key to the symbols in Appendix 2.

Table 8.8. *Intraethnic genetic diversity of the five Brazilian geographic regions*

Ethnic category	Region				
	North	Northeast	Center-West	Southeast	South
European-derived	0.193 ± 0.052	0.258 ± 0.054	0.556 ± 0.222	0.475 ± 0.089	0.395 ± 0.071
African-derived	0.205 ± 0.048	0.227 ± 0.054	0.151 ± 0.112	0.411 ± 0.081	0.394 ± 0.073
Mixed	0.305 ± 0.069	0.278 ± 0.045	—	0.257 ± 0.055	0.408 ± 0.073

Genetic systems considered:

North
European-derived: PGM1, PGM2, PGD, ACP, AK, TF, CP, ESD, ABO, MN, RH, HB, G6PD, HP
African-derived: PGM1, PGM2, PGD, ACP, AK, TF, CP, ESD, ABO, MN, RH, HB, G6PD, HP
Mixed: PGM1, PGM2, PGD, ACP, AK, TF, CP, ESD, ABO, MN, FY, RH, HB, G6PD, HP, HLA-A, HLA-B

Northeast
European-derived: ABO, RH, HB, G6PD, HP, PGM1, PGM2, PGD, ACP, TF, CP, ESD, MN, GC
African-derived: ABO, RH, HB, G6PD, HP, PGM1, PGM2, PGD, ACP, TF,CP, ESD, AK
Mixed: ABO, RH, HB, G6PD, HP, PGM1, PGM2, PGD, ACP, TF, CP, ESD, MN, GC

Center-West
European-derived: RH, HB, HLA-A, HLA-B
African-derived: RH, HB, ABO
Mixed: ABO, RH

Southeast
European-derived: ABO, MN, RH, FY, HB,HP, ACP, ESD, HLA-A, HLA-B
African-derived: ABO, MN, RH, FY, HB,HP, ACP, ESD, HLA-A
Mixed: ABO, RH, ACP, ESD, G6PD, KELL, PGM1, PGD, TF

Table 8.8. (cont.)

South
European-derived: ABO, RH, MN, FY, GC, HB, G6PD, HP, PGM1, PGM2, PGD, ACP, AK, TF, CP, ESD, HLA-A, HLA-B, HLA-C, HLA-DQ, HLA-DR
African-derived: ABO, RH, MN, GC, HB, G6PD, HP, PGM1, PGM2, PGD, ACP, AK, TF, CP, ESD, HLA-A, HLA-B, HLA-C, HLA-DQ, HLA-DR
Mixed: ABO, RH, GC

References. North: Montenegro (1960); Junqueira *et al.* (1962); De Lucca (1975a); Meira *et al.* (1984b); Rosa *et al.* (1984); Santos *et al.* (1987); Schneider *et al.* (1987); Guerreiro and Chautard-Freire-Maia (1988); Bortolini *et al.* (1992); Guerreiro *et al.* (1993); Ribeiro-dos-Santos *et al.* (1995); Melo-dos-Santos *et al.* (1996). *Northeast:* Torres (1930); Da Silva (1948); Novais (1953); Pedreira (1954); Lima and Walter (1955); Shansis and Carpilovsky (1956); Junqueira *et al.* (1962); Nunes Moreira (1963); Marques Ferreira *et al.* (1973); Azevêdo (1980); Azevêdo *et al.* (1980b); Franco *et al.* (1981, 1982); Oliveira *et al.* (1983); Loiola *et al.* (1986/1987); Conceição *et al.* (1987); Bortolini *et al.* (1992, 1997a,b, 1998); Pedrosa *et al.* (1995a,b). *Center–West:* Leme Lopes and Lopes da Costa (1951); Serra *et al.* (1964); Trachtenberg *et al.* (1988); Alvares Filho *et al.* (1995). *Southeast:* Bier and Machado (1933); Seiler and Bier (1935); Godoy (1937); Lacaz (1939); Lacaz *et al.* (1946); Ottensooser and Pasqualin (1946); Mesquita and Leite-Ribeiro (1947); Patiño Salazar and Mello (1948); Faria *et al.* (1951); Lacaz and Maspes (1951); Lacaz *et al.* (1955); Pinto *et al.* (1955); Saldanha (1956c); Shansis and Carpilovsky (1956); Memória and Barbosa (1957); Junqueira and Wishart (1958); Frota and de Paula (1960); Russo (1964); Vallada *et al.* (1967); Fragoso (1970); Bortolozzi and de Lucca (1971); Rocha *et al.* (1973); Itskan and Saldanha (1975); Curi and Kroll (1976); Mestriner (1976); Mourant *et al.* (1976); Oliveira (1977); Bernardo *et al.* (1982); Palatnik *et al.* (1982); Saldanha (1982); Barretto *et al.* (1983); Conti and Krieger (1985); Rodini and Bortolozzi (1986); Venturelli and Moraes (1986); Trachtenberg *et al.* (1988); Salaru and Otto (1989); Engrácia *et al.* (1990); Rosales *et al.* (1992); Alvares Filho *et al.* (1995). *South:* Osório *et al.* (1957); Mota *et al.* (1963); Salzano (1963); Salzano and Hirschfeld (1965); Salzano *et al.* (1967); Maranhão (1968); Franco *et al.* (1981); Beiguelman and Sevá-Pereira (1983); Culpi and Salzano (1984); Culpi and Corção (1984); Culpi and Lourenço (1984); Stueber-Odebrecht *et al.* (1984); Kvitko and Weimer (1988); Trachtenberg *et al.* (1988b); Bortolini *et al.* (1992, 1997b); Rainha de Souza *et al.* (1995); Dornelles *et al.* (1999); Petzl-Erler (1999).

Table 8.9. *Total genetic diversity and (in parentheses, per cent) its fraction due to interethnic differences, Brazilian regions*

| Genetic system | Region | | | | |
	North	Northeast	Center-West	Southeast	South
ABO	0.373 (0.4)	0.457 (2.9)	0.341 (1.7)	0.483 (3.2)	0.473 (0.4)
MN	0.467 (0.0)			0.495 (0.1)	
RH	0.339 (0.5)	0.416 (0.4)	0.483 (3.2)	0.437 (0.5)	
PGM1	0.293 (0.0)	0.351 (0.0)			0.449 (1.4)
PGM2	0.025 (0.4)	0.067 (6.3)			
PGD	0.026 (0.2)	0.085 (0.2)			
ACP	0.289 (0.2)	0.420 (0.0)		0.328 (0.0)	
AK	0.013 (0.3)				
TF	0.047 (1.6)	0.018 (0.5)			
CP	0.027 (0.7)	0.034 (0.7)			
ESD	0.264 (1.5)	0.188 (0.1)		0.247 (2.3)	
HB	0.021 (0.5)	0.053 (0.6)			
G6PD	0.134 (2.9)	0.303 (3.0)			
HP	0.529 (0.1)	0.520 (0.1)			
GC					0.319 (3.8)
All loci	0.202 (0.7)	0.243 (1.0)	0.412 (2.6)	0.394 (0.6)	0.415 (1.6)

Table 8.10. *Intraethnic genetic diversity of four Brazilian geographic regions for a constant set (12) of systems*[a]

Ethnic category	Region			
	North	Northeast	Southeast	South
European-derived	0.222 ± 0.056	0.268 ± 0.058	0.247 ± 0.060	0.259 ± 0.064
African-derived	0.240 ± 0.050	0.283 ± 0.055	0.275 ± 0.055	0.273 ± 0.056
Mixed	0.240 ± 0.052	0.283 ± 0.054	0.277 ± 0.054	—

[a]Genetic systems considered: ABO, MN, RH, HB, G6PD, HP, PGM1, PGD, ACP, TF, CP, ESD.

Table 8.11. *Total genetic diversity and (in parentheses, per cent) its fraction due to inter-region differences, considering a constant set (12) of systems, Brazilian populations*

Genetic systems	European-derived	African-derived	Mixed
ABO	0.447 (0.6)	0.435 (1.0)	0.435 (0.4)
MN	0.492 (0.4)	0.486 (0.7)	0.485 (0.9)
RH	0.433 (1.2)	0.376 (1.6)	0.408 (1.1)
HB	0.011 (0.6)	0.050 (0.3)	0.043 (0.3)
G6PD	0.082 (2.2)	0.335 (1.4)	0.231 (4.8)
HP	0.508 (0.4)	0.535 (0.4)	0.513 (0.0)
PGM1	0.345 (0.8)	0.329 (0.3)	0.338 (0.6)
PGD	0.053 (0.9)	0.049 (0.3)	0.075 (0.7)
ACP	0.388 (1.9)	0.352 (0.8)	0.331 (1.5)
TF	0.003 (0.2)	0.048 (1.3)	0.026 (0.1)
CP	0.016 (0.3)	0.022 (2.3)	0.035 (0.4)
ESD	0.228 (0.3)	0.196 (0.1)	0.292 (2.3)
All loci	0.251 (0.9)	0.268 (0.8)	0.267 (1.1)

Table 8.12. *Total genetic diversity and (in parentheses, per cent) its fraction due to interethnic differences, considering a constant set (12) of systems, Brazilian populations*

Genetic systems	North	Northeast	Southeast	South
ABO	0.373 (0.5)	0.457 (0.2)	0.464 (1.0)	0.473 (0.2)
MN	0.467 (0.1)	0.492 (0.6)	0.495 (0.1)	0.492 (0.0)
RH	0.339 (0.5)	0.416 (0.6)	0.437 (0.7)	0.431 (0.5)
HB	0.018 (0.7)	0.064 (1.2)	0.053 (0.6)	0.028 (0.9)
G6PD	0.130 (4.9)	0.301 (2.7)	0.303 (3.0)	0.217 (8.6)
HP	0.520 (0.2)	0.520 (1.0)	0.525 (1.7)	0.510 (1.0)
PGM1	0.293 (0.0)	0.351 (0.0)	0.349 (0.2)	0.360 (0.2)
PGD	0.026 (0.2)	0.085 (0.2)	0.064 (0.5)	0.053 (0.1)
ACP	0.289 (0.2)	0.420 (0.1)	0.328 (0.0)	0.408 (1.6)
TF	0.047 (2.5)	0.018 (0.7)	0.024 (0.8)	0.008 (0.2)
CP	0.027 (1.0)	0.034 (1.0)	0.024 (0.1)	0.001 (0.0)
ESD	0.264 (2.2)	0.188 (0.1)	0.247 (2.3)	0.234 (0.1)
All loci	0.233 (0.7)	0.279 (0.2)	0.243 (1.0)	0.268 (1.1)

Table 8.13. *Information on DNA genetic systems available for European-derived Latin American populations*

Region and country	Genetic systems	References
Continental Central America		
Costa Rica	G6PD	Beutler *et al.* (1991*b*)
South America		
Argentina	DXS52; HLA-B35	Salle *et al.* (1990); Satz *et al.* (1995)
	THO1, FABP, vWFA, FES/FPS, HPRTB, F13A1, CSF1PO, D6S366	Sala *et al.* (1998)
Brazil	Y-27H39, vWF1, vWF2, G6PD	Baronciani *et al.* (1993); Santos *et al.* (1993, 1998); Weimer *et al.* (1993, 1998); Pena *et al.* (1994); Sonati *et al.* (1994); Saad *et al.* (1997*a*,*b*)
	EST00444	Haddad and Pena (1993)
	HLA-DRB1, DQA1, DQB1, DPB1, B39	Moraes *et al.* (1993); Maertens *et al.* (1998)
	D1S80	Heidrich *et al.* (1995)
	ABO	Franco *et al.* (1995); Zago *et al.* (1996); Olsson *et al.* (1997, 1998)
	D13S71, D13S193, D13S124, FLT1, D13S121, D13S118, D13S197, D13S122, PLA2A, THO1, CSF1R, F13A1, CYP19, LPL, DM-CTG, SCA, DRPLA, HD-CAG, D20S473, D20S604, D20S481, D21S1435, D21S1446	Kimmel *et al.* (1996); Deka *et al.* (1999)
	DYS19, YAP, DYS199, DYS392, DYS393, DYS390	Rodriguez-Delfin *et al.* (1997)
	ALB	Arai *et al.* (1989); Franco *et al.* (1999*a*)
	ALDH2, ADH2	Santos *et al.* (1997)
	Prothrombin (G20210A)	Franco (1998*a*)
	MTHFR	Arruda *et al.* (1998*a*); Franco *et al.* (1998*b*)
	CBS	Franco *et al.* (1998*c*)
	GSTM1, GSTT1	Arruda *et al.* (1998*b*)
	BCHE	Souza *et al.* (1998)
	mtDNA, DYS199, SRY, YAP	Alves-Silva *et al.* (1999*a*,*b*); Pena *et al.* (2000)

	Protein C	Mendes et al. (1999)
	DRD4	Roman et al. (1999)
	HPA	Castro et al. (1999)
	Factor V Leiden	Franco et al. (1999)
	mtDNA	Alves-Silva et al. (2000)
Colombia	DYS19, DYS199, DYS271, DYS287 (YAP), DYS388, DYS390, DYS391, DYS392, DYS393, SRY-2627, 92R7, mtDNA	Carvajal-Carmona et al. (2000)
Venezuela	ALB	Weimer et al. (1999)

Table 8.14. *Information on DNA genetic systems available for African-derived Latin American populations*

Region and country	Genetic systems	References
Continental Central America		
Belize	mtDNA	Monsalve and Hagelberg (1997)
Caribbean Islands		
Bahamas	D4S139, D2S44, D1S7, D10S28, D1S80	Duncan et al. (1993, 1996)
Costa Rica	G6PD	Beutler et al. (1991b)
	mtDNA	Stine et al. (1992)
Jamaica	D4S139, D2S44, D1S7, D10S28	Duncan et al. (1993)
	APO, AT3-I/D, GC, FY-null, ICAM-1, LPL, OCA2, RB2300, Sb19.3, mtDNA, YAP	Parra et al. (1998)
Trinidad	D4S139, D2S44, D1S7, D10S28	Duncan et al. (1993)

Table 8.14. (*cont.*)

Region and country	Genetic systems	References
South America		
Brazil	HLA-DRB1, DQA1, DQB1, DPB1	Moraes *et al.* (1993)
	FIX, FVIII	Figueiredo (1994); Figueiredo *et al.* (1994c), Silva and Figueiredo (1994)
	D1S80	Heidrich *et al.* (1995)
	ABO	Franco *et al.* (1995); Zago *et al.* (1996); Olsson *et al.* (1997, 1998)
	DYS19, YAP, DYS199, DYS392, DYS393, DYS390	Rodriguez-Delfin *et al.* (1997)
	mtDNA	Bortolini *et al.* (1997a,c, 1999)
	ALB	Arai *et al.* (1989), Nielsen *et al.* (1997)
	ALDH2, ADH2	Santos *et al.* (1997)
	Prothrombin (G20210A)	Franco *et al.* (1998a)
	MTHFR	Arruda *et al.* (1998a); Franco *et al.* (1998b)
	CBS	Franco *et al.* (1998c)
	vWF1, F13A1, D1S80, D4S43, APO-B, PAH, DXS52, D4S19	Bortolini *et al.* (1998, 1999); Silva *et al.* (1999)
	vWF1, vWF2	Arpini-Sampaio *et al.* (1999)
	GSTM1, GSTT1	Arruda *et al.* (1998b)
	BCHE	Souza *et al.* (1998)
	Protein C	Mendes *et al.* (1999)
	DRD4	Roman *et al.* (1999)
	HPA	Castro *et al.* (1999)
	DM-CTG	Deka and Chakraborty (1999)
	D13S71, D13S193, D13S124, FLT1, D13S121, D13S118, D13S197, D13S122, PLA2A, THO1, CSF1R, F13A1, CYP19, LPL, DM-CTG, SCA, DRPLA, HD-CAG, D20S473, D20S604, D20S481, D21S1435, D21S1446	Kimmel *et al.* (1996); Deka *et al.* (1999)

Colombia	HLA-DRB1, DQA1, DQB1, DPA1, DPB1, IGHA1, IGHA2	Keyeux (1993); Keyeux and Bernal (1996); Jiménez et al. (1996); Trachtenberg et al. (1996)
Ecuador	HLA-DQA1	Zimmerman et al. (1995)
Uruguay	mtDNA, Yαh, YAP, DYS19	Bravi et al. (1997)
Venezuela	DXS52, DYS19, F13A1, vWF1, APO-B, D4S43, D1S80, PAH	Bortolini et al. (1998)
No country information	HLA, DRB1, DRB5, DQA1, DQB1	Fernandez-Viña et al. (1991a,b)

Table 8.15. *Information on DNA genetic systems available for Mestizo or unclassified Latin American populations*

Region and country	Genetic systems	References
Continental Central America		
México	G6PD	Beutler and Kuhl (1990a,b); Beutler (1991); Beutler et al. (1991a,b, 1992); Medina et al. (1997)
	HLA-B35, mtDNA	Stoneking et al. (1991); Fernandez-Viña et al. (1995); Satz et al. (1995); Green et al. (2000)
Panama	G6PD	Beutler (1994)
Caribbean Islands		
Cuba	G6PD	Beutler (1989); Beutler et al. (1996)
Puerto Rico	G6PD	Beutler and Kuhl (1990b)
South America		
Argentina	DXS52	Salle et al. (1990)
	D4S139, D2S44, D1S7, D5S110, D10S28, D17S26	Sala et al. (1997)
	mtDNA, Yαh, DYS199, DYS19, YAP, pSRY	Dipierri et al. (1998)
	D1S7, D2S44, DYS139, D5S110, D8S358, D10S28, D17S26, D1S80, THO1, FABP, D6S366, CSF1PO, TPOX, F13A1, FES/FPS, vWF1, MBPA/B, D16S539, D7S820, D13S317, RENA4, HPRTB, DYS385, DYS389I, DYS389II, DYS19, DYS390, DYS391, DYS392, DYS393, YCAII	Sala et al. (1999)

Table 8.15. (cont.)

Region and country	Genetic systems	References
Brazil	ADH, ALDH	Santos et al. (1995a)
	HLA-DRB, DQA1, DRB1, DQB1	Goldberg and Kalil (1989a,b,c); Moraes et al. (1997); Goldberg et al. (1998b)
	ALB	Arai et al. (1989); Kragh-Hansen et al. (1996); Nielsen et al. (1997); Franco et al. (1999)
	Yαh, DYS19, YAP	Santos et al. (1996)
	G6PD	Weimer et al. (1993, 1998); Baronciani et al. (1993); Sonati et al. (1994); Saad et al. (1997a); Marques et al. (1998)
	ACE	Miranda et al. (1998)
	CKR-5	Pareira et al. (1998)
	ASA-PD	Pedron et al. (1999)
	HPA	Chiba et al. (1998)
	FcγRIIIB	Kuwano et al. (1998)
	vWF1, APOB, D4S43, D1S80, F13A1, DYS19, DYS199, mtDNA	Rodrigues (1999); Santos et al. (1999)
Chile	G6PD	Beutler et al. (1992)
	FMR1, FRAXAC1, DXS548	Jara et al. (1998)
Colombia	IGHA1, IGHA2	Keyeux and Bernal (1996)
	HLA-A, B, C	Fleischhauer et al. (1999)
Peru	HLA-Cw	Sanz et al. (1999)
No country information	KELL	Lee et al. (1997)

Table 8.16. *Gene diversity analysis considering the D1S80 locus in different world populations*

Populations	No. of populations (No. of chromosomes)	Gene diversity			Gene flow N_{mFM}[c]	References
		Total	Intrapopulational	Interpopulational absolute[a] (relative in %)[b]		
Amerindians	16 (539)	0.790	0.700	0.096 (12.1)	1.8	1–3
Europeans	10 (2153)	0.797	0.792	0.005 (0.6)	41.4	4–12
Sub-Saharan Africans	2 (68)	0.872	0.857	0.015 (1.7)	14.4	13
Neo-Latin Americans	14 (619)	0.877	0.850	0.029 (3.3)	7.2	2,13–15
African-derived	11 (455)	0.884	0.859	0.027 (3.0)	8.1	1,2,13–15
European-derived	2 (125)	0.805	0.800	0.005 (0.5)	49.7	2,14
All	42 (3379)	0.859	0.780	0.081 (9.4)	2.4	—

[a] Dm, Average minimum genetic distance among populations (Nei, 1973).

[b] G_{ST}: coefficient of gene differentiation (Nei, 1986).

[c] N_{mFM}, female/male gene flow/generation (Slatkin, 1995). In the present case it was estimated as $1/4\,[(1/G_{ST}) - 1]$.

References: 1. Duncan *et al.* (1996); 2. Zago *et al.* (1996); 3. Hutz *et al.* (1997); 4. Sajantila *et al.* (1992); 5. Schnee-Greise *et al.* (1993); 6. Gené *et al.* (1995); 7. Kádasi *et al.* (1994); 8. Deka *et al.* (1995); 9. Pinheiro *et al.* (1996); 10. Rosé *et al.* (1996); 11. Lorente *et al.* (1997); 12. Katsuyama *et al.* (1998); 13. Silva *et al.* (1999); 14. Heidrich *et al.* (1995); 15. Bortolini *et al.* (1998).

Table 8.17. *Mitochondrial DNA studies performed in Latin American non-Indian populations*

Ethnic category and country	Methods of study			Sequencing		References
	RFLP	No. of sites	SSO	HVS-I	HVS-II	
European-derived						
Colombia	Yes	4	No	No	No	Carvajal-Carmona et al. (2000)
Brazil	Yes	29	No	Yes	Yes	Alves-Silva et al. (2000)
African-derived						
Jamaica	Yes	6	No	No	No	Stine et al. (1992)
Belize	No	—	No	Yes	No	Monsalve and Hagelberg (1997)
Brazil	Yes	3	No	Yes	No	Bortolini et al. (1997a,c, 1999)
Uruguay	Yes	19	No	No	No	Bravi et al. (1997)
Mestizo or unclassified						
México	No	—	Yes	Yes	Yes	Stoneking et al. (1991)
	Yes	8	No	Yes	No	Green et al. (2000)
Brazil	Yes	6	No	No	No	Santos et al. (1999)
Argentina	Yes	5	No	No	No	Dipierri et al. (1998)

SSO, Sequence-specific oligonucleotide probes; RFLP, restriction fragment length polymorphisms.

Table 8.18. *Interpopulation variability (%) considering the 10 most variable specific HVS-I mtDNA sites in four ethnic categories*

Site	Africans	Europeans	Amerindians	African-derived Latin Americans
16051				37
111			32	
126		9		
129	19			20
144		8		
162		7		
166				21
183	12			
187	14			31
189		8		31
193		9		20
209	15		31	
215	11			
217			40	
218	14			
219	17	11		
223	17		33	
235		7		
239				20
257		7		
264				21
278		10		
291		7		
292			74	
298			46	
300	11			
309				36
311				36
325			38	
327			40	
357			33	
362	12		40	

Sources and populations, Africans: Watson *et al*. (1997): West Africans, Mandenka (MAN), Songhai (SON), Yoruba (YOR), Hausa (HAU), Fulbe (FUL), Kanuri (KAN). Mateu *et al*. (1997): Bantu-speakers, Bubi (BUB). Pinto *et al*. (1996): Bantu-speakers, Fang (FAN). Europeans: Richards *et al*. (1996): Finland (FIN), Germany (GER), Bavaria (BAV), Denmark (DEN), Iceland (ICE), Wales/UK (WAL), Cornwall/UK (COR), Sardinia (SAR), Northern Spain (SPA), Viscaya/Spain (VIS), Portugal (POR), Caucasian/UK (UK). Amerindians: Bortolini *et al*. (1998): Arara (ARA), Yanomama (YAN), Wayampi (WAY), Kayapo (KAY), Xavante (XAV), Zoró (ZOR), Gavião (GAV), Mapuche (MAP), Wai-Wai (WAI), Suruí (SUR). African-derived Latin Americans: Bortolini *et al*. (1997*a*): Porto Alegre (POA), Salvador (SAL), Paredão (PAR). Monsalve and Hagelberg (1997): Black Carib, Belize (CAR).

Table 8.19. *Gene diversity analysis using the DYS19 data set in four ethnic categories*

Population	No. of populations (No. of chromosomes)	Gene diversity			Gene flow N_{mM}[c]	Variance among populations[d]		References
		Total	Intrapopulational	Interpopulational absolute[a] (relative in %)[b]		Absolute variance	Relative variance (ϕ_{st}, in %)	
Amerindians	15 (256)	0.389	0.268	0.130 (32.6)	2.7	0.055	25.0**	1–4
Europeans	35 (2972)	0.659	0.611	0.049 (7.4)	12.5	0.011	3.2**	3,5
Sub-Saharan Africans	20 (344)	0.660	0.581	0.083 (12.5)	7.0	0.016	4.3**	5
Neo-Latin Americans								
African-derived	9 (208)	0.758	0.647	0.125 (16.2)	5.2	0.011	3.0*	1,7
European-derived	3 (393)	0.627	0.621	0.009 (1.4)	70.4	− 0.0004	− 0.001	1,3,6
Neo-Brazilians	8 (470)	0.736	0.628	0.123 (16.4)	5.1	0.029	8.3***	1,6,7
Neo-Venezuelans	4 (45)	0.719	0.673	0.061 (8.3)	11.0	− 0.002	− 0.008	7
All	14 (669)	0.742	0.642	0.108 (14.4)	6.9	0.024	6.9**	1,3,6,7
Grand total	84 (4241)	0.750	0.548	0.204 (26.0)	2.8	0.037	10.6**	—

[a] Dm, average minimum genetic distance among populations (Nei, 1973).

[b] G_{ST}', coefficient of gene differentiation (Nei, 1986).

[c] N_{mM}, male gene flow/generation (Slatkin, 1995). In the present case it was estimated as $[(1/G_{ST}') - 1]$.

[d] The significance levels of the statistics were obtained by comparison of the actual values with the distribution of 100 values obtained at random. $*P < 0.05$; $**P < 0.01$; $***P < 0.001$. Negative values suggest that the population divisions explain less variance than random permutations of the pooled samples (Excoffier and Slatkin, 1995).

References: 1. Zago *et al.* (1996); 2. Ruiz-Linares *et al.* (1996); 3. Krijff *et al.* (1997); 4. Vallinoto *et al.* (1999); 5. Hammer *et al.* (1997); 6. Santos *et al.* (1995b); 7. Bortolini *et al.* (1998).

Table 8.20. *Comparison between estimates of total and interpopulation variability obtained using protein, blood group and DNA markers among African-derived South Americans*

Country (Population)	Gene diversity		Genetic system	No. loci	References
	Total (absolute)	Interpopulation (%)			
Brazil (POA, SAL)	0.209	0.9	Protein and blood group	10	Bortolini *et al.* (1997*b,c*)
Brazil (CAM, TRO, PAR, CAJ)	0.223	5.2	Protein and blood group	11	Bortolini *et al.* (1995*b*, 1998)
Brazil (CAM, TRO, PAR, CAJ)	0.810	5.6	VNTR/STR	8	Bortolini *et al.* (1998)
Brazil (POA, SAL, PAR)	0.129	19.0	mtDNA (HVS-I)	1	Bortolini *et al.* (1997*a,c*)
Venezuela (GAG, PAT)	0.499	1.6	Protein and blood group	6	Bortolini *et al.* (1995*a*)
Venezuela (CUR, BIR, SOT, PAN)	0.215	2.8	Protein and blood group	11	Bortolini *et al.* (1998)
Venezuela (CUR, BIR, SOT, PAN)	0.785	3.7	VNTR/STR	8	Bortolini *et al.* (1998)

Key to populations. Brazil (state of the federation), rural samples: CAM, Cametá (Pará); PAR, Paredão (Rio Grande do Sul); TRO, Trombetas (Pará); CAJ, Cajueiro (Maranhão). Urban samples: POA, Porto Alegre (Rio Grande do Sul); SAL, Salvador (Bahia). Venezuela (state of the federation), rural samples: GAG, Ganga (Miranda); PAT, Patanemo (Carabobo); CUR, Curiepe (Miranda); BIR, Birongo (Miranda); SOT, Sotillo (Miranda); PAN, Panaquire (Miranda).

Table 8.21. *Glucose-6-phosphate variants observed in Latin America*

Variant	Molecular substitution	WHO class	Ethnic classification	Country	References
Lages	14 Gly → Arg	4	E	Brazil	1
A-(Distrito Federal, Castilla, Tepic)	68 Val → Met, 126 Asn → Asp	3	A, E, M	Brazil, Costa Rica, México, Puerto Rico	2
São Borja	113 Asp → Asn	4	E	Brazil	1
Santa Maria	181 Asp → Val, 126 Asn → Asp	2	U, E	Brazil, Costa Rica	3
Mediterranean (Panama)	188 Ser → Phe	2	E, M	Brazil, México, Panama	2, 4
Santiago	198 Arg → Pro	1	U	Chile	5
Mexico City	227 Arg → Glu	3	M	México	6
Corur	274 Glu → Lys	1	E	Brazil	7
Wexham	278 Ser → Phe	1	E	Brazil	7
Seattle	282 Asp → His	3	E, M	Brazil, México	1, 6
A-(Unnamed, Guantanamo)	323 Leu → Pro, 126 Asn → Asp	3	A, M	Cuba, México	6, 8
Farroupilha	326 Pro → His	4	E	Brazil	1
Chatham	335 Ala → Thr	3	M	Brazil	2
Guadalajara	387 Arg → Cys	1	M	México	6
Anaheim	393 Arg → His	1	E	Brazil	1
Puerto Limon	398 Glu → Lys	1	E	Costa Rica	3
Sumaré	431 Val → Gly	1	E	Brazil	9
Santiago de Cuba	447 Gly → Arg	1	U	Cuba	10
Campinas	488 Gly → Val	1	E	Brazil	11

WHO class: 1, nonspherocytic hemolytic anemia; 2, severe deficiency; 3, moderate deficiency; 4, no deficiency.

Ethnic classification: A, African-derived; E, European-derived; M, mixed ancestry; U, unknown.

References: 1. Weimer et al. (1993, 1998); 2. Beutler and Kuhl (1990a,b), Beutler et al. (1991a,b), Saad et al. (1997a); 3. Beutler et al. (1991b), Marques et al. (1998); 4. E. Beutler and W. Kuhl (unpublished, 1991); 5. Beutler et al. (1992); 6. Medina et al. (1997); 7. Sonati et al. (1994); 8. P.J. Mason, M. Estrada, C. Perázzio et al. (1995); 10. Beutler and Kuhl (1995); 11. Perázzio et al. (1993).

Table 8.22. *Albumin variants detected among Latin American populations*

Variant	Synonyms	Molecular substitution	Ethnic classification	Geographical distribution
Yanomama 2	—	114 Arg → Gly	Am	Restricted to the Indians of Brazil and Venezuela
Vera Cruz	Tradate 2	225 Lys → Gln	E	Brazil, Italy
Tovar	Reading, Tagliacozzo, Cooperstown, New Guinea, Canterbury	313 Lys → Asn	E	Venezuela, Sweden, Ireland, England, Italy, USA, New Guinea, New Zealand
Porto Alegre 1	Coari 1, Belém 4	358 Glu → Lys	E	Brazil, Uruguay
Passo Fundo	Nagasaki 2	375 Asp → Asn	E, As	Brazil, Japan
Porto Alegre 2	Coari 2, Manaus 1, Fortaleza, Vancouver, Birmingham, Adana	501 Glu → Lys	Af/E, As	Brazil, India, Turkey
Maku	Makiritare 2, Wapishana, Mura 1, Belém 3, Belém 5, Oriximiná 1	541 Lys → Glu	Am	Brazil, Venezuela
México	Belém 2	550 Asp → Gly	Am	Brazil, México, USA
B	—	570 Glu → Lys	E	Venezuela

Ethnic classification: Af, African-derived; Am, Amerindian; As, Asiatic; E, European-derived.
Source: Franco *et al.* (1999); Weimer *et al.* (1999).

9 Gene dynamics

> The human species, according to the best theory I can form of it, is
> composed of two distinct races, the men who borrow, and the men who
> lend   *Charles Lamb, quoted by Jonathan Marks*

Variability is ubiquitous, but explanations difficult

The fantastic development of genetic techniques has disclosed a previously
unimagined amount of variability in humans, as well as in all the organic
world generally. Scientific attempts to understand this variation began
with the seminal work of Darwin (1859), but the harmonious fusion of
genetic and evolutionary concepts occurred in the first half of the twentieth
century only, with the empirical contributions of Dobzhansky (1937),
Mayr (1942), Simpson (1944, 1949) and Stebbins (1950), while the mathematical foundations of what had been called the synthetic theory were
established by Fisher (1930), Wright (1930, 1931), and Haldane (1932).
According to these scholars, whom Mayr (1980) called 'bridge builders'
due to their ability to cross disciplinary boundaries, the basic facts of
evolution are now known. Mutations (including any type of change in the
genetic material) provide the variability that will be tested by natural
selection. The latter establishes which part of this variation should be
maintained or favored, and which part eliminated.

Besides these primary factors, several others should be considered, the
most important being related to the constraints developed by population
structure (size, subdivisions, mobility). In humans cultural influences cannot be overemphasized, since key technologic improvements may be of
paramount importance for the fate of a given community or society.

Having said that, it is important to emphasize that in humans the
specific roles of each of these factors can seldom be established in specific
situations. The action of natural selection in the elimination of deleterious
variants (for instance, hereditary diseases) is of course well known. But
only one clearly established instance of positive selection is recognized in
our species, that related to the hemoglobinopathies and malaria. Presently
a series of evolutionary inferences can be drawn related to the type
(whether changes occurred in synonymous or nonsynonymous DNA sites)
or level (less variation than that expected, suggesting past 'bottleneck'

301

effects) of the genetic variability. But they are just that, inferences.

Taking for granted that biologic and cultural factors certainly should be different in populations isolated for large periods of time, it is not surprising that people living in different continents should develop diverse genetic pools. The question that remains is how, quantitatively, these interpopulation differences relate to the intrapopulation, interindividual variability.

Ethnic and 'race' classifications

Inhabitants of distinct continents can be usually distinguished morphologically. This fact was variously considered in ancient times. Karl von Linné's (1707–78) classification had already separated Amerindians, Europeans, Asians and Africans, and it was one of his contemporaries, George Louis Leclerc de Buffon (1707–88) who first applied the term race to different varieties of *Homo sapiens* (Comas, 1966).

The problem is that race is also a social concept, and biologically different people may have different cultures; also, a culturally defined unit is often called an ethnic group (from the Greek *ethnos*, meaning nation). An ethnic group can be a nation, or groups defined by language, social customs or religion. Since the term is intentionally vague, Ashley Montagu (see, for instance, Montagu, 1960) was one of the first scholars to suggest that the word race should be substituted for ethnic group. This suggestion has been followed by us throughout this book.

Independently of names, the question remains: are human 'races' (biologically defined groups) real entities or simple social constructs? This question cannot be considered at length here, but it is sufficient to mention that (a) most of the genetic variability (around 85%) occurs at the intrapopulation level. Therefore, only 15% can be ascribed to interethnic differences; but (b) this fact does not imply that there are no specific clear biologic differences among them. Given a sufficient number of genetic markers, it is relatively easy to ascertain if a person has European, African or Asian ancestry (details in Salzano, 1997).

Latin American ethnic categories

In certain societies people classify themselves or are classified by others in terms of morphologic features, such as skin color, type and color of the hair, nose and lip characteristics. In others the descent rule prevails; if a person has, for instance, an ancestor, no matter how remote, who came from Africa, he/she will be classified as a Negro independently of physical features.

In communities where interethnic marriages would be rigidly controlled through customs or laws, and segregation of residence and other aspects of social life would be enforced, such classifications could be relatively easy to apply. This situation was generally not true in Latin America, resulting in complex classifications. An example would be the 16 categories developed in eighteenth century Spanish America that included *castizo, morisco, cambujo* and *ahí te estás*, none, of course, socially or legally functional (Wagley, 1971).

In Brazil, traditionally, the census officers used four categories related to skin color (*branco, pardo, preto, amarelo*, or white, brown, black, yellow); but when the National Research on Sampling by Domicile tried to establish an open question for self-classification, not less than 190 different denominations appeared (PNAD, 1976). Besides this multiplicity of designations there is also the question of variation between interviewer evaluations, as compared with self-characterization. This problem was considered by Silva (1994), based on data collected in a sample of 573 individuals from São Paulo, Brazil (Table 9.1). Some extreme differences, such as the evaluation as white of five people who classified themselves as black and vice versa could be due to errors of understanding or of codification of the answers. But there are other discrepancies worth noting. For instance, there was only 79% of agreement in the classification of whites, and the agreement between the mixed and black categories was even lower (69% and 48%, respectively). Silva (1994), using data on the socioeconomic status of these individuals, obtained evidence that part of these discrepancies is probably due to this factor, that is 'money whitens the skin'. This, however, would occur only among individuals not much different in the continuum of skin color distribution since, as was stressed by Wagley (1952) no Brazilian would be so fool as to designate a black person, even if rich, as white.

This problem of ethnic definition is not restricted to Latin America. Goldstein and Morning (2000) estimated that the decision to allow respondents in the 2000 USA census to mark one or more 'races', instead of the traditional single-'race' approach, may lead to complications in the implementation of 'race'-based policies. They estimated that 3.1–6.6% of the US population is likely to mark multiple 'races'.

Taking into consideration all these aspects, we decided in the present work to group as African-derived, European-derived or Mestizo populations those that were classified as such by the investigators or by themselves. By acting in such a noncomital way we are not endorsing any type of ethnic classification, although we believe that ignoring them would lead to the loss of an important aid in the understanding of the full spectrum of human variation.

Racism and racial paradises

Oliveira (see Oliveira and Castro Faria, 1971) has developed a concept that, although it was based on Indian/white relationships, has a wider application. According to him, an *interethnic system* is formed by the relations between two populations dialectically joined by diametrically opposed, though interdependent interests. Thus, in contact situations involving European-derived and African-derived persons, there are patterns of dominance relationships that date back to slavery times; they generate opposite goals for the two groups, although economically they may depend on each other.

Racism, the assumption of inherent biologic superiority of a certain group in relation to others, and the consequent discrimination against them, is a widespread phenomenon that up to now has defied all attempts at eradication. Complex sociologic and socioeconomic factors are certainly involved in its maintenance, and a selected bibliography about these factors in relation to Latin America is presented in Table 9.2. They cover a period of 36 years and provide a useful overview of this question. Additional references can be found in Salzano and Freire-Maia (1970) and Salzano (1997). Brazil was for a long time considered a 'racial paradise', but this concept was challenged in several of the studies listed in Table 9.2. The sheer existence of a Brazilian law (enacted in 1951) to punish discrimination acts indicates that such acts are being practiced. Presently infractors of this law are not protected by bail.

Assortative mating

Mating choice is a complex behavioural characteristic, which involves psychologic, cultural, socioeconomic and biologic variables (Trachtenberg *et al.*, 1985). The concept of beauty varies between cultures; Jones (1996), for instance, quantified clear differences between Brazilians, Americans and Russians in relation to standards of attractiveness (see Chapter 5). It is not surprising, therefore, that mating does not occur at random in relation to ethnicity, since individuals from different ethnic groups may differ not only in physical traits, but also in social class and customs.

Data on this question are presented in Table 9.3, which gives information about 15 555 Brazilian interethnic marriages. As can be seen, marriage is far from random when subjects are classified according to skin color and related characteristics. The diagonal in the table lists the homogamic matings; while only 6980 would be expected if marriages were independent of these traits, actually 12 294 were observed, that is 1.76 times more. There is also a pronounced asymmetry in relation to the matings listed above or below the diagonal. In the first case they are represented by unions in which the wife is darker than the husband, while in the second case the opposite is true (wife lighter). There are clearly many more cases of this latter type (1875) than of the former (1386). Therefore, the direction of the gene flow depends also on the sex of the persons involved.

Historical genetics: admixture quantification

Bernstein (1931) and Ottensooser (1944) were the first to use allele frequency data in admixed and parental populations to estimate the accumulated proportional contributions of the ancestral groups to that admixed population. The derived formula is based in a model of simple dilution and can be used when just a system (or a particularly informative one) is available and there are just two progenitors. Some years later Ottensooser (1962) provided an extension that could deal with three parental stocks.

From this simple start many other researchers have developed methods which vary in precision and complexity. All of them depend on two basic assumptions: (a) there is no error in the choice of parental groups, or in their gene frequencies; and (b) no changes in allele frequencies are occurring independently of the gene flow.

Three much used methodologies are those of maximum likelihood (Krieger *et al.*, 1965), weighted least squares (Roberts and Hiorns, 1962, 1965; Elston, 1971), and gene identity (Chakraborty, 1975). The method of weighted averages, developed by Cavalli-Sforza and Bodmer (1971), has the limitation that it cannot be used for mutiallelic loci (Long and Smouse, 1983). An important addition to these evaluations was provided by the latter, as well as by Long (1991), and Long *et al.* (1991), through the multiple regression coefficient (R^2), which estimates the proportion of the allele frequencies that can be explained by the admixture event *per se*. Chakraborty (1985) also changed the gene identity method to allow adjustment of the model in terms of R^2.

All these methods use allele frequencies that would be present in the parental populations, just weighting them to provide the estimates. A different approach can be employed taking into consideration markers that are distinctive of a given ethnic group. In this case either direct counts in the admixed population, or the method of Szathmáry and Reed (1978) should be used.

In an extensive review about this subject Chakraborty (1986) stressed the difficulties in reaching a consensus about the best method to be used, but he suggested that methods incorporating the R^2 parameter should be preferred. In the present work we followed his suggestion, using the gene identity method developed by him (Chakraborty, 1975, 1985), and the weighted least squares method of Long and Smouse (1983), Long (1991), and Long *et al.* (1991).

Admixture studies are also important to detect linkage disequilibrium in linkage studies (see, for instance, McKeigue, 1998; Lautenberger *et al.*, 2000), but this approach will not be discussed here.

Parental contributions to Latin American populations

Previous studies

Tables 9.4–9.6 present the previous estimates obtained about the proportion of European, African and Amerindian genes present respectively in populations classified as European-derived, African-derived or Mestizo (in the latter case combined with those unclassified in relation to ethnic categories).

Let us first consider the groups classified as European-derived (Table 9.4). Results concerning four countries are available, and in relation to Brazil they were subdivided by region. They have been obtained using five different methods, and sample sizes from the 18 urban and rural communities of four countries vary widely. Be that as it may, the results obtained in Cuba, Venezuela and Uruguay basically confirm the previous phenotypic classification of the subjects. This is not true for the Brazilian samples, however, in which a considerable degree of non-European heritage has been estimated. In the northeastern town of Natal subjects classified as European-derived could have as much as 20% of African and 13% of Amerindian ancestry (Franco *et al.*, 1982).

Data concerning Afro-Americans living on 10 countries, from which 12 urban and 30 rural samples were derived, are presented in Table 9.5. The admixture estimates were based on six different methods, and sample sizes varied from 18 to 3429. In Jamaica and one Carib community from Belize the genetic estimates agreed with the morphologic classification, indicating

80% to 93% of African parentage. In the other populations, however, this contribution was estimated at much lower levels, in several instances lower than 50%. It is difficult to explain these cases. Sampling effects may be involved, or problems in the application of the admixture programs. Independently of these cases, what is clear is the generalized presence of European and Amerindian components in these groups.

Table 9.6 lists the information on Mestizo or unclassified populations. It derives from 68 urban and nine rural samples originated from nine countries. Sample sizes vary, sometimes reaching considerable numbers. In these cases, however, just one or two markers were used (generally the ABO and Rh systems), and therefore the estimates are subjected to considerable error. México and Brazil were more extensively studied, but geographic trends are not apparent. As expected, the major contribution is of Amerindian origin in México, but of European origin in Brazil. The Bolivian sample (from La Paz) displays, as the Mexican ones, a high Amerindian component (83%). In the samples from the other countries the relative contributions of Europeans, Africans and Amerindians are variously distributed, without clear patterns.

New estimates

With the data surveyed in Chapters 7 and 8 it was possible to obtain a large number of new assessments about the interethnic admixture in Latin America. The advantage of this investigation is that two methods of assessment were uniformly employed in all samples, thus providing an internal control. Standard errors were also calculated, therefore indicating the reliability of the figures given; while with both methods it was possible to calculate R^2 values, which provided indications about the applicability of the admixture assumptions. On the negative side, since the genetic information available for each country was very uneven, the weight to be given to the different estimates also varies. Frequencies for the parental populations were based on our extensive data bank and will be supplied on request.

The values obtained for Latin America outside Brazil are listed in Tables 9.7–9.9. Those related to European-derived groups, shown in Table 9.7, should be compared with the early values given in Table 9.4. The numbers now presented for Cuba basically confirm the previous estimates, furnishing, however, values for the Amerindian contribution (10–14%), which had not been calculated before. For Venezuela, the only previous investigation (Weimer *et al.*, 1999) was restricted to an isolate of German ancestry. The new values (0–16% of African, 12–22% of Amerindian parentage) are certainly more representative of Venezuela's general population. As for Uruguay, the present numbers suggest more admixture, but since the early results were based on a vast array of comparisons they are probably better. There is a clear distinction between Montevideo and Tacuarembó, the latter showing a higher degree of non-European ancestry.

For Puerto Rico, St Barthélemy, Costa Rica, Ecuador, Peru and Argentina these are the first estimates for samples classified as being of European descent. They indicate variable amounts of African and Amerindian components, but should be viewed with caution, since they are based on few markers.

The data for African-derived populations are much more extensive. (Tables 9.5 and 9.8). Previously samples from nine countries (outside Brazil) had been evaluated, and of these seven have been considered here also. The results from Cuba, Belize and Venezuela mainly confirmed the previous estimates, but those from Guatemala and Nicaragua suggest more European influence, those from Jamaica more admixture in general, and those from Ecuador a diminished Amerindian component. For nine countries these are the first estimates. Particularly high African heritages were found for Haiti, Panama and Surinam.

Tables 9.6 and 9.9 present respectively the previous and new data on samples classified as Mestizo or unclassified in relation to ethnic category. Eight countries (besides Brazil) had been included in previous studies, and we had made reevaluations for seven of them. Basic agreement between the two sets was found for México, Uruguay and Chile, but this was not true for the remainder. Less Amerindian heritage was found for Venezuela and Argentina, less European and more African for Peru, and more European, with less African and Amerindian components for Colombia. Results concerning 18 countries are presented here for the first time. Predominantly European ancestry was found for samples from the Dominican Republic, Puerto Rico, El Salvador, Costa Rica, Panamá and French Guiana; predominantly African from those of Haiti, Jamaica, Trinidad and Surinam; important European and African components in those from Cuba, Guadeloupe, Martinique, Netherland Antilles and Honduras; important European and Amerindian fractions in Guatemala and Paraguay; while the results from Ecuador yielded too many differences between the two methods of evaluation, so that no firm conclusion could be made.

The Brazilian new data are presented in Tables 9.10–9.12. For the samples classified as being European derived (Table 9.4 for previous and 9.10 for present estimates) the two groups of values agree reasonably well for the northeast and south regions. The present results, however, suggest more Amerindian admixture in the north region than the previous estimates. For the center-west and southeast regions these are the first estimates for samples of this classification. As expected, the Amerindian component is higher in the center-west. In general, the genetic values agree with the morphologic classification, with somewhat less European contribution in the north. However, suggestions of admixture are universal.

The agreement between the morphologic and genetic data for the African-derived Brazilian groups (Tables 9.5 and 9.11) are less marked than in the previous category, the two sets indicating also, in all regions, considerable non-African contributions.

Results concerning the admixed or unclassified in relation to ethnic category samples (Tables 9.6 and 9.12) show basic agreement between the two sets of values. For the present evaluation, in the south the weighted least squares method provides a better agreement with the numbers expected than does the gene identity calculation.

Asymmetries due to sex

In the previous chapter gene flow estimates using an autosome marker (D1S80, Table 8.16) and a Y chromosome polymorphism (DYS19, Table 8.19) were obtained. Similar values can be calculated for mtDNA using the data discussed in the text there. In this way, since D1S80 is a biparental marker, while mtDNA and DYS19 have respectively exclusive matrilineal and patrilineal descent, it is possible to verify whether the interpopulation relationships occur differentially when the variable of gender is taken into consideration.

Table 9.13 presents the results obtained. Theoretically, the numbers obtained with the autosome polymorphism divided by two should be equal to the sum of the isolated female and male estimates. This does not happen because the estimates were obtained with data derived from different samples. Therefore, only relative degrees of magnitude should be considered, rather than the strict absolute numbers. The interethnic differences observed have already been considered in Chapter 8. Here we shall examine the sex differences. As is shown in the table, there is a dichotomy in this regard; while Europeans and Africans show around 50% more female than male gene flow, in Amerindians and African-derived Latin Americans the contribution of females and males is about even. More striking differences between female and male migration rates were obtained by Seielstad *et al.* (1998) at the worldwide level. They found that the Y chromosome-derived value they calculated was about eight times higher than those obtained using mtDNA or autosome data. Results similar to those obtained here were found, however, by Mesa *et al.* (2000) in Amerindians.

Another question refers to the differential contribution of males and females to the gene pool of ethnically admixed populations. The data available refer mainly to African-derived Latin Americans, and they are displayed in Table 9.14. A total of 12 populations, both urban and rural and distributed through three South American countries could be assembled. Despite the fact that the admixture estimates based on autosomes varied considerably among them, a consistent pattern was found (with five exceptions in the 24 comparisons). The European component was mostly contributed by males, while the opposite was true for the Amerindian fraction (mainly contributed by females). The African heritage, however, was furnished in about equal terms by males and females. Carvajal-Carmona *et al.* (2000) found a similar distribution in a Mestizo sample from Medellín, Colombia.

Coda

Human biologic variation is *structured*. Many factors which influence the biologic fate of other organisms are also important for humans, such as population size, mobility, or patterns of fertility and mortality. In addition, however, cultural agents play a decisive role in the patterning of such variability. Social constructs may be important for the self-characterization of a group and the way in which its members are accepted or rejected as friends by neighbors. The dialectics of interindividual and interpopulation interactions is complex; despite stereotypes and discriminatory policies, the picture that emerges from Latin America is one of universal gene flow. Its rate varies from place to place, and in a given community with the circumstances of daily contact, but it is there. Fortunately, genetic techniques and concepts can now be used to quantify the amount of interchange or isolation that occurred in the past, therefore contributing to the elucidation of the history of the study population. Inferences can also be made about mating choices and their consequences. The asymmetry in relation to sex observed for the introduction of genes characteristic of certain ethnic groups clearly reflect patterns of social dominance or submission. In colonial times the European-derived males would introduce genes of their ethnic group into their African-derived slave females. More recently, the association between socioeconomic status and morphologic traits indicative of ethnicity may also be involved in the differential introduction of genes in interethnic marriages. It is also important to emphasize that the measures adopted to prevent discrimination are leading to increased possibilities of mating choices, and changes regarding the preferred partner. With the more uniform distribution of genes eventually connected with ethnicity it is possible to visualize good prospects for the solution of the critical problems of interethnic relations.

Table 9.1. *Ethnic classification (as assessed by skin color): comparison between interviewer assessment and the self-classification made by the subject of the investigation, São Paulo, 1986*

Interviewer classification		Self-classification					
		White	Brown	Black	Yellow	Others or no data	Total
White	n	322	73	5	3	5	408
	%	78.9	17.9	1.2	0.8	1.2	100
Brown	n	14	85	19	0	6	124
	%	11.3	68.6	15.3	0.0	4.8	100
Black	n	5	8	13	0	1	27
	%	18.5	29.6	48.2	0.0	3.7	100
Yellow	n	0	0	0	14	0	14
	%	0.0	0.0	0.0	100.0	0.0	100.0
Total	n	341	166	37	17	12	573
	%	59.5	29.0	6.4	3.0	2.1	100.0

Source: Silva (1994).

Table 9.2. *Selected bibliography about sociocultural aspects of the interethnic relations in Latin America*

Author	Characteristics of the work
Freyre (1930)	Trained by Franz Boas, Freyre was one of the first intellectuals to separate biology and culture in the examination of the socioeconomic condition of African-derived Brazilians. He praised Brazil's 'racial democracy'.
Wagley (1952); Bastide and Fernandes (1959); Fernandes (1965); Bastide (1973)	These UNESCO-sponsored studies challenged the myth of Brazilian 'racial democracy' by documenting the ethnic discrimination in this country, demonstrating also a strong correlation between blackness and poverty.
Hoetink (1973)	Considered interethnic relations in slavery and post-slavery times in Curaçao and Surinam
Solaun and Kronus (1973)	Discrimination is not always accompanied by interethnic violence.
Tolentino (1974)	Questions the efforts to relate as a single entity New World and Old World slavery and prejudices.
Toplin (1974)	Specialized essays within strict regional boundaries. Emphasized the brutal aspects of slavery in Latin America.
Lavrin (1989)	Examined the politics of reproduction adopted by different Latin American countries in Colonial times, beginning with the *Real Pragmática* of Carlos III, which in 1778 established norms for marriages in the Colonies.
Stepan (1991)	Considered the eugenics movement in Latin America in the period 1900–40, with special emphasis in Argentina, Brazil, and México. Clear regional variation could be discerned.
Hellwig (1992)	Presented 27 contributions by African American scholars, analyzing the US black perception of Brazilian interethnic relations, which evolved from enthusiastic endorsement to outright repudiation.
Schwarcz (1993)	Analyzed the interethnic questions in Brazil in the period 1870–1930, especially in relation to scientists and institutions.
Skidmore (1993)	Examined interethnic relationships in Brazil criticizing the 'racial democracy' concept.
Wade (1993)	Considered 'racial' identity in Colombia, drawing an analytic distinction between 'race' and class.
Maio and Santos (1996)	Investigated the relationships between 'race', science and society in Brazil from the nineteenth to the present century in a multidisciplinary way.

Table 9.3. *Interethnic marriages (as assessed by skin color) in a large sample of the Brazilian population*

			Wife's color				
Husband's color			White	Brown	Black	Yellow	Total
White	*n*	Obs.	6361	1092	92	3	7548
		Exp.	3844	3312	365	27	
Brown	*n*	Obs.	1421	5425	193	9	7048
		Exp.	3589	3093	341	25	
Black	*n*	Obs.	123	304	466	1	894
		Exp.	456	392	43	3	
Yellow	*n*	Obs.	17	5	1	42	65
		Exp.	33	29	3	0	
Total	*n*		7922	6826	752	55	15 555

Source: Silva (1987), based on the National Research on Sampling by Domicile (PNAD, 1976).
Obs., observed; Exp., expected.

Table 9.4. *Previous estimates on the proportion of European, African, and Amerindian components present in Latin American populations classified as European-derived (autosome markers)*

Country and region	No. and nature of the samples	Sample sizes	No. of genetic systems considered	Method	Parental contribution (%) European	African	Amerindian	References
Cuba	1 urban	155	1	1	92	8	ND	1
Venezuela	1 rural	102–108	8	2	100	0	0	2
Brazil								
North	3 urban	171	1–10	3,4	69–76	2–14	13–22	3
Northeast	1 urban	64–136	8	4	67	20	13	4
South	10 urban and rural	24–1261	1–17	1,2,4,5	80–92	3–19	0–13	5
Uruguay	2 urban	37–504	6–10	2,4	65–90	9–15	1–20	6

Methods: 1, Bernstein (1931), Ottensooser (1944); 2, Chakraborty (1975, 1985); 3, Ottensooser (1962); 4, Krieger *et al.* (1965); 6, Szathmáry and Reed (1978).

References: 1. Garcia *et al.* (1982); 2. Weimer *et al.* (1999); 3. Schneider and Salzano (1979); Santos *et al.* (1983, 1987); 4. Franco *et al.* (1982); 5. Franco *et al.* (1982), Culpi and Salzano (1984); Dornelles *et al.* (1999); 6. Sans *et al.* (1993, 1997, 2001). ND, not determined. When more than one number is given in the column of sample sizes, they indicate that they differed in relation to the systems or populations considered.

Table 9.5. *Previous estimates on the proportion of European, African, and Amerindian components present in Latin American populations classified as African-derived (autosome markers)*

Country and region No. and nature of the samples		Sample sizes	No. of genetic systems considered	Method	Parental contribution (%)			References
					European	African	Amerindian	
Cuba	1 urban	194	1	1	57	43	ND	1
Jamaica	1 urban	102	10	4,5	7	93	ND	2
Belize								
Caribs	4 rural	187–277	1	2	0.5–5	71–80	17–24	3
Creoles	1 urban, 2 rural	48–115	1	2	17–43	52–75	5–9	3
Guatemala								
Caribs	1 rural	131–205	1–7	2,3	1,2	70–75	22–29	3
Nicaragua	1 rural	80–86	7	3	1	66	33	4
St Vincent								
Caribs	2 rural	70–147	1	2	10–17	41–58	32–42	3
Creoles	1 rural	98	1	2	16	79	5	3
Venezuela	7 rural	18–246	5–11	4,5	0–37	46–79	0–26	5
Ecuador	2 rural	78–177	17	6	11–14	58–75	14–28	6
Brazil								
North	1 urban, 5 rural	40–453	1–14	2,4–6	18–51	18–62	0–34	7
Northeast	2 urban, 4 rural	35–1131	3–17	1,4–6	9–38	47–72	0–44	8
Southeast	3 urban	100–3429	1–5	1,5	22–56	44–73	0–5	9
South	2 urban, 1 rural	20–1,000	3–13	1,4–6	18–62	38–80	0–13	10
Uruguay	1 urban	33	12	5	35	52	13	11

Methods: 1. Bernstein (1931), Ottensooser (1944); 2, Direct count of ethnic-specific markers; 3, Elston (1971); 4, Long and Smouse (1983); 5, Chakraborty (1975, 1985); 6, Krieger et al. (1965).

References: 1. Garcia et al. (1982); 2. Parra et al. (1998); 3. Crawford et al. (1981); 4. Biondi et al. (1988); 5. Castro de Guerra et al. (1996); Bortolini et al. (1998, 1999); 6. Martinez-Labarga et al. (1999); 7. Schneider and Salzano (1979); Santos et al. (1987); Bortolini et al. (1992, 1997a–c, 1998); Guerreiro et al. (1999); 8. Saldanha (1957); Krieger et al. (1965); Bortolini et al. (1992, 1997a–c, 1998); Arpini-Sampaio et al. (1999); 9. Saldanha (1957); Saldanha (1982); Bortolini et al. (1992, 1997a–c, 1998); 10. Franco et al. (1982); Culpi and Salzano (1984); Bortolini et al. (1992, 1997a–c, 1998); 11. Sans et al. (2001).

ND, not determined. When more than one number is given in the column of sample sizes, they indicate that they differed in relation to the systems or populations considered.

Table 9.6. *Previous estimates on the proportion of European, African and Amerindian components present in Latin American populations classified as Mestizo or unclassified in relation to ethnic categories (autosome markers)*

Country and region	No. and nature of the samples	Sample sizes	No. of genetic systems considered	Method	Parental contribution (%)			References
					European	African	Amerindian	
México								
North	2 urban	202–511	7–9	1,2	40–79	8	21–52	1
Central	9 urban, 1 rural	104–1212	4–9	1,3,4	22–71	1–18	28–70	2
South	1 urban	220	8	1	31	2	67	3
East	6 urban	109–228	7	1	28–43	6–40	31–51	4
Venezuela	4 urban	8211–58 339	2	1	35–61	0–19	39–46	5
Colombia	2 urban	4093–19 303	2	1	0–42	11–84	16–47	5
Peru	1 urban	36 613	2	1	44	0	56	5
Bolivia	1 urban	29 960	2	1	17	<1	83	5
Brazil								
North	18 urban	93–1002	1–14	1,2,5	19–67	2–52	13–55	6
Northeast	9 urban	68–2129	2–17	1–6	25–62	16–74	1–38	7
Southeast	2 urban	5795–72 678	2	1	40–56	30–52	8–13	5
South	2 urban, 2 rural	131–33 710	2,7	1,7	61–83	0–39	17	8
Argentina	6 urban, 2 rural	48–105 307	1,2	1,7	<1–66	0–17	29–>99	9
Uruguay	1 urban	76–64 023	2–21	1,2	54–65	15	20–31	10
Chile	4 urban, 4 rural	24–47 130	1–2	1,6	20–65	23	35–80	11

Methods: 1, Krieger et al. (1965); 2, Chakraborty (1975, 1985); 3, Elston (1971); 4, Roberts and Hiorns (1962); 5, Ottensooser (1962); 6, Bernstein (1931), Ottensooser (1944); 7, Long and Smouse (1983).

References: 1. Lisker et al. (1990), Cerda-Flores et al. (1991); 2. Crawford et al. (1974, 1976); Tiburcio et al. (1978); Lisker et al. (1986, 1988, 1990, 1995); 3. Lisker et al. (1990); 4. Lisker and Babinsky (1986); Lisker et al. (1990); 5. Lopez-Camelo et al. (1996); 6. Ottensooser (1962); Ayres et al. (1968); Schneider and Salzano (1979); Schüler et al. (1982); Santos et al. (1983); Rosa et al. (1984); Hamel et al. (1984); Santos et al. (1987); Guerreiro and Chautard-Freire-Maia (1988); Guerreiro et al. (1993); Santos and Guerreiro (1995); Ribeiro-dos-Santos et al. (1995); Santos et al. (1996); 7. Ottensooser (1962); Saldanha (1962b); Krieger et al. (1965); Roberts and Hiorns (1965); Elston (1971); Franco et al. (1982); Chakraborty (1985); Conceição et al. (1987); Weimer et al. (1991); Lima et al. (1996); 8. Saldanha et al. (1998), Lopez-Camelo et al. (1996); 9. Lopez-Camelo et al. (1996); Dipierri et al. (2000); 10. Lopez-Camelo et al. (1996); Sans et al. (1997); 11. Rothhammer (1987); Lopez-Camelo et al. (1996); Harb et al. (1998). ND, not determined. When more than one number is given in the column of sample sizes, they indicate that they differed in relation to the systems or populations considered.

Table 9.7. *New estimates obtained on the proportion of European, African and Amerindian components present in Latin American populations (except Brazil) classified as European-derived*

Country or region	Parental contribution (% ± SD) and adequacy of the admixture model (R^2)								Genetic systems considered[c]	References
	Gene identity[a]				Weighted least squares[b]					
	European	African	Amerindian	R^2	European	African	Amerindian	R^2		
Caribbean Islands										
Cuba (n = 146–4622)	86±16	0±19	14±3	99	80±6	10±6	10±5	100	ABO, KELL, MN, RH, Hb	1
Puerto Rico (n = 179–3988)	57±0	37±0	6±0	100	60±8	33±10	7±4	100	ABO, RH	2
St Barthélemy (n = 74)	100±25	0±17	0±17	94	100±13	0±6	0±16	98	HLA-A, HLA-B, HLA-C	3
Central America										
Costa Rica (n = 6911)	94±ND	6±ND	0±ND	ND	91±ND	9±ND	0±ND	84	HB	4
South America										
Venezuela (n = 100–131)	78±5	0±10	22±7	99	73±19	15±23	12±13	96	RH, ACP, AK, ESD, PGM1(iso), HP, TF	5
Ecuador (n = 200)	67±ND	7±ND	26±ND	ND	68±ND	7±ND	25±ND	ND	ABO	6
Peru (n = 98–102)	0±0	3±0	97±1	100	0±0	38±18	62±18	99	GC, TF	7
Argentina (n = 1000)	82±10	3±5	15±7	100	84±1	0±0	16±1	100	RH (3 sera)	8
Uruguay (n = 82–604)	40±0	30±0	30±0	99	52±15	24±12	24±11	94	ABO, FY, KELL, RH (3 sera), HLA-C	9

[a]Chakraborty (1975, 1985). [b]Long and Smouse (1983). [c]For details see Appendix 2.
References: 1. Sáenz et al. (1993), Hidalgo et al. (1995); 2. Torregrosa (1945/46); Mourant et al. (1976); 3. Fumeron et al. (1981); 4. Sáenz (1988); 5. Mourant et al. (1976); 6. Weimer et al. (1999); 7. Mourant et al. (1976); 8. Etcheverry (1949); 9. Alvarez et al. (1993); Sans et al. (1993).
SD, standard deviation; ND, not determined. When more than one number is given in relation to sample sizes (*n*), they indicate that they differed in relation to the systems or populations considered.

Table 9.8. *New estimates obtained on the proportion of European, African, and Amerindian components present in Latin American countries (except Brazil) classified as African-derived*

Country or region	Parental contribution (% ± SD) and adequacy of the admixture model (R^2)								Genetic systems considered[c]	References
	Gene identity[a]				Weighted least squares[b]					
	European	African	Amerindian	R^2	European	African	Amerindian	R^2		
Caribbean Islands										
Cuba (n = 138–6996)	42±5	41±4	17±0	99	36±10	47±10	17±7	100	ABO, KELL, MN, HB	1
Dominican Republic (n = 67–798)	6±7	66±8	28±1	99	14±21	58±26	28±15	98	ABO, RH, HB	2
Haiti (n = 323–2066)	0±9	100±7	0±3	99	0±32	100±33	0±16	99	ABO, CP	3
Jamaica (n = 98–10180)	18±2	68±2	14±1	99	20±27	75±29	5±19	85	RH, HB, HLA-C, HLA-DR	4
Dominica (n = 67–1000)	0±48	49±36	51±12	99	0±10	52±18	48±39	90	ABO, MN, KELL, HB	5
Guadeloupe (n = 152–1000)	36±5	60±6	4±1	99	39±15	61±13	0±11	99	ABO, HP, TF	6
Martinique (n = 685)	30±ND	63±ND	7±ND	ND	29±ND	64±ND	7±ND	ND	ABO	7
Continental Central America										
Belize (n = 547–1844)	26±4	69±3	5±2	99	29±6	66±5	5±4	100	DI, FY, KELL, MN, RH, CP, HP, TF	8
Guatemala (n = 199–439)	15±3	63±0	22±0	99	14±11	66±12	20±10	99	ABO, HB, ACP, AK, ESD, PGM1, PGM2, PGD, CP, HP, TF	9
Honduras (n = 151–1431)	69±8	31±7	0±3	99	59±24	41±13	0±14	97	ABO, FY, KELL, MN, RH, HB	10
Nicaragua (n = 163–173)	40±1	52±1	8±1	100	33±11	45±5	22±12	99	HB, ACP, ESD, G6PD, GLOI, PGM1, PGD, CP, HP	11

Population								Systems	Ref.
Panama (n = 402–7064)	0±3	100±5	0±2	99	7±11	92±12	1±6	99	ACP, ESD, PGM1, PGM2, PGD, HP, TF	12
South America										
Venezuela (n = 75–2370)	49±3	49±3	2±2	99	40±17	45±15	15±12	90	ABO, FY, KELL, MN, RH, RH (3 sera), HB, ACP, AK, ESD, PGM1, PGD, HP, TF, HLA-C, HLA-DQ	13
Surinam (n = 322–1951)	4±5	93±5	3±3	99	0±18	89±9	11±11	95	ABO, KELL, MN, RH, HB, GC, HP, TF	14
Colombia (n = 78–1441)	13±1	71±2	16±0	99	8±9	83±13	9±9	97	ABO, HB, RH, HP, HLA-C	15
Ecuador (n = 156)	37±7	55±5	8±5	98	15±13	80±12	5±9	97	HB, ACP, ADA, CA2, ESD, GLOI, G6PD, PGM1, PGM1 (iso), PGD	16

[a]Chakraborty (1975, 1985). [b]Long and Smouse (1983). [c]For details see Appendix 2.
References: 1. Colombo and Martínez (1985); Hidalgo et al. (1995); 2. Alvarez (1951), Mourant et al. (1976), Sáenz et al. (1993); 3. Mourant et al. (1976); 4. Went (1957); Went and Mac Iver (1958), Mourant et al. (1976); Blank et al. (1995); 5. Mourant et al. (1976); Sáenz et al. (1993); 6. Languillon (1951); Sauvain-Dugerdil (1989); 7. Languillon (1951); 8. Sauvain-Dugerdil (1989); 9. Scrimshaw et al. (1961), Tejada et al. (1965); Crawford et al. (1981); Sáenz (1988); 10. Mourant et al. (1976); Sauvain-Dugerdil (1989); 11. Biondi et al. (1988); 12. Ferrell et al. (1978); Sáenz et al. (1993); 13. Tejada et al. (1965); Mourant et al. (1976); Arends et al. (1978); Arends (1984); Castro de Guerra (1991); Bortolini et al. (1992); Castro de Guerra et al. (1996); Bortolini et al. (1998); 14. Kahn (1936); Butts (1955); Liachowitz et al. (1958); Jonxis (1959); Nijenhuis and Gemser- Runia (1965); Peetoom et al. (1965); Smink and Prins (1965); Mourant et al. (1976); Sáenz et al. (1993); 15. Restrepo and Gutierrez (1968); Restrepo (1971); Monsalve et al. (1987); 16. Martinez-Labarga et al. (1999).
SD, standard deviation. ND, not determined. When more than one number is given in relation to sample sizes (n), they indicate that they differed in relation to the systems or populations considered.

Table 9.9. *New estimates obtained on the proportion of European, African, and Amerindian components present in Latin American countries (except Brazil) classified as Mestizo or unclassified in relation to ethnic categories*

Country or region	Parental contribution (% ± SD) and adequacy of the admixture model (R²)								Genetic systems considered[c]	References
	Gene identity[a]				Weighted least squares[b]					
	European	African	Amerindian	R²	European	African	Amerindian	R²		
Caribbean Islands										
Cuba (n = 62–159 389)	43 ± 2	39 ± 2	18 ± 2	99	38 ± 9	39 ± 8	23 ± 7	97	ABO, KELL, MN, RH, HB, GM	1
Dominican Republic										
(n = 520–10 461)	53 ± 12	33 ± 15	14 ± 3	99	59 ± 7	25 ± 8	16 ± 4	100	ABO, MN, RH	2
Haiti (n = 92–477)	13 ± 1	87 ± 1	0 ± 1	99	13 ± 7	87 ± 3	0 ± 4	99	ABO, FY, KELL, LE, MN, HB, ACP, AK, PGM1, PGD, GC, HP, TF	3
Jamaica (n = 24–99 821)	0 ± 16	92 ± 20	8 ± 5	99	1 ± 9	96 ± 8	3 ± 9	100	ABO, HB, GC	4
Puerto Rico (n = 110–1147)	82 ± 44	18 ± 29	0 ± 17	99	56 ± 25	38 ± 16	6 ± 19	98	ABO, HB, HP, TF	5
Guadeloupe (n = 884–26 163)	57 ± 9	43 ± 9	0 ± 12	96	50 ± 7	43 ± 7	7 ± 10	97	ABO, LE, RH, HB, GM	6
Martinique (n = 1488)	45 ± ND	51 ± ND	4 ± ND	ND	44 ± ND	49 ± ND	7 ± ND	ND	ABO	7
Dutch Antilles (n = 90)	0 ± 15	88 ± 17	12 ± 2	99	51 ± 18	41 ± 16	8 ± 6	100	ABO, RH	8
Trinidad (n = 4866)	0 ± 1	100 ± 2	0 ± 0	99	0 ± 39	100 ± 64	0 ± 28	97	ABO, RH	9
Continental Central America										
Mexico (n = 84–115 438)	49 ± 4	21 ± 5	30 ± 4	97	37 ± 11	33 ± 11	30 ± 8	96	ABO, KELL, LE, MN, RH, PGM1, HB, FY, AK, ESD, PGM1 (iso), PGD, GC, HP, TF, HLA-C	10
Guatemala (n = 1600)	41 ± ND	22 ± ND	37 ± ND	ND	49 ± ND	15 ± ND	36 ± ND	ND	ABO	11
Honduras (n = 247)	59 ± ND	41 ± ND	0 ± ND	ND	58 ± ND	42 ± ND	0 ± ND	ND	FY	12
El Salvador (n = 25–931)	96 ± 54	0 ± 56	4 ± 10	97	90 ± 13	0 ± 11	10 ± 24	75	ABO, HB, HLA-C, HLA-DR	13
Costa Rica (n = 4877–30 699)	67 ± 1	25 ± 0	8 ± 0	99	64 ± 4	29 ± 3	7 ± 3	100	ABO, HB	14
Panama (n = 100)	100 ± 82	0 ± 99	0 ± 70	87	87 ± 86	13 ± 86	0 ± 1	41	HLA-C	15

South America

Population									Systems considered	No.
Venezuela (n = 102–20975)	55±5	24±3	21±3	99	52±15	21±10	27±13	95	HB, G6PD, HP, TF, HLA-C, HLA-DQ	16
Colombia (n = 42–199 972)	58±7	33±8	9±2	99	49±19	38±19	13±12	93	ABO, KELL, MN, RH, HB, HLA-C, HLA-DQ	17
Surinam (n = 243–4762)	0±67	100±61	0±7	99	0±41	100±39	0±16	99	ABO, MN	18
French Guiana (n = 440)	58±ND	31±ND	11±ND	ND	65±ND	24±ND	11±ND	ND	ABO	19
Ecuador (n = 400–635)	53±28	22±27	25±10	98	11±32	47±31	42±23	79	ABO, HLA-C, HLA-DQ	20
Peru (n = 1000–196 988)	21±0	32±0	47±0	99	32±6	19±7	49±4	100	ABO, KELL, MN, RH	21
Paraguay (n = 50)	46±3	5±4	49±1	99	54±17	0±19	46±9	99	ABO, RH	22
Argentina (n = 1542–341 852)	67±18	26±19	7±5	99	71±14	19±17	10±8	94	ABO, KELL, MN, RH	23
Uruguay (n = 96–76 865)	59±1	28±1	13±1	99	67±16	22±12	11±11	93	ABO, FY, KELL, MN, RH, ACP, AK, ESD, PGM1, PGM2, G6PD, PGD, HP, HLA-C, HLA-DQ	24
Chile (n = 83–42 776)	28±0	12±0	60±0	99	32±13	14±10	54±11	97	ABO, FY, KELL, MN, RH, ACP, ESD, PGM1, PGM2, HP	25

[a]Chakraborty (1975, 1985). [b]Long and Smouse (1983). [c]For details see Appendix 2.

References: 1. Soto-Pradera et al. (1955); Colombo and Martinez (1985); Hidalgo et al. (1995); 2. Alvarez (1951); Mourant et al. (1976); 3. Basu et al. (1976); Mourant et al. (1976); Roychoudhury and Nei (1988); 4. Snyder (1929); Serjeant et al. (1986); Roychoudhury and Nei (1988); 5. van der Sar (1959); Tills et al. (1983); 6. Montestruc and Ragusin (1944); Languillon (1951); Fabritius et al. (1980); Darlu et al. (1990); 7. Montestruc and Ragusin (1944); 8. Rife (1972); 9. Arnaud and Young (1955); 10. Salazar-Mallen and Portilla (1944); Salazar-Mallen and Arteaga (1951); Lisker et al. (1955, 1967, 1986, 1988, 1990); Ruffié et al. (1969); Crawford et al. (1974); Mourant et al. (1976); Garza-Chapa et al. (1978); Tiburcio et al. (1978); Grunbaum et al. (1980); Tills et al. (1983); Roychoudhury and Nei (1988); Ruiz Reyes (1983); Cerda-Flores et al. (1994); 11. Cabrera (1950); 12. Spencer et al. (1978); 13. Fleischhacker (1959); Sáenz (1986); Santiago-Delpin (1991); 14. Echandi (1953); Sáenz (1988); Madrigal (1989); 15. Santiago-Delpin (1991); 16. Mourant et al. (1976); Boada and Yates (1979); Arends (1984); Santiago-Delpin (1991); 17. Paez Perez and Freudenthal (1944); Soriano-Lleras (1954); Restrepo et al. (1965); Fajardo and Lavalle (1966); Caraballo et al. (1992); 18. Collier et al. (1952); Nijenhuis and Gemser-Runia (1965); 19. Floch and Lajudie (1948); 20. Hoffstetter (1949); Santiago-Delpin (1991); 21. Rahm (1931); Valdez (1965); Mourant et al. (1976); Tills et al. (1983); 22. Mourant et al. (1976); 23. Rahm (1931); Gargiulo et al. (1939); Pezzi et al. (1951); Mourant et al. (1976); 24. Mazzela (1949); Gispert et al. (1950); Soto Pradera et al. (1955); Mourant et al. (1976); Tills et al. (1983); Santiago-Delpin (1991); Sans et al. (1995); 25. Rahm (1931); Schepeler and von Dessauer (1945); Sandoval (1961); Sandoval and Hidalgo (1961); Alvial and Henckel (1963); Cruz-Coke et al. (1967); Pinto-Cisternas et al. (1971); Harb et al. (1998) Acuña et al. (1988).

SD, standard deviation. ND, not determined. When more than one number is given in relation to sample sizes (n), they indicate that they differed in relation to the systems or populations considered.

Table 9.10. *New estimates obtained on the proportion of European, African, and Amerindian components present in Brazilian populations classified as European-derived*

| Region | Parental contribution (% ± SD) and adequacy of the admixture model (R²) | | | | | | | | Genetic systems considered[c] | References |
| | Gene identity[a] | | | | Weighted least squares[b] | | | | | |
	European	African	Amerindian	R²	European	African	Amerindian	R²		
North (n = 48–2054)	52±5	0±4	48±2	99	54±7	7±3	39±7	99	ABO, MN, RH, HB, ACP, AK, G6PD, ESD, PGM1, PGD, CP, HP, TF	1
Northeast (n = 64–27 604)	69±1	26±1	5±0	100	75±7	20±5	5±7	99	ABO, MN, RH, HB, ACP, ESD, G6PD, PGM1, PGD, CP, GC, HP, TF	2
Center-West (n = 1316–1948)	64±1	19±1	17±0	100	73±14	13±13	14±11	99	ABO, RH, HB	3
Southeast (n = 89–60 270)	52±1	44±2	4±1	100	60±10	34±10	6±8	99	ABO, FY, KELL, MN, RH, HB, ACP, ESD, HP, TF	4
South (n = 107–5527)	75±0	18±0	7±0	100	70±7	10±5	20±6	98	ABO, FY, MN, RH, HB, ACP, AK, ESD, G6PD, HP, PGM1, PGD, CP, GC, TF, HLA-C, HLA-DQ	5

[a]Chakraborty (1975, 1985). [b]Long and Smouse (1983). [c]For details see Appendix 2.

References: 1. De Lucca (1975a,b); Rosa et al. (1984); Santos et al. (1987); Guerreiro and Chautard Freire-Maia (1988); 2. Torres (1930); Novais (1953); Pedreira (1954); Lima and Walter (1955); Shansis and Carpilovsky (1956); Nunes-Moreira (1963); Marques-Ferreira et al. (1973); Azevêdo (1980); Azevêdo et al. (1980); Franco et al. (1981, 1982); Conceição et al. (1987); 3. Leme Lopes and Lopes da Costa (1951); Alvares Filho et al. (1995); Serra et al. (1964); 4. Bier and Machado (1933); Godoy (1937); Ottensooser and Pasqualin (1946); Lacaz et al. (1946); Mesquita and Leite-Ribeiro (1947); Patiño-Salazar and Mello (1948); Lacaz and Maspes (1951); Pinto et al. (1955); Shansis and Carpilovsky (1956); Saldanha (1956c); Memoria and Barbosa (1957); Junqueira and Wishart (1958); Frota and de Paula (1960); Russo (1964); Bortolozzi and De Lucca (1971); Rocha et al. (1973); Itskan and Saldanha (1976); Mestriner (1976); Curi and Kroll (1976); Saldanha (1982); Palatnik et al. (1982); Barreto et al. (1983); Venturelli and Moraes (1986); Salaru and Otto (1989); Engrácia et al. (1990); Alvares Filho et al. (1995); 5. Mota et al. (1963); Salzano (1963); Salzano and Hirschfeld (1965); Salzano et al. (1967); Franco et al. (1981); Culpi and Salzano (1984); Culpi and Lourenço (1984); Osório et al. (1957); Kuitko and Weimer (1988); Rainha de Souza et al. (1995); Muniz et al. (1998); Petzl-Erler (1999); Dornelles et al. (1999).

SD, standard deviation. When more than one number is given in relation to sample sizes (n), they indicate that they differed in relation to the systems or populations considered.

Table 9.11. *New estimates obtained on the proportion of European, African, and Amerindian components present in Brazilian populations classified as African-derived*

Region	Parental contribution (% ± SD) and adequacy of the admixture model (R^2)								Genetic systems considered[c]	References
	Gene identity[a]				Weighted least squares[b]					
	European	African	Amerindian	R^2	European	African	Amerindian	R^2		
North (n = 18–482)	30 ± 3	25 ± 2	45 ± 1	99	21 ± 10	42 ± 9	37 ± 10	98	ABO, MN, RH, HB, ACP, AK, ESD, G6PD, PGM1, PGD, CP, HP, TF	1
Northeast (n = 70–18 898)	38 ± 0	55 ± 0	7 ± 0	100	38 ± 7	54 ± 5	8 ± 7	99	ABO, RH, HB, ACP, AK, ESD, G6PD, PGM1, PGD, CP, HP, TF	2
Center-West (n = 455–2624)	28 ± 1	34 ± 1	37 ± 0	100	32 ± 8	28 ± 10	40 ± 6	100	ABO, RH, HB	3
Southeast (n = 178–11 534)	23 ± 1	77 ± 1	0 ± 0	100	22 ± 11	78 ± 10	0 ± 8	98	ABO, FY, MN, RH, HB, ACP, ESD, HP	4
South (n = 53–1236)	36 ± 1	51 ± 1	13 ± 0	100	36 ± 5	50 ± 6	14 ± 5	99	ABO, MN, RH, HB, ACP, AK, ESD, G6PD, PGM1, PGD, CP, GC, HP, TF, HLA-C, HLA-DQ	5

[a]Chakraborty (1975, 1985). [b]Long and Smouse (1983). [c]For details see Appendix 2.
References: 1. Rosa *et al.* (1984); Schneider *et al.* (1987); Guerreiro and Chautard Freire-Maia (1988); Bortolini *et al.* (1992); 2. Torres (1930); Da Silva (1948); Pedreira (1954); Lima and Walter (1955); Shansis and Carpilovsky (1956); Nunes-Moreira (1963); Azevêdo (1980); Azevêdo *et al.* (1980); Conceição *et al.* (1987); Pedrosa *et al.* (1995a,b); Bortolini *et al.* (1997b; 1998); 3. Leme Lopes and Lopes da Costa (1951); Alvares Filho *et al.* (1995); Serra *et al.* (1964); 4. Bier and Machado (1933); Godoy (1937); Patiño-Salazar and Mello (1948); Pinto *et al.* (1955); Shansis and Carpilovsky (1956); Junqueira and Wishart (1958); Frota and de Paula (1960); Russo (1964); Rocha *et al.* (1973); Itskan and Saldanha (1975); Mourant *et al.* (1976); Mestriner (1976); Saldanha (1982); Palatnik *et al.* (1982); Barretto *et al.* (1983); Venturelli and Moraes (1986); Salaru and Otto (1989); Engrácia *et al.* (1990); Alvares Filho *et al.* (1995); 5. Osório *et al.* (1957); Mota *et al.* (1963); Salzano (1963); Salzano and Hirschfeld (1965); Salzano *et al.* (1967); Franco *et al.* (1981); Culpi and Salzano (1984); Culpi and Lourenço (1984); Kuitko and Weimer (1988); Bortolini *et al.* (1992, 1997b); Petzl-Erler (1999).
SD, standard deviation. When more than one number is given in relation to sample sizes (*n*), they indicate that they differed in relation to the systems or populations considered.

Table 9.12. *New estimates obtained on the proportion of European, African, and Amerindian components present in Brazilian populations classified as admixed or without ethnic classification*

Region	Parental contribution (% ± SD) and adequacy of the admixture model (R^2)								Genetic systems considered[c]	References
	Gene identity[a]				Weighted least squares[b]					
	European	African	Amerindian	R^2	European	African	Amerindian	R^2		
North ($n = 47$–5647)	43±5	19±4	38±2	99	43±7	22±5	35±6	99	ABO, FY, MN, RH, HB, ACP, AK, ESD, G6PD, PGD, CP, HP, TF	1
Northeast ($n = 371$–47 119)	46±2	44±2	10±1	99	45±7	49±6	6±6	100	ABO, MN, RH, HB, ACP, ESD, G6PD, PGM1, PGD, CP, GC, HP, TF	2
Center-West ($n = 42$–3396)	63±1	37±2	0±0	100	84±41	0±49	16±25	95	ABO, RH	3
Southeast ($n = 67$–29 684)	57±8	41±8	2±2	99	53±17	31±10	16±10	98	ABO, KELL, RH, ACP, ESD, G6PD, PGM1, PGD, TF	4
South ($n = 188$–18 963)	27±33	73±58	0±25	99	54±23	46±31	0±17	98	ABO, RH, GC	5

[a]Chakraborty (1975, 1985). [b]Long and Smouse (1983). [c]For details see Appendix 2.

References: 1. Montenegro (1959, 1960); Junqueira et al. (1962); Rosa et al. (1984); Santos et al. (1987); Guerreiro and Chautard Freire-Maia (1988); Guerreiro et al. (1993); Ribeiro-dos-Santos et al. (1995); Melo-dos-Santos et al. (1996); Rosa et al. (1984); Santos et al. (1987); Guerreiro et al. (1954); Lima and Walter (1955); Shansis and Carpilovsky (1956); Junqueira et al. (1962); Nunes-Moreira (1963); Franco et al. (1982); Oliveira et al. (1983); Loiola et al. (1986/87); Conceição et al. (1987); 3. Serra et al. (1964); 4. Lacaz (1939); Patiño-Salazar and Mello (1943); Faria et al. (1951); Lacaz et al. (1955); Saldanha (1956c); Shansis and Carpilovsky (1956); Junqueira and Wishart (1958); Frota and de Paula (1960); Russo (1964); Vallada et al. (1967); Itskan and Saldanha (1975); Curi and Kroll (1976); Mestriner (1976); Oliveira (1977); Bernardo et al. (1982); Barretto et al. (1983); Conti and Krieger (1985); Rodini and Bortolozzi (1986); Engrácia et al. (1990); 5. Osório et al. (1957); Mota et al. (1963); Salzano and Hirschfeld (1965); Maranhão (1968); Beiguelman and Sevá-Pereira (1983); Stueber-Odebrecht et al. (1984).

SD, standard deviation. When more than one number is given in relation to sample sizes (*n*), they indicate that they differed in relation to the systems or populations considered.

Table 9.13. *Differences in gene flow according to sex in African-derived Latin Americans and their putative ancestors*

Group	Gene flow estimates			
	Female + Male	Female	Male	Female/Male
Parental stocks:				
European	41.4	19.0	12.5	1.5
African	14.4	11.5	7.0	1.6
Amerindian	1.8	2.2	2.7	0.8
African-derived Latin Americans	8.1	5.2	5.2	1.0

Source: Chapter 8 (text plus Tables 8.16 and 8.18). Markers considered: D1S80 (Female + Male), mtDNA (Female) and DYS19 (Male).

Table 9.14. *Gene flow estimates evaluated independently or differentiated by sex in African-derived South American populations: female and male contributions were calculated using uniparental genetic markers*

Country and region	No. of populations	Genetic markers	Parental contribution (%)		
			European	African	Amerindian
Venezuela					
Rural, north	4	Autosomes	17–29	53–68	3–26
		Female contribution	0	51–100	0–88
		Male contribution	100	0–49	12–100
Brazil					
Urban, northeast	1	Autosomes	38	58	4
		Female contribution	29	57	100
		Male contribution	71	43	0
Urban, southeast	1	Autosomes	22	73	5
		Female contribution	83	46	71
		Male contribution	17	54	29
Urban, south	1	Autosomes	54	46	< 1
		Female contribution	17	79	100
		Male contribution	83	21	0
Rural, north	2	Autosomes	21–28	51–57	15–28
		Female contribution	14–56	48–62	63–71
		Male contribution	44–86	38–52	29–37
Rural, northeast	1	Autosomes	29	58	13
		Female contribution	34	45	96
		Male contribution	66	55	4
Rural, south	1	Autosomes	30	64	6
		Female contribution	40	45	100
		Male contribution	60	55	0
Uruguay	1	Autosomes	38	46	15
		Female contribution	30	73	82
		Male contribution	70	27	18

Source: Bortolini *et al.* (1999); Sans *et al.* (2001).

10 *Synthesis*

The task of a mind is to produce future   *Paul Valéry*

Integration or disintegration?

The first dilemma faced by a student of science is the choice of the problem and of the method of its investigation. He or she can be interested in general or specific questions, and it is not unreasonable to suppose that the final decision may be influenced, at least indirectly, by his/her genetic constitution. Even a superficial perusal of the present biologic literature reveals that the emphasis is now on a reductionist approach to the problems, and not on more synthetic evaluations.

As was emphasized by Mayr (1982), the term 'reduction' has been used with at least three different meanings, which he classified as constitutive, explanatory, or theory reductionism. A systems approach to biology, however, suggests that its phenomena have the peculiarity (termed by many as emergence) that the characteristics of the whole cannot be deduced from the simple knowledge of its components.

This idea is especially important when we are considering problems of human evolution, since it is difficult to dismiss any aspect of our biology or environment as not being important. This interdisciplinary approach was asserted as most important, for instance, by two groups of specialists assembled by the World Health Organization some years ago (WHO, 1964, 1968).

Attempts to evaluate the genetics and evolution of human populations at the continental or world level are scarce. The most recent, encyclopedic effort, was that of Cavalli-Sforza *et al.* (1994), in which they covered all continents in a systematic way, establishing afterwards main generalizations. In their treatment of the Americas, however, they restricted their attention to Amerindians only. Since, as was indicated in Chapter 2, individuals of this ethnic classification represent only a fraction of the Latin American total population (Middle America: 21%; South America: 6%), we decided to concentrate our attention in the large majority. Another reason for our choice was that the Amerindians had been also considered in two other relatively recent evaluations (Salzano and Callegari-Jacques, 1988; Crawford, 1998).

327

Previous attempts to consider the Brazilian population and Latin America as a whole (Salzano and Freire-Maia, 1970; Salzano, 1971b) occurred three decades ago; it was time, therefore, to try a new synthesis.

The land and the people

In Chapter 3 we indicated that the term 'Latin America' was probably invented by the French, in their attempts in the nineteenth century to colonize the Americas at the south of the Rio Grande River. But we also adduced information suggesting that the area and the people are sufficiently distinctive to be considered as a whole.

The region is vast, comprising about 21 million km^2, with a wide latitude and longitude range. Along it practically all types of soil, flora, fauna and climate exist, providing, therefore, a most appropriate target for human colonization. This probably started about 30 to 20 thousand years ago, leading to a diverse range of human occupation. Before the European discovery the cultural development of its Asiatic-derived inhabitants occurred in a process that Ribeiro (1970, 1977) classified as emergent, that is, in certain places they developed from the tribal to the nation level independently of outside influences.

This situation changed drastically from the sixteenth century onwards, with the extensive immigration of predominantly Europeans and Africans, but also of Eastern Asiatics, Eastern Indians and Pakistanis. They established themselves throughout the area, but in a discontinuous, heterogeneous way; and despite rules and discriminatory customs interbred extensively. As a result, around half of the 486 million individuals now living in Latin America cannot be clearly classified in relation to ethnic origin, and are identified as creole, mestizo or mixed.

Interactions between these diversified people, however, were generally not peaceful, as exemplified by the wars of conquest and, centuries later, of the independence of the newly established nations. The process has not yet ended, since there are countries which have not yet achieved complete political independence from outside governments.

The problems of the region are not only political, but also economic. In this sense, there is no nation which has not been influenced, in varying degrees, by foreign agencies. The modernization process occurred basically through historical incorporation, not by endogenous factors, and took place independently of Ribeiro's (1970, 1977) characterization of these societies as the Witness, New or Transplanted Peoples. The chilling reality is that presently there is in the region an increase in the number of poor people at a rate of two every minute!

Population characteristics

The *structure* of a population depends of its history, intrinsic factors such as fertility and mortality, the mobility of its members, and patterns of mating (whether it occurs locally or with people from elsewhere). A large number of studies was conducted in Latin America in relation to these aspects, and they are covered in Chapter 3. Archival studies furnished precious information about the spatial distribution of historical populations, components of their demographic growth, and specific aspects of their fertility and mortality. In this area of knowledge it is important to combine the empirical information with mathematical models and simulation experiments, to test specific hypotheses about the main factors involved in the determination of these variables.

Migration can be characterized at the macro or micro levels. Those that contributed to the formation of the Latin American populations have been briefly mentioned in the previous section. Levels of internal mobility have been variously considered (such as the rural to urban movements), but more specific, community investigations are probably more rewarding for the establishment of connections between demography and genetics. Several measures for the quantification of these factors have been devised (individual migration, marital distance, parent–offspring distance, exogamy index, mean matrimonial radius, inbreeding) and the appropriate data are presented in Chapter 3. Wide differences occur among and within countries, which are related to marriage customs and other socioeconomic variables, such as availability of modern means of transportation and job opportunities.

Life histories

In evolutionary terms, an organism has to establish a balance between how much it should invest in personal survival as opposed to reproductive costs. In humans, with their long maturation phase, the dynamics of physical growth should be carefully monitored. Essential for the maintenance of adequate physiologic homeostasis, especially in extreme environments such as those at high altitudes, are nutritional factors, as well as cultural adaptations. Latin America has three main food cultures, involving respectively emphasis on maize, beans and potatoes, starchy roots and tubers, or animal products. Under- and overnutrition are heterogeneously distributed through the area, and here again, cultural beliefs such as those listed in Chapter 4 should be considered.

The relative roles of genetic and environmental agents in processes such

as physical growth, working capacity, patterns of activity, motor performance, blood pressure and adaptation to high altitude are not clearly
established. They probably interact in complex ways, making generalizations difficult. The concept of *plasticity* should be emphasized, instead of
rigidly controlled physiologic and morphologic boundaries.

Aging characterizes the later stages of human life. Evolutionary reasons
for this process have been variously considered. It is possible that the
phenomenon is just a by-product of processes which put youth quality
first, independently of effects which would occur afterwards. As the number of old people increases, there is a need for more investigations both at
the cellular and organismal level of the aging condition that, as obesity, is
becoming a public health problem.

In its simplest terms, life consists of energy capture and its conversion.
Patterns of energy flow, therefore, characterize different aspects of life
histories, and should be duly considered. The dynamics of physical
growth, for instance, differ by region of the body; and there is clear
heterogeneity, in Latin America, in relation to secular trends in stature.

Form and function

The analysis of the structure of the human body can be quantified in terms
of size and shape. As is true for the characteristics considered in the
previous section, here again the final product is determined by genetic–
environmental interactions. Latin American populations do not present
especially tall subjects, but there are some internation differences. There is
also ethnic diversity, but the offspring of interethnic crosses does not
uniformly present intermediate values, as would be expected for multifactorial traits.

Morphologic comparisons of migrants with people who remained at
their home country have a respected past in anthropology, and some of
these studies were conducted in Latin America. The fact that migrants are
not a random sample of the original population should be always considered; therefore, ideally biologically related individuals should be examined in
the parent and the new home. This is not, however, an easy task. Diverse
traits are also expressed differently in these comparisons. These studies can
throw light on the question of the plasticity of morphologic traits.

In paired organs, symmetry should be considered, since it gives an
indication of the normality of the ontogenetic development. More symmetrical individuals may be favored in situations of assortative mating,
and may also show better fitness (measured, for instance, in terms of lower

morbidity or better quantity and/or quality of offspring). But different parts of the body may differ in relation to this characteristic, so that analyses are not easy. Ontogenic fields should be investigated, also, for patterns of palm and finger dermatoglyphics. A special situation concerns analyses at the cellular level; in this case the dynamics of cell division and function should be considered.

Well-being and sickness

The relationships between disease and evolution are numerous. The sheer question of why sickness was not eliminated by natural selection is import-ant, stressing the limitations of this main evolutionary factor. A conveni-ent subdivision for further inquiries is that related to the infectious and noninfectious diseases. In Latin America millions of people still suffer from illnesses due to infectious agents and parasites which have been drastically reduced or eliminated in First World countries. This is a reflec-tion of low socioeconomic levels and generalized lack of emphasis, by governments, on health or education (as opposed to military) expenses. But host–parasite interactions should (and have been) considered also. The adaptive systems of humans and their pathogenic agents display dialectic relationships of opposition and dependence that need close inves-tigation. Cultural concepts of illness and well-being should also be con-sidered.

Chapter 6 lists a series of data related to diseases in Latin America, namely: (a) a total of 82 specific studies, conducted in eight countries, concerning 17 infectious diseases; (b) nonmolecular investigations on 73 pathologic entities, plus three groups of conditions, performed in six countries; (c) molecular studies on 37 pathologic conditions, studied in 14 countries; (d) linkage data on 11 pathological entities from five countries; (e) twenty-nine associations between diseases and non-HLA genetic markers, as well as 34 between different conditions and HLA alleles. Especially studied diseases or groups of diseases were the ectodermal dysplasias, hemophilias, von Willebrand and Huntington conditions, cys-tic fibrosis, muscular dystrophies, and fragile X syndrome. Specific pro-grams considered congenital malformations (ECLAMC), endemic goiter and atherosclerosis.

In the area of mutagenesis, teratogenesis, and carcinogenesis, mention should be made of the investigations in an area of high natural background radiation, and of the Goiânia accident, which released high levels of radiation. The National System of Information on Teratogenic Agents has been functioning for 10 years now, and cases involving thalidomide and misoprostol exposure have been described. Finally, a multicenter ambitious project which has the objective of generating one million gene sequences from the most common cancers found in Brazil should be mentioned.

Genetic screening for several genetic conditions is being performed in Cuba, Brazil, and México. Information about genetic counseling, provided in 13 countries, has been referred to, as well as the participation of Latin American medical geneticists in a cross-cultural world project.

All this illustrates the high level of interest and the large number of studies being conducted in Latin America in the area of medical genetics.

Hemoglobin variation

The hemoglobin genetic system is probably the best investigated in humans, and research on it is extensive in Latin America also. More than half a million people had been tested in relation to the three most common alleles, and neonatal screening for the hemoglobinopathies has been carried out in Cuba for several decades now. Also of note are the detailed clinical investigations of these conditions performed in Jamaica. African-derived populations can have as much as 8% of the $Hb\beta*S$ mutation, but the frequency of $Hb\beta*C$ is much lower (around 1%). As expected, these frequencies diminish as the amount of non-African ancestry increases. But other, rare variants, occur in the region; 30 concerning the α locus, 42 related to β, and 14 involving other chains are listed in Chapter 7. There is also information there about the 3.7-kb deletion in α-thalassemia and of 22 types of β-thalassemias.

The investigation of mutations along the β S and β A chromosomes can be evaluated through specific arrangements, called haplotypes. Five β S haplotypes show a distinctive distribution in Africa, and their prevalences in Latin America are characterized by a marked difference between Central America and Surinam on one hand, and Brazil on the other. While in the former the Benin arrangement is the most frequent (with the exception of México), Bantu is the haplotype most prevalent in Brazil. This probably reflects the different sources of Africans who were brought to these different regions.

Not less than 16 β A haplotypes can be distinguished, showing wide interpopulation variation. The data summarized in Chapter 7 included 861 individuals living in three countries. The variability present in African-derived Latin Americans is not much different from that observed in Africa.

Selected sickle cell disease studies performed in 13 countries or regions and covering a period of 69 years, besides the classical Jamaican investigations, are also listed. The data for thalassemia major, however, are much more restricted.

A long series of events separate the primary action of a given genetic variant and the organismal phenotype. Therefore, although essential, the molecular information cannot predict the final picture. An example was provided by Zago *et al.* (1982), who verified that two abnormalities, an α chain mutation associated with β⁰-thalassemia, *improved* the clinical picture of a patient.

Protein and DNA variability

Extensive studies related to protein markers were performed in Latin American groups, but they were very unevenly distributed (European-derived populations, 25 genetic systems investigated in nine countries; African-derived, 69 in 22; Mestizo or unclassified, 56 in 26). The main result obtained in these studies was that the variability is distributed in a continuous way, independently of ethnic categories. Nor could clear geographic trends be detected on the alleles of the ABO system, the most studied in the area and the world.

The amount of information available in relation to the DNA markers is more restricted, but there are data concerning 62 loci in European-derived, 74 in African-derived, and 65 in Mestizo or ethnically unclassified groups. We have chosen to evaluate the distribution of an autosome marker (D1S80), as well as of those having matrilineal (mtDNA) or patrilineal (DYS19) inheritance. Despite some peculiarities, the amount of variability found in Latin Americans is equivalent to that present in their putative ancestors. In the protein–DNA comparisons the most interesting result found was the high (21%) interpopulation variability found for the mtDNA variants.

Systematic molecular investigations in two important proteins (G6PD and albumin) disclosed results worthy of note. In relation to G6PD, 19 variants were found in Latin America, distributed all over the molecule (from position 14 to 488). It is remarkable that up to now no definite G6PD variant of Amerindian origin has been found. In relation to albumin, nine variants were encountered. Yanomama 2 was found (in high prevalence) among Amerindians of this tribe only. In contrast, Porto Alegre 2 was observed in southern, northeastern and northern Brazil, as well as in subjects living or derived from Turkey and India.

Classification systems sometimes are ignored

The question whether there are real human races continues to be discussed, with no end in sight. People inhabiting distinct continents can be usually distinguished morphologically, and given a sufficient number of genetic markers it is possible to ascertain if a person had European, African or Asian ancestry. Since in many instances the predominance of a given ancestry is associated with a certain socioeconomic status, discrimination based on ethnic characteristics inevitably occurs. In Latin America three points related to these questions should be emphasized: (a) classification into neat categories is generally difficult and sometimes impossible; (b) despite this fact there is preference for homogamic matings; and (c) interethnic crossings are, however, very common.

The eventual quantification of the gene flow that has occurred and is still occurring between these categories is mainly of historical interest; and presently there are relatively sophisticated genetic–statistical methods which can furnish estimates in cases of di- or trihybrid ancestry.

In Chapter 9 we reviewed the past studies in which such quantifications were made, and derived two sets of evaluations for samples related to 14 European-derived, 21 African-derived and 30 Mestizo or unclassified groups living in an equivalent number of countries or regions. The main conclusion is clear: despite variations from place to place, gene flow is universal.

Recently, with the use of exclusive matrilineal or patrilineal genetic markers, it has become possible to evaluate the influence of gender in these crossings. In Latin Americans of mixed ancestry the European component was mainly contributed by males, while the Amerindian fraction is mostly derived from females. This is a reflection of unions that occurred in the Colonial period and which could be detected independently of the demographic, cultural, and biologic changes that occurred afterwards.

The future

Predictions for the future are notoriously difficult, especially when Third World countries or populations are involved. Some trends that we expect will occur in Latin America in the future, however, are listed in Table 10.1. In general they do not differ from those expected for the world population, but the list could serve as a program for the future. Its items present an optimistic view of what might happen. It would be easy to develop what would be, perhaps, a more realistic picture. Certainly, the rate of change will differ according to the regions considered, and obstacles to the envisaged development are many. But let us finish with the hope for the maximum of happiness for the largest number of people.

Table 10.1. *Future trends for the Latin American populations*

1. Demography
  1.1. Declining mortality
  1.2. Declining fertility
  1.3. Increase in life expectancy
  1.4. Higher percentage of old people
  1.5. Higher divorce rates, increased frequency of consensual unions, higher proportion of women without partners
  1.6. Increased mobility

2. Health and disease
  2.1. Better nutrition
  2.2. Increase in size
  2.3. Decreased incidence of infectious diseases
  2.4. Increased incidence of chronic, degenerative diseases
  2.5. Increased exposure to noxious agents
  2.6. Increased control and treatment of hereditary diseases and congenital malformations

3. Genetic variability
  3.1. More homogeneous distribution of normal and abnormal genetic variants

4. Socioeconomic and cultural variables
  4.1. Increased levels of education
  4.2. Increased penetration of the regional and world economies
  4.3. Decreased differences between rich and poor
  4.4. Less ethnic and gender discrimination
  4.5. More political freedom
  4.6. Globalization of decisions at the individual and population levels

Appendix 1

Publications related to the genetics of Latin American populations (except Brazil) not mentioned in the text but which were considered in the compilation of Tables 8.1–8.3

Acuña, M. and Eaton, L. (1996). Serum butyrylcholinesterase (CHE1) polymorphism in a Chilean population of mixed ancestry. *Gene Geography*, **10**, 161–5.

Acuña, M. and Eaton, L. (1998). Interaction between ABO and haptoglobin systems in a Chilean population of mixed ancestry. *Genetics and Molecular Biology*, **21**, 439–41.

Allen, F.H. Jr. and Milch, R.A. (1961). Blood groups of Dominicans. *American Journal of Physical Anthropology*, **19**, 98.

Alvarez, I., Vila, V., Perez, H., Sosa, M., *et al.* (1981). Frecuencia de los antigenos HLA-ABC en la poblacion uruguaya. *I Congresso e Workshop, Sociedade Latinoamericana de Histocompatibilidade*, São Paulo, SP.

Arce Larreta, J. (1941). Características serológicas de la primera infancia, en niños peruanos nativos de los Andes. *Crónica Médica* (Lima), **58**, 29–34.

Arellano, J., Vallejo, M., Estrada, H.G. and Kretschmer, R.R. (1981). HLA profile of the Mexican Mestizo population. *Tissue Antigens*, **18**, 242–6.

Arenas, D., Coral, R., Cisneros, B., *et al.* (1996). Carrier detection in Duchenne and Becker muscular dystrophy using dinucleotide repeat polymorphisms. A study in Mexican families. *Archives of Medical Research*, **27**, 151–6.

Arends, T. and Gallango, M.L. (1972). Aloalbuminemia: su distribucion en poblaciones venezolanas. *Acta Científica Venezolana*, **22** (Suppl. 3), 191–5.

Balanza, E. and Taboada, G. (1985). The frequency of lactase phenotypes in Aymara children. *Journal of Medical Genetics*, **22**, 128–30.

Barrios García, B., Alé Suárez, O. and Tarano Cartaya, G. (1984). El sistema proteasa inhibidor y sus variantes en Cuba. *Memorias, VI Congreso Latinoamericano de Genética*, 322–3.

Barros, L. and Witkop, C.J. Jr. (1963). Oral and genetic study of Chileans 1960. III. Periodontal disease and nutritional factors. *Archives of Oral Biology*, **8**, 195–206.

Bartholomew, C. and Pong, O.Y. (1976). Lactose intolerance in East Indians of Trinidad. *Tropical and Geographical Medicine*, **28**, 336–8.

Battaglia, A. (1949). Grupos sanguineos y factor Rh en la poblacion de Buenos Aires. *Revista de la Sociedad Argentina de Hematologia e Hemoterapia*, **1**,

169–72.

Battistuzzi, G., Biondi, G., Rickards, O., Astolfi, P. and De Stefano, G.F. (1986). Historical and demographic factors and the genetic structure of an Afroamerican community of Nicaragua. In Roberts, D.F. and De Stefano, G.F., editors. *Genetic Variation and its Maintenance*. Cambridge: Cambridge University Press, pp. 181–90.

Bernal, J.E., Papiha, S.S., Keyeux, G., Lanchbury, J.S. and Mauff, G. (1985). Complement polymorphism in Colombia. *Annals of Human Biology*, **12**, 261–5.

Bernal, J.E., Duran, C. and Papiha, S.S. (1988). HLA antigens in the Uitoto Indians and an urban population of Colombia. *Human Heredity*, **38**, 337–40.

Bonaiti-Pellié, C., Dugoujon, J.M., Bois, E., Blanc, M. and Rivat-Péran, L. (1988). Studies on an isolated West Indies population. VI. Immunoglobulin (Gm and Km) allotypes. *Gene Geography*, **2**, 15–21.

Bravo Aguiar, M.L.J. (1995). Estudio de la estructura genética de la población de Antioquia, Colombia. I. *Antropologia Biológica*, **3**, 15–28.

Bravo, M.L.J., Olarte, G.A., Rothhammer, F., Gómez, J.R., Pineda, B.E. and Pineda, H. (1994). Composición genética de algunas poblaciones urbanas y rurales de Colombia. *Antropologia Biológica*, **2**, 59–66.

Bravo, M.L., Valenzuela, C.Y. and Arcos-Burgos, O.M. (1996). Polymorphisms and phyletic relationships of the Paisa community from Antioquia (Colombia). *Gene Geography*, **10**, 11–17.

Campusano, C., Zambra, E., Figueroa, H., Lazo, B., Reyes, J. and Pinto-Cisternas, J. (1984). Distribución de la fosfatasa acida (AcP) y la fosfoglucomutasa I (PGMI) en la ciudad de Valparaiso, Chile. *Memorias, VI Congreso Latinoamericano de Genética*, 293–4.

Campusano, C., Lazo, B. and Medina, M.C. (1996). ACP and PGM1 polymorphisms in a Chilean population. *Gene Geography*, **10**, 167–70.

Cantú, J.M. and Ibarra, B. (1982). Phosphoglucomutase: Evidence for a new locus expressed in human milk. *Science*, **216**, 639–40.

Cerda-Flores, R.M. and Garza-Chapa, R. (1989). Variation in the gene frequencies of three generations of humans from Monterrey, Nuevo León, México. *Human Biology*, **61**, 249–61.

Cerda-Flores, R.M., Ramirez-Fernandes, E. and Garza-Chapa, R. (1987). Genetic admixture and distances between populations from Monterrey, Nuevo León, Mexico and their putative ancestral populations. *Human Biology*, **59**, 31–49.

Chauvenet, P.H., Anderson, S.A. and Banowsky, L.H. (1981). HLA frequencies in a Mexican American population. *Tissue Antigens*, **17**, 323–31.

Cox, D.W., Andrews, B.J. and Wills, D.E. (1986). Genetic polymorphism of α2HS-glycoprotein. *American Journal Human Genetics*, **38**, 699–706.

Crawford, M.H. (1983). The anthropological genetics of the Black Caribs (Garifuna) of Central America and the Caribbean. *Yearbook of Physical Anthropology*, **26**, 161–92.

Crawford, M.H. (1984). *Current Developments in Anthropological Genetics*, vol. 3. *Black Caribs: a Case Study in Biocultural Adaptation*. New York: Plenum

Press.

De Leo, C., Castelan, N., Lopez, M., *et al.* (1997). HLA Class I and Class II alleles and haplotypes in Mexican Mestizos established from serological typing of 50 families. *Human Biology*, **69**, 809–18.

Diaz, J.W. and Cheredeer, A.N. (1977). Distribution of HLA antigens in a Cuban population. *Tissue Antigens*, **9**, 71–9.

Díaz-Narváez, V.P., Hidalgo, P. and Pérez, M.R. (1995). Polimorfismo genético del sistema de la haptoglobina (Hp) en la Provincia de Pinar del Rio (Cuba). *Antropologia Biológica*, **3**, 13–20.

Dipierri, J.E., Gonzalez, L.K., D'Agostino, S.P. and Guerra, V.Q. (1983). Deficiencia primaria de lactasa tipo adulto en la población autóctona de Jujuy (Rep. Argentina). *Mendeliana*, **6**, 17–22.

Dipierri, J.E., Alfaro, E., Bejarano, I.F. and Etchart, A.A. (1998). Acetylator phenotypes: allele frequency in northwestern Argentina and review of acetylator distribution in the Americas. *Human Biology*, **70**, 959–64.

Duque Gomez, L. (1944). Grupos sanguineos entre los indigenas del Departamento de Caldas. *Revista del Instituto Etnológico Nacional*, **1**, 623–53.

Dykes, D.D., Crawford, M.H. and Polesky, H.F. (1983). Population distribution in North and Central America of PGM_1 and Gc subtypes as determined by isoelectric focusing (IEF). *American Journal of Physical Anthropology*, **62**, 137–45.

Elizondo, J., Sáenz, G.F., Páez, C.A., *et al.* (1982). G6PD-Puerto Limón: A new deficient variant of glucose-6-phosphate dehydrogenase associated with congenital nonspherocytic hemolytic anemia. *Human Genetics*, **62**, 110–12.

Escajadillo, T. (1948). La población mestiza del Peru y los grupos sanguineos. *II Curso Sudamericano de Transfusión y Hematologia*, Santiago, Chile, 58–67.

Escobar-Gutiérrez, A., Gorodezky, C. and Salazar-Mallén, M. (1973). Distribution of some of the HL-A system lymphocyte antigens in Mexicans. *Vox Sanguinis*, **25**, 151–5.

Estrada, J.F., Hidalgo Calcines, P., Gomes, B.D., Castellano, S.B. and Crespo, A.M. (1989). Frecuencia fenotípica y génica del antígeno Kell en la región central de Cuba (Villa Clara). *Revista Cubana de Hematologia, Inmunologia y Hemoterapia*, **5**, 575–9.

Fernandez, M.P. (1948). Contribución al estudio de los grupos sanguineos en la Provincia de Trujillo. *Revista de la Universidad Nacional de Trujillo*, **2**, 76–85.

Fonseca-Pérez, T., González-Coira, M. and Arias, S. (1996). PI locus (alpha–1-antitrypsin) allele frequencies in an Andean Venezuelan population. *Gene Geography*, **10**, 65–74.

Gallango, M.L. (1969). Gc polymorphism in Venezuelan Mestizos. *Human Heredity*, **19**, 530–3.

Gallango, M.L. and Arends, T. (1968). Lp and Ag systems in Venezuelan populations. *Proceedings of the 11th Congress of the International Society of Blood Transfusion* (Sydney), **29**, 299–302.

Gallango, M.L. and Arends, T. (1969). Phenotypical variants of pseudocholinesterase in myeloma patients. *Humangenetik*, **7**, 104–8.

Gallango, M.L. and Castillo, O. (1974). α_2-Macroglobulin polymorphism: a new genetic system detected by immunoelectrophoresis. *Journal of Immunogenetics*, **1**, 147–52.

Garrocho, S.C. and Castillo, E.R. (1973). Grupos sanguineos y nivel socioeconomico en San Luis Potosi. *Medicina* (México), **53**, 520–6.

Garza-Chapa, R., Davila-Rodriguez, M.I., Leal-Garza, C.H., Gonzalez-Quiroga, G. and Rojas-Alvarado, M.A. (1995). Gene frequencies and admixture estimates for ABO, $Rh_0(D)$, and MN blood groups in persons with mono- and polyphyletic surnames in Monterrey, N.L., Mexico. *American Journal of Human Biology*, **7**, 65–75.

Gibbs, W.N., Ottey, F. and Dyer, H. (1972). Distribution of glucose-6-phosphate dehydrogenase phenotypes in Jamaica. *American Journal of Human Genetics*, **24**, 18–23.

Gibbs, W.M., Gray, R. and Lowry, M. (1979). Glucose-6-phosphate dehydrogenase deficiency and neonatal jaundice in Jamaica. *British Journal of Haematology*, **43**, 263–74.

Gibbs, W.N., Wardle, J. and Serjeant, G.R. (1980). Glucose-6-phosphate dehydrogenase deficiency and homozygous sickle cell disease in Jamaica. *British Journal of Haematology*, **45**, 73–80.

Giraudo, C., Gomez, V. and Marcellino, J.A. (1982). Estudio inmunogenético en un semiaislado humano de la Sierra de Comechingones (Cordoba, Argentina). *Medicina*, **42**, (Suppl. 1), 51–5.

González, R., Estrada, M. and Colombo, B. (1975). G-6-PD polymorphism and racial admixture in the Cuban population. *Humangenetik*, **26**, 75–8.

González, R., Ballester, J.M., Estrada, M., *et al.* (1976). A study of the genetical structure of the Cuban population: red cell and serum biochemical markers. *American Journal of Human Genetics*, **28**, 585–96.

González, R., Wade, M., Estrada, M., Svarch, E. and Colombo, B. (1977). G6PD Pinar del Rio: a new variant discovered in a Cuban family. *Biochemical Genetics*, **15**, 909–13.

González, R., Estrada, M., Garcia, M. and Gutierrez, A. (1980). G6PD Ciudad de la Habana: a new slow variant with deficiency found in a Cuban family. *Human Genetics*, **55**, 133–6.

González-Quiroga, G., Ramirez-del Rio, J.L., Cerda-Flores, R.M. and Garza-Chapa, R. (1994). Frequency and origin of G6PD deficiency among icteric newborns in the metropolitan area of Monterrey, Nuevo León, México. *Gene Geography*, **8**, 157–64.

González Vallarino, Z. (1958). Frecuencia de los factores ABO y Rh en pobladores del Perú. *Sangre*, **3**, 31–3.

Gorodezky, C., Terán, L. and Gutiérrezn A.E. (1979). HLA frequencies in a Mexican Mestizo population. *Tissue Antigens*, **14**, 347–52.

Guderian, R.H. and Vargas, J.G. (1986). Duffy blood group distribution and the incidence of malaria in Ecuador. *Transactions of the Royal Society of Tropical Medicine and Hygiene*, **80**, 162–3.

Guerreiro, J.F., Ribeiro-dos-Santos, A.K.C, Santos, E.J.M., *et al.* (1999).

Genetical–demographic data from two Amazonian populations composed of descendants of African slaves: Pacoval and Curiau. *Genetics and Molecular Biology*, **22**, 163–7.

Gutiérrez, A., Garcia, M., Estrada, M., Quintero, I. and González, R. (1987). Glucose-6-phosphate dehydrogenase (G6PD) Guantánamo and G6PD Caujerí: two new glucose-6-phosphate-deficient variants found in Cuba. *Biochemical Genetics*, **25**, 231–8.

Helmuth, R., Fildes, N., Blake, E. *et al.* (1990). HLA-DQα allele and genotype frequencies in various human populations, determined by using enzymatic amplification and oligonucleotide probes. *American Journal of Human Genetics*, **47**, 515–23.

Herzog, P. and Gonzales Corona, P.O. (1967). Hp-, Gm-, Inv- und Tf-typen in Santiago de Cuba (Cuba). *Folia Haematologica*, **87**, 260–6.

Hidalgo, P.C. (1998). Consideraciones sobre la constitución genética de la población cubana. *Revista Española de Antropologia Biológica*, **19**, 5–20.

Hidalgo, P.C. and Machado, M.J. (1991). Polimorfismo genetico del pepsinogeno A urinario en la Provincia de Villa Clara (Cuba). *Revista Cubana de Iinvestigaciones Biomédicas*, **10**, 109–12.

Ibarra, B., Vaca, G., Sanchez-Corona, J. *et al.* (1979). Los Angeles variant of galactose-1-phosphate uridyltransferase (E.C.2.7.7.12) in a Mexican family. *Human Genetics*, **48**, 121–4.

Joe, J., Tjin, T., Prins, H.K. and Nijenhuis, L.E. (1965). Hereditary and acquired blood factors in the negroid population of Surinam. I. Origin, collection and transport of the blood samples; characteristic blood groups. *Tropical and Geographical Medicine*, **17**, 56–60.

Kostyu, D.D., Amos, D.B. and Hinostroza, S. (1975). An analysis of the 4c complex of HL-A based on Indian populations. *Tissue Antigens*, **5**, 420–30.

Lanchbury, J.S., Bernal, J.E. and Papiha, S.S. (1984). Genetic polymorphism of glutamate-pyruvate transaminase and glyoxalase I in Colombia. *Human Heredity*, **34**, 222–5.

Lastra, E.T. and Klappenbach, M. (1945). Estudio del factor Rh y su importancia en la practica obstetrica. *Obstetricia y Ginecologia Latino-Americanas*, **3**, 911–20.

Layrisse, M., Wilbert, J. and Arends, T. (1958). Frequency of blood group antigens in the descendants of Guayquerí Indians. *American Journal of Physical Anthropology*, **16**, 307–18.

Layrisse, Z., Rodríguez, M.P., Rodríguez-Iturbe, B., García, E., Stoikow, Z. and Salas, G. (1976). Genetics of the HL-A system in a Venezuelan heterogeneous population. *Vox Sanguinis*, **31**, 37–47.

Lehmann, H., Araujo, A.C. and Chaves, M. (1946). Grupos sanguineos entre los indios 'Kwaiker'. *Boletin de Arqueologia* (Bogotá), **2**, 227–30.

Liau, S.W., Mickey, R., Romano, P. and Lee, T.D. (1984). Study of the HLA system in the Haitian population. *Tissue Antigens*, **23**, 308–13.

Lindo-Haines, G. (1980). Blood group distribution and ABO haemolytic disease of the newborn in Jamaica. *Medical and Laboratory Sciences*, **37**, 263–6.

Lisker, R. (1982). Deficiencia de lactasa en poblaciones mexicanas. *Actas, V*

Congreso Latinoamericano de Genética, 61–6.

Lisker, R., López-Habib, G., Daltabuit, M., Rostenberg, I. and Arroyo, P. (1974). Lactase deficiency in a rural area of Mexico. *American Journal of Clinical Nutrition*, **27**, 756–9.

Lisker, R., Gonzalez, B. and Daltabuit, M. (1975). Recessive inheritance of the adult type of intestinal lactase deficiency. *American Journal of Human Genetics*, **27**, 662–4.

Lisker, R., Briceno, R.P., Zavala, C., Navarrette, J.I., Wessels, M. and Yoshida, A. (1977). A glucose-6-phosphate dehydrogenase GD(-) Castilla variant characterized by mild deficiency associated with drug-induced hemolytic anemia. *Journal of Laboratory and Clinical Medicine*, **90**, 754–9.

Lisker, R., Pérez-Briceño, R., Granados, J. and Babinsky, V. (1988). Gene frequencies and admixture estimates in the State of Puebla, Mexico. *American Journal of Physical Anthropology*, **76**, 331–5.

Lisker, R., Ramirez, E., Perez Briceño, R., Granados, J. and Babinsky, V. (1990). Gene frequencies and admixture estimates in four Mexican urban centers. *Human Biology*, **62**, 791–801.

Lisker, R., Ramírez, E., Peñaloza, R. and Salamanca, F. (1994). Red cell acid phosphatase types and GC polymorphisms in Mérida, Oaxaca, León, and Saltillo, Mexico. *Human Biology*, **66**, 1103–9.

Lisker, R., Ramirez, E., González-Villalpando, C. and Stern, M.P. (1995). Racial admixture in a Mestizo population from Mexico City. *American Journal of Human Biology*, **7**, 213–16.

Lisker, R., Ramirez, E. and Babinsky, V. (1996). Genetic structure of autochthonous populations of Meso-America: México. *Human Biology*, **68**, 395–404.

Machado Kano, M. and Opolsky, A.F. (1984). Distribution and frequency of glucose-6-phosphate deficiency in the Central Region of Cuba. *Genetika*, **20**, 864–7.

McCurdy, P.R., Maldonado, N., Dillon, D.E. and Conrad, M.E. (1973). Variants of glucose-6-phosphate dehydrogenase (G6PD) associated with G6PD deficiency in Puerto Ricans. *Journal of Laboratory and Clinical Medicine*, **82**, 432–7.

McGowan, V.M., Reed, T.E., Schanfield, M.S., Goliah, S. and Poon-King, T. (1990). Genetic variation at the immunoglobulin allotype loci in Creoles of Trinidad. *American Journal of Physical Anthropology*, **81**, 555–62.

Meera Khan, P., Verma, C., Wijnen, L.M.M., Wijnen, J. Th., Prins, H.K. and Nijenhuis, L.E. (1986). Electrotypes and formal genetics of red cell glutathione peroxidase (GPX1) in the Djuka of Surinam. *American Journal of Human Genetics*, **38**, 712–23.

Monplaisir, N., Valette, I., Lepage, V, et al. (1985). Study of HLA antigens of the Martinican population. *Tissue Antigens*, **26**, 1–11.

Monplaisir, N., Valette, I. and Bach, J.F. (1986). HLA antigens in 88 cases of rheumatic fever observed in Martinique. *Tissue Antigens*, **28**, 209–13.

Morganti, G. (1949). Il problema genetico del sistema Rh. *Annales de Biologia Normal y Pathologica*, **1**, 387–429.

Müller, A. and Arends, T. (1971). Electrophoretic phenotypes of adenylate kinase in Venezuelan populations. *American Journal of Human Genetics*, **23**, 507–9.

Peñaloza, R., Villalobos, H., Salamanca, F. and Zavala, C. (1985). Variantes electroforéticas de la fosfatase ácida eritrocítica en muestras de individuos del Distrito Federal y algunos Estados de la República Mexicana. *Gaceta Médica de Mexico*, **121**, 189–93.

Pik, C., Loos, J.A., Jonxis, J.H.P. and Prins, H.K. (1965). Hereditary and acquired blood factors in the negroid population of Surinam. II. The incidence of haemoglobin anomalies and the deficiency of glucose-6-phosphate dehydrogenase. *Tropical and Geographical Medicine*, **17**, 61–8.

Rivers, H., Montes, E. and Barrientos, R. (1973). Newborn hematological normal values, and Rh and ABO system distribution from a group of Salvadorean women. *Revista del Instituto de Investigaciones Medicas*, **2**, 349–51.

Sáenz, G.F., Brilla, E., Arroyo, G., Valenciano, E. and Jiménez, J. (1971). Deficiencia de la dehidrogenasa de la glucosa-6-fosfato (G-6-PD) eritrocítica en Costa Rica. I. Generalidades sobre el defecto y hallazgos en población de raza negra. *Revista Médica del Hospital Nacional de Niños Dr. C.S. Herrera*, **6**, 129–46.

Sáenz, G.F., Chaves, M., Berrantes, A., Elizondo, J., Montero, A.G. and Yoshida, A. (1984). A glucose-6-phosphate dehydrogenase variant, GD (-) Santamaria found in Costa Rica. *Acta Haematologica*, **72**, 37–40.

Saha, N. and Samuel, A.P.W. (1987). A genetic study of Blacks from Trinidad. *Human Heredity*, **37**, 365–70.

Sandoval, L. (1965). El sistema de grupos sanguíneos Kidd en la población de Santiago. *Sangre*, **10**, 257–60.

Serra, A. (1932). Isohemagglutination and blood grouping. Their relation to transfusion. Report of 5135 blood groups in Porto Rico. *Puerto Rico Journal of Public Health*, **7**, 443–54.

Serra, J.L., Séger, J., Lanset, S., *et al.* (1987). Studies on an isolated West Indies population (V): genetic differentiation, evidence for founder effect and drift. *Gene Geography*, **1**, 81–92.

Serrano, C. (1977). Distribución de los grupos sanguíneos (sistemas ABO y Rh) en un contingente militar mexicano. *Anales de Antropologia*, **14**, 373–80.

Serre, J.L. and Babron, M.C. (1985). Polymorphism and genetic evolution in an isolate in the Antilles: Saint-Berthélemy. *Annals of Human Biology*, **12**, 413–20.

Solá, J.E. (1931). Los grupos sanguíneos en las diferentes formas de alienación mental. *La Semana Médica*, **18**, 699–706.

Suarez, R.M., Olavarrieta, S., Buso, R., Meyer, L.M. and Suarez, R.M. Jr. (1961). Glucose-6-phosphate dehydrogenase deficiency among certain Puerto Rican groups. *Boletin de la Asociación Médica de Puerto Rico*, **53**, 41–8.

Sutton, R.N.P. (1963). Erythrocyte glucose-6-phosphate dehydrogenase deficiency in Trinidad. *The Lancet*, 1, 855.

Taboada, G. and Balanza, E. (1983). Aspectos hereditarios de la deficiencia de lactasa intestinal de tipo adulto. *Salud Boliviana*, 2, 173–6.

Umetsu, K., Yuasa, I., Yamashita, T., *et al.* (1989). Genetic polymorphisms of orosomucoid and alpha-2-HS-glycoprotein in Thai, Sri Lankan and Paraguayan populations. *Japanese Journal of Human Genetics*, 34, 195–202.

Vaca, G., Ibarra, B., Garcia Cruz, D., *et al.* (1985). G6PD Jalisco and G6PD Morelia: Two new Mexican variants. *Human Genetics*, 71, 82–5.

Valenzuela, C.Y. and Harb, Z. (1977). Socioeconomic assortative mating in Santiago, Chile: a demonstration using stochastic matrices of mother–child relationships applied to ABO blood groups. *Social Biology*, 24, 225–33.

Van Der Sar, A., Schouten, H. and Boudier, A.M.S. (1964). Glucose-6-phosphate dehydrogenase deficiency in red cells. Incidence in the Curaçao population, its clinical and genetic aspects. *Enzymologia*, 27, 289–310.

van Loghem, E., Shuster, J., Fudenberg, H.H. and Franklin, E.C. (1969). Phylogenetic studies of immunoglobulins: evolution of Gm factors in Primates. *Annals, New York Academic of Sciences*, 162(1), 161–9.

Vu Tien, J., Pison, G., Lévy, D., Darcos, J.-C., Constans, J. and Mauran-Sendrail, A. (1975). Le phénotype HpO dans quelques populations d'Afrique et d'Amérique Centrale. *Comptes Rendus de l'Academie de Sciences de Paris, Série D*, 280, 2281–4.

Appendix 2

αGLU	Alpha glucosidase
α2M	Alpha-2-macroglobulin
ABH	ABH secretor
ABO	ABO blood group
ACAH-C5	C5 cholinesterase
ACE	Angiotensin-converting enzyme
ACONM	Mitochondrial aconitase
ACONS	Soluble aconitase
ACP	Acid phosphatase
ADA	Adenosine deaminase
ADH	Alchool dehydrogenase
ADH2	Alcohol dehydrogenase-2
ADH3	Alcohol dehydrogenase-3
AHSG	Alpha-2-HS-glycoprotein
AK	Adenylate kinase
ALADH	Delta-aminolevulinate dehydrogenase
ALB	Albumin
ALDH	Acetaldehyde dehydrogenase
ALDH2	Aldehyde dehydrogenase-2
ALPP	Placental alkaline phosphatase
APO	Apolipoprotein
APOB	Apolipoprotein B
APOC2	Apolipoprotein C-II
ASA-PD	Arylsulfatase A pseudodeficiency
AT3–I/D	Antithrombin III insertion-deletion polymorphism
BF	Properdin factor B
C3	Complement component 3
C4	Complement component 4
C5	Complement component 5
C6	Complement component 6
C7	Complement component 7
CA2	Carbonic anhydrase
CBS	Cystathionine β-synthase

CHE1	Serum cholinesterase 1
CHE2	Serum cholinesterase 2
CKR-5	Beta-chemokine receptor gene
CP	Ceruloplasmin
CSF1	Macrophage colony stimulating factor
CYP19	Cytochrome P450 aromatization of androgen
DHA1	Alcohol dehydrogenase, locus 1
DHA2	Alcohol dehydrogenase, locus 2
DHA3	Alcohol dehydrogenase, locus 3
DI	Diego blood group
DM	Myotonic dystrophy
DRD4	Dopamine D4 receptor gene
DRPLA	Dentatorubral-pallidoluysian atrophy
ESD	Esterase D
EST	Expressed sequence tag
FABP	Fatty acid binding protein
FcγRIIIB	Gene that encodes the glucose phosphate isomerase-linked receptor on neutrophils
FES	Oncogene FES, feline sarcoma virus
FLT	Oncogene FLT
FY	Duffy blood group
FVIII	Coagulation factor VIII
FIX	Coagulation factor IX
F13A	Coagulation factor 13A
F13B	Coagulation factor 13B
FMR1	Gene causing the fragile X syndrome
FRAXAC1	Microsatellite flanking the fragile X site
GC	Group-specific component
GC (iso)	Group-specific component (isoelectric focusing)
G6PD	Glucose-6-phosphate dehydrogenase
GALT	Galactose 1-phosphate uridyl-transferase
GLOI	Glyoxalase I
GM	GM immunoglobulin
Go^a	Gonzales blood group
GPI	Glucose phosphate isomerase
GPX1	Glutathione peroxidase
GPT	Glutamic pyruvate transaminase
GRM1	Alpha-1-acid glycoprotein 1
GRM1 (iso)	Alpha-1-acid glycoprotein 1 (isoelectric focusing)
GSTM1	Glutathione S-transferase mu
GSTT1	Glutathione S-transferase theta

HB	Hemoglobin
HD	Huntington disease
HLA-A	Histocompatibility locus A
HLA-B	Histocompatibility locus B
HLA-C	Histocompatibility locus C
HLA-D	Histocompatibility locus D
HLA-DP	Histocompatibility locus DP
HLA-DQ	Histocompatibility locus DQ
HLA-DR	Histocompatibility locus DR
HP	Haptoglobin
HP (iso)	Haptoglobin (isoelectric focusing)
HPA	Human platelet antigen
HPRT	Hypoxanthine-guanine phosphoribosyltransferase
ICAM-1	Intercellular adhesion molecule-1
IGHA	Immunoglobulin switch alpha region
KELL	Kell blood group
KIDD	Kidd blood group
LAC-DF	Lactase deficiency
LDH	Lactate dehydrogenase
LE	Lewis blood group
LPL	Lipoprotein lipase
LU	Lutheran blood group
MBPA/B	Myelin basic protein A/B
MN	MN blood group
MNSs	MNSs blood group
MTHFR	Methylenetetrahydrofolate reductase
NADH-MR	NADH metahemoglobin reductase
OCA-2	Oculocutaneous albinism-2
ORM1	Orosomucoid 1
P	P blood group
PAH	Phenylalanine hydroxylase
PEPA	Peptidase A
PEPB	Peptidase B
PEPC	Peptidase C
PGD	Phosphogluconate dehydrogenase
PGI	Phosphoglucose isomerase
PGM1	Phosphoglucomutase 1
PGM1 (iso)	Phosphoglucomutase 1 (isoelectric focusing)
PGM2	Phosphoglucomutase 2
PGM3	Phosphoglucomutase 3
PGM4	Phosphoglucomutase 4

PGP	Phosphoglycolate phosphatase
PI	Alpha-1-antitrypsin
PLA	Plasminogen activator
PLAP	Placental alkaline phosphatase
RB	Retinoblastoma
RENA4	Renin A4
RH	Rhesus blood group
Sb19.3	*Alu* insertion, region 19.3
SCA	Spinocerebellar ataxia
SODA	Superoxide dismutase A
SRY	Sex-determining region of the Y
TF	Transferrin
THO1	Tyrosine hydroxylase
TPOX	Thyroid peroxidase X
UMPK	Uridine monophosphate kinase
vWF	von Willebrand Factor
Yαh	Alphoid heteroduplex haplotypes
YAP	Polymorphic *Alu* element of the Y (DYS287)
YCAII	Y chromosome short tandem repeat CAII

Note: For anonymous DNA sequences, the convention is to use D (= DNA) followed by 1–22, X or Y to indicate the chromosome location, then S for a unique segment, Z for a chromosome-specific repetitive DNA family or F for a multilocus DNA family, and finally a serial number. The letter E following an anonymous clone number indicates that the clone is known to be expressed.

Appendix 3

Publications related to the genetics of Brazilian populations not mentioned in the text but which were considered in the compilation of Table 8.7

Aben-Athar, J. (1927). Isso-aglutininas do sangue dos brasileiros. *Scientia Medica* (Rio de Janeiro), **5**, 145–53.

Alcântara, V.M., Chautard-Freire-Maia, E.A., Picheth, G. and Vieira, M.M. (1990). Frequency of the *CHE1*K* allele of serum cholinesterase in a sample from southern Brazil. *Human Heredity*, **40**, 386–90.

Alcântara, V.M., Lourenço, M.A.C. de., Salzano, F.M., *et al.* (1995). Butyrylcholinesterase polymorphisms (BCHE and CHE2 loci) in Brazilian Indian and admixed populations. *Human Biology*, **67**, 717–26.

Alter, A.A., Gelb, A.G., Chown, B., Rosenfield, R.E. and Cleghorn, T.E. (1967). Gonzales (Goᵃ), a new blood group character. *Transfusion*, **7**, 88–91.

Alves, M.G.L., Pitombeira, M.S., Lima, L.M.A., Fujita, M.N. and Martins, J.M. (1986). Antígeno Diego (Diᵃ) na população do Ceará: estudo realizado em dois serviços de hemoterapia em Fortaleza. *Revista de Medicina da Universidade Federal do Ceará* (Fortaleza), **26/27**, 1–7.

Anonymous. (1944). Blood typing of Brazilian soldiers. *JAMA* (*Journal of the American Medical Association*), **126**, 186.

Arpini-Sampaio, Z.A. (1998). Estrutura genética de três isolados Afro-Brasileiros. Ph.D. thesis, University of São Paulo, Ribeirão Preto.

Azevêdo, E.E.S. (1969). Estudo eletroforético dos polimorfismos adenilato-quinase e 6-fosfogluconato desidrogenase no Nordeste brasileiro. *Ciência e Cultura*, **21**, 234.

Azevêdo, E.S. (1970). Adenylate kinase and phosphogluconate dehydrogenase polymorphisms in northeastern Brazil. *American Journal of Human Genetics*, **22**, 1–6.

Azevêdo, E.S. (1976). Polimorfismos: da bioquímica às populações. *Ciência e Cultura*, **28**, 298–303.

Azevêdo, E.S. (1979). Características antropogenéticas da população da Bahia, Brasil. *Ciencia Interamericana*, **19**, 28–35.

Azevêdo, E.S. and Azevedo, T.F.S. (1974). Glucose-6-phosphate dehydrogenase deficiency and neonatal jaundice in Bahia, Brazil. *Ciência e Cultura*, **26**, 1044–7.

Azevêdo, E.S. and Yoshida, A. (1969). Brazilian variant of glucose-6-phosphate

350   *Appendix 3*

dehydrogenase (Gd Minas Gerais). *Nature*, **222**, 380–2.
Azevêdo, E., Krieger, H. and Morton, N.E. (1969). Ahaptoglobinemia in
northeastern Brazil. *Human Heredity*, **19**, 609–12.
Azevêdo, E.S., Silva, M.C.B.O. and Tavares-Neto, J. (1975). Human alcohol
dehydrogenase ADH$_1$, ADH$_2$ and ADH$_3$ loci in a mixed population of
Bahia, Brazil. *Annals of Human Genetics*, **39**, 321–7.
Azevêdo, E.S., Silva, M.C.B.O. and Tavares-Neto J. (1977). Studies on human
alcohol dehydrogenase *locus* ADH$_3$ in Bahia, Brazil. *Revista Médica da
Bahia*, **23**, 129–36.
Azevêdo, W.C., Silva, M.L.F., Grassi, M.C.B. and Azevêdo, E.S. (1978).
Deficiência de glicose-6-fosfato desidrogenase em pacientes de um Hospital
Geral de Salvador, Bahia, Brasil. *Revista Brasileira de Pesquisas Médicas e
Biológicas*, **11**, 49–52.
Azevêdo, E.S., Silva, M.C.B.O., Lima, A.M.V.M.D., Fonseca, E.F. and
Conceição, M.M. (1979). Human aconitase polymorphism in three samples
from northeastern Brazil. *Annals of Human Genetics*, **43**, 7–10.
Azevêdo, E.S., Silva, K.M.C., Silva, M.C.B.O., Lima, A.M.V.M.D., Fortuna,
C.M.M. and Santos, M.G. (1981). Genetic and anthropological studies in
the island of Itaparica, Bahia, Brazil. *Human Heredity*, **31**, 353–7.
Bacila, M. (1946). Contribuição ao estudo do fator Rh em Curitiba. Ph.D. thesis,
Curitiba, Tipografia João Haupt e Cia.
Balazs, I. (1993). Population genetics of 14 ethnic groups using phenotypic data
from VNTR loci. In Pena, S.D.J., Chakraborty, R., Epplen, J.T. and
Jeffreys, A.J., editors. *DNA Fingerprinting: State of the Science*, Basel:
Birkhäuser Verlag, pp. 193–210.
Balthazar, P.A., Gabriel Jr. A., Oliveira, D.P., *et al.* (1985). HLA-DR em
portadores de lúpus eritematoso sistêmico com anticorpos
anti-ribonucleoproteína e na doença mista do tecido conectivo. *Revista do
Hospital de Clínicas da Faculdade de Medicina* (São Paulo), **40**(6), 249–53.
Barraviera, B. (1984). Malária causada pelo *Plasmodium falciparum* e a redução
da metahemoglobina pela via das pentoses. M.Sc. dissertation, Universidade
Estadual Paulista Julio de Mesquita Filho, Botucatu, SP.
Barraviera, B., Meira, D.A., Machado, P.E.A. and Curi, P.R. (1987). Malária no
município de Humaitá, Estado do Amazonas. XXI. Prevalência da
deficiência de glicose-6-fosfato desidrogenase (G6PD) em amostra da
população e em doentes com malária causada pelo *Plasmodium falciparum*.
Revista do Instituto de Medicina Tropical (São Paulo), **29**, 374–80.
Barreto, J.M. (1936). Grupos sangüíneos. *Revista Médica do Paraná*, **5**(8), 291–6.
Barretto, O.C.O. (1970). Erythrocyte glucose-6-phosphate dehydrogenase
deficiency in São Paulo, Brazil. *Revista Brasileira de Pesquisas Médicas e
Biológicas*, **3**, 61–5.
Barretto, O.C.O.P., Enokihara, M.Y., Mazar Jr. W., Ziwian, Z.L.J. and Ferreira,
J.L.M.S. (1983). Distribuição do sistema ABO e RH, destacando-se a
pesquisa do antígeno D^u, em Santo André, SP. *Revista do Hospital de
Clínicas da Faculdade de Medicina* (São Paulo), **38**(3), 111–14.
Beiguelman, B. and Marchi, A. (1962). The frequency of blood groups among

Japanese immigrants in Brazil. *American Journal of Physical Anthropology*, 20, 29–31.

Carlos, L.M.B., Silva, V.F.P., Pitombeira, M.S. and Campos, O.R. (1992). Incidência do caráter secretor em doadores de sangue. *Revista de Medicina da Universidade Federal do Ceará*, 32, 11–18.

Carvalho, J.R. (1929). Fixidez dos isoaglutinogeneos e das isoaglutininas (contribuição para o seu estudo e aplicação na investigação de paternidade). Thesis, Faculdade de Medicina de São Paulo, São Paulo.

Cerqueira, A.J.B., Junqueira, P.C. and Tsuno, T. (1968). Grupos sangüíneos dos sistemas ABO, Rh e Diego em Japoneses. *Folha Médica*, 57, 567–71.

Chautard-Freire-Maia, E.A., Primo-Parmo, S.L., Canever, M.A. and Culpi, L. (1979). Sistema da colinesterase do soro: microtécnica e freqüências de fenótipos não usuais em amostras de Curitiba. *Ciência e Cultura*, 31, 624–5.

Chautard-Freire-Maia, E.A., Carvalho, R.D.S., Silva, M.C.B.O., Souza, M.G.F. and Azevêdo, E.S. (1984). Frequencies of atypical serum cholinesterase in a mixed population of northeastern Brazil. *Human Heredity*, 34, 364–70.

Chautard-Freire-Maia, E.A., Lourenço, M.A.C. and Jugend, R.M. (1984). Phenotype frequencies of the *CHE2* locus of serum cholinesterase in a sample collected in Curitiba. *Revista Brasileira de Genética*, 7, 709–15.

Chautard-Freire-Maia, E.A., Primo-Parmo, S.L., Lourenço, M.A.C. and Culpi, L. (1984). Frequencies of atypical serum cholinesterase among Caucasians and Negroes from southern Brazil. *Human Heredity*, 34, 388–92.

Chaves, M. and Azevêdo, E.S. (1987). O alelo D^u em Salvador, Bahia: freqüência gênica e identificação de tipo. *Ciência e Cultura*, 39, 316–17.

Corvelo, T.C.O. and Salzano, F.M. (1984). New genetic data from an Amazonian town. *Interciencia*, 9, 236–8.

Corvelo, T.C.O. and Schneider, H. (1996). A subgroups of the ABO blood system. *Ciência e Cultura*, 48, 184–8.

Costa, E.J., Estalote, A.C. and Palatnik, M. (1996). Electrophoretic variation of hair proteins. *Brazilian Journal of Medical and Biological Research*, 29, 1427–9.

Costa Ferreira, H. (1945). O fator Rh e sua importância clínica. *Arquivos de Cirurgia e Clínica Experimental*, 6, 60.

Cuadrado, R.R. and Davenport, F.M. (1970). Antibodies to influenza viruses in military recruits from Argentina, Brazil and Colombia: Their relation to ABO blood group distribution. *Bulletin of the World Health Organization*, 42, 873–84.

De Lucca, E.J. (1974). Grupos sangüíneos ABO e Rh em Humaitá (AM). *Folha Médica*, 68, 451–4.

Deka, R., Miki, T., Yin, S.-J., *et al.* (1995). Normal CAG repeat variation at the DRPLA locus in world populations. *American Journal of Human Genetics*, 57, 508–11.

Donin, C. and Lipinski-Figueiredo, E. (1984). Frequency of serum cholinesterase variants in a sample of Brazilian newborns. *Revista Brasileira de Genética*, 7, 593–6.

Duarte, A. (1926). Contribuição ao estudo dos grupos sangüíneos na Bahia.

Thesis, Universidade da Bahia, Salvador.

Duarte, A. (1935). Grupos sangüíneos da raça negra. In: *Estudos Afro-Brasileiros. Trabalhos apresentados no 1º. Congresso Afro-Brasileiro* (Rio de Janeiro).

Duarte, E. and Oliveira, J.E.D. (1978). Intolerância à lactose em adultos. *Revista Brasileira de Pesquisas Médicas e Biológicas*, **11**, 105–9.

Engrácia, V., Mestriner, M.A., Cabello, P.H. and Krieger, H. (1991). Association between the acid phosphatase 1 and adenosine deaminase systems in a Brazilian sample. *Human Heredity*, **41**, 147–50.

Faria, R. and Ottensooser, F. (1951). Grupos ABO e tipos de Rh em pretos e mulatos de São Paulo. *Seara Medica*, **6**, 497–503.

Faria, R., Mello, N.R. and Murat, L.G. (1951). Contribuição ao estudo médico e social do doador. *Arquivos de Clínica*, **12**, 408–17.

Favero, F. (1935). Contribuição do Instituto Oscar Freire para o estudo dos tipos sangüíneos. *Arquivos de Medicina Legal e Identificação*, **5**(12), 231–43.

Favero, F. and Novah, E. (1936). Predominância dos tipos sangüíneos no meio universitário de São Paulo. *Boletim do Instituto Oscar Freire* (São Paulo), **3**, 9–11.

Favero, F. and Novah, E. (1937). Os tipos sangüíneos no meio universitário de São Paulo. *Arquivos da Sociedade de Medicina Legal e Criminologia*, **8**, 19–21.

Favero, F. and Novah, E. (1945). As verificações do Instituto Oscar Freire quanto aos tipos sangüíneos nos universitários paulistas. *Anais da Faculdade de Medicina da Universidade de São Paulo*, **21**, 227–9.

Ferreira, E., Ward, F.E. and Amos, D.B. (1975). HL-A in a Brazilian population: Evidence for new HL-A specificities. In: *Histocompatibility Testing 1975*. Copenhagen: Munksgaard, pp. 226–32.

Ferreira, H.C. and Toledo, R. (1953). Moléstia hemolítica de recém-nascidos atribuída aos fatores A e P; freqüência dos gens P-p na cidade de São Paulo. *Revista Paulista de Medicina*, **43**, 235–42.

Ferreira, H.C., Lacaz, C.S. and Mellone, O. (1946). Resultados de 260 determinações do fator Rh na cidade de São Paulo. *Revista Brasileira de Medicina*, **3**, 89–91.

Figueiredo, E.L. (1982). Deficiência de alfa$_1$-antitripsina: avaliação de técnica semiquantitativa e freqüência em amostra de Curitiba. M.Sc. dissertation, Federal University of Paraná, Curitiba.

Franco, M.H.L.P. and Salzano, F.M. (1985). A comparative study of albumin variants found in Brazil. *Human Heredity*, **35**, 34–8.

Franco, M.H.L.P., Moreira, D.M., Salzano, F.M., Santos, S.E.B., Conceição, M.M. and Schneider, H. (1986). New data on the association between the glyoxalase I and haptoglobin loci. *Human Heredity*, **36**, 126–8.

Garlipp, C.R. and Ramalho, A.S. (1988). Aspectos clínicos e laboratoriais da deficiência de desidrogenase de 6-fosfato de glicose (G-6-PD) em recém nascidos brasileiros. *Revista Brasileira de Genética*, **11**, 717–28.

Godoy, O. (1937). Grupos sangüíneos nos criminosos de São Paulo. *Arquivos da Sociedade de Medicina Legal e Criminologia*, **8**, (Suppl.), 185–91.

Gomes, M.B., Ruzany, F., Quadra, A.A., Sarno, E.N. and Arduino, F. (1981). Histocompatibility antigens and insulin-dependent diabetes: A study of 20 Brazilian families. *Brazilian Journal of Medical and Biological Research*, **14**, 379–81.

Grisi, M.E.S. and Azevêdo, E.S. (1976). Hypohaptoglobinemia in rural and urban populations of Bahia State, Brazil. *Human Biology*, **48**, 653–7.

Guerreiro, J.F. and Chautard-Freire-Maia, E.A. (1984). Studies on serum cholinesterase (*CHE1* locus) in a sample from northern Brazil. *Revista Brasileira de Genética*, **7**, 717–25.

Guerreiro, J.F., Ribeiro-dos-Santos, A.K.C., Santos, E.J.M., *et al.* (1999). Genetical-demographic data from two Amazonian populations composed of descendants of African slaves: Pacoval and Curiau. *Genetics and Molecular Biology*, **22**, 163–7.

Hinrichsen, R. (1929). Estado atual da questão dos grupos hemáticos. *Actas e Trabalhos do 1º. Congresso Brasileiro de Eugenia* (Rio de Janeiro), 169.

Horta de Figueiredo, L. (1951). Resumo sobre 3000 classificações de grupos sangüíneos. *Arquivos de Clínica*, **12**, 445–9.

Hubner, M. (1952). Investigação do fator Rh e grupos sangüíneos em 2000 enfermas. *Acta Ginecologica*, **6**, 130.

Hutz, M.H., Yoshida, A. and Salzano, F.M. (1977). Three rare G-6-PD variants from Porto Alegre, Brazil. *Human Genetics*, **39**, 191–7.

Imanishi, T., Akaza, T., Kimura, A., Tokunaga, K. and Gojobori, T. (1992). Allele and haplotype frequencies of HLA and complement loci in various ethnic groups. In Tsuji, K., Aizawa, M. and Sasazuki, T., editors. *HLA 1991*, Vol. 1. Oxford: Oxford University Press, pp. 1065–220.

Junqueira, P.C. and Wishart, P.J. (1953). Incidência do fator Cʷ na população do Rio de Janeiro. *Arquivos Brasileiros de Medicina*, **43**, 199–204.

Kelso, A.J., Siffert, T. and Maggi, W. (1992). Association of ABO phenotypes and body weight in a sample of Brazilian infants. *American Journal of Human Biology*, **4**, 607–11.

Kelso, A.J., Maggi, W. and Beals, K.L. (1994). Body weight and ABO blood types: are AB females heavier? *American Journal of Human Biology*, **6**, 385–7.

Kelso, A.J., Siffert, T. and Thieman, A. (1995). Do type B women have more offspring? An instance of asymmetrical selection at the ABO blood group locus. *American Journal of Human Biology*, **7**, 41–4.

Krieger, H. and Barbosa, C.A.A. (1979). Smallpox and the ABO system association: A critical review. *Revista Brasileira de Biologia*, **39**, 195–9.

Kvitko, K. and Weimer, T.A. (1990). Phosphoglucomutase proteins in human milk. *Revista Brasileira de Genética*, **13**, 125–31.

Lacaz, C.S. (1951). Novos dados estatísticos sobre a incidência do fator Rh na cidade de São Paulo (Brasil). *Revista Paulista de Medicina*, **38**, 17–20.

Lacaz, C.S. and Maspes, V. (1951). Distribuição do fator P na cidade de São Paulo. *Folia Clinica et Biologica*, **17**, 165–8.

Lacaz, C.S., Ferreira, H.C. and Mellone, O. (1947). Dados estatísticos sobre o fator Rh. *Anais Paulistas de Medicina e Cirurgia*, **53**, 319–23.

Lacaz, C.S., Pinto, D.O., Borges, S.O., Mellone, O. and Yahn, O. (1955). Incidência do fator Kell em São Paulo. *Folia Clinica et Biologica*, **23**, 43–8.

Lacaz, C.S., Pinto, D.O., Borges, S.O., Mellone, O. and Yahn, O. (1955). The incidence of the Kell factor in São Paulo, Brazil. *American Journal of Physical Anthropology*, **13**, 349–50.

Lai, L., Nevo, S. and Steinberg, A.G. (1964). Acid phosphatases of human red cells: predicted phenotype conforms to a genetic hypothesis. *Science*, **145**, 1187–8.

Lefèvre, J.A. (1927). Da hereditariedade dos grupos sangüíneos e sua aplicação na investigação da paternidade. Thesis, São Paulo, Irmãos Ferraz.

Leme Lopes, M.B. (1948). O fator Rh na população do Rio de Janeiro. Dados estatísticos. *Revista Brasileira de Medicina*, **5**, 479–82.

Leme Lopes, M.B. (1951). Cinco anos de experiência em Banco Central de Sangue. *Arquivos de Clínica*, **12**, 533–47.

Leme Lopes, M.B. and Junqueira, P.C. (1951). O sistema Rh no Rio de Janeiro. Dados estatísticos. *Seara Médica*, **6**, 483–5.

Lewgoy, F. and Salzano, F.M. (1968). G-6-PD deficiency gene dynamics in a Brazilian population. *Acta Geneticae Medicae et Gemellologiae*, **17**, 595–606.

Lima, P. (1932). Grupos sangüíneos em cadáveres. Da imutabilidade dos grupos sangüíneos 'Post-mortem'. Thesis, Faculdade de Medicina, São Paulo.

Magna, L.A., Morandin, R.C., Pinto Jr. W. and Beiguelman, B. (1980). Frequency of the atypical serum cholinesterase in southeastern Brazilian Caucasoids. *Revista Brasileira de Genética*, **3**, 329–37.

Magna, L.A., Moura, T.J.A., Pinto Jr. W. and Beiguelman, B. (1983). Atypical pseudocholinesterase in Northeastern Brazilian Caucasoids. *Revista Brasileira de Genética*, **6**, 381–4.

Marques, J. and Campos, J.O. (1975). Incidência da deficiência de glicose-6-fosfato deidrogenase em negros de Minas Gerais. *Revista da Associação Médica Brasileira*, **21**(4), 111–12.

Marques Ferreira, C.S., Souza, M.M.M. and Azevedo, E.S. (1973). ABO gene frequencies in a mixed sample of 9391 blood donors in Bahia, Brazil. *Ciência e Cultura*, **25**, 573–4.

Massud, M., Andrade, S.M.G.M. and Uribe, M.R.N. (1978). Genética e tratamento da tuberculose. *Revista Médica*, **7**(8), 18–26.

Mellone, O., Ludovicci, J., Maluf, M. and Macruz, R. (1952). Incidência dos grupos sangüíneos do sistema ABO no Serviço de Transfusão do Hospital das Clínicas de São Paulo. *Revista Paulista de Medicina*, **40**, 287–8.

Melo, M.C.A., Lima, A.M.V.M.D., Silva, M.C.B.O., Sousa, M.G.F. and Azevêdo, E.S. (1988). PGM3 polymorphism in human placenta samples from Bahia, Brazil. *Gene Geography*, **2**, 65–9.

Mesquita, M.P. and Leite-Ribeiro, V. (1947). Pesquisas sobre o fator Rh na cidade do Rio de Janeiro. *O Hospital*, **32**(4), 505–22.

Mestriner, M.A. and Salzano, F.M. (1998). Monomorphic and polymorphic enzyme genetic markers of the Waiãpi Indians of Amapá and inhabitants of Manaus, Amazonas. *Genetics and Molecular Biology*, **21**, 311–14.

Montenegro, L. (1959). Grupos sanguíneos em amostra da população do

Amazonas. *Anais, I Reunião Brasileira de Genética Humana*, 92–4.

Morton, N.E. (1964). Genetic studies of northeastern Brazil. *Cold Spring Harbor Symposia on Quantitative Biology*, **29**, 69–79.

Moura, M.M.F. and Krieger, H. (1994). Studies on polymorphism segregation in northeastern Brazil. II. Pairs of relatives. *Revista Brasileira de Genética*, **17**, 109–12.

Moura-Neto, R.S., Silva, R., Carvalho, E.F. and Zorio, D.R. (1993). Comparison of Rio de Janeiro DNA typing data with the FBI worlwide study. *Ciência e Cultura*, **45**, 258–62.

Naoum, P.C., Megid, J.C.C., Cruz Filho, N.A. and Cury, P.R. (1977). Deficiência de glicose-6-fosfato deidrogenase em população hospitalar. Comparação com caracteres, nível de hemoglobina, e determinação de tamanho de amostra. *Revista Brasileira de Patologia Clínica*, **13**, 265–70.

Novah, E. (1949). Estatística de tipos sanguíneos do Instituto Oscar Freire. *Boletim do Instituto Oscar Freire*, **5**, 3–4.

Osorio, F., Amaral, M. and Maulaz, P. (1957). Indice bioquimico racial em Pelotas. *Revista da Faculdade de Odontologia de Pelotas*, **3**, 101–8.

Ottensooser, F. (1944). Cálculo do grau de mistura racial através dos grupos sangüíneos. *Revista Brasileira de Biologia*, **4**, 531–7.

Ottensooser, F. and Cavalcanti, M.A.A. (1966). O genótipo paterno como elemento prognóstico na doença hemolítica por Rh: influências raciais. *Revista Paulista de Medicina*, **68**, 11–19.

Ottensooser, F. and Pasqualin, R. (1944). Aspectos sorológicos da transfusão de sangue. III. Acidentes hemolíticos após transfusão, eritroblastose fetal e o novo fator sanguíneo Rh. *Arquivos de Biologia*, **27**(261), 49–56.

Ottensooser, F. and Rosales, T. (1970). Os subgrupos de A em São Paulo. *Revista do Hospital de Clínicas da Faculdade de Medicina* (São Paulo), **25**, 161–4.

Ottensooser, F., Lacaz, C.S., Ferreira, H.C. and Mellone, O. (1947). Os 8 tipos de Rh. Aplicações clínicas e antropológicas. *Arquivos de Biologia*, **31**, 36–42.

Ottensooser, F., Lacaz, C.S., Ferreira, H.C. and Mellone, O. (1948). Distribution of the Rh types in São Paulo (Brazil). *Blood*, **3**, 696–8.

Ottensooser, F., Leon, N. and Cunha, A.B. (1961). Beiträge zur Kenntnis der Gm-Gruppen. *Zeitschrift für Immuntätsforschung und experimentelle Therapie*, **122**, 165–78.

Ottensooser, F., Leon, N., Sato, M. and Saldanha, P.H. (1963). Blood groups of a population of Ashkenazi Jews in Brazil. *American Journal of Physical Anthropology*, **21**, 41–8.

Ottensooser, F., Leon, N. and Almeida, T.V. (1975). ABO blood groups and isoagglutinins in systemic lupus erythematosus. *Revista Brasileira de Pesquisas Médicas e Biológicas*, **8**, 421–6.

Otto, P.A. and Bozóti, M.M. (1968). Digital dermatoglyphics and blood-groups. *The Lancet*, **II**, 1250–1.

Palatnik, M. (1984). A and AB ABO blood groups variants in Brazil. *Revista Brasileira de Genética*, **7**, 727–33.

Palatnik, M., Sá e Benevides, Maria J.F. and Salzano, F.M. (1969). ABH salivary secretion and White/Negro gene flow in a Brazilian population. *Human*

Biology, **41**, 83–96.

Palatnik, M., Soares, M.B.M., Junqueira, P.C., Corvelo, T.C.O., Soares, V.M.F.C. and Faria, G.M.F. (1980). Weak reacting A antigen in an AB Brazilian Negroid donor. *Revista Brasileira de Biologia*, **40**, 721–7.

Palatnik, M., Rosenblit, J., Junqueira, P.C., El Khoury, A.B., Alves, Z.M.S. and Santos, A.A. (1981). El fenotipo A_x, Un estudio serológico y genético. *Sangre*, **26**, 324–34.

Palatnik, M., Junqueira, P.C. and Alves, Z.M.S. (1982). Fs: antigenic determinant possibly related to the Duffy blood group. *Blood Transfusion and Immunohaematology*, **25**, 629–37.

Palatnik, M., Laranjeira, N.S.M., Malajovich, M.A.M., *et al.* (1985). The A_3 phenotype: serologic and anthropologic aspects. *Revista Brasileira de Genética*, **8**, 115–22.

Palatnik, M., Laranjeira, N.S.M., Simões, M.L.M.S. and Loureiro, J.B. (1987). Genetic heterogeneity of the blood group O, Le(a-b-) phenotype. *Revista Brasileira de Genética*, **10**, 761–8.

Palatnik, M., Laranjeira, N.S.M., Simões, M.L.M.S. and Loureiro, J.B. (1988). Distribution of the Lewis and secretor traits in Rio de Janeiro, Brazil. *Revista Brasileira de Genética*, **11**, 187–92.

Palatnik, M., Simões, M.L.M.S., Alves, Z.M.S. and Laranjeira, N.S.M. (1990). The 60 and 63 kDa proteolytic peptides of the red cell membrane band–3 protein: their prevalence in human and non-human primates. *Human Genetics*, **86**, 126–30.

Palatnik, M., Simões, M.L.M.S., Guinsburg, S.S. and Lopes, H. (1992). Genetic polymorphism of red cell membrane band 3 in Japanese Brazilians. *Gene Geography*, **6**, 17–20.

Pasqualin, R. (1941). Grupos sangüíneos e índice bioquímico racial em São Paulo. *Arquivos de Biologia*, **25**, 179–81.

Pena, S.D.J., Santos, P.C., Campos, M.C.B.N. and Macedo, A.M. (1993). Paternity testing in Brazil by DNA fingerprinting with the multilocus probe F10. *Ciência e Cultura*, **45**, 236–40.

Pena, S.D.J., Santos, P.C., Campos, M.C.B.N. and Macedo, A.M. (1993). Paternity testing with the F10 multilocus DNA fingerprinting probe. In Pena, S.D.J., Chakraborty, R., Epplen, J.T. and Jeffreys, A.J., editors. *DNA Fingerprinting: State of the Science*. Basel: Birkhäuser Verlag, pp. 237–47.

Penalva da Silva, F. and Krieger, H. (1981). The maintenance of ABO polymorphism. *Revista Brasileira de Genética*, **4**, 705–12.

Penalva da Silva, F. and Krieger, H. (1982). Haptoglobin segregation and interaction with ABO blood groups. *Revista Brasileira de Genética*, **5**, 157–63.

Primo-Parmo, S.L., Chautard-Freire-Maia, E.A., Lourenço, M.A.C., Salzano, F.M. and Freitas, M.J.M. (1986). Studies on serum cholinesterase (CHE1 and CHE2) in Brazilian Indian and admixed populations. *Revista Brasileira de Genética*, **9**, 467–78.

Ramalho, A.S. and Beiguelman, B. (1976). Deficiência de desidrogenase de 6-fosfato de glicose (G6-PD) em doadores de sangue brasileiros. *Folha*

Médica, **73**, 281–3.

Resende, J. (1944). A eritroblastose fetal. Problema obstétrico. Thesis, Escola de Medicina e Cirurgia, Rio de Janeiro.

Ribeiro, E.B. (1963). Os grupos sangüíneos ABO e o câncer do estômago em São Paulo. *Anais Paulistas de Medicina e Cirurgia*, **86**(2), 87–95.

Ribeiro, V.R.L. (1949). Estudo estatístico sobre a incidência do fator Rh no Distrito Federal. *Vida Médica*, **17**(2), 24–6.

Rieger, T.T., Weimer, T.A., Salzano, F.M., Franco, M.H.L.P. and Moreira, D.M. (1988). ESD, ACP and GLO frequencies in Southern Brazil, with a note on heterozygosity levels. *Revista Brasileira de Genética*, **11**, 155–63.

Robinson, W.M., Hickmann, A.C., Weimer, T.A., Franco, M.H.L.P., Geiger, C.J. and Salzano, F.M. (1995). Protein genetics and paternity determinations in southern Brazil. *Indian Journal of Human Genetics*, **1**, 105–10.

Russo, E. (1964). Estudo de 4390 classificações de fatores Rh e de grupos sangüíneos com relação ao sexo e à cor. 4 casos de incompatibilidade sangüínea materno-fetal. *O Hospital*, **65**, 141–60.

Saldanha, P.H. (1956). ABO blood groups, age and disease. *Revista Brasileira de Biologia*, **16**, 349–53.

Saldanha, S.G. and Itskan, S.B. (1971). Grupos sangüíneos e atividade da glucose-6-fosfato desidrogenase em japoneses da cidade de São Paulo. *Revista Brasileira de Biologia*, **31**, 337–40.

Saldanha, P.H., Frota-Pessoa, O., Eveleth, P., Ottensooser, F., Cunha, A.B. and Cavalcanti, M.A.A. (1960). Estudo genético e antropológico de uma colonia de holandeses do Brasil. *Revista de Antropologia* (São Paulo), **8**, 1–42.

Saldanha, P.H., Maia, J.C.C. and Nóbrega, F.G. (1969). Distribution of erythrocyte glucose-6-phosphate dehydrogenase activity and electrophoretic variants among different racial groups in Brazil. *Revista Brasileira de Pesquisas Médicas e Biológicas*, **2**, 327–33.

Saldanha, S.G. (1979). ABO blood groups and salivary secretion of ABH antigens in patients with congenital heart defects. *Revista Brasileira de Genética*, **2**, 57–68.

Salzano, F.M., Franco, M.H.L.P. and Ayres, M. (1974). Alloalbuminemia in two Brazilian populations: a possible new variant. *American Journal of Human Genetics*, **26**, 54–8.

Santos, F.R., Epplen, J.T. and Pena, S.D.J. (1993). Testing deficiency paternity cases with a Y-linked tetranucleotide repeat polymorphism. In Pena, S.D.J., Chakraborty, R., Epplen, J.T. and Jeffreys, A.J., editors. *DNA Fingerprinting: State of the Science.* Basel: Birkhäuser Verlag, pp. 261–5.

Santos, R.P. (1944). Classificação de sangue das Forças Expedicionárias Brasileiras. *Arquivos do Instituto de Biologia do Exército*, **5**, 61–2.

Santos, S.E.B., Salzano, F.M., Franco, M.H.L.P. and Melo e Freitas, M.J. (1983). Mobility, genetic markers, susceptibility to malaria and race mixture in Manaus, Brazil. *Journal of Human Evolution*, **12**, 373–81.

Schüler, L., Salzano, F.M., Franco, M.H.L.P., Melo e Freitas, M.J., Mestriner, M.A. and Simões, A.L. (1982). Demographic and blood genetic characteristics in an Amazonian population. *Journal of Human Evolution*,

358   *Appendix 3*

11, 549–58.
Schüler, L., Chardosim, N.G.O. and Salzano, F.M. (1986). Electrophoretic
salivary genetic variation and patterns of dispersion in a Brazilian
population. *International Journal of Anthropology*, 1, 229–38.
Sena, L.L.A. and Ramalho, A.S. (1985). Clinical evaluation of
glucose-6-phosphate dehidrogenase (G-6-PD) deficiency in a Brazilian
population. *Revista Brasileira de Genética*, 8, 89–96.
Sena, L.L.A., Ramalho, A.S., Barreto, O.C.O. and Lima, F.A.M. (1986).
Deficiência de desidrogenase de 6-fosfato glicose (G6-PD): dados de
prevalência e de morbidade na região de Natal, RN. *Revista da Associação
Médica Brasileira*, 32, 17–20.
Sessa, P.A., Lima Pereira, F.E., Barros, G.C., Mattos, C.A. and Daher, V.R.
(1985). Blood groups of the ABO and Rh systems and muco-cutaneous
leishmaniasis. *Revista Brasileira de Genética*, 8, 183–5.
Sevá-Pereira, A., Magalhães, A.F.N., Pereira-Filho, R.A. and Beiguelman, B.
(1983). Primary adult lactose malabsorption, a common genetic trait among
southeastern Brazilians. *Revista Brasileira de Genética*, 6, 747–59.
Silva, M.C.B.O., Lima, A.M.V.M.D., Santiago, C.A.S. and Azevêdo, E.S.
(1985). Alkaline phosphatase polymorphism of the human placenta: a study
in Brazilians. *Revista Brasileira de Genética*, 8, 79–87.
Silva, M.C.B.O., Azevêdo, E.S., Sousa, M.G.F., Carvalho, R.D.S. and
Passos-Bueno, M.R. (1995). Alkaline phosphatase (ALPP) in placental
extract and in the serum of black mixed parturients from Bahia, Brazil.
Revista Brasileira de Genética, 18, 305–9.
Silva, M.I.A.F., Salzano, F.M. and Lima, F.A.M. (1981). Migration, inbreeding,
blood groups and hemoglobin types in Natal, Brazil. *Studies in Physical
Anthropology*, 7, 3–11.
Silva, R.S. and Salzano, F.M. (1991). Placental polymorphisms in Southern
Brazil and their relationship to race and mother-child interactions. *Revista
Brasileira de Genética*, 14, 185–95.
Silva, R.S., Weimer, T.A. and Salzano, F.M. (1981). Rare and common types of
phosphoglucomutase in two Brazilian populations. *Human Biology*, 53,
227–38.
Simões, J. Jr. (1959). Incidência do fator Rh e grupos sangüíneos no meio
universitário. *Iª. Reunião Brasileira de Genética Humana*, 150–6.
Sousa, M.G.F., Passos-Bueno, M.R., Silva, M.C.B.O., Duarte, A.F.B.G. and
Azevêdo, E.S. (1990). The contribution of sex, electrophoretic phenotype,
pregnancy and race to the variability of delta-aminolevulinate dehydrase
(ALADH) levels in human erythrocytes. A study in Black mixed Brazilians.
Clinica Chimica Acta, 194, 229–34.
Stueber-Odebrecht, N. (1985). Studies on the CHE1 locus of serum cholinesterase
and surnames in a sample from Santa Catarina (southern Brazil). *Revista
Brasileira de Genética*, 8, 535–43.
Tavares-Neto, J. and Azevêdo, E.S. (1978). Family names and ABO blood group
frequencies in a mixed population of Bahia, Brazil. *Human Biology*, 50,
361–7.

Tondo, C.V., Carmen Mundt and F.M. Salzano. (1963). Haptoglobin types in Brazilian Negroes. *Annals of Human Genetics*, **26**, 325–31.

Valette, I.V.C. and Salzano, F.M. (1981). Genetic and physico-chemical studies in quantitative haptoglobin variants found in Porto Alegre. *Revista Brasileira de Biologia*, **41**, 81–90.

Varella-Garcia, M. (1977). Blood groups and dermatoglyphics in a Brazilian population of Arabian origin. *Ciência e Cultura*, **29**, 826–9.

Versiani, V. (1946). O fator Rh em Belo Horizonte. *Brasil Médico*, **60**, 367–70.

Villaça, J.A.F., Sad, W.E., Lima, L.M.A. and Rouger, P. (1982). Fenótipo Bm com fraca atividade sérica anti-B. *Revista da Associação Médica Brasileira*, **28**, 148–50.

Waldrigues, C.P.N. (1982). Populational study of three polymorphisms of erythrocytic enzymes (ESD, CAII and GLO) in the population of Bambuí, MG, Brazil. *Revista Brasileira de Genética*, **5**, 236–7.

Weimer, T.A., Salzano, F.M. and Hutz, M.H. (1981). Erythrocyte isozymes and hemoglobin types in a southern Brazilian population. *Journal of Human Evolution*, **10**, 319–28.

Weimer, T.A., Corvello, C.M. and Salzano F.M. (1984). The acid phosphatase polymorphism: new data and review of results in Amerindians. *Interciencia*, **9**, 401–3.

Weimer, T.A., Schüler, L., Beutler, E. and Salzano, F.M. (1984). Gd(+) Laguna, a new rare glucose-6-phosphate dehydrogenase variant from Brazil. *Human Genetics*, **65**, 402–4.

Weimer, T.A., Rieger, T.T. and Salzano, F.M. (1987). Occurrence of a rare variant of superoxide dismutase in Brazil. *Human Heredity*, **37**, 26–9.

Weissmann, J. (1981). Seltene Varianten im Adenosindesaminase-System. Hinweis auf die Phänotypen ADA 4-1, 4-2, 5-1, 5-2 und 6-1. *Ärztlich Laboratorium*, **27**, 44–7.

Weissmann, J., Vollmer, M. and Pribilla, O. (1981). Phänotypenverteilung und Genfrequenzen der Adenosindesaminase in Schleswig-Holstein im Vergleich mit Stichproben aus Portugal, Brasilien und Südafrika (Bantu-Xhosa und Weiße). *Zeitschrift für Rechtsmedizin*, **86**, 227–32.

Weitkamp, L.R., Salzano, F.M., Neel, J.V., Porta, F., Geerdink, R.A. and Tárnoky, A.L. (1973). Human serum albumin: twenty-three genetic variants and their population distribution. *Annals of Human Genetics*, **36**, 381–92.

References

Abel, L., Demenais, F., Prata, A., Souza, A.E. and Dessein, A. (1991). Evidence for the segregation of a major gene in human susceptibility/resistance to infection by *Schistosoma mansoni*. *American Journal of Human Genetics*, **48**, 959–70.

ABEP (Associação Brasileira de Estudos Populacionais) (1988). *Demografia da População Negra*. Anais, VI Encontro Nacional de Estudos Populacionais, Vol. 3. Belo Horizonte: Associação Brasileira de Estudos Populacionais.

Abramowicz, M. (1969). Contribuição para o estudo comparado das impressões dactiloscópicas em brancos e amarelos. *Revista da Faculdade de Odontologia de São Paulo*, **7**, 165–73.

Abreu, M.S. and Peñalver, M.S.A. (1992). Hemoglobinopatias S en la Argentina. *Medicina* (Buenos Aires), **52**, 341–6.

Acosta, A.X., Peres, L.C., Mazzucatto, L.F. and Pina-Neto, J.M. (1998). A viable fetus presenting 68, XX[73]/69, XXX[27] triploid mosaicism. *Genetics and Molecular Biology*, **21**, 307–10.

Acuña, M.P., Rothhammer, F., Llop, E., Harb, Z. and Palomino, H. (1988). Composición genética de las poblaciones rurales de la cuenca del Rio Elqui, Chile. *Evolución Biológica*, **2**, 149–65.

Acurcio, F.A., Cesar, C.C. and Guimarães, M.D.C. (1998). Health care utilization and survival among patients with AIDS in Belo Horizonte, Minas Gerais, Brazil. *Cadernos de Saúde Pública*, **14**, 811–20.

Adams, J. and Smouse, P.E. (1985). Genetic consequences of demographic changes in human populations. In Chakraborty, R. and Szathmáry, E.J.E., editors, *Diseases of Complex Etiology in Small Populations. Ethnic Differences and Research Approaches*. New York: Alan R. Liss, pp. 283–99.

Adams, J., Hermalin, A., Lam, D. and Smouse, P. (1990). *Convergent Issues in Genetics and Demography*. Oxford: Oxford University Press.

Agostini, J.M. and Meirelles-Nasser, C. (1986). Consanguineous marriages in the archdiocese of Florianópolis, south Brazil. *Revista Brasileira de Genética*, **9**, 479–86.

Agostini, J.M.S., Cavalli, I.J., Erdtmann, B. and Mattevi, M.S. (1989). Population and familial studies on the variability of the heterochromatic and euchromatic segments of the human Y chromosome. *Revista Brasileira de Genética*, **12**, 405–18.

Aguiar, G.F.S., Silva, H.P. and Marques, N. (1991). Patterns of daily allocation of sleep periods: a case study in an Amazonian riverine community. *Chronologia*, **18**, 9–19.

361

Aguilar, L., Lisker, R., Hernández-Peniche, J. and Martínez-Villar, C. (1978). A new syndrome characterized by mental retardation, epilepsy, palpebral conjunctival telangiectasias and IgA deficiency. *Clinical Genetics*, **13**, 154–8.

Aguilar, L., Lisker, R., Ruz, L. and Mutchinik, O. (1981). Constitutive heterochromatin polymorphisms in patients with malignant diseases. *Cancer*, **47**, 2437–9.

Ahern, E.J., Jones, R.T., Brimhall, B. and Gray, R.H. (1970). Haemoglobin F Jamaica ($\alpha_2\eta_2^{61\text{Lys}\rightarrow\text{Glu};\ 136\ \text{Ala}}$). *British Journal of Haematology*, **18**, 369–75.

Albarus, M.H., Salzano, F.M. and Goldraich, N.P. (1997). Genetic markers and acute febrile urinary tract infection in the first year of life. *Pediatric Nephrology*, **11**, 691–4.

Alberto, F.L., Figueiredo, M.S., Zago, M.A., Araújo, A.G. and dos-Santos, J.E. (1999). The Lebanese mutation as an important cause of familial hypercholesterolemia in Brazil. *Brazilian Journal of Medical and Biological Research*, **32**, 739–45.

Albuquerque, M.F.P.M. (1993). Urbanização, favelas e endemias: A produção da filariose no Recife, Brasil. *Cadernos de Saúde Pública*, **9**, 487–497.

Alexandre, C.O.P. and Roisenberg, I. (1985). A genetic and demographic study of hemophilia A in Brazil. *Human Heredity*, **35**, 250–4.

Alexandre, C.O.P., Camargo, L.M.A., Mattei, D., *et al.* (1997). Humoral immune response to the 72 kDa heat shock protein from *Plasmodium falciparum* in populations at hypoendemic areas of malaria in western Brazilian Amazon. *Acta Tropica*, **64**, 155–66.

Alfaro, E. and Dipierri, J.E. (1996). Isonimia, endogamia, exogamia y distancia marital en la Provincia de Jujuy. *Revista Argentina de Antropología Biológica*, **1**, 41–56.

Alfaro, E. and Dipierri, J.E. (1997). Consanguinidad y uniones matrimoniales en poblaciones jujeñas de altura. *Revista Española de Antropología Biológica*, **18**, 57–71.

Alho, C.S., Salzano, F.M. and Zatz, M. (1995). Analysis of deletions and their relationship with clinical severity, family recurrence, and intelligence in Duchenne and Becker muscular dystrophy patients from southern Brazil. *Brazilian Journal of Genetics*, **18**, 617–22.

Allain, M., Figueiredo, G., Rivorêdo, C. and Vieira, L. (1983). Estudo da mortalidade neonatal no Hospital Agamenon Magalhães. *Revista de Pediatria de Pernambuco*, **1**, 103–70.

Allamand, V., Broux, O., Richard, I., *et al.* (1995). Preferential localization of the limb-girdle muscular dystrophy type 2A gene in the proximal part of a 1-cM 15q15.1-q15.3. *American Journal of Human Genetics*, **56**, 1417–30.

Almeida, L.E.A., Barbieri, M.A., Gomes, U.A., *et al.* (1992). Peso ao nascer, classe social e mortalidade infantil em Ribeirão Preto, São Paulo. *Cadernos de Saúde Pública*, **8**, 190–8.

Almeida-Melo, N. and Azevêdo, E.S. (1977). Beckman's scoring method and racial admixture in normal individuals. *Human Heredity*, **27**, 310–13.

Alonso, M.E., Yescas, P., Cisneros, B., *et al.* (1997). Analysis of the (CAG)n repeat causing Huntington's disease in a Mexican population. *Clinical*

Genetics, **51**, 225–30.

Alvares Filho, F., Naoum, P.C., Moreira, H.W., Cruz, R., Manzanato, A.J. and Domingos, C.R.B. (1995). Distribución geográfica etaria y racial de la hemoglobina S en Brasil. *Sangre*, **40**, 97–102.

Alvarez, I., Sans, M., Toledo, R., Sosa, M., Bengochea, M. and Salzano, F.M. (1993). HLA gene and haplotype frequencies in Uruguay. *International Journal of Anthropology*, **8**, 163–8.

Alvarez, J.J. (1951). Studies on the A-B-O, M-N, and Rh-Hr blood factors in the Dominican Republic, with special reference to the problem of admixture. *American Journal of Physical Anthropology*, **9**, 127–48.

Alvarez, O.T., Haces, O.J., Diaz, J.G. and Gutierrez, S. (1989). Consideraciones sobre el somatotipo de atletas nacionales de beisebol. In Serrano, C. and Salas, M.E., editors, *Estudios de Antropología Biológica (IV Coloquio de Antropología Física Juan Comas)*. México: Universidad Nacional Autónoma de México, pp. 407–15.

Alvarez-Leal, M., Perez-Vazquez, J.C., Gaspar-Belmonte, J.A., Castillo-Solis, T. and Zuñiga-Charles, A. (1997). Carrier detection of hemophilia by immunological and coagulation assays in a selected sample of the Mexican population. *American Journal of Human Biology*, **9**, 173–8.

Alves, A.F.P. and Azevêdo, E.S. (1977). Recessive form of Freeman–Sheldon's syndrome or 'whistling face'. *Journal of Medical Genetics*, **14**, 139–41.

Alves, A.F.P., Dourado, A.S. and Azevêdo, E.S. (1976). The effects of some biological characteristics on Legg Calvé Perthes disease in Bahia, Brazil. *Ciência e Cultura*, **28**, 952–4.

Alves-Silva, J., Guimarães, P.E.M., Rocha, J., Pena, S.D.J. and Prado, V.F. (1999*a*). Identification in Portugal and Brazil of a mtDNA lineage containing a 9-bp triplication of the intergenic COII/tRNA[Lys] region. *Human Heredity*, **49**, 56–8.

Alves-Silva, J., Santos, M.S., Guimarães, P.E.M., *et al.* (2000). The ancestry of Brazilian mtDNA lineages. *American Journal of Human Genetisc*, **67**, 444–61.

Alves-Silva, J., Santos, M.S., Pena, S.D.J. and Prado, V.F. (1999*b*). Multiple geographic sources of region V 9-bp deletion haplotypes in Brazilians. *Human Biology*, **71**, 245–59.

Alvial, B.I. and Henckel, C. (1963). Estudio de los grupos sanguíneos en la población hospitalaria de Concepción. *Boletin de la Sociedad de Biologia de Concepción, Chile*, **38**, 35–40.

Ammerman, A.J. and Cavalli-Sforza, L.L. (1984). *The Neolithic Transition and the Genetics of Population in Europe*. Princeton, NJ: Princeton University Press.

Anderson, M., Went, L.N., MacIver, J.E. and Dixon, H.G. (1960). Sickle-cell disease in pregnancy. *The Lancet*, ii, pp. 516–21.

Andrade, F.L., Annichino-Bizzacchi, J.M., Saad, S.T.O., Costa, F.F. and Arruda, V.R. (1998). Prothrombin mutant, factor V Leiden, and thermolabile variant of methylenetetrahydrofolate reductase among patients with sickle cell disease in Brazil. *American Journal of Hematology*, **59**, 46–50.

Andrade, L.O., Machado, C.R.S., Chiari, E., Pena, S.D.J. and Macedo, A.M. (1999). Differential tissue distribution of diverse clones of *Trypanosoma cruzi* in infected mice. *Molecular and Biochemical Parasitology*, **100**, 163–72.

Andrade, M.C. (1991). O Brasil e a América Latina. São Paulo: Editora Contexto.

Andreazza, M.L. and Nadalin, S.O. (1994). O cenário da colonização no Brasil Meridional e a família imigrante. *Revista Brasileira de Estudos de População*, **11**(1), 61–87.

Andrew, L.J., Brancolini, V., dela Pena, L.S. *et al.* (1999). Refinement of the chromosome 5p locus for familial calcium pyrophosphate dihydrate deposition disease. *American Journal of Human Genetics*, **64**, 136–45.

Anjos, L.A., Boileau, R.A. and Misner, J.E. (1992). Maximal mechanical aerobic and anaerobic power output of low-income Brazilian schoolchildren as a function of growth. *American Journal of Human Biology*, **4**, 647–56.

Anjos, L.A., Meirelles, E., Knackfuss, I., Cardoso, C. and Costa, S.G. (1989). Indicadores de gordura corporal em crianças de 7 a 11 anos de idade vivendo em condições sócio-ambientais diferentes no Rio de Janeiro, Brasil. *Ciência e Cultura*, **41**, 1179–88.

Anonymous (1989). Brazil's singular AIDS patterns. *International Family Planning Perspectives*, **15**, 50.

Anonymous (1999). Genética. Descobertas em ritmo acelerado. *Pesquisa* (Fundação de Amparo à Pesquisa do Estado de São Paulo), **47**, 18–21.

Antonarakis, S.E., Boehm, C.D., Serjeant, G.R., Theisen, C.E., Dover, G.J. and Kazazian, H.H. Jr. (1984). Origin of the β^S-globin gene in Blacks: the contribution of recurrent mutation or gene conversion or both. *Proceedings of the National Academy of Sciences, USA*, **81**, 853–6.

Antunes, J.L.F. and Waldman, E.A. (1999). Tuberculosis in the twentieth century: Time-series mortality in São Paulo, Brazil, 1900–97. *Cadernos de Saúde Pública*, **15**, 463–76.

Aragón, L.E. and Imbiriba, M.N.O. (1989). *Populações Humanas e Desenvolvimento Amazônico*. Belém: Universidade Federal do Pará.

Arai, K., Huss, K., Madison, J., *et al.* (1989). Amino acid substitutions in albumin variants found in Brazil. *Proceedings of the National Academy of Sciences, USA*, **86**, 1821–5.

Arango, C., Concha, M., Zaninovic, V., *et al.* (1988). Epidemiology of tropical spastic paraparesis in Colombia, and associated HTLV-I infection. *Annals of Neurology*, **23** (Suppl.), S161–5.

Araújo, A.M. and Salzano, F.M. (1974). Marital distances and inbreeding in Porto Alegre, Brazil. *Social Biology*, **21**, 249–55.

Araújo, A.M. and Salzano, F.M. (1975a). Parental characteristics and birthweight in a Brazilian population. *Human Biology*, **47**, 37–43.

Araújo, A.M. and Salzano, F.M. (1975b). Congenital malformations, twinning and associated variables in a Brazilian population. *Acta Geneticae Medicae et Gemellologiae*, **24**, 31–9.

Araújo, M.N.T., Silva, N.P., Andrade, L.E.C., Sato, E.I., Gerbase-de-Lima, M. and Leser, P.G. (1997). C2 deficiency in blood donors and lupus patients:

prevalence, clinical characteristics and HLA-associations in the Brazilian population. *Lupus*, **6**, 462–6.

Arechabaleta, G.Z. (1982). Peso y talla de escolares caraqueños. In Villanueva, M. and Serrano, C., editors, *Estudios de Antropología Biológica (I Coloquio de Antropología Física Juan Comas)*. México: Universidad Nacional Autónoma de México, pp. 389–418.

Arellanes-Garcia, L., Bautista, N., Mora, P., Ortega-Larrocea, G., Burguet, A. and Gorodezky, C. (1998). HLA-DR is strongly associated with Vogt–Koyanagi–Harada disease in Mexican Mestizo patients. *Ocular Immunology and Inflammation*, **6**, 93–100.

Arellano, J., Vallejo, M., Jimenez, J., Mintz, G. and Kretschmer, R.R. (1984). HLA-B27 and ankylosing spondylitis in the Mexican Mestizo population. *Tissue Antigens*, **23**, 112–16.

Arena, J.F.P. (1976*a*). Estudo biométrico de recém-nascidos de uma população brasileira. *Revista Paulista de Medicina*, **88**, 95–101.

Arena, J.F.P. (1976*b*). Instabilidade congênita do quadril e sinal de estalido na articulação coxofemoral em recém-nascidos de uma população brasileira. *Revista da Associação Médica Brasileira*, **22**, 276–80.

Arena, J.F.P. (1977). Incidência de malformações em uma população brasileira. *Revista Paulista de Medicina*, **89**, 42–9.

Arenas, D., Coral, R., Cisneros, B., *et al.* (1996). Carrier detection in Duchenne and Becker muscular dystrophy using dinucleotide repeat polymorphisms. A study in Mexican families. *Archives of Medical Research*, **27**, 151–6.

Arends, A., Alvarez, M., Velázquez, D., *et al.* (2000). Determination of β-globin gene cluster haplotypes and prevalence of α-thalassemia in sickle cell anemia patients in Venezuela. *American Journal of Hematology*, **64**, 87–90.

Arends, T. (1961*a*). El problema de las hemoglobinopatias en Venezuela. *Revista Venezolana de Sanidad y Asistencia Social,*, **26**, 61–8.

Arends, T. (1961*b*). Frecuencia de las hemoglobinas anormales en Venezuela. *Archivos del Hospital Vargas,*, **3**, 225–36.

Arends, T. (1966). Haemoglobinopathies, thalassaemia and glucose-6-phosphate deficiency in Latin America and the West Indies. *New Zeland Medical Journal* (Suppl.), **65**, 831–44.

Arends, T. (1967). High concentrations of haemoglobin A2 in malaria patients. *Nature*, **215**, 1517–18.

Arends, T. (1971*a*). Epidemiology of hemoglobin variants in Venezuela. In Arends, T., Bemski, G. and Nagel, R.L., editors, *Genetical, Functional, and Physical Studies of Hemoglobins*. Basel: S. Karger, pp. 82–98.

Arends, T. (1971*b*). Hemoglobinopathies and enzyme deficiencies in Latin American populations. In Salzano, F.M., editor, *The Ongoing Evolution of Latin American Populations*. Springfield, IL: C.C. Thomas, pp. 509–59.

Arends, T. (1984). Epidemiologia de las variantes hemoglobínicas en Venezuela. *Gaceta Médica de Caracas*, **92**, 189–224.

Arends, T. and Layrisse, M. (1956). Investigación de las hemoglobinas anormales en Venezuela. Primeros casos de Hemoglobina C. *Memorias del VI Congreso Venezolano de Ciencias Médicas*, **2**, 777–86.

Arends, T., Bemski, G. and Nagel, R.L. (1971). *Genetical Functional and Physical Studies of Hemoglobins.* Basel: S. Karger.

Arends, T., Castillo, O., Garlin, G., Maleh, Y., Anchustegui, M. and Salazar, R. (1987). Hemoglobin Alamo [$\alpha^2\beta^2$ 19(B1)ASN→ASP] in a Venezuelan family. *Hemoglobin,* **11**, 135–8.

Arends, T., Gallango, M.L., Muller, A., Gonzalez-Marrero, M. and Bandez, O.P. (1978). Tapipa: a negroid Venezuelan isolate. In: Meier, R.J., Otter, C.M. and Abdel-Hameed, F., editors, *Evolutionary Models and Studies in Human Diversity.* The Hague: Mouton, pp. 201–14.

Arends, T., Garlin, G., de Perera, A.P. and Castillo, O. (1984). Hemoglobin Deer Lodge ($\alpha^2\beta^2$ 2(NA2)HIS→ARG) in a Venezuelan family. *Hemoglobin,* **8**, 621–6.

Arends, T., Garlin, G., Pérez-Bández, O. and Anchustegui, M. (1982). Hemoglobin variants in Venezuela. *Hemoglobin,* **6**, 243–6.

Arends, T., Salazar, R., Anchustegui, M. and Garlin, G. (1990). Hemoglobin variants in the northeastern region of Venezuela. *Interciencia,* **15**, 36–41.

Arévalo, M.D.P., Carvalho, K.B.L. and Motta, P.A. (1974). Dermatoglifos em distúrbios mentais: Esquizofrenia. *A Folha Médica,* **69**, 599–605.

Arias, S. (1974). Inherited congenital profound deafness in a genetic isolate. *Birth Defects Original Article Series,* **X**(10), 230–43.

Arias-Cazorla, S. and Rodríguez-Larralde, A. (1987). Metacarpophalangeal pattern profiles in Venezuelan and northern Caucasoid samples compared. *American Journal of Physical Anthropology,* **73**, 71–80.

Arnaldi, L.A.T., Polimeno, N.C., Arruda, V.R. and Annichino-Bizzacchi, J.M. (1999). A novel splice site mutation in a Brazilian patient with hereditary antithrombin deficiency type I. *Human Heredity,* **49**, 119–20.

Arneaud, J.D. and Young, O. (1955). A preliminary survey of the distribution of ABO and Rh blood groups in Trinidad, B.W.I. *Documenta Medica Geographica Tropica,* **7**, 375–8.

Arnhold, I.J.P., Osorio, M.G.F., Oliveira, S.B., *et al.* (1998). Clinical and molecular characterization of Brazilian patients with growth hormone gene deletions. *Brazilian Journal of Medical and Biological Research,* **31**, 491–7.

Arpini-Sampaio, Z., Costa, M.C.B., Melo, A.A., Carvalho, M.F.V.A., Deus, M.S.M. and Simões, A.L. (1999). Genetic polymorphisms and ethnic admixture in African-derived Black communities of northeastern Brazil. *Human Biology,* **71**, 69–85.

Arruda, V.R., Eid, K.A.B., Zen, G.C., Gonçalves, M.S., Saad, S.T.O. and Costa, F.F. (1993). Hepatitis C antibody (anti-HCV) prevalence in Brazilian patients with sickle cell diseases. *Vox Sanguinis,* **65**, 247.

Arruda, V.R., Grignolli, C.E., Gonçalves, M.S., *et al.* (1998*b*). Prevalence of homozygosity for the deleted alleles of glutathione S-transferase mu (GSTM1) and theta (GSTT1) among distinct ethnic groups from Brazil. Relevance to environmental carcinogenesis? *Clinical Genetics,* **54**, 210–14.

Arruda, V.R., Siqueira, L.H., Gonçalves, M.S., *et al.* (1998*a*). Prevalence of the mutation C677→T in the methylene tetrahydrofolate reductase gene among distinct ethnic groups in Brazil. *American Journal of Medical Genetics,* **78**,

332–5.

Arzimanoglou, I., Tuchman, A., Li, Z., *et al.* (1995). Cystic fibrosis screening in Hispanics. *American Journal of Human Genetics*, **56**, 544–7.

Asensio, A.S., Beiguelman, B., Magna, L.A., Rossi, E. and Krieger, H. (1986). Alpha 1-antitripsina e artrite reumatóide. *Revista Brasileira de Reumatologia*, **26**, 19–22.

Ashcroft, M.T. (1971). Some aspects of growth and development in different ethnic groups in the Commonwealth West Indies. In Salzano, F.M., editor, *The Ongoing Evolution of Latin American Populations.* Springfield, IL: C.C. Thomas, pp. 281–309.

Ashton-Prolla, P. and Félix, T.M. (1997). Say syndrome: a new case with cystic renal dysplasia in discordant monozygotic twins. *American Journal of Medical Genetics*, **70**, 353–6.

Ashton-Prolla, P., Ashley, G.A., Giugliani, R., Pires, R.F., Desnick, R.J. and Eng, C.M. (1999). Fabry disease: comparison of enzymatic, linkage, and mutation analysis for carrier detection in a family with a novel mutation (30 del G). *American Journal of Medical Genetics*, **84**, 420–4.

Assis, A.M., Santos, L.M.P., Martins, M.C., *et al.* (1997). Distribuição da anemia em pré-escolares do semi-árido da Bahia. *Cadernos de Saúde Pública*, **13**, 237–43.

Assunção, R.M., Barreto, S.M., Guerra, H.L. and Sakurai, E. (1998). Mapas de taxas epidemiológicas, Uma abordagem Bayesiana. *Cadernos de Saúde Pública*, **14**, 713–23.

Ayres, M., Salzano, F.M. and Ludwig, O.K. (1966). Blood group changes in leukaemia. *Journal of Medical Genetics*, **3**, 180–5.

Ayres, M., Salzano, F.M. and Ludwig, O.K. (1967). Multiple antigenic changes in a case of acute leukaemia. *Acta Haematologica*, **37**, 150–8.

Ayres, M., Salzano, F.M., Castro, I.V. and Barros, R.M.S. (1968). Componentes raciais da população de Belém, PA. Primeiros dados. *Ciência e Cultura*, **20**, 188–9.

Ayres, M., Salzano, F.M., Franco, M.H.L.P. and Barros, R.M.S. (1976). The association of blood groups, ABH secretion, haptoglobins and hemoglobins with filariasis. *Human Heredity*, **26**, 105–9.

Azevêdo, E.S., Morton, N.E., Miki, C. and Yee, S. (1969). Distance and kinship in Northeastern Brazil. *American Journal of Human Genetics*, **21**, 1–22.

Azevêdo, E.S. (1973). Historical note on inheritance of sickle cell anemia. *American Journal of Human Genetics*, **25**, 457–8.

Azevêdo, E.S. (1980). Subgroup studies of black admixture within a mixed population of Bahia, Brazil. *Annals of Human Genetics*, **44**, 55–60.

Azevêdo, E.S. (1984). Alguns fatores genéticos na evolução da infecção pelo *S. mansoni.* In Rocha, H., editor, *Aspectos Peculiares da Infecção por* S. mansoni. Salvador: Universidade Federal da Bahia, pp. 187–99.

Azevêdo, E.S. and Santos, M.C.N. (1982). Joint mobility in children: a population study. *Acta Anthropogenetica*, **6**, 33–43.

Azevêdo, E.S., Alves, A.F.P., Silva, M.C.B.O., Souza, M.G.F., Lima, A.M.V.M.D. and Azevêdo, W.C. (1980*b*). Distribution of abnormal

368   *References*

hemoglobins and glucose-6-phosphate dehydrogenase variants in 1200
school children of Bahia, Brazil. *American Journal of Physical Anthropology*,
53, 509–12.

Azevêdo, E.S., Assemany, S., Souza, M.M.M. and Santana, M.E.A. (1971).
Ahaptoglobinemia and predisposition to iron-deficiency anaemia. *Journal of
Medical Genetics*, **8**, 140–2.

Azevêdo, E.S., Biondi, J. and Ramalho, L.M. (1973). Cryptophthalmos in two
families from Bahia, Brazil. *Journal of Medical Genetics*, **10**, 389–92.

Azevêdo, E.S., Chautard-Freire-Maia, E.A., Freire-Maia, N., *et al.* (1986).
Mating types in a mixed and multicultural population of Salvador, Brazil.
Revista Brasileira de Genética, **9**, 487–96.

Azevêdo, E.S., Freire-Maia, N., Azevedo, M.C.C., Weimer, T.A. and Souza,
M.M.M. (1980*a*). Inbreeding in a Brazilian general hospital. *Annals of
Human Genetics*, **43**, 255–64.

Azevêdo, E.S., Krieger, H., Mi, M.P. and Morton, N.E. (1965). PTC taste
sensitivity and endemic goiter in Brazil. *American Journal of Human
Genetics*, **17**, 87–90.

Azevêdo, E.S., Tavares-Neto, J., Carvalho, R.E. and Alves, M.G.H. (1979).
Further studies on the association of Chagas disease and race. *Ciência e
Cultura*, **31**, 671–5.

Bachega, T.A.S.S., Billerbeck, A.E.C., Madureira, G., *et al.* (1999). Low
frequency of CYP21B deletions in Brazilian patients with congenital adrenal
hyperplasia due to 21-hydroxylase deficiency. *Human Heredity*, **49**, 9–14.

Baena de Moraes, M.H., Beiguelman, B. and Krieger, H. (1989). Decline of the
twinning rate in Brazil. *Acta Geneticae Medicae et Gemellologiae*, **38**, 57–63.

Bailey, S., Campos, H., Schosinsky, K. and Mata, L. (1987). Relationship of
upper body fat distribution to serum glucose and lipids in a Costa Rican
population. *American Journal of Physical Anthropology*, **73**, 111–17.

Baker, P.T. (1978). *The Biology of High Altitude Peoples.* Cambridge: Cambridge
University Press.

Baldwin, C.T., Hoth, C.F., Amos, J.A., Silva, E.O. and Milunsky, A. (1992). An
exonic mutation in the *HuP2* paired domain gene causes Waardenburg's
syndrome. *Nature*, **355**, 637–8.

Balemans, W., van den Ende, J., Paes-Alves, A.F., *et al.* (1999). Localization of
the gene for sclerosteosis to the van Buchem disease-gene region on
chromosome 17q12-q21. *American Journal of Human Genetics*, **64**, 1661–9.

Ballew, C. and Haas, J.D. (1986). Altitude differences in body composition
among Bolivian newborns. *Human Biology*, **58**, 871–82.

Balthazar, P.A., Gabriel, A. Jr., Oliveira, D.P., *et al.* (1985). HLA-DR em
portadores de lupus eritematoso sistêmico com anticorpos
anti-ribonucleoproteína e na doença mista do tecido conectivo. *Revista do
Hospital de Clínicas da Faculdade de Medicina da Universidade de São Paulo*,
40, 249–53.

Bar, M.E., Damborsky, M.P., Oscherov, E.B., Alvarez, B.M., Mizdraji, G. and
Avalos, G. (1997). Infestación domiciliaria por triatominos y
seroprevalencia humana en el Departamento Empredado, Corrientes,

Argentina. *Cadernos de Saúde Pública*, **13**, 305–12.

Baraibar, B.C. (1961). Análisis bioestadístico de las determinaciones de los grupos sanguíneos A-B-O, en nuestra población. *Revista de la Facultad de Humanidades y Ciencias, Universidad de la República* (Montevideo), **19**, 5–13.

Barbier, D., Demenais, F., Lefait, J.F., *et al.* (1987). Susceptibility to human cutaneous leishmaniasis and HLA, Gm, Km markers. *Tissue Antigens*, **30**, 63–7.

Barbosa, A.A.L., Cavalli, I.J., Abé, K., Santos, M.G. and Azevêdo, E.S. (1997). Family names and the length of the Y chromosome in Brazilian blacks. *Brazilian Journal of Genetics*, **20**, 93–6.

Barbosa, A.S., Ferraz-Costa, T.E., Semer, M., Liberman, B. and Moreira-Filho, C.A. (1995). XY gonadal dysgenesis and gonadoblastoma: a study in two sisters with a cryptic deletion of the Y chromosome involving the SRY gene. *Human Genetics*, **95**, 63–6.

Barbosa, C.A.A. (1993). Analysis of mixture in the distributions of immunoglobin levels in a Chagasic population. *Revista Brasileira de Genética*, **16**, 1043–8.

Barbosa-Coutinho, L.M., Assis-Brasil, B.M., Drachler, M.L., Rotta, N.T. and Giugliani, R. (1987). Gangliosidose GM1-Tipo 1. Estudo anátomo-clínico de um caso. *Arquivos de Neuro-Psiquiatria*, **45**, 60–6.

Barbujani, G., Bertorelle, G. and Chikhi, L. (1998). Evidence for Paleolithic and Neolithic gene flow in Europe. *American Journal of Human Genetics*, **62**, 488–91.

Barbujani, G., Sokal, R.R. and Oden, N.L. (1995). Indo-European origins: a computer-simulation test of five hypotheses. *American Journal of Physical Anthropology*, **96**, 109–32.

Bardakdjian, J., Kister, J., Rhoda, M.D., *et al.* (1988). Hb J-Cordoba [$\alpha^2 A\beta^2$ 95(FG2)LYS→MET]. A new Hb variant found in Argentina. *Hemoglobin*, **12**, 1–11.

Barnola, J., Tovar-Escobar, G. and Potenza, L. (1953). Enfermedad por celulas falciformes. *Archivos Venezolanos de Puericultura e Pediatria*, **16**, 293–376.

Baronciani, L., Tricta, F. and Beutler, E. (1993). G6PD 'Campinas': a deficient enzyme with a mutation at the far 3′ end of the gene. *Human Mutation*, **2**, 77–8.

Barrantes, R. (1975). Endogamia y flujo génico en una población humana, Parroquia de Santa Maria de Dota, Costa Rica, 1888–1962. Lic. Thesis, San José: University of Costa Rica.

Barrantes, R. (1978). Estructura poblacional y consanguinidad en Dota, Costa Rica, 1888–1962. *Revista de Biología Tropical*, **26**, 347–57.

Barrantes, R. (1980). Las malformaciones congénitas en Costa Rica. I. Mortalidad, registro y vigilancia. *Acta Médica Costarricense*, **23**, 119–31.

Barreiro, C. (1997). Situación de los servicios de genética médica en Argentina. *Brazilian Journal of Genetics*, **20** (Suppl. 1), 5–10.

Barrera, H.A., Gonzalez, M.L., Rivera, J.A., Rojas, A. and Vázquez, R.M. (1991). Genética molecular humana en México. *Ciencia y Desarrollo*,

17(101), 68–80.

Barreto, G. and Salzano, F.M. (1992). *Alu I* and *Hae III* restriction endonucleases, human chromosomes and infertility. *Revista Brasileira de Genética*, 15, 149–60.

Barretto, O.C.O.P., Enokihara, M.Y., Mazar, W.Jr., Ziwian, Z.L.J. and Ferreira, J.L.M.S. (1983). Distribuição do sistema ABO e RH, destacando-se a pesquisa do antígeno Dᵘ, em Santo André, SP. *Revista do Hospital Clínico da Faculdade de Medicina de São Paulo*, 38, 111–14.

Barretto, O.C.O.P., Mansur, A.J., Mizukami, S. and Grinberg, M. (1987). Haptoglobinas em população normal e em portadores de cardiopatias valvares. *Revista Paulista de Medicina*, 105, 128–9.

Barros, F.C. and Victora, C.G. (1996). Saúde materno-infantil em Pelotas, Rio Grande do Sul, Brasil, 1982–93: uma década de transição. *Cadernos de Saúde Pública*, 12 (Suppl. 1), 4–92.

Barros, F.C., Victora, C.G., Granzoto, J.A., Vaughan, J.P. and Lemos, A.V., Jr. (1984). Saúde perinatal em Pelotas, RS, Brasil. Fatores sociais e biológicos. *Revista de Saúde Pública*, 18, 301–12.

Barros, L. and Witkop, C.J. Jr. (1963). Oral and genetic study of Chileans 1960. III. Periodontal disease and nutritional factors. *Archives of Oral Biology* 8, 195–206.

Barth, M.L., Giugliani, R., Goldenfum, S.L., *et al.* (1990). Application of a flowchart for the detection of lysosomal storage diseases in 105 high-risk Brazilian patients. *American Journal of Medical Genetics*, 37, 534–8.

Bassanezi, M.S.C.B. (1997). Repensando a demografia histórica. *Revista Brasileira de Estudos de População*, 14(1/2), 97–100.

Bassi, R.A. (1983). Casamentos consangüíneos em populações brasileiras e alguns parâmetros migracionais e etários, associados a casamentos, em Curitiba. M.Sc. thesis, Curitiba: Federal University of Paraná.

Bassi, R.A. and Freire-Maia, N. (1985). Marriage age and inbreeding in Curitiba, southern Brazil. *Revista Brasileira de Genética*, 8, 199–203.

Bassit, L., Kleter, B., Ribeiro-dos-Santos, G., *et al.* (1998). Hepatitis G virus: prevalence and sequence analysis in blood donors of São Paulo, Brazil. *Vox Sanguinis*, 74, 83–7.

Bastide, R. (1973). *Estudos Afro-Brasileiros*. São Paulo: Editora Perspectiva.

Bastide, R. and Fernandes, F. (1959). *Brancos e Negros em São Paulo*. São Paulo: Companhia Editora Nacional.

Bastos, F.I., Boschi-Pinto, C., Telles, P.R. and Lima, E. (1993). O não-dito da AIDS. *Cadernos de Saúde Pública*, 9, 90–6.

Basu, A., Namboodiri, K.K., Weitkamp, L.R., Brown, W.H., Pollitzer, W.S. and Spivey, M.A. (1976). Morphology, serology, dermatoglyphics, and microevolution of some village populations in Haiti, West Indies. *Human Biology*, 48, 245–69.

Batista, D.A.S., Campos, M.T.G.R., Vianna-Morgante, A.M. and Otto, P.A. (1985). Dermatoglyphics in patients with the Prader–Willi syndrome and normal karyotypes. *Revista Brasileira de Genética*, 8, 107–14.

Bau, C.H.D., and Salzano, F.M. (1995). Alcoholism in Brazil: the role of personality and susceptibility to stress. *Addiction*, **90**, 693-8.

Bau, C.H.D., Roman, T., Almeida, S. and Hutz, M.H. (1999). Dopamine D4 receptor gene and personality dimensions in Brazilian male alcoholics. *Psychiatric Genetics*, **9**, 139-43.

Beall, C.M. (1981). Optimal birthweights in Peruvian populations at high and low altitudes. *American Journal of Physical Anthropology*, **56**, 209-16.

Beet, E.A. (1949). The genetics of the sickle cell trait in a Bantu tribe. *Annals of Eugenics*, **14**, 279-84.

Beiguelman, B. (1962). Estudo genético e antropológico de imigrantes japoneses e seus descendentes não miscigenados. *Revista de Antropologia* (São Paulo), **10**, 109-42.

Beiguelman, B. (1963*a*). A somatometric study on Japanese immigrants and Japanese unmixed descendants in Brazil. *Zeitschrift für Morphologie und Anthropologie*, **53**, 296-9.

Beiguelman, B. (1963*b*). Grupos sangüíneos e lepra. *Revista Brasileira de Leprologia*, **31**, 34-44.

Beiguelman, B. (1964*a*). Sistema ABO e epidemiologia de lepra. *Revista Paulista de Medicina*, **65**, 80-6.

Beiguelman, B. (1964*b*). Taste sensitivity to phenylthiourea and leprosy. *Acta Geneticae Medicae et Gemellologiae*, **13**, 193-6.

Beiguelman, B. (1971). Genetics of ab and A'-d ridge counts. *Revista Brasileira de Pesquisas Médicas e Biológicas*, **4**, 337-42.

Beiguelman, B. (1975). Terapeutica de la lepra y farmacogenética. *Revista Médica de Chile*, **103**, 344-9.

Beiguelman, B. and Pinto, W. Jr. (1971). A new approach to dermatoglyphic studies. *Revista Brasileira de Pesquisas Médicas e Biológicas*, **4**, 305-9.

Beiguelman, B. and Prado, D. (1963). Recessive juvenile glaucoma. *Journal de Génétique Humaine*, **12**, 53-4.

Beiguelman, B. and Sevá-Pereira, A. (1983). Deficiência de lactase intestinal e intolerância ao leite. *Ciência e Cultura*, **35**(6), 722-34.

Beiguelman, B. and Villarroel-Herrera, H.O. (1993). Factors influencing the decline of twinning incidence in a southeastern Brazilian population. *Revista Brasileira de Genética*, **16**, 793-801.

Beiguelman, B., Colletto, G.M.D.D., Franchi-Pinto, C. and Krieger, H. (1998*a*). Birth weight of twins. I. The fetal growth patterns of twins and singletons. *Genetics and Molecular Biology*, **21**, 151-4.

Beiguelman, B., Colletto, G.M.D.D., Franchi-Pinto, C. and Krieger, H. (1998*b*). Birth weight of twins. II. Fetal genetic effect on birth weight. *Genetics and Molecular Biology*, **21**, 155-8.

Beiguelman, B., Franchi-Pinto, C. and Magna, L.A. (1997). Biological and social traits associated with twinning among Caucasoids and Negroids. *Brazilian Journal of Genetics*, **20**, 311-18.

Beiguelman, B., Franchi-Pinto, C., Dal Colletto, G.M.D. and Krieger, H. (1995). Annual variation of sex ratio in twin births and in singletons in Brazil. *Acta Geneticae Medicae et Gemelollogiae*, **44**, 163-8.

Beiguelman, B., Krieger, H. and Marques da Silva, L. (1996). Maternal age and

372   References

Down syndrome in southeastern Brazil. *Brazilian Journal of Genetics*, **19**, 637–40.

Beiguelman, B., Pinto, W. Jr., Dall'Aglio, F.F., Silva, E. and Vozza, J.A. (1968). G–6PD deficiency among lepers and healthy people in Brazil. *Acta Genetica et Statistica Medica*, **18**, 159–62.

Beiguelman, B., Ramalho, A.S., Arena, J.F.P. and Garlipp, C.R. (1977). A acetilação da isoniazida em brasileiros caucasóides e negróides com tuberculose pulmonar. *Revista Paulista de Medicina*, **89**, 12–15.

Bejarano, I., Dipierri, J., Junqueira, C. and Alfaro, E. (1997). Causas de muerte en la Puna de Atacama (periodo 1890–1950), Distribución por sexos y edades. *Revista Española de Antropología Biológica*, **18**, 247–59.

Bejarano, I., Dipierri, J.E. and Ocampo, S.B. (1996). Variación regional de la tendencia secular de la talla adulta masculina en la Provincia de Jujuy. *Revista Argentina de Antropología Biológica*, **1**, 7–18.

Beltrão, K.I. and Migon, H.S. (1989). Migrações anuais rural-urbano-rural período 70/80. *Revista Brasileira de Estudos de População*, **6**, 63–94.

Benedito-Silva, A.A., Mena-Barreto, L., Alam, M.F., *et al.* (1998). Latitude and social habits as determinants of morning and evening types in Brazil. *Biological Rhythm Research*, **29**, 591–7.

Benítez, J.P. (1964). Paraguai. Independência e organização do estado (1811–1870). In: Levene, R., editor, *História das Américas*, Vol. 5. Rio de Janeiro: W.M. Jackson, pp. 295–332.

Benoist, J. 1966. Du social au biologique: Étude de quelques interactions. *L'Homme*, **6**, 5–26.

Benoist, J. and Dansereau, G. (1972). Données qualitatives et quantitatives sur les dermatoglyphes digitaux et palmaires de Saint-Barthélémy (Antilles Françaises). *Bulletins et Mémoires de la Societé d'Anthropologie de Paris,*, **9**, ser., **12**, 165–76.

Bentley, G.R. (1999). Aping our ancestors: comparative aspects of reproductive ecology. *Evolutionary Anthropology*, **7**, 175–85.

Berg, M.A., Guevara-Aguirre, J., Rosenbloom, A.L., Rosenfeld, R.G. and Francke, U. (1992). Mutation creating a new splice site in the growth hormone receptor genes of 37 Ecuadorean patients with Laron syndrome. *Human Mutation*, **1**, 24–34.

Bergmann, M. (1977). *Nasce um Povo*. Petrópolis: Editora Vozes.

Bern, Z. (1995). As três Américas e o gerenciamento contínuo do diverso. In Bern, Z. and de Grandis, R., editors, *Imprevisíveis Américas: Questões de Hibridação Cultural nas Americas,* Porto Alegre, Sagra-D.C.: Luzzato, pp. 51–60.

Bernal, J.E., Duran de Rueda, M.M. and Brigard, D. (1988). Human lymphocyte antigen in actinic prurigo. *Journal of the American Academy of Dermatology*, **18**, 310–12.

Bernal, J.E., Ortega, G. and Umana, A. (1983). The contribution of genetic disease to paediatric mortality in a University Hospital in Bogotá. *Journal of Biosocial Science*, **15**, 465–71.

Bernand, C. and Gruzinski, S. (1997). *História do Novo Mundo. Da Descoberta à*

Conquista, uma Experiência Européia (1492–1550). São Paulo: Editora da Universidade de São Paulo.

Bernardo, M.A., Aquino, L.V., Souza, R.A., Santos, M.C.T. and Lirio, A.S. (1982). Tipos de transferrina na população do Estado do Rio de Janeiro. *Ciência e Cultura*, **34** (7, Suppl.), 760.

Bernstein, F. (1931). Die geographische Verteilung der Blutgruppen und ihre anthropologische Verteilung. In *Comitato Italiano per lo Studio dei Problemi della Populazione*. Roma: Istituto Poligrafico dello Stato.

Berquó, E. (1989). A família no século XXI. *Ciência Hoje*, **10**(58), 58–65.

Berti, P.R. and Leonard, W.R. (1998). Demographic and socioeconomic determinants of variation in food and nutrient intake in an Andean community. *American Journal of Physical Anthropology* **105**, 407–17.

Berti, P.R., Leonard, W.R. and Berti, W.J. (1998). Stunting in an Andean community: prevalence and etiology. *American Journal of Human Biology*, **10**, 229–40.

Bérubé, N.G., Smith, J.R. and Pereira-Smith, O.M. (1998). The genetics of cellular senescence. *American Journal of Human Genetics*, **62**, 1015–19.

Bethell, L. (1984). *The Cambridge History of Latin America*. Cambridge, Cambridge University Press. (Brazilian version, 1997, Editora da Universidade de São Paulo, São Paulo, and Fundação Alexandre Gusmão, Brasília.)

Beutler, E. (1989). Glucose-6-phosphate dehydrogenase: new perspectives. *Blood*, **73**, 1397–401.

Beutler, E. (1991). Glucose-6-phosphate dehydrogenase deficiency. *New England Journal of Medicine*, **324**, 169–74.

Beutler, E. (1994). G6PD deficiency. *Blood*, **84**, 3613–36.

Beutler, E. and Kuhl, W. (1990a). Linkage between a *Pvu II* restriction fragment length polymorphism and G6PD A-$^{202A/376G}$: evidence for a single origin of the common G6PD A- mutation. *Human Genetics*, **85**, 9–11.

Beutler, E. and Kuhl, W. (1990b). The NT 1311 polymorphism of G6PD: G6PD Mediterranean mutation may have originated independently in Europe and Asia. *American Journal of Human Genetics*, **47**, 1008–12.

Beutler, E., Kuhl, W., Ramirez, E. and Lisker, R. (1991a). Some Mexican glucose-6-phosphate dehydrogenase variants revisited. *Human Genetics*, **86**, 371–4.

Beutler, E., Kuhl, W., Sáenz, G.F. and Rodríguez, W. (1991b). Mutation analysis of glucose-6-phosphate dehydrogenase (G6PD) variants in Costa Rica. *Human Genetics*, **87**, 462–4.

Beutler, E., Vulliamy, T. and Luzzatto, L. (1996). Hematologically important mutations: glucose-6-phosphate dehydrogenase. *Blood Cells, Molecules, and Diseases*, **22**, 49–56.

Beutler, E., Westwood, B., Prchal, J.T., Vaca, G., Bartsocas, C.S. and Baronciani, L. (1992). New glucose-6-phosphate dehydrogenase mutations from various ethnic groups. *Blood*, **80**, 255–6.

Bevilaqua, L.R.M., Mattevi, V.S., Ewald, G.M., *et al.* (1995). Beta-globin gene cluster haplotype distribution in five Brazilian Indian tribes. *American*

Journal of Physical Anthropology, **98**, 395–401.

Bier, O.G. and Machado, J.C. (1933). Blood groups in São Paulo. *Folia Clinica et Biologica*, **5**, 41–4.

Bina, J.C., Tavares-Neto, J., Prata, A. and Azevêdo, E.S. (1978). Greater resistance to development of severe schistosomiasis in Brazilian Negroes. *Human Biology*, **50**, 41–9.

Biondi, G., Battistuzzi, G., Rickards, O., *et al.* (1988). Migration pattern and genetic marker distribution of the Afro-American population of Bluefields, Nicaragua. *Annals of Human Biology*, **15**, 399–412.

Biondi, G., Rickards, O., Guglielmino, C.R. and De Stefano, G.F. (1993). Marriage distances among the Afroamericans of Bluefields, Nicaragua. *Journal of Biosocial Science*, **25**, 523–30.

Blackwell, J.M., Black, G.F., Peacock, C.S., *et al.* (1997). Immunogenetics of leishmanial and mycobacterial infections: The Belém Family Study. *Philosophical Transactions of the Royal Society of London, Series B*, **352**, 1331–45.

Blakemore, H. (1985). Chile. In Blakemore, H. and Smith, C.T., editors, *Latin America. Geographical Perspectives*, 2nd edn. London: Methuen, pp. 457–531.

Blakemore, H. and Smith, C.T. (1985*a*). *Latin America. Geographical Perspectives.* 2nd edn. London: Methuen.

Blakemore, H. and Smith, C.T. (1985*b*). Conclusion, unity and diversity in Latin America. In Blakemore, H. and Smith, C.T., editors, *Latin America. Geographical Perspectives.* 2nd edn. London: Methuen, pp. 533–43.

Blanco, R. and Chakraborty, R. (1975). Consanguinity and demography in some Chilean populations. *Human Heredity*, **25**, 477–87.

Blanco, R. and Covarrubias, E. (1971). Estudio genético y demográfico en Caleu. *Revista Médica de Chile*, **99**, 139–44.

Blanco, R., Arcos-Burgos, M., Paredes, M., *et al.* (1998). Complex segregation analysis of nonsyndromic cleft lip/palate in a Chilean population. *Genetics and Molecular Biology*, **21**, 139–44.

Blanco, R., Rothhammer, F., Olarte, G., Palomino, H. and Justiniano, M. (1973). Análisis genético-cuantitativo de cinco rasgos morfológicos dentarios. *Revista Médica de Chile*, **101**, 223–6.

Blank, M., Blank, A., King, S., Yashiki, S., *et al.* (1995). Distribution of HLA and haplotypes of Colombian and Jamaican black populations. *Tissue Antigens*, **45**, 111–16.

Blau, N., Niederwieser, A., Curtius, H.C., *et al.* (1989). Prenatal diagnosis of atypical phenylketonuria. *Journal of Inherited Metabolic Diseases*, **12** (Suppl. 2), 295–8.

Bloch, M. and Rivera, H. (1969). Hemoglobinas anormales y deficiencia de glucose-6-fosfato dehidrogenasa en El Salvador. *Sangre*, **14**, 121–4.

Boada, J.J. and Yates, A.P. (1979). Phenotypes of glucose-6-phosphate dehydrogenase in a Mestizo population. *Acta Científica Venezolana*, **30**, 172–4.

Boëtsch, G., Sauvain-Dugerdil, C. and Roberts, D.F. (1996). Continuity, collapse or metamorphosis? Demographic anthropology and the study of change within human populations. Special issue. *International Journal of Anthropology*, **11**(2–4), 3–191.

Bogin, B. (1978). Seasonal pattern in the rate of growth in height of children living in Guatemala. *American Journal of Physical Anthropology*, **49**, 205–10.

Bogin, B. and Keep, R. (1999). Eight thousand years of economic and political history in Latin America revealed by anthropometry. *Annals of Human Biology*, **26**, 333–51.

Bogin, B. and MacVean, R.B. (1978). Growth in height and weight of urban Guatemalan primary school children of low and high socioeconomic class. *Human Biology*, **50**, 477–87.

Bogin, B. and MacVean, R.B. (1981). Body composition and nutritional status of urban Guatemalan children of high and low socioeconomic class. *American Journal of Physical Anthropology*, **55**, 543–51.

Bogin, B. and MacVean, R.B. (1982). Ethnic and secular influences on the size and maturity of seven-year-old children living in Guatemala City. *American Journal of Physical Anthropology*, **59**, 393–8.

Boia, M.N., Motta, L.P., Salazar, M.S.P., Muttis, M.P.S., Coutinho, R.B.A. and Coura, J.R. (1999). Estudo das parasitoses intestinais e da infecção chagásica no município de Novo Airão, Estado do Amazonas, Brasil. *Cadernos de Saúde Pública*, **15**, 497–504.

Bolzan, A., Guimarey, L. and Frisancho, A.R. (1999). Study of growth in rural school children from Buenos Aires, Argentina using upper arm muscle area by height and other anthropometric dimensions of body composition. *Annals of Human Biology*, **26**, 185–93.

Bompeixe, E.P., Costa, S.M.C.M., Arruda, W.O. and Petzl-Erler, M.L. (1999). Lack of association between parenchymal neurocysticercosis and HLA Class I and Class II antigens. *Genetics and Molecular Biology*, **22**, 7–11.

Bonatto, S.L. and Salzano, F.M. (1997a). A single and early migration for the peopling of the Americas supported by mitochondrial DNA sequence data. *Proceedings of the National Academy of Sciences, USA*, **94**, 1866–71.

Bonatto, S.L. and Salzano, F.M. (1997b). Diversity and age of the four major mtDNA haplogroups, and their implications for the peopling of the New World. *American Journal of Human Genetics*, **61**, 1413–23.

Bonatto, S.L., Redd, A.J., Salzano, F.M. and Stoneking, M. (1996). Lack of ancient Polynesian–Amerindian contact. *American Journal of Human Genetics*, **59**, 253–6.

Bonaventura, J. and Riggs, A. (1967). Polymerization of hemoglobins of mouse and man: structural basis. *Science*, **158**, 800–2.

Bordin, S., Martins, J.T., Gonçalves, M.S., Melo, M.B., Saad, S.T.O. and Costa, F.F. (1998). Haplotype analysis and $^{A}\gamma$ gene polymorphism associated with the Brazilian type of hereditary persistence of fetal hemoglobin. *American Journal of Hematology*, **58**, 49–54.

Borges-Osório, M.R.L. and Salzano, F.M. (1985). Language disabilities in three twin pairs and their relatives. *Acta Geneticae Medicae et Gemellologiae*, **34**, 95–100.

Borges-Osório, M.R.L. and Salzano, F.M. (1986). Laterality and language disability. *Interciencia*, **11**, 84–5.

Borges-Osório, M.R.L. and Salzano, F.M. (1987). Frequencies of language disabilities and their family patterns in Porto Alegre, Brazil. *Behavior Genetics*, **17**, 53–69.

Bortolini, M.C. and Salzano, F.M. (1999). β^S haplotype diversity in Afro-Americans, Africans, and Euro-Asiatics: an attempt at a synthesis. *Ciência e Cultura*, **51**, 175–80.

Bortolini, M.C., Castro de Guerra, D., Salzano, F.M. and Weimer, T.A. (1995a). Inter and intrapopulational genetic diversity in Afro-Venezuelan and African populations. *Interciencia*, **20**, 90–3.

Bortolini, M.C., Salzano, F.M., Zago, M.A., Silva, W.A. Jr. and Weimer, T.A. (1997c). Genetic variability in two Brazilian ethnic groups: A comparison of mitochondrial and protein data. *American Journal of Physical Anthropology*, **103**, 147–56.

Bortolini, M.C., Silva, W.A. Jr., Castro de Guerra, D., *et al.* (1999). African-derived South American populations: a history of symmetrical and asymmetrical matings according to sex revealed by bi- and uniparental genetic markers. *American Journal of Human Biology*, **11**, 551–63.

Bortolini, M.C., Silva, W.A. Jr., Weimer, T.A., *et al.* (1998). Protein and hypervariable tandem repeat diversity in eight African-derived South American populations: Inferred relationships do not coincide. *Human Biology*, **70**, 443–61.

Bortolini, M.C., Weimer, T.A., Franco, M.H.L.P., *et al.* (1992). Genetic studies in three South American Black populations. *Gene Geography*, **6**, 1–16.

Bortolini, M.C., Weimer, T.A., Salzano, F.M., *et al.* (1995b). Evolutionary relationships between black South American and African populations. *Human Biology*, **67**, 547–59.

Bortolini, M.C., Weimer, T.A., Salzano, F.M., Moura, L.B. and Silva, M.C.B.O. (1997b). Genetic structure of two urban Afro-Brazilian populations. *International Journal of Anthropology*, **12**, 5–16.

Bortolini, M.C., Zago, M.A., Salzano, F.M., *et al.* (1997a). Evolutionary and anthropological implications of mitochondrial DNA variation in African Brazilian populations. *Human Biology*, **69**, 141–59.

Bortolozzi, J. and de Lucca, E.J. (1971). Distribuição dos grupos sangüíneos ABO e Rh em Botucatu (SP). *Revista Paulista de Medicina*, **78**, 173–6.

Boussiou, M., Lokopoulos, D., Christaki, J. and Fessas, Ph. (1991). The origin of the sickle mutation in Greece: evidence from the β^S globin gene polymorphisms. *Hemoglobin*, **15**, 459–67.

Bowles, G.T. (1984). China, Mongolia, Korea. In Schwidetzky, I., editor, *Rassengeschichte der Menschheit*, No. 10. München: R. Oldenbourg Verlag, pp. 39–105.

Boy, R., Pimentel, M.M.G., Hemerly, A.P., Silva, M.P.S., Barreiro, A.P. and Cabral de Almeida, J.C. (1998a). Chromosome 6q deletion: report of a new case and review of the literature. *Genetics and Molecular Biology*, **21**, 145–9.

Boy, R., Llerena, J., Pimentel, M.M.G. and Cabral de Almeida, J.C. (1998b).

Geleophysic dysplasia: report on two sibs. *Genetics and Molecular Biology*, **21**, 159–62.

Brading, D.A. (1997). A Espanha dos Bourbons e seu Império Americano. In: Bethell, L., editor, *História da América Latina*, Vol. 1, *América Latina Colonial.* São Paulo: Editora da Universidade de São Paulo, pp. 391–445.

Braga, J.C., Freire-Maia, N., Abdala, H. and Tartuce, N. (1977). Estudo clinicogenético de uma amostra de retardados mentais. I. Aspectos psicológicos, psiquiátricos e neurológicos. *Ciência e Cultura*, **29**, 985–91.

Braga, M.C.C., Otto, P.A. and Spinelli, M. (1999). Recurrence risks in cases of nonsyndromic deafness. *Brazilian Journal of Dysmorphology and Speech–Hearing Disorders*, **2**, 33–40.

Branco, P.P.M. (1991). *A População Idosa e o Apoio Familiar.* São Paulo: Fundação Sistema Estadual de Análise de Dados.

Brandon, W. (1961). *The American Heritage Book of Indians.* New York: American Heritage.

Bravi, C.M., Sans, M., Bailliet, G., *et al.* (1997). Characterization of mitochondrial DNA and Y-chromosome haplotypes in a Uruguayan population of African ancestry. *Human Biology*, **69**, 641–52.

Brittain, A.W. (1992). Birth spacing and child mortality in a Caribbean population. *Human Biology*, **64**, 223–41.

Brittain, A.W., Morrill, W.T. and Kurland, J.A. (1988). Parental choice and infant mortality in a West Indian population. *Human Biology*,, **60**, 679–92.

Britten, R.J. and Kohne, D.E. (1968). Repeated sequences in DNA. *Science*, **161**, 529–40.

Brown, L.R. (1980). Food or fuel, new competition for the world's cropland. *Interciencia*, **5**, 365–72.

Bruno, Z.V., Tavares, Z.M.V.B. and Bruno, Z.V. (1981/1982). Estudo comparativo dos níveis hematológicos e proteinêmicos em pré-escolares, portadores e não-portadores de enteroparasitosis, no município de Fortaleza, Ceará, 1977/80. *Revista de Medicina da Universidade Federal do Ceará*, **21/22**, 55–60.

Buchalter, M.S., Wannmacher, C.M.D. and Wajner, M. (1983). Tay–Sachs disease: screening and prevention program in Porto Alegre. *Revista Brasileira de Genética*, **6**, 539–47.

Bueno, E. (1998*a*). *História do Brasil.* Porto Alegre: Zero Hora/RBS Jornal.

Bueno, E. (1998*b*). *Náufragos, Traficantes e Degredados. As Primeiras Expedições ao Brasil, 1500–1531.* Rio de Janeiro: Objetiva.

Burki, S.J. and Edwards, S. (1996). A América Latina e a crise mexicana, Novos desafios. In Langoni, C.G., editor, *A Nova América Latina.* Rio de Janeiro: Fundação Getulio Vargas, pp. 1–55.

Busch, C.P., Ramdath, D.D., Ramsewak, S. and Hegele, R.A. (1999). Association of *PON2* variation with birth weight in Trinidadian neonates of South Asian ancestry. *Pharmacogenetics*, **9**, 351–6.

Buschang, P.H. and Malina, R.M. (1980). Brachymesophalangia–V in five samples of children: a descriptive and methodological study. *American Journal of Physical Anthropology*, **53**, 189–95.

Buschang, P.H. and Malina, R.M. (1983). Growth in height and weight of mild to moderately undernourished Zapotec school children. *Human Biology*, **55**, 587–97.

Buschang, P.H., Malina, R.M. and Little, B.B. (1986). Linear growth of Zapotec schoolchildren: growth status and yearly velocity for leg length and sitting height. *Annals of Human Biology*, **13**, 225–34.

Buss, D.M. (1985). Human mate selection. *American Scientist*, **73**, 47–51.

Buss, D.M. (1989). Sex differences in human mate preferences: Evolutionary hypotheses tested in 37 cultures. *Behavioral and Brain Sciences*, **12**, 1–49.

Buss, D.M. (1994). The strategies of human mating. *American Scientist*, **82**, 238–49.

Butts, D.C.A. (1955). Blood groups of the Bush Negroes of Surinam. *Documenta Medica Geographica Tropica*, **7**, 43–9.

Byard, P.J. and Lees, F.C. (1982). Skin colorimetry in Belize. II. Inter- and intrapopulation variation. *American Journal of Physical Anthropology*, **58**, 215–19.

Bydlowski, S.P., Novak, E.M., Issa, J.S., Forti, N., Giannini, S.D. and Diament, J. (1996). DNA polymorphisms of apolipoprotein B and AI-CIII-AIV genes in a Brazilian population: A preliminary report. *Brazilian Journal of Medical and Biological Research*, **29**, 1269–74.

Caballero, E.C. (1990). La población esclava en ciudades puertos del Rio de la Plata, Montevideo y Buenos Aires. In Nadalin, S.O., Marcílio, M.L. and Balhana, A.P., editors, *História e População. Estudos sobre a América Latina*. São Paulo: Fundação Sistema Estadual de Análise de Dados, pp. 218–25.

Cabello, F. (1991). Una visita a un antigo paradigma en Chile: deterioro económico y social y epidemias. *Interciencia*, **16**, 176–81.

Cabello, G.M.K., Moreira, A.F., Horovitz, D., *et al.* (1999). Cystic fibrosis: low frequency of DF508 mutation in two population samples from Rio de Janeiro, Brazil. *Human Biology*, **71**, 189–96.

Cabello, P.H. and Krieger, H. (1991). Note on estimates of the inbreeding coefficient through study of pedigrees and isonymous marriages. *Human Biology*, **63**, 719–23.

Cabello, P.H., Hatagima, A., Lima, A.M.D., Azevêdo, E.S. and Krieger, H. (1997). Algunos aspectos de la estructura genética de una población del nordeste brasileño, Parentesco y mezcla racial relacionados con el origen de los apellidos. *Revista Española de Antropología Biológica*, **18**, 7–18.

Cabello, P.H., Lima, A.M.V.M., Azevêdo, E.S. and Krieger, H. (1995). ABO blood groups and *Leishmania donovani chagasi* infection: an apparent association. *Revista Brasileira de Genética*, **18**, 297–9.

Cabral de Almeida, J.C., Reis, D.F., Llerena, J.C. Jr. and Pereira, E.T. (1989*a*). 'Pure' partial trisomy 3 p due to the malsegregation of a balanced maternal translocation t (X; 3) (p22.3; p21). *Annales de Génétique*, **32**, 180–3.

Cabral de Almeida, J.C., Reis, D.F. and Martins, R.R. (1989*b*). Interstitial deletion of (17) (p11.2). A microdeletion syndrome. Another example. *Annales de Génétique*, **32**, 184–6.

Cabral, D.F., Maciel-Guerra, A.T. and Hackel, C. (1998). Mutations of androgen receptor gene in Brazilian patients with male pseudohermaphroditism. *Brazilian Journal of Medical and Biological Research*, 31, 775–8.

Cabral-Alexandre, I. and Meirelles-Nasser, C. (1995). Familial resemblance of facial measurements in Florianópolis, South-Brazil. *Revista Brasileira de Genética*, 18, 301–4.

Cabrera, M., Shaw, M.A., Sharples, C., *et al*. (1995). Polymorphism in tumor necrosis factor genes associated with mucocutaneous leishmaniasis. *Journal of Experimental Medicine*, 182, 1259–64.

Cabrera, M.A. (1950). Breve estudio sobre la repartición de los grupos sanguíneos entre nuestra población y la evolución de la investigación de la paternidad por medio de ellos en nuestro país. *Salubridad y Asistencia, Guatemala*, 3, 16–17.

Cabrera Llano, J., González, M.C., Hernández, N.L. and Hidalgo, P.C. (1992). Pesquisaje de hemoglobinas anormales en 1149 muestras de sangre de cordon umbilical. *Medicentro*, 8, 13–18.

Callegari-Jacques, S.M., Salzano, F.M. and Peña, H.F. (1977). Palmar dermatoglyphic patterns in twins. *Human Heredity*, 27, 437–43.

Calmon, P. (1964). Fundação do Império do Brasil. In Levene, R., editor, *História das Américas*, Vol. 8. Rio de Janeiro: W.M. Jackson, pp. 219–309.

Camarano, A.A. (1986). Migração e estrutura produtiva, O caso das regiões metropolitanas nordestinas. *Revista Brasileira de Estudos de População*, 3(2), 23–46.

Camargo, L. (1998). *Almanaque Abril 98*. São Paulo: Editora Abril.

Camargo-Neto, E., Schulte, J., Lewis, E. and Giugliani, R. (1991). Two-year report from a comprehensive multi-regional Brazilian screening program. *Proceedings, 8th International Neonatal Screening Symposium*, pp. 45–6.

Camargo-Neto, E., Schulte, J., Silva, L.C.S. and Giugliani, R. (1993). Cromatografia em camada delgada para a detecção neonatal de fenilcetonúria e outras aminoacidopatias. *Revista Brasileira de Análises Clínicas*, 25, 81–2.

Campana, M.A. and Roubicek, M.M. (1996). Maternal and neonatal variables in twins, an epidemiological approach. *Acta Geneticae Medicae et Gemellologiae*, 45, 461–9.

Candotti, E., Velho, O., Lent, R., Guimarães, A.P. Filho and Almeida, D.F. (1988). Autos de Goiânia. *Ciência Hoje*, 7 (Suppl.), 1–48.

Canto-Lara, S.B., Cardenas-Maruffo, M.F., Vargas-Gonzalez, A. and Andrade-Narvaez, F. (1998). Isoenzyme characterization of *Leishmania* isolated from human cases with localized cutaneous leishmaniasis from the State of Campeche, Yucatán Peninsula, México. *American Journal of Tropical Medicine and Hygiene*, 58, 444–7.

Caraballo, L.R., Marrugo, J., Erlich, H. and Pastorizo, M. (1992). HLA alleles in the population of Cartagena (Colombia). *Tissue Antigens*, 39, 128–33.

Carakushansky, G., Aguiar, M.B., Gonçalves, M.R., Berthier, C.O., Kahn, E., Carakushansky, M. and Pena, S.D.J. (1996). Identical twin discordance for the Brachmann–de Lange syndrome revisited. *American Journal of Medical Genetics*, **63**, 458–60.

Carakushansky, G., Rosembaum, S., Ribeiro, M.G., Kahn, E. and Carakushansky, M. (1998). Achondroplasia associated with Down syndrome. *American Journal of Medical Genetics*, **77**, 168–9.

Caratini, A., Carnese, F.R. and Gómez, P. (1996). Endogamia-exogamia grupal de los inmigrantes españoles en la ciudad de Buenos Aires, Su variación en el espacio y en el tiempo. *Revista Española de Antropología Biológica*, **17**, 63–75.

Carbonell, J.M. (1964). A luta pela independência em Cuba (1810–1898). In: Levene, R., editor, *História das Américas*, Vol. 7. Rio de Janeiro: W.M. Jackson, pp. 263–374.

Cardenas Barahona, E. and Peña Reyes, M.E. (1989). Capacidad vital y composición corporal bajo entrenamiento deportivo. In Serrano, C. and Salas, M.E., editors, *Estudios de Antropología Biológica (IV Coloquio de Antropología Física Juan Comas)*. México: Universidad Nacional Autónoma de México, pp. 329–44.

Carmenate, M.M., Díaz, M.E., Toledo, E.M., Martinez, A.J., Martinez, C.P., Wong, I., Moreno, R. and Moreno, V. (1997). Ciclo reproductivo y factores sociales en mujeres cubanas. *Antropología Física Latinoamericana*, **1**, 47–66.

Carmona, C., Perdomo, R., Carbo, A., *et al.* (1998). Risk factors associated with human cystic echinococcosis in Florida, Uruguay: results of a mass screening study using ultrasound and serology. *American Journal of Tropical Medicine and Hygiene*, **58**, 599–605.

Carneiro, J.I. (1945). Los grupos sanguineos de la Provincia de Tucuman. *Archivos de Farmacia y Bioquimica* (Tucuman), **2**, 169–79.

Carneiro, T.A., Pessoa, E.P., Stambowsky, B., Montenegro, R.M. and Carneiro, D.A. (1970). Estudo antropométrico realizado em Fortaleza no grupo etário de 7 e 8 anos de idade. *Revista da Faculdade de Medicina da Universidade Federal do Ceará*, **10**, 63–6.

Carnevale, A., Hernández, M., Reyes, R., Paz, F. and Sosa, C. (1985). The frequency and economic burden of genetic disease in a pediatric hospital in México City. *American Journal of Medical Genetics*, **20**, 665–75.

Carnevale, A., Lisker, R., Villa, A.R. and Armendares, S. (1998). Attitudes of Mexican geneticists towards prenatal diagnosis and selective abortion. *American Journal of Medical Genetics*, **75**, 426–31.

Carrera de Boscán, B. and Rodríguez, A. (1984). Datos estadísticos sobre partos gemelares durante 1981 en la maternidad Concepción Palacios, Caracas, Venezuela. *Memorias, VI Congreso Latinoamericano de Genética*, 309–10.

Carvajal-Carmona, L.G., Soto, I.D., Pineda, N., *et al.* (2000). Strong Amerind/white sex bias and a possible Sephardic contribution among the founders of northwest Colombia. *American Journal of Human Genetics*, **67**, 1287–95.

Carvalho, E.M., Acioli, M.D., Branco, M.A.F., *et al.* (1998). Evolução da esquistossomose na Zona da Mata Sul de Pernambuco. Epidemiologia e situação atual: controle ou descontrole? *Cadernos de Saúde Pública*, **14**,

787–95.

Carvalho, F.M., Silvany, A.M. Neto, Paim, J.S. and Mello, A.M.C. (1988). Morbidade referida e utilização de consulta médica em cinco populações do Estado da Bahia. *Ciência e Cultura*, **40**, 853–8.

Carvalho, J.A.M. (1985). Estimativas indiretas e dados sobre migrações, Uma avaliação conceitual e metodológica das informações censitárias recentes. *Revista Brasileira de Estudos de População*, **2**(1), 31–73.

Carvalho, J.A.M. (1996). Os saldos dos fluxos migratórios internacionais do Brasil na década de 80 – uma tentativa de estimação. *Revista Brasileira de Estudos de População*, **13**, 3–14.

Carvalho, M.R., Giugliani, R., Fiori, R.M., Barcellos, L. and Costa, J.C. (1990). Zellweger (cerebro-hepato-renal) syndrome: first Brazilian case with confirmed peroxisomal defect. *Revista Brasileira de Genética*, **13**, 363–70.

Carvalho, T.B., Safatle, H.P.N., Gonçalves, A., Padovani, C.R. and Ferrari, I. (1999). Chromosome disorders in a hospital for musculoskeletal diseases. *Brazilian Journal of Dysmorphology and Speech–Hearing Disorders*, **2**, 11–16.

Casadei, F.M., Motonaga, S.M., Cardoso, A.C.V., Villani, A.C., Richieri-Costa, A. and Giachetti, C.M. (1998). Branchio-Oto-Renal (BOR syndrome): report of a Brazilian family. *Brazilian Journal of Dysmorphology and Speech–Hearing Disorders*, **2**, 3–8.

Casillas, L.E. and Vargas, L.A. (1987). Una gráfica para uso de la comunidad en la detección de alteraciones del crecimiento y la nutrición de escolares. In Faulhauber, M.E.S. and Lizárraga Cruchaga, X., editors, *Estudios de Antropología Biológica (III Coloquio de Antropología Física Juan Comas)*. México: Universidad Nacional Autónoma de México, pp. 11–27.

Castilho, E.M., Naoum, P.C., Graciano, R.A.S. and Silva, R.A. (1987). Prevalências de talassemia alfa em pacientes com anemia e em pessoas sem anemia. *Revista Brasileira de Patologia Clínica*, **23**, 131–4.

Castilla, E.E. and Adams, J. (1990). Migration and genetic structure in an isolated population in Argentina, Aicuña. In Adams, J., Lam, D.A., Hermalin, A.I. and Smouse, P.E., editors, *Convergent Issues in Genetics and Demography*. Oxford: Oxford University Press, pp. 45–62.

Castilla, E.E. and Orioli, I.M. (1983). El estudio colaborativo latinoamericano de malformaciones congénitas: ECLAMC/Monitor. *Interciencia*, **8**, 271–8.

Castilla, E.E. and Parreiras, I.M.O. (1982). Malformations and inbreeding. *Boletin Genético* (Castelar), **11**, 19–24.

Castilla, E.E. and Villalobos, H. (1977). *Malformaciones Congénitas*. Maracaibo: Universidad de Zulia.

Castilla, E.E., Ashton Prolla, P., Barreda-Mejia, E., *et al.* (1996*b*). Thalidomide, a current teratogen in South America. *Teratology*, **54**, 273–7.

Castilla, E.E., Dutra, M.G., Fonseca, R.L. and Paz, J.E. (1997*a*). Hand and foot postaxial polydactyly. Two different traits. *American Journal of Medical Genetics*, **73**, 48–54.

Castilla, E.E., Gomez, M.A., Lopez-Camelo, J.S. and Paz, J.E. (1991*a*). Frequency of first-cousin marriages from civil marriage certificates in Argentina. *Human Biology*, **63**, 203–10.

382   *References*

Castilla, E.E., Lopez-Camelo, J.S. and Paz, J.E. (1995). *Atlas Geográfico de las Malformaciones Congénitas en Sudamérica.* Rio de Janeiro: Fundação Oswaldo Cruz.

Castilla, E.E., Lopez-Camelo, J.S., Dutra, G.P. and Paz, J.E. (1991*b*). Birth defects monitoring in underdeveloped countries: an example from Uruguay. *International Journal of Risk & Safety in Medicine*, **2**, 271–88.

Castilla, E.E., Lopez-Camelo, J.S., Dutra, M.G., Queenan, J.T., Simpson, J.L. and NFP-ECLAMC Group (1997*b*). The frequency and spectrum of congenital anomalies in natural family planning users in South America. No increase in a case–control study. *Advances in Contraception*, **13**, 395–404.

Castilla, E.E., Lopez-Camelo, J.S., Orioli, I.M., Sánchez, O. and Paz, J.E. (1988). The epidemiology of conjoined twins in Latin America. *Acta Geneticae Medicae et Gemellologiae*, **37**, 111–18.

Castilla, E.E., Lopez-Camelo, J.S., Paz, J.E. and Orioli, I.M. (1996*a*). *Prevención Primaria de los Defectos Congénitos.* Rio de Janeiro: Fundação Oswaldo Cruz.

Castilla, E.E., Rittler, M., Dutra, M.G., Lopez-Camelo, J.S., Campaña, H., Paz, J.E., Orioli, I.M. and the ECLAMC-Downsurv Group (1998). Survival of children with Down syndrome in South America. *American Journal of Medical Genetics*, **79**, 108–11.

Castillo-Taucher, S. (1997). Los servicios de genética médica en Chile. *Brazilian Journal of Genetics*, **20** (Suppl. 1), 25–31.

Castro de Guerra, D. (1991). Factors condicionantes de la estructura genética en dos poblaciones negras Venezolanas. Ph.D. thesis, Instituto Venezolano de Investigaciones Científicas, Caracas.

Castro de Guerra, D., Arvello, H., Larralde, A.R. and Salzano, F.M. (1996). Genetic study in Panaquire, a Venezuelan population. *Human Heredity*, **46**, 323–8.

Castro de Guerra, D., Arvelo, H. and Pinto-Cisternas, J. (1993). Estructura de población y factores influyentes en dos pueblos negros venezolanos. *América Negra*, **5**, 37–47.

Castro de Guerra, D., Arvelo, H. and Pinto-Cisternas, J. (1999). Population structure of two black Venezuelan populations studied through their mating structure and other related variables. *Annals of Human Biology*, **26**, 141–50.

Castro de Guerra, D., Hutz, M.H. and Salzano, F.M. (1997). Beta-globin gene cluster haplotypes in an admixed Venezuelan population. *American Journal of Human Biology*, **9**, 323–7.

Castro de Guerra, D., Pinto-Cisternas, J. and Rodríguez-Larralde, A. (1990*b*). Inbreeding as measured by isonymy in two Venezuelan populations and its relationship to other variables. *Human Biology*, **62**, 269–78.

Castro de Guerra, D., Rodríguez-Larralde, A. and Pinto-Cisternas, J. (1990*a*). Distribución de los apellidos y estructura de población en algunas poblaciones de origen negro de la zona costera norcentral de Venezuela. *Acta Científica Venezolana*, **41**, 241–9.

Castro, V., Origa, A.F., Annichino-Bizzacchi, J.M., *et al.* (1999). Frequencies of platelet-specific alloantigen systems 1–5 in three distinct ethnic groups in

Brazil. *European Journal of Immunogenetics*, **26**, 355–60.

Castro Faria, L. (1952). Pesquisas de Antropologia Física no Brasil. História. Bibliografia. *Boletim do Museu Nacional, Antropologia*, **13**, 1–106.

Cat, I., Marinoni, L.P., Giraldi, D.J., Furtado, V.P., Pasquini, R., Freire-Maia, N. and Braga, H. (1967). Epidermolysis bullosa dystrophica, hypoplastic type, associated with Pelger–Huet anomaly. *Journal of Medical Genetics*, **4**, 302–3.

Cavalcanti, D.P., Ferrari, I. and Pinto, W. Jr. (1988). Chromosome analysis of 52 spontaneous abortuses in Brazil. *Revista Brasileira de Genética*, **11**, 149–54.

Cavalli, I.J., Mattevi, M.S., Erdtmann, B., Sbalqueiro, I.J. and Maia, N.A. (1984). Quantitative analysis of C bands in chromosomes 1, 9, 16 and Y in Caucasian and Japanese males. *Human Heredity*, **34**, 62–4.

Cavalli, I.J., Mattevi, M.S., Erdtmann, B., Sbalqueiro, I.J. and Maia, N.A. (1985). Equivalence of the total constitutive heterochromatin content by an interchromosomal compensation in the C band sizes of chromosomes 1, 9, 16, and Y in Caucasian and Japanese individuals. *Human Heredity*, **35**, 379–87.

Cavalli-Sforza, L.L. and Bodmer, W.F. (1971). *The Genetics of Human Populations*. San Francisco: W.H. Freeman.

Cavalli-Sforza, L.L., Menozzi, P. and Piazza, A. (1994). *The History and Geography of Human Genes*. Princeton: NJ: Princeton University Press.

Cavalli-Sforza, L.L., Menozzi, P. and Piazza, A. (1994). *The History and Geography of Human Genes*. Princeton, NJ: Princeton University Press.

Centeno, N.A., Bello, A.H. and Beltrán, J.S. (1996). Sistemas de salud de las comunidades indígenas y negras de Colombia estudiadas por la Gran Expedición Humana. In Bernal, J., editor, *Terrenos de la Gran Expedición Humana*, Vol. 9. Bogotá: Pontificia Universidad Javeriana, pp. 1–164.

Cerda-Flores, R.M., Kshatriya, G.K., Barton, S.A., et al. (1991). Genetic structure of the populations migrating from San Luis Potosi and Zacatecas to Nuevo León in México. *Human Biology*, **63**, 309–27.

Cerda-Flores, R.M., Barton, S.A., Hanis, C.L. and Chakraborty, R. (1994). Genetic variation by birth cohorts in Mexican Americans of Starr County, Texas. *American Journal of Human Biology*, **6**, 669–74.

Ceroni, M., Piccardo, P., Rodgers-Johnson, P., et al. (1988). Intrathecal synthesis of IgG antibodies to HTLV-I supports the etiological role for HTLV-I in tropical spastic paraparesis. *Annals of Neurology*, **23** (Suppl.), S188–91.

Cézar, P.C., Mizusaki, K., Pinto, W. Jr., Opromolla, D.W.A. and Beiguelman, B. (1974). Hemoglobina S e lepra. *Revista Brasileira de Pesquisas Médicas e Biológicas*, **7**, 151–67.

Chacin-Bonilla, L., Guanipa, N., Cano, G., Parra, A.M., Estevez, J. and Raleigh, X. (1998). Epidemiological study of intestinal parasitic infections in a rural area from Zulia State, Venezuela. *Interciencia*, **23**, 241–7.

Chakraborty, R. (1975). Estimation of race admixture: a new method. *American Journal of Physical Anthropology*, **42**, 507–11.

Chakraborty, R. (1985). Gene identity in racial hybrids and estimation of admixture rates. In Neel, J.V. and Ahuja, Y., editors, *Genetic*

Microdifferentiation in Man and Other Animals. New Delhi: Indian
Anthropological Association, pp. 171–80.

Chakraborty, R. (1986). Gene admixture in human populations: models and
predictions. *Yearbook of Physical Anthropology*, **29**, 1–43.

Chalar, E.M., Galli, A.L., Milone, C.A. and Pintos, L.B. (1966). Frecuencia de
los sistemas Kell Cellano, MN en dadores de la Ciudad de La Plata.
Sumários, II Congreso Argentino de Hematologia y Hemoterapia (Córdoba,
Argentina), 48.

Chan, S-Y., Ho, L., Ong, C-K., *et al*. (1992). Molecular variants of human
papillomavirus type 16 from four continents suggest ancient pandemic
spread of the virus and its coevolution with humankind. *Journal of Virology*,
66, 2057–66.

Chávez, E.R. (1996). A crise migratória do verão de 1994. Balanço e perspectivas
do fluxo emigratório cubano, 1984–1996. *Revista Brasileira de Estudos de
População*, **13**(2), 135–67.

Chediak, M., Cabrera Calderin, J. and Prado y Vargas, G. (1939). Anemia a
hematies falciformes. Contribucion a su estudio en Cuba. *Archivos de
Medicina Interna*, **5**, 313–70.

Chiarella, J.M., Goldberg, A.C., Abel, L., Carvalho, E.M., Kalil, J. and Dessein,
A. (1998). Absence of linkage between MHC and a gene involved in
susceptibility to human schistosomiasis. *Brazilian Journal of Medical and
Biological Research*, **31**, 665–70.

Chiavenato, J.J. (1986). *O Negro no Brasil. Da Senzala à Guerra do Paraguai*. São
Paulo: Brasiliense.

Chiba, A.K., Kuwano, S.T., Carvalho, K.I., *et al*. (1998). Gene frequencies of
human platelet antigens in South American Indians and Brazilian blood
donors. *Transfusion*, **38** (Suppl.), 28S.

Chikhi, L., Destro-Bisol, G., Bertorelle, G., Pascali, V. and Barbujani, G. (1998).
Clines of nuclear DNA markers suggest a largely Neolithic ancestry of the
European gene pool. *Proceedings of the National Academy of Sciences, USA*,
95, 9053–8.

Chouza, C., Ketzoian, C., Caamaño, J.L., Cáceres, R., Coirolo, G., Dieguez, E.
and Rega, I. (1996). Prevalence of Parkinson's disease in a population of
Uruguay. Preliminary results. *Advances in Neurology*, **69**, 13–17.

Christensen, A.F. (1998). Ethnohistorical evidence for inbreeding among the
pre-Hispanic Mixtec royal caste. *Human Biology*, **70**, 563–77.

Clavijo, H.A., Schüler, L., Sanseverino, M.T. and Giugliani, R. (1992). An
information service on teratogenic agents in Brazil. *World Health Forum*, **13**,
196.

Clevelario, J., Jr. (1997). A participação da imigração na formação da população
brasileira. *Revista Brasileira de Estudos de População*, **14**(1/2), 51–71.

Coelho, J.C., Scalco, F.B., Tobo, P. and Giugliani, R. (1995). Diagnóstico
laboratorial da doença de Niemann-Pick tipo C: Relato de uma experiência
pioneira no Brasil. *Revista do Hospital de Clínicas de Porto Alegre*, **15**,
249–51.

Coelho, J.C., Wajner, M., Burin, M.G., Vargas, C.R. and Giugliani, R. (1997).

Selective screening of 10 000 high-risk Brazilian patients for the detection of inborn errors of metabolism. *European Journal of Pediatrics*, **156**, 650–4.

Coimbra, A.M.F.C., Ayres, M. and Salzano, F.M. (1971). Distância marital em gerações sucessivas da população de Belém, Pará. *Ciência e Cultura*, **23**(Suppl.), 95.

Colantonio, S.E. (1998). Estructura poblacional a partir de apellidos y migración. Departamento Pocho (Provincia de Córdoba, Argentina). *Revista Española de Antropología Biológica*, **19**, 45–63.

Colantonio, S.E. and Celton, D.E. (1996). Estructura de una población semiaislada actual, Reproducción, selección natural y deriva genética. *Revista Española de Antropologia Biológica*, **17**, 105–27.

Collazo, T., Magarino, C., Chavez, R., Suardiaz, B., Gispert, S., Gomez, M., Rojo, M. and Heredero, L. (1995). Frequency of delta-F508 mutation and XV2C/KM19 haplotypes in Cuban cystic fibrosis families. *Human Heredity*, **45**, 55–7.

Collet, A. 1958. Recherche des hémoglobines anormales chez les lépreux. *Archives du Institut Pasteur de Martinique*, **11**, 16.

Colletto, G.M.D., Krieger, H. and Magalhães, J.R. (1981). Estimates of the genetical and environmental determinants of serum lipid and lipoprotein concentrations in Brazilian twins. *Human Heredity*, **31**, 232–7.

Colletto, G.M.D., Krieger, H. and Magalhães, J.R. (1983). Genetic and environmental determinants of 17 serum biochemical traits in Brazilian twins. *Acta Geneticae Medicae et Gemellologiae*, **32**, 23–9.

Collier, W.A., Wolff, A.E. and Zaal, A.E.G. (1952). Contributions to the geographical pathology of Surinam. I. Blood groups of the Surinam population. *Documenta Medica Geographica Tropica*, **4**, 92–5.

Colombo, B. and Martínez, G. (1981). Haemoglobinopathies including thalassaemia. 2. Tropical America. *Clinics in Haematology*, **10**, 730–56.

Colombo, B. and Martínez, G. (1985). Hemoglobin variants in Cuba. *Hemoglobin*, **9**, 415–22.

Colombo, C. (1984). Diários da Descoberta da América. Porto Alegre: L & PM.

Colonia, V.J. and Roisenberg, I. (1979). Investigation of associations between ABO blood groups and coagulation, fibrinolysis, total lipids, cholesterol, and triglycerides. *Human Genetics*, **48**, 221–30.

Comas, J. (1959). El índice cnémico en tibias prehispanicas y modernas del Valle de México. *Cuadernos del Instituto de História, Série Antropológica*, **5**, 1–55.

Comas, J. (1966). *Manual de Antropología Física*. México: Universidad Nacional Autónoma de México.

Conceição, M.M., Lyra, L.G., Azevêdo, E.S., Almeida-Melo, N. and Fonseca, E.F. (1979). Association between HbsAg and race in a mixed population of northeastern Brazil. *Revista Brasileira de Pesquisas Médicas e Biológicas*, **12**, 405–9.

Conceição, M.M., Salzano, F.M., Franco, M.H.L.P., Weimer, T.A. and Krieger, H. (1987). Demography, genetics, and race admixture in Aracaju, Brazil. *Revista Brasileira de Genética*, **10**, 313–31.

Confalonieri, U., Araújo, A.J.G. and Ferreira, L.F. (1981). *Trichuris trichiura* infection in Colonial Brazil. *Paleopathology Newsletter*, **35**, 13–14.

Confalonieri, U., Ferreira, L.F. and Araújo, A. (1991). Intestinal helminths in lowland South American Indians: some evolutionary interpretations. *Human Biology*, **63**, 863–73.

Conti, F. and Krieger, H. (1985). PGM₁, PGM₂, G6PD, 6PGD e SOD_A, Freqüências gênicas e possíveis efeitos sobre características métricas na população de Bambuí (MG). *Ciência e Cultura*, **37** (Suppl.), 766.

Coope, E. and Roberts, D.F. (1971). Dermatoglyphic studies of populations in Latin America. In Salzano, F.M., editor, *The Ongoing Evolution of Latin American Populations*. Springfield, IL: C.C. Thomas, pp. 405–53.

Coral-Vázquez, R., Arenas, D., Cisneros, B., *et al.* (1997). Pattern of deletions of the dystrophin gene in Mexican Duchenne/Becker muscular dystrophy patients: the use of new designed primers for the analysis of the major deletion 'hot spot' region. *American Journal of Medical Genetics*, **70**, 240–6.

Cormand, B., Harboe, T.L., Gort, L., *et al.* (1998). Mutation analysis of Gaucher disease patients from Argentina: High prevalence of the *RecNci* I mutation. *American Journal of Medical Genetics*, **80**, 343–51.

Corredor, A., Nicholls, R.S., Duque, S., Munoz de Hoyos, P., Alvarez, C.A., Guderian, R.H., Lopez, H.H. and Palma, G.I. (1998). Current status of onchocerciasis in Colombia. *American Journal of Tropical Medicine and Hygiene*, **58**, 594–8.

Costa, D. Jr. and Borborema, R.D.P. (1974). Hemoglobinopatias. Estudos realizados em pacientes do Hospital dos Servidores do Estado do Pará. *Revista Brasileira de Patologia Clínica*, **10**, 25–8.

Costa, E.B. (1998). *História Ilustrada do Rio Grande do Sul*. Porto Alegre: Já Porto Alegre Editores.

Costa, F.F. and Zago, M.A. (1986). Bone marrow and peripheral blood globin chain synthesis in sickle cell β° thalassaemia. *Journal of Medical Genetics*, **23**, 252–5.

Costa, F.F., Figueiredo, M.S., Sonati, M.F., Kimura, E.M. and Martins, C.S.B. (1992). The IVS-I-110 (G→T) and codon 39 (C→T) β-thalassemia mutations in association with a α-thal-2 (–3.7 kb) and Hb Hasharon [α 47 (CE5) ASP→HIS] in a Brazilian patient. *Hemoglobin*, **16**, 525–9.

Costa, F.F., Sonati, M.F. and Zago, M.A. (1991a). Hemoglobin Stanleyville II (α78 Asn→Lys) is associated with a 3.7-kb α-globin gene deletion. *Human Genetics*, **86**, 319–20.

Costa, F.F., Tavella, M.H. and Zago, M.A. (1989). Deletion type α-thalassemia among Brazilian patients with sickle cell anemia. *Revista Brasileira de Genética*, **12**, 605–11.

Costa, F.F., Tavella, M.H. and Zago, M.A. (1991b). β-thalassemia intermedia and IVS-1 nt6 homozygosis in Brazil. *Brazilian Journal of Medical and Biological Research*, **24**, 157–61.

Costa, F.F., Zago, M.A., Sonati, M.F. and Bottura, C. (1987). The association of Hb Stanleyville II with α thalassemia and Hb S. *Nouvelle Revue Française d'Hématologie*, **29**, 387–90.

Costa, H.C., Opromolla, D.V.A., Virmond, M. and Beiguelman, B. (1993).

Influence of the rapid acetylator phenotype on the emergence of DDS resistant *Mycobacterium leprae*. *Revista Brasileira de Genética*, 16, 1029–34.

Costa, H.C., Souza, L.C.D. and Leirião, J.A. (1985). Correlações entre os níveis hematimétricos e os diferentes graus de desnutrição em crianças portadoras de parasitas intestinais da periferia de Bauru, SP. *Salusvita*, 4, 64–72.

Costa, I. (1983). *Economia escravista Brasileira*. Special issue. *Estudos Econômicos*, 13, 1–287.

Costa, S.H., Martin, I.R., Freitas, S.R.S. and Pinto, C.S. (1990). Family planning among low-income women in Rio de Janeiro, 1984–1985. *International Family Planning Perspectives*, 16, 16–28.

Costa-Macedo, L.M., Esteves da Costa, M.C. and Almeida, L.M. (1999). Parasitismo por *Ascaris lumbricoides* em crianças menores de dois anos: estudo populacional em comunidade do Estado do Rio de Janeiro. *Cadernos de Saúde Pública*, 15, 173–8.

Costa-Macedo, L.M., Machado-Silva, J.R., Rodrigues-Silva, R., Oliveira, L.M. and Vianna, M.S.R. (1998). Enteroparasitoses em pré-escolares de comunidades favelizadas da cidade do Rio de Janeiro, Brasil. *Cadernos de Saúde Pública*, 14, 851–5.

Costanzi, E., Erwenne, C.M. and Armelin, M.C.S. (1993). PCR detection of *XbaI* polymorphism in the human *Rb* gene of retinoblastoma patients. *Brazilian Journal of Medical and Biological Research*, 26, 1031–6.

Costanzi, E., Silva, M.E., Farah, L.M.S., Andrade, J.A.D. and Farah, C.A. (1987). Alfafetoproteína em líquido amniótico de gestantes brasileiras. *Revista Paulista de Medicina*, 105, 75–80.

Costanzi, E., Silva-Fernandes, M.E. and Erwenne, C.M. (1989*b*). Esterase D analysis in familial retinoma and retinoblastoma. *Ophthalmic Paediatrics and Genetics*, 10, 157–60.

Costanzi, E., Silva-Fernandes, M.E., D'Almeida, V. and Erwenne, C.M. (1989*a*). Esterase D assay in Brazilian retinoblastoma families. *American Journal of Medical Genetics*, 34, 391–6.

Coura-Filho, P. (1997). Distribuição da esquistossomose no espaço urbano. 1. O caso da região metropolitana de Belo Horizonte, Minas Gerais, Brazil. *Cadernos de Saúde Pública*, 13, 245–55.

Coura-Filho, P., Farah, M.W.C., Rezende, D.F., Lamartine, S.S., Carvalho, O.S. and Katz, N. (1995). Determinantes ambientais e sociais da esquistossomose mansoni em Ravena, Minas Gerais, Brasil. *Cadernos de Saúde Pública*, 11, 254–65.

Covarrubias, E. (1965). Microevolución en poblaciones humanas chilenas. I. Flujo genético y siete rasgos en dos poblaciones contrastantes. *Biologica*, 37, 62–77.

Crawford, M.H. (1976). *The Tlaxcaltecans*, Prehistory, Demography, Morphology and Genetics. Lawrence, KS: University of Kansas Publications in Anthropology, pp. 1–208.

Crawford, M.H. (1983). The anthropological genetics of the Black Caribs (Garifuna) of Central America and the Caribbean. *Yearbook of Physical Anthropology*, 26, 161–92.

388   *References*

Crawford, M.H. (1998). *The Origins of Native Americans. Evidence from Anthropological Genetics.* Cambridge: Cambridge University Press.
Crawford, M.H., Gonzalez, N.L., Schanfield, M.S., Dykes, D.D., Skradski, K. and Polesky, H.F. (1981). The Black Caribs (Garifuna) of Livingston, Guatemala: genetic markers and admixture estimates. *Human Biology*, **53**, 87–103.
Crawford, M.H., Leyshon, W.C., Brown, K., Lees, F. and Taylor, L. (1974). Human biology in Mexico, II: a comparison of blood group, serum and red cell enzyme frequencies, and genetic distances of the Indian populations of Mexico. *American Journal of Physical Anthropology*, **41**, 251–68.
Crawford, M.H., Lisker, R. and Perez Briceno, R. (1976). Genetic microdifferentiation of two transplanted Tlaxcaltecan populations. In Crawford, M.H., editor, *The Tlaxcaltecans: Prehistory, Demography, Morphology and Genetics.* Lawrence, KS: University of Kansas Publications in Anthropology, pp. 169–75.
Crispim, J., Trigueiro, K.G., Benevides-Filho, F.R.S. and Salzano, F.M. (1972). Third molar agenesis in a trihybrid Brazilian population. *American Journal of Physical Anthropology*, **37**, 289–92.
Crooks, D.L. (1994). Relationship between environment and growth for Mopan children in Belize. *American Journal of Human Biology*, **6**, 571–84.
Crossley, J.C. (1985). The River Plate countries. In Blakemore H. and Smith, C.T., editors, *Latin America. Geographical Perspectivas.* London: Methuen, pp. 383–455.
Crow, J.F. (1958). Some possibilities for measuring selection intensities in man. *Human Biology*, **30**, 1–13.
Crow, J.F. (1966). The quality of people, human evolutionary changes. *Bioscience*, **16**, 863–7.
Crow, J.F. (1980). The estimation of inbreeding by isonymy. *Human Biology*, **52**, 1–14.
Crow, J.F. (1986). *Basic Concepts in Population, Quantitative, and Evolutionary Genetics.* New York: W.H. Freeman.
Crow, J.F. and Mange, A.P. (1965). Measurement of inbreeding from the frequency of marriages between persons of the same surname. *Eugenics Quarterly*, **12**, 199–203.
Cruz-Coke, R. (1965*a*). Consanguinidad parental en una población hospitalizada. *Revista Médica de Chile*, **93**, 583–7.
Cruz-Coke, R. (1965*b*). Asociación de defectos de visión de colores y cirrosis hepática. *Revista Médica de Chile*, **93**, 519–21.
Cruz-Coke, R. (1970). *Color Blindness.* Springfield, IL: C.C. Thomas.
Cruz-Coke, R. (1971). Consecuencias genéticas de la anticoncepción en Santiago. *Revista Médica de Chile*, **99**, 190–4.
Cruz-Coke, R. (1997). Patterns of morbidity in prehistoric and historic Andean populations. In Barton, S.A., Rothhammer, F. and Schull, W.J., editors, *Patterns of Morbidity in Andean Aboriginal Populations: 8000 Years of Evolution.* Santiago: Amphora Editores, pp. 85–102.
Cruz-Coke, R. and Biancani, F. (1966). Demografia genética de Arica. *Revista

Médica de Chile, **94**, 63–70.
Cruz-Coke, R. and Brncic, D. (1982). *Genética. V Congreso Latinoamericano.* Santiago: Asociación Latinoamericana de Genética.
Cruz-Coke, R. and Covarrubias, E. (1962). Factors influencing blood pressure in a rural Chilean community. *The Lancet*, **I**, 1138–40.
Cruz-Coke, R. and Covarrubias, E. (1963). Herencia multifactorial de la presión arterial en una población endogámica. *Revista Médica de Chile*, **91**, 252–8.
Cruz-Coke, R. and Iglesias, R. (1963). Asociación de alelo *T* com fenotipos patológicos renales. *Revista Médica de Chile*, **91**, 829–33.
Cruz-Coke, R. and Moreno, R.S. (1994). Genetic epidemiology of single gene defects in Chile. *Journal of Medical Genetics*, **31**, 702–6.
Cruz-Coke, R. and Rivera, L. (1980). Genetic characteristics of hemophilia A in Chile. *Human Heredity*, **30**, 161–70.
Cruz-Coke, R. and Silva, J.C.N. (1969). Estudio médico genético de muestra universal de nacidos vivos mediante computación. I. Introducción al método. *Revista Médica de Chile*, **97**, 320–31.
Cruz-Coke, R. and Valenzuela, C.Y. (1973). Enfermedades hereditarias en un hospital general. *Revista Médica de Chile*, **101**, 212–15.
Cruz-Coke, R. and Valenzuela, C.Y. (1975). Enfermedades genéticas en Chile. *Revista Médica de Chile*, **103**, 327–30.
Cruz-Coke, R., Cristoffanini, A., Barrera, R., Montenegro, A. and Villagra, R. (1967). Flujo genico del sistema ABO en la poblacion de Arica. *Revista Médica de Chile*, **95**, 614–18.
Cruz-Coke, R., Etcheverry, R. and Nagel, R. (1964*a*). Influence of migration on blood pressure of islanders. *The Lancet*, **I**, 697–9.
Cruz-Coke, R., Laval, E., Mendoza, A. and Montenegro, A. (1968). Asociación del cromosoma rhesus Cde (*R1*) com *Salmonela typhi* in vivo. *Archivos de Biologia Médica y Experimentación*, **4**, 252–7.
Cruz-Coke, R., Nagel, R. and Etcheverry, R. (1964*b*). Effects of locus MN on diastolic blood pressure in a human population. *Annals of Human Genetics*, **28**, 39–48.
Cruz-Coke, R., Rivera, L., Kattan, L. and Mardones, J. (1971). Defectos de visión de colores en mujeres alcoholicas y sus parientes. *Revista Médica de Chile*, **99**, 118–24.
Cruz-Coke, R., Valenzuela, C.Y. and Navarro, J.C. (1972). La morbidité génétique: Methodes de mesures et résultats obtenus a Santiago. *Population*, **27**, 1045–52.
Cuadrado, R.R. and Davenport, F.M. (1970). Antibodies to influenza viruses in military recruits from Argentina, Brazil and Colombia. Their relation to ABO blood group distribution. *Bulletin of the World Health Organization*, **42**, 873–84.
Cuadrado, R.R. and Kagan, I.G. (1967). The prevalence of antibodies to parasitic diseases in sera of young army recruits from the United States and Brazil. *American Journal of Epidemiology*, **86**, 330–40.
Cullen, T.L. and Penna Franca, E. (1977). *International Symposium on Areas of High Natural Radioactivity.* Rio de Janeiro: Academia Brasileira de Ciências.

Culpi, L. and Corção, G. (1984). Polimorfismo da haptoglobina em Curitiba. *Ciência e Cultura*, **36** (Suppl.), 843.

Culpi, L. and Lourenço, M.A.C. (1984). Esterase D in a sample of Caucasian and Black individuals from the population of Curitiba, State of Paraná, Brazil. *Revista Brasileira de Genética*, **7**, 313–19.

Culpi, L. and Salzano, F.M. (1984). Migration, genetic markers and race admixture in Curitiba, Brazil. *Journal of Biosocial Sciences*, **16**, 127–35.

Cunha, J.M.P. (1987). A migração nas regiões administrativas do Estado de São Paulo segundo o censo de 1980. *Revista Brasileira de Estudos de População*, **4**(2), 87–111.

Curi, P.R. (1991). Agrupamento de países segundo indicadores básicos e econômicos. *Revista Brasileira de Estudos de População*, **8**(1/2), 112–24.

Curi, P.R. and Kroll, L.B. (1976). Determinação do tamanho da amostra e distribuição dos grupos sanguíneos ABO e Rh em Botucatu, São Paulo, Brasil. *Revista Brasileira de Pesquisas Médicas e Biológicas*, **9**, 129–36.

Curtin, P., Feierman, S., Thompson, L. and Vansina, J. (1978). *African History*. London: Longman.

D'Aloja, A. (1987). Determinación dinamométrica. In Faulhaber, M.E.S. and Lizarraga Cruchaga, X., editors, *Estudios de Antropologia Biológica (III Coloquio de Antropologia Física Juan Comas)*. México: Universidad Nacional de México, pp. 255–65.

D'Aloja, A. (1989). Función pulmonar en mineros de la Sierra de Juarez, Oaxaca. In Serrano, C. and Salas, M.E., editors, *Estudios de Antropología Biológica (IV Coloquio de Antropología Física Juan Comas)*. México: Universidad Nacional Autónoma de México, pp. 305–28.

Da Rocha, F.J., Salzano, F.M., Peña, H.F. and Callegari, S.M. (1972). New studies on the heritability of anthropometric characteristics as ascertained from twins. *Acta Geneticae Medicae et Gemellologiae*, **21**, 125–34.

Da Silva, E.M. (1948). Blood groups of Whites, Negroes and Mulattoes from the State of Maranhão, Brazil. *American Journal of Physical Anthropology*, **6**, 423–8.

Dalalio, M.M.O. (1994). Análise de Associação entre Cardiopatia Chagásica Crônica e Antigenos de Classe I (HLA-A, -B, -C) e de Classe II (HLA-DR e – DQ). M.Sc. thesis, State University of Maringá.

Daltabuit, M. and Leatherman, T.L. (1998). The biocultural impact of tourism on Mayan communities. In Goodman, A.H. and Leatherman, T.L., editors, *Building a New Biocultural Synthesis. Political-Economic Perspectives on Human Biology*. Ann Arbor, MI: University of Michigan Press, pp. 317–37.

Daltabuit, M. and Tronick, E. (1987). Microambiente del infante en el Altiplano peruano. In Faulhaber, M.E.S. and Lizarraga Cruchaga, X., editors, *Estudios de Antropología Biológica (III Coloquio de Antropología Física Juan Comas)*. México: Universidad Nacional de México, pp. 147–70.

Darlu, P., Sagnier, P.P. and Bois, E. (1990). Genealogical and genetical African admixture estimations, blood pressure and hypertension in a Caribbean community. *Annals of Human Biology*, **17**, 387–97.

Darwin, C. (1859). *On the Origin of Species by Means of Natural Selection, or the*

Preservation of Favored Races in the Struggle for Life. London: John Murray.

David, M.H., Wilke, S.M., Silva, N.S.G. and Xavier-Filho, E. (1985). Estudo do desenvolvimento psico-motor em crianças de 0–3 anos de idade de diferentes níveis socioeconômicos (Porto Alegre, RS). *Arquivos de Medicina Preventiva* (Porto Alegre), **7**, 71–9.

De Boni, L.A. (1987). *A Presença Italiana no Brasil*, Vol. I. Porto Alegre: Escola Superior de Teologia, and Torino, Fondazione Agnelli.

De Boni, L.A. (1990). *A Presença Italiana no Brasil*, Vol. II. Porto Alegre: Escola Superior de Teologia, and Torino, Fondazione Agnelli.

De Boni, L.A. (1996). *A Presença Italiana no Brasil*, Vol. III. Porto Alegre: Escola Superior de Teologia, and Torino, Fondazione Agnelli.

De Boni, L.A. and Costa, R. (1991). *Far la Mérica. A Presença Italiana no Rio Grande do Sul.* Porto Alegre: Riocell.

De Lucca, E.J. (1975a). Pesquisas genéticas na população de Humaitá (AM). I. Grupos sangüíneos ABO, Rh e MN. *Revista Brasileira de Medicina,* **32**, 397–400.

De Lucca, E.J. (1975b). Pesquisas genéticas na população de Humaitá (AM). II. Dermatoglifos digitais e palmares. *Folha Médica,* **71**, 287–90.

De Lucca, M., Pérez, B., Desviat, L.R. and Ugarte, M. (1998). Molecular basis of phenylketonuria in Venezuela: presence of two novel null mutations. *Human Mutation,* **11**, 354–9.

De Messias, I.J., Reis, A., Brenden, M., Queiroz-Telles, F. and Mauff, G. (1991). Association of major histocompatibility complex class III complement components C2, BF, and C4 with Brazilian paracoccidioidomycosis. *Complement and Inflammation,* **8**, 288–93.

De Messias, I.J., Santamaria, J., Brenden, M., Reis, A. and Mauff, G. (1993). Association of C4B deficiency (C4B*0) with erythema nodosum in leprosy. *Clinical and Experimental Immunology,* **92**, 284–7.

De Piñango, C.L.A. and Arends, T. (1965). Hemoglobinas anormales en donantes de sangre nativos del Estado Bolívar. *Acta Científica Venezolana,* **16**, 215–18.

De Santis, L. (1997). Infectious/communicable diseases in the Caribbean. In Halberstein, R.A., editor, *Health and Disease in the Caribbean.* Lexington, KY: Association of Caribbean Studies, pp. 23–56.

De Stefano, G.F. and Calicchia, M.C. (1986). Digital dermatoglyphics in two Afro-American communities of Central America. *International Journal of Anthropology,* **1**, 25–38.

De Torregrosa, M.V., Ortiz, A. and Vargas, D. (1956). Sickle-cell-spherocytosis associated with hemolytic anemia. *Blood,* **11**, 260–6.

De Vitto, L.P.M., Costa, O.A., Bevilaqua, M.C., Passerotti, S. and Richieri-Costa, A. (1997). New autosomal recessive syndrome of progressive sensorineural hearing loss and cataracts: report on two Brazilian patients. *American Journal of Medical Genetics,* **70**, 247–9.

De Walt, B.R. (1998). The political ecology of population increase and malnutrition in southern Honduras. In Goodman, A.H. and Leatherman,

T.L., editors, *Building a New Biocultural Synthesis. Political-Economic Perspectives on Human Biology.* Ann Arbor, MI: University of Michigan Press, pp. 295–316.

De-Araujo, M., Sanches, M.R., Suzuki, L.A., Guerra, G. Jr., Farah, S.B. and de-Mello, M.P. (1996). Molecular analysis of CYP21 and C4 genes in Brazilian families with the classical form of steroid 21-hydroxylase deficiency. *Brazilian Journal of Medical and Biological Research,* **29**, 1–13.

Debaz, H., Olivo, A., Garcia, M.N.V., *et al.* (1998). Relevant residues of the DRβ1 third hypervariable region contributing to the expression and to severity of rheumatoid arthritis in Mexicans. *Human Immunology,* **59**, 287–94.

Deghaide, N.H., Dantas, R.O. and Donadi, E.A. (1998). HLA Class I and II profiles of patients presenting with Chagas' disease. *Digestive Diseases Science,* **43**, 246–52.

Deka, R. and Chakraborty, R. (1999). Trinucleotide repeats, genetic instability and variation in the human genome. In Papiha, S.S., Deka, R. and Chakraborty, R., editors, *Genomic Diversity. Applications in Human Population Genetics.* New York: Kluwer Academic/Plenum, pp. 53–64.

Deka, R., Shriver, M.D., Yu, L.M., Ferrell, R.E. and Chakraborty, R. (1995). Intra and interpopulation diversity at short tandem repeat loci in diverse populations of the world. *Electrophoresis,* **16**, 1659–64.

Deka, R., Shriver, M.D., Yu, L.M., *et al.* (1999). Genetic variation at twenty-three microsatellite loci in sixteen human populations. *Journal of Genetics,* **78**, 99–121.

Del Brutto, O.H., Granados, G., Talamas, O., Soteb, J. and Gorodezky, C. (1991). Genetic pattern of the HLA system: HLA A, B, C, DR, and DQ antigens in Mexican patients with parenchymal brain cysticercosis. *Human Biology,* **63**, 85–93.

Delgado, H., Habicht, J-P., Yarbrough, C., *et al.* (1975). Nutritional status and the timing of deciduous tooth eruption. *American Journal of Clinical Nutrition,* **28**, 216–24.

Demarchi, D.A. and Marcellino, A.J. (1996). El uso de dermatoglifos en la discriminación interpoblacional. *Revista Argentina de Antropología Biológica,* **1**, 246–58.

Dennell, R. (1983). *European Economic Prehistory: A New Approach.* London, Academic Press.

Devoto, F.C.H. (1969). Fenotipos y genotipos del complexo de Carabelli en la población contemporánea del Departamento de los Andes (Salta, Argentina). *Revista de la Asociación Odontológica Argentina,* **57**, 3–13.

Dewey, R.A. and Vidal-Rioja, L. (1993). Differential methylation of the human apolipoprotein AI (APO AI) gene in liver and leucocyte cell DNA. *Revista Brasileira de Genética,* **16**, 1057–64.

Diamond, J.M. (1990). The talk of the Americas. *Nature,* **344**, 589–90.

Dias Neto, E., Souza, C.P., Rollinson, D., Katz, N., Pena, S.D.J. and Simpson, A.J.G. (1993). The random amplification of polymorphic DNA allows the identification of strains and species of schistosomes. *Molecular and*

Biochemical Parasitology, **57**, 83–8.

Díaz, M.E., Carmenate, M.M., Prado, C., *et al.* (1998). Estado nutricional y estilo de vida en mujeres pre y postmenopáusicas cubanas. *Revista Española de Antropología Biológica*, **19**, 77–92.

Díaz, M.E., Montero, M., Wong, I., Moreno, V., Toledo, E.M. and Moreno, R. (1994). Fat distribution in Cuban infants. *American Journal of Human Biology*, **6**, 687–92.

Díaz, M.E., Norat, T., Toledo, E.M., Wong, I. and Moreno, V. (1989). Distribución relativa de la grasa corporal en lactantes cubanos mediante el analisis de los componentes principales. In Serrano, C. and Salas, M.E., editors, *Estudios de Antropologia Biológica (IV Coloquio de Antropologia Física Juan Comas)*, México: Universidad Autónoma de México, pp. 361–75.

Dickinson, F. (1994). Potential independent factors of variability of biological status and reproductive history in Yucatecan women. *Studies in Human Ecology*, **11**, 31–54.

Dickinson, F., Cervera, M.D., Murguia, R. and Uc, L. (1991). Growth, nutritional status, and environmental change in Yucatan, México. *Studies in Human Ecology*, **9**, 135–49.

Dickinson, F., Murguia, R., Cervera, M.D., Hernandez, H., Kim, M. and Leon, F. (1989). Antropometria de una población en crecimiento en la costa de Yucatan. In Serrano, C. and Salas, M.E., editors, *Estudios de Antropología Biológica (IV Coloquio de Antropología Física Juan Comas)*. México: Universidad Nacional Autónoma de México, pp. 123–50.

Dipierri, J.E. and Ocampo, S.B. (1985). Anencefalia en la Provincia de Jujuy (República Argentina). *Mendeliana*, **7**, 49–55.

Dipierri, J.E., Alfaro, E., Martínez-Marignac, V.L., *et al.* (1998). Paternal directional mating in two Amerindian subpopulations located at different altitudes in Northwestern Argentina. *Human Biology*, **70**, 1001–10.

Dipierri, J.E., Alfaro, E., Peña, J.A., Constans, J. and Dugoujon, J-M. (2000). GM, KM immunoglobulin allotypes and other serum genetic markers (HP, GC, PI, and TF) among South American populations living at different altitudes (Jujuy Province, Argentina): Admixture estimates. *Human Biology*, **72**, 305–19.

Dobbins, D.A. and Kindick, C.M. (1972). *Anthropometry of the Latin-American Armed Forces*. Forth Clayton, Canal Zone: United States Army Tropic Test Center.

Dobzhansky, Th. (1937). *Genetics and the Origin of Species*. New York: Columbia University Press.

Dominguez, M.G., Rivera, H., Vásquez, A.I. and Ramos, A.L. (1999). Single cell chromosome rearrangements in individuals with reproductive failure. *Genetics and Molecular Biology*, **22**, 21–3.

Donadi, E.A., Carvalho, I.F. and Falcão, R.P. (1989). Circulating immune complexes in sickle cell-β^0 thalassemia. *Brazilian Journal of Medical and Biological Research*, **22**, 1255–7.

Donadi, E.A., Voltarelli, J.C., Paula-Santos, C.M., Kimachi, T. and Ferraz, A.S.

(1998). Association of Alport's syndrome with HLA-DR2 antigen in a group of unrelated patients. *Brazilian Journal of Medical and Biological Research*, **31**, 533–7.

Dornelles, C.L., Callegari-Jacques, S.M., Robinson, W.M., *et al.* (1999). Genetics, surnames, grandparents' nationalities, and ethnic admixture in southern Brazil: do the patterns of variation coincide? *Genetics and Molecular Biology*, **22**, 151–61.

Dressler, W.W., Balieiro, M.C. and Santos, J.E. (1999). Culture, skin color, and arterial blood pressure in Brazil. *American Journal of Human Biology*, **11**, 49–59.

Drets, M.E. (1997). La importancia de un laboratorio de investigación básica para el desarrollo y el entrenamiento de recursos humanos en citogenética en el Uruguay – '36 años de actividad de investigación' (1960–96). *Brazilian Journal of Genetics*, **20** (Suppl. 1), 123–31.

Droulers, M. and Maury, P. (1981). Colonização da Amazônia maranhense. *Ciência e Cultura*, **33**, 1033–50.

Du Toit, B.M. (1997). Ethnomedical (folk) healing in the Caribbean. In Halberstein, R.A., editor, *Health and Disease in the Caribbean*. Lexington, KY: Association of Caribbean Studies, pp. 95–109.

Dufour, D.L., Reina, J.C. and Spurr, G.B. (1999). Food and macronutrient intake of economically disadvantaged pregnant women in Colombia. *American Journal of Human Biology*, **11**, 753–62.

Dufour, D.L., Staten, L.K., Reina, J.C. and Spurr, G.B. (1994). Anthropometry and secular changes in stature of urban Colombian women of differing socioeconomic status. *American Journal of Human Biology*, **6**, 749–60.

Duncan, G., Thomas, E., Gallo, J.C., Baird, L.S., Garrison, J. and Herrera, R.J. (1996). Human phylogenetic relationships acoording to the D1S80 locus. *Genetica*, **98**, 277–87.

Duncan, G.T., Noppinger, K., Carey, J. and Tracey, M. (1993). Comparison of VNTR allele frequencies and inclusion probabilities over six populations. *Genetica*, **88**, 51–7.

Dyke, B. (1971). Potential mates in a small human population. *Social Biology*, **18**, 28–39.

Echandi, C.A. (1953). Grupos sanguíneos en Costa Rica. *Revista de Biologia Tropical*, **1**, 15–16.

Echavarria, A. and Molina, C. (1971). Thalassemia syndromes and abnormal hemoglobins in Colombia. In Arends, T., Bemski, G. and Nagel, R.L., eds., *Genetical, Functional, and Physical Studies of Hemoglobins*. Basel: S. Karger, pp. 65–81.

Eckhardt, R.B. and Melton, T.W. (1992). *Population Studies on Human Adaptation and Evolution in the Peruvian Andes*. University Park, Occasional Papers in Anthropology 14, University Park, PA: Pennsylvania State University.

Economou, E.P., Antonarakis, S.E., Dowling, C.C., Ibarra, B., de la Mora, E. and Kazazian, H.H. Jr. (1991). Molecular heterogeneity of β-thalassemia in Mestizo Mexicans. *Genomics*, **11**, 474.

Effros, R.B. (1998). Replicative senescence in the immune system: impact of the Hayflick limit on T-cell function in the elderly. *American Journal of Human Genetics*, **62**, 1003–7.

Eizirik, D.L., Monteiro, C.M.C., Voltarelli, J.C. and Foss, M.C. (1987). Frequency of HLA antigens in a Brazilian type I diabetic population. *Brazilian Journal of Medical and Biological Research*, **20**, 533–7.

Elizaga, J.C. (1970). *Migraciones a las Areas Metropolitanas de América Latina.* Santiago: Centro Latinoamericano de Demografia.

Elizondo, J. and Zomer, M. 1970. Hemoglobinas anormales en la población asegurada costarricense. *Acta Médica Costarricense*, **13**, 249–55.

Ell, E., Camacho, L.A.B. and Chor, D. (1999). Perfil antropométrico de funcionários de banco estatal no Estado do Rio de Janeiro/Brasil. I. Índice de massa corporal e fatores sócio-demográficos. *Cadernos de Saúde Pública*, **15**, 113–21.

Elliot, J.H. (1997*a*). A conquista espanhola e a colonização da América. In: Bethell, L., editor, *História da América Latina*, Vol. 1, *América Latina Colonial.* São Paulo: Editora da Universidade de São Paulo, pp. 135–94.

Elliot, J.H. (1997*b*). A Espanha e a América nos séculos XVI e XVII. In: Bethell, L., editor, *História da América Latina*, Vol. 1, *América Latina Colonial.* São Paulo: Editora da Universidade de São Paulo, pp. 283–337.

Ellis, N.A., Ciocci, S., Proytcheva, M., Lennon, D., Groden, J. and German, J. (1998). The Ashkenazic Jewish Bloom syndrome mutation blm^{Ash} is present in non-Jewish Americans of Spanish ancestry. *American Journal of Human Genetics*, **63**, 1685–93.

Elston, R.C. (1971). The estimation of admixture in racial hybrids. *Annals of Human Genetics*, **35**, 9–17.

Englert, S. and Cruz-Coke, R. (1975). Estructura genealógica de la población tribal de Isla de Pascua. *Revista Médica de Chile*, **103**, 340–3.

Engrácia, V., Mestriner, M.A. and Krieger, H. (1990). Human red cell enzymes distribution in a population sample from Bambui (Minas Gerais). *Revista Brasileira de Genética*, **13**, 115–23.

Engstrom, E.M. and Anjos, L.A. (1999). Déficit estatural nas crianças brasileiras: Relação com condições sócio-ambientais e estado nutricional materno. *Cadernos de Saúde Pública*, **15**, 559–67.

Enquist, M. and Arak, A. (1994). Symmetry, beauty and evolution. *Nature*, **372**, 169–72.

Enríquez, J. and Martín, L. (1998). Son necesarias curvas de muestra condicional para el estudio de los recién nacidos de bajo-peso (BPN)? *Revista Española de Antropología Biológica*, **19**, 93–103.

Erdtmann, B., Gomes de Freitas, A.A., Souza, R.P. and Salzano, F.M. (1971). Klinefelter's syndrome and G. trisomy. *Journal of Medical Genetics*, **8**, 364–8.

Erdtmann, B., Salzano, F.M. and Mattevi, M.S. (1975). Chromosome studies in patients with congenital malformations and mental retardation. *Humangenetik*, **26**, 297–306.

Erdtmann, B., Salzano, F.M. and Mattevi, M.S. (1981*b*). Size variability of the Y chromosome distal C band in Brazilian Indians and Caucasoids. *Annals of Human Biology*, **8**, 415–24.

Erdtmann, B., Salzano, F.M. and Mattevi, M.S. (1982). Quantitative analysis of C-band size in human chromosomes. *Acta Anthropogenetica*, **6**, 151–62.

Erdtmann, B., Salzano, F.M., Mattevi, M.S. and Flores R.Z. (1981*a*). Quantitative analysis of C bands in chromosomes 1, 9, and 16 of Brazilian Indians and Caucasoids. *Human Genetics*, **57**, 58–63.

Escamilla, M.A., Demille, M.C., Benavides, E., *et al.* (2000). A minimalist approach to gene mapping: locating the gene for acheiropodia, by homozygosity analysis. *American Journal of Human Genetics*, **66**, 1995–2000.

Escamilla, M.A., McInnes, L.A., Spesny, M., *et al.* (1999). Assessing the feasibility of linkage disequilibrium methods for mapping complex traits: an initial screen for bipolar disorder loci on chromosome 18. *American Journal of Human Genetics*, **64**, 1670–8.

Escamilla, M.A., Spesny, M., Reus, V.I., *et al.* (1996). Use of linkage disequilibrium approaches to map genes for bipolar disorder in the Costa Rican population. *American Journal of Medical Genetics*, **67**, 244–53.

Escobar-Gutiérrez, A., Gorodezky, C. and Salazar-Mallén, M. (1973). Distribution of some of the HL-A system lymphocyte antigens in Mexicans. II. Studies in atopics and lepers. *Vox Sanguinis*, **25**, 151–5.

Esperon, L.C.M. (1978). *Erros Inatos do Metabolismo dos Aminoácidos.* Rio Grande: Fundação Universidade do Rio Grande.

Etcheverry, M.A. (1949). Grupo sanguineo y factor Rh en los Vascos. *Revista de la Sociedad Argentina de Hematologia e Hemoterapia*, **1**, 114–18.

Evans, A.S., Casals, J., Opton, E.M., Borman, E.K., Levine, L. and Cuadrado, R.R. (1969). A nationwide survey of Colombian military recruits, 1966. I. Description of sample and antibody patterns with arboviruses, polioviruses, respiratory viruses, tetanus, and treponematosis. *American Journal of Epidemiology*, **90**, 292–303.

Eveleth, P.B. (1972). An anthropometric study of northeastern Brazilians. *American Journal of Physical Anthropology*, **37**, 223–32.

Excoffier, L. and Slatkin, M. (1995). Maximum-likelihood estimation of molecular haplotype in a diploid population. *Molecular Biology and Evolution*, **12**, 921–7.

Fabritius, H., Millan, J., Blot, M., *et al.* (1980). Hémoglobines rares en Guadeloupe et aux Petites Antilles. *Nouvelle Revue Française d'Hématologie*, **22**, 243–48.

Fajardo, L.F. and Lavalle, Z.N. (1966). Distribution of blood antigens A, B and D in the population of Bogotá (Analysis of 30 000 samples). *American Journal of Physical Anthropology*, **24**, 257–60.

Falcão-Conceição, D.N., Gonçalves-Pimentel, M.M., Moreira, C.P., Kneppers, A.L.J. and Bakker, E. (1992). Use of two multiplex (PCR) reactions as an initial deletion screening method for DMD and BMD patients. *Revista Brasileira de Genética*, **15**, 657–66.

Falcão-Conceição, D.N., Kneppers, A.L.J. and Bakker, E. (1994). DNA analysis of Brazilian Duchenne muscular dystrophy families using (CA)n microsatellite markers. *Revista Brasileira de Genética*, **17**, 113–20.

Falcão-Conceição, D.N., Pereira, M.C.G., Gonçalves, M.M. and Baptista, M.L. (1983). Familial occurrence of heterozygous manifestations in X-linked muscular dystrophies. *Revista Brasileira de Genética*, **6**, 527–38.

Falcão-Conceição, D.N., Santos, D.M. and Gonçalves-Pimentel, M. (1988). Detection rates for Duchenne muscular dystrophy gene carriers using logistic discrimination: creatine-kinase and hemopexin. *Revista Brasileira de Genética*, **11**, 995–1008.

Falkner, F. and Tanner, J.M. (1986). *Human Growth. A Comprehensive Treatise.* (3 volumes.) New York: Plenum Press.

Farah, S.B., Garmes, H.M., Cavalcanti, D.P., *et al.* (1991). Use of Y-chromosome-specific DNA probes to evaluate an XX male. *Brazilian Journal of Medical and Biological Research*, **24**, 149–56.

Faria, M.A.M. (1988). Saúde e trabalho industrial: condições de saúde dos operários brasileiros. *Ciência e Cultura*, **40**, 967–75.

Faria, R., Mello, N.R. and Murat, L.G. (1951). Contribuição ao estudo médico e social do doador. *Arquivos Clínicos*, **12**, 408–17.

Farid-Coupal, N., Contreras, M.L. and Castellano, H.M. (1981). The age at menarche in Carabobo, Venezuela with a note on the secular trend. *Annals of Human Biology*, **8**, 283–8.

Faulhaber, J. (1976). El crecimiento de un grupo de niños normales de la Ciudad de México. *Anales de Antropología*, **13**, 275–88.

Faulhaber, J. (1982). Variaciones en la velocidad del crecimiento en el transcurso del año en niños de la Ciudad de México. In Villanueva, M. and Serrano, C., editors, *Estudios de Antropología Biológica (I Coloquio de Antropología Física Juan Comas)*. México: Universidad Nacional Autónoma de México, pp. 363–88.

Faulhaber, J. (1987). Peso, talla y menarquia en niñas adolescentes de la Ciudad de México. In Faulhaber, M.E.S. and Lizárraga Cruchaga, X., editors, *Estudios de Antropología Biológica (III Coloquio de Antropologia Física Juan Comas)*. México: Universidad Nacional Autónoma de México, pp. 85–107.

Faulhaber, J. (1989). La dentición en adolescentes de la Ciudad de México. In Serrano, C. and Salas, M.E., editors, *Estudios de Antropología Biológica (IV Coloquio de Antropología Física Juan Comas)*. México: Universidad Nacional Autónoma de México, pp. 179–201.

Faulhaber, J. and Schwidetzky, I. (1986). México. In Schwidetzky, I., editor, *Rassengeschichte der Menschheit, 11. Lieferung. Amerika I: Nordamerika, México*. Munich: R. Oldenbourg Verlag, pp. 81–151.

Faulhaber, M.E.S. (1982). Relación entre el grado de maduración osea y tres variables antropométricas en niñas de distinto nivel socioeconomico de la Ciudad de México. In Villanueva, M. and Serrano, C., editors, *Estudios de Antropología Biológica (I Coloquio de Antropología Física Juan Comas)*. México: Universidad Nacional Autónoma de México, pp. 419–32.

Febres, F., Scaglia, H., Lisker, R., *et al.* (1975). Hypothalamic-pituitary-gonadal function inpatients with myotonic dystrophy. *Journal of Clinical Endocrinology and Metabolism*, **41**, 833–40.

Feitosa, M.F. and Krieger, H. (1992). Demography of the human sex ratio in some Latin American countries, 1967–1986. *Human Biology*, **64**, 523–30.

Feitosa, M.F. and Krieger, H. (1993). Some factors affecting the secondary sex ratio in a Latin American sample. *Human Biology*, **65**, 273–8.

Feitosa, M.F., Azevêdo, E.S., Lima, A.M. and Krieger, H. (1999). Genetic causes involved in *Leishmania chagasi* infection in northeastern Brazil. *Genetics and Molecular Biology*, **22**, 1–5.

Feitosa, M.F., Krieger, H., Borecki, I., Beiguelman, B. and Rao, D.C. (1996). Genetic epidemiology of the Mitsuda reaction in leprosy. *Human Heredity*, **46**, 32–5.

Felsenstein, J. (1985). Confidence limits on phylogenies: an approach using bootstrap. *Evolution*, **39**, 783–91.

Fernandes, F. (1965). *A Integração do Negro na Sociedade de Classes*. São Paulo: Dominus.

Fernandes, O., Souto, R.P., Castro, J.A., *et al.* (1998). Brazilian isolates of *Trypanosoma cruzi* from humans and triatomines classified into two lineages using mini-exon and ribosomal RNA sequences. *American Journal of Tropical Medicine and Hygiene*, **58**, 807–11.

Fernández, M.E., Schmidt, A. and Basauri, V. (1976). *La Población de Costa Rica*. San José: Editorial Universitario de Costa Rica.

Fernandez, R., Covarrubias, E., Benado, M., Castelli, G. and Lamborot, L. (1966). Microevolución en poblaciones humanas chilenas. II. Consanguinidad e inmigración en los matrimonios del area de Ñuñoa, Santiago 1850–1960. *Biológica*, **29**, 32–55.

Fernandez, T.S., Silva, M.L.M., Souza, J.M., Tabak, D. and Abdelhay, E. (1997). Cytogenetic study of 50 Brazilian patients with primary myelodysplastic syndrome. *Brazilian Journal of Genetics*, **20**, 87–91.

Fernandez-Mestre, M.T., Layrisse, Z., Montagnani, S., *et al.* (1998). Influence of the HLA Class II polymorphism in chronic Chagas' disease. *Parasite Immunology*, **20**, 197–203.

Fernandez-Viña, M.A., Gao, X., Moraes, M.E., *et al.* (1991*a*). Alleles at four HLA class II loci determined by oligonucleotide hybridization and their associations in five ethnic groups. *Immunogenetics*, **34**, 299–312.

Fernandez-Viña, M.A., Moraes, J.R., Moraes, M.E., Miller, S. and Stastny, P. (1991*b*). HLA Class II haplotypes in Amerindians and in black North and South Americans. *Tissue Antigens*, **38**, 235–7.

Fernandez-Viña, M.A., Lazaro, A.M., Sun, Y., Miller, S., Forero, L. and Stastny, P. (1995). Population diversity of B-locus alleles observed by high-resolution DNA typing. *Tissue Antigens*, **45**, 153–68.

Ferrari, N. (1988). Hemophilia in Florianópolis: frequency and carrier detection. *Revista Brasileira de Genética*, **11**, 975–80.

Ferraroni, J.J. and Hayes, J. (1979). Aspectos epidemiológicos da malária no Amazonas. *Acta Amazonica*, **9**, 471–9.

Ferraroni, J.J., Ferraroni, M.J.R., Ushijima, R.N. and Frade, J.M. (1983). Prevalence of antibodies to nine human pathogens in a remote Amazonia population. *Ciência e Cultura*, **35**, 981–7.

Ferreira, A.H.B. (1996). Os movimentos migratórios e as diferenças de renda *per capita* entre os estados no Brasil (1970–1980). *Revista Brasileira de Estudos de População*, **13**(1), 67–78.

Ferreira, A.P.S., Acosta, A.X., Pina-Neto, J.M. and Ramos, E.S. (1995). Coffin-Siris syndrome in a child with consanguineous parents. *Revista Brasileira de Genética*, **18**, 339–41.

Ferreira, C.E.C. and Flores, L.P.O. (1997/1998). The dimensions of infant mortality in São Paulo. *Brazilian Journal of Population Studies*, **1**, 145–64.

Ferreira, D.M., Pagnan, N.A.B. and Otto, P.A. (1985). The A'-d ridge count in Ullrich–Turner syndrome: study in a large sample. *Revista Brasileira de Genética*, **8**, 193–8.

Ferrell, R.E., Nunez, A., Bertin, T., Labarthe, D.R. and Schull, W.J. (1978). The Blacks of Panama: their genetic diversity as assessed by 15 inherited biochemical systems. *American Journal of Physical Anthropology*, **48**, 269–75.

Ferrera, A., Olivo, A., Alaez, C., Melchers, W.J.G. and Gorodezky, C. (1999). HLA DQA1 and DQB1 loci in Honduran women with cervical dysplasia and invasive cervical carcinoma and their relationship to human papillomavirus infection. *Human Biology*, **71**, 367–79.

Fett-Conte, A.C. and Richieri-Costa, A. (1990). Acheiropodia: report on four new Brazilian patients. *American Journal of Medical Genetics*, **36**, 341–4.

Fialkow, P.J., Lisker, R., Detter, J., Giblett, E.R. and Zavala, C. (1969). 6-Phosphogluconate dehydrogenase: hemizygous manifestation in a patient with leukemia. *Science*, **163**, 194–5.

Fiedel, S.J. (1996). *Prehistoria de América*. Barcelona: Crítica.

Field, L.W. (1994). Who are the Indians? Reconceptualizing indigenous identity, resistance, and the role of social science in Latin America. *Latin America Research Review*, **29**, 237–56.

Figueiredo, M.S. (1993). Molecular analysis of hemophilia B in Brazilian patients. *Brazilian Journal of Medical and Biological Research*, **26**, 919–31.

Figueiredo, M.S. (1994). Location and rapid analysis of the intragenic *Bam HI* polymorphic site of the factor IX gene. *Brazilian Journal of Medical and Biological Research*, **27**, 1117–21.

Figueiredo, M.S., Bernardi, F. and Zago, M.A. (1992*a*). A novel deletion of FVIII gene associated with variable levels of FVIII inhibitor. *European Journal of Haematology*, **48**, 152–4.

Figueiredo, M.S., Bowen, D.J., Silva, W.A. Jr. and Zago, M.A. (1994*c*). Factor IX gene haplotypes in Brazilian Blacks and characterization of unusual DdeI alleles. *British Journal of Haematology*, **87**, 789–96.

Figueiredo, M.S., Kerbauy, J., Gonçalves, M.S., *et al.* (1996). Effect of α-thalassemia and β-globin gene cluster haplotypes on the hematological and clinical features of sickle-cell anemia in Brazil. *American Journal of Hematology*, **53**, 72–6.

Figueiredo, M.S., Santos, J.E., Alberto, F.L. and Zago, M.A. (1992*b*). High frequency of the Lebanese allele of the *LDLr* gene among Brazilian patients with familial hypercholesterolaemia. *Journal of Medical Genetics*, **29**,

813–15.

Figueiredo, M.S., Silva, M.C.B.O., Guerreiro, J.F., *et al.* (1994*b*). The heterogeneity of the β^S cluster haplotypes in Brazil. *Gene Geography*, **8**, 7–12.

Figueiredo, M.S., Tavella, M.H. and Simões, B.P. (1994*a*). Large DNA inversions, deletions, and *Taq I* site mutations in severe haemophilia A. *Human Genetics*, **94**, 473–8.

Firshein, I.L. (1961). Population dynamics of the sickle-cell trait in the Black Caribs of British Honduras, Central America. *American Journal of Human Genetics*, **13**, 233–54.

Fischer, R.R., Giddings, J.C. and Roisenberg, I. (1988). Hereditary combined deficiency of clotting factors V and VIII with involvement of von Willebrand factor. *Clinical and Laboratory Haematology*, **10**, 53–62.

Fischer, R.R., Lerner, C., Bandinelli, E., Fonseca, A.S.K. and Roisenberg, I. (1989). Inheritance and prevalence of von Willebrand's disease severe form in a Brazilian population. *Journal of Inherited Metabolic Diseases*, **12**, 293–301.

Fischer, R.R., Lucas, E.M., Pereira, A.M.B. and Roisenberg, I. (1996). Preparation of heterologous antiserum for the determination of von Willebrand factor in human plasma. *Brazilian Journal of Medical and Biological Research*, **29**, 1641–4.

Fischer, R.R., Pereira, W.V., Pereira, D.V. and Roisenberg, I. (1984). Inherited Factor V deficiency. *Human Heredity*, **34**, 226–30.

Fischmann, A. and Guimarães, J.J.L. (1986). Risco de morrer no primeiro ano de vida entre favelados e não favelados no município de Porto Alegre, RS (Brasil), em 1980. *Revista de Saúde Pública*, **20**, 219–26.

Fisher, R.A. (1930). *The Genetical Theory of Natural Selection.* Oxford: Clarendon Press.

Fisher, R.A. (1935). The sheltering of lethals. *American Naturalist*, **69**, 446–55.

Fisher, R.A. (1943). The relation between the number of species and the number of individuals in a random sample of an animal population. Part 3. *Journal of Animal Ecology*, **12**, 42–58.

Fix, A.G. (1996). Gene frequency clines in Europe: demic diffusion or natural selection? *Journal of the Royal Anthropological Institute*, **2**, 625–43.

Fleischhacker, H. (1959). Estadistica de los grupos de sangre y mezcla de razas en El Salvador (Centroamerica). *Comunicaciones* (San Salvador), **80**, 21–9.

Fleischhauer, K., Agostino, A., Zino, E., *et al.* (1999). Molecular characterization of HLA class I in Colombians carrying HLA-A2: High allelic diversity and frequency of heterozygotes at the HLA-B locus. *Tissue Antigens*, **53**, 519–26.

Fleming, L.E., Bean, J.A., Baden, D. and Brams, E. (1997). Environmental health in the Caribbean. In Halberstein, R.A., editor, *Health and Disease in the Caribbean.* Lexington, KY: Association of Caribbean Studies, pp. 6–22.

Flinn, M.V. and England, B.G. (1995). Childhood stress and family environment. *Current Anthropology*, **36**, 854–66.

Floch, H. and Lajudie, P. (1948). Répartition des groupes sanguins en Guyane Française. *Bulletin de la Société de Pathologie Exotique*, **41**, 241–3.

Flores, M. (1971). Nutritional studies in Central America and Panama. In

Salzano, F.M., editor, *The Ongoing Evolution of Latin American Populations.* Springfield, IL: C.C. Thomas, pp. 311–31.

Flores, R.Z., Mattos, L.F.C. and Salzano, F.M. (1998). Incest: frequency, predisposing factors, and effects in a Brazilian population. *Current Anthropology*, **39**, 554–8.

Florey, C.V., Cuadrado, R.R., Henderson, J.R. and Goes, P. (1967). A nationwide survey of Brazilian military recruits, 1964. I. Method and sampling results. *American Journal of Epidemiology*, **86**, 314–18.

Folberg, A., Barth, M.L., Dutra, J.C., Munarski, R. and Giugliani, R. (1990). Dosagem de ácido siálico urinário em crianças normais e sua aplicação para a determinação de sialidoses e sialúrias. *Revista Brasileira de Análises Clínicas*, **22**, 2–5.

Folberg, A., Giugliani, R., Andrade, H.H.R. and Pinto-Junior, W. (1992). Study of fibroblast and leucocyte beta-galactosidase activity in a family with multiple cases of GM1 ganghosidosis. *Revista Brasileira de Genética*, **15**, 675–86.

Fonseca, L.G. and Freire-Maia, N. (1970). Further data on inbreeding levels in Brazilian populations. *Social Biology*, **17**, 324–8.

Fonseca, M.J.M., Chor, D. and Valente, J.G. (1999). Hábitos alimentares entre funcionários de um banco estatal: Padrão de consumo alimentar. *Cadernos de Saúde Pública*, **15**, 29–39.

Fontoura-da-Silva, S.E. and Chautard-Freire-Maia, E.A. (1996). Butyrylcholinesterase variants (*BCHE* and *CHE2* loci) associated with erythrocyte acetylcholinesterase inhibition in farmers exposed to pesticides. *Human Heredity*, **46**, 142–7.

Forster, P., Harding, R., Torroni, A. and Bandelt, H-J. (1996). Origin and evolution of native American mtDNA variation: a reappraisal. *American Journal of Human Genetics*, **59**, 935–45.

Fox, D.J. (1985). México. In Blakemore, H. and Smith, C.T., editors, *Latin America. Geographical Perspectives.* London: Methuen, pp. 25–75.

Fragoso, C., Lima, P.R.M., Nogueira, L.M., *et al.* (1974). Homosexuality and inbreeding. *Revista Brasileira de Genética*, **17**, 443–5.

Fragoso, S.C. (1970). Haptoglobinas: determinação de seus tipos. Thesis, Escola de Medicina e Cirurgia, Rio de Janeiro.

França, E., Araújo, A.S., Nunesmaia, H.G. and Hutz, M.H. (1998). Haplótipos do gene *HBB*S* em pacientes com anemia falciforme na população do Recife/PE. *Genetics and Molecular Biology*, **21** (Suppl.), 358.

Franchi-Pinto, C., Colletto, G.M.D., Krieger, H. and Beiguelman, B. (1999). Genetic effects on Apgar score. *Genetics and Molecular Biology*, **22**, 13–16.

Franchi-Pinto, C., Pinto, W. Jr. and Beiguelman, B. (1994). Meconium-like substance in midtrimester amniotic fluid: significance for the neuropsychomotor evolution of the infant. *Revista Brasileira de Genética*, **17**, 105–8.

Franco, G.R., Adams, M.D., Soares, M.B., Simpson, A.J.G., Venter, J.C. and Pena, S.D.J. (1995*a*). Identification of new *Schistosoma mansoni* genes by the EST strategy using a directional cDNA library. *Gene*, **152**, 141–7.

Franco, G.R., Simpson, A.J.G. and Pena, S.D.J. (1995*b*). Sequencing and identification of expressed *Schistosoma mansoni* genes by random selection of cDNA clones from a directional library. *Memórias do Instituto Oswaldo Cruz*, **90**, 215–16.

Franco, M.H.L.P., Brennan, S.O., Chua, E.K.M., *et al.* (1999). Albumin genetic variability in South America: Population distribution and molecular studies. *American Journal of Human Biology*, **11**, 359–66.

Franco, M.H.L.P., Salzano, F.M. and Maia de Lima, F.A. (1981). Blood groups and serum protein types in two Brazilian populations. *Revista Brasileira de Genética*, **4**, 689–704.

Franco, M.H.L.P., Weimer, T.A. and Salzano, F.M. (1982). Blood polymorphisms and racial admixture in two Brazilian populations. *American Journal of Physical Anthropology*, **58**, 127–32.

Franco, R.F., Araújo, A.G., Guerreiro, J.F., Elion, J. and Zago, M.A. (1998*b*). Analysis of the 677 C→T mutation of the methylenetetrahydrofolate reductase gene in different ethnic groups. *Thrombosis and Haemostasis*, **79**, 119–21.

Franco, R.F., Elion, J., Lavinha, J., Krishnamoorthy, R., Tavella, M.H. and Zago, M.A. (1998*c*). Heterogeneous ethnic distribution of the 844 ins68 in the cystathionine β-synthase gene. *Human Heredity*, **48**, 338–42.

Franco, R.F., Elion, J., Santos, S.E.B., Araújo, A.G., Tavella, M.H. and Zago, M.A. (1999*b*). Heterogeneous ethnic distribution of the Factor V Leiden mutation. *Genetics and Molecular Biology*, **22**, 143–5.

Franco, R.F., Santos, S.E.B., Elion, J., Tavella, M.H. and Zago, M.A. (1998*a*). Prevalence of the G20210A polymorphism in the 3'-untranslated region of the prothrombin gene in different human populations. *Acta Haematologica*, **100**, 9–12.

Franco, R.F., Simões, B.P. and Zago, M.A. (1995). Relative frequencies of the two 0 alleles of the histo-blood ABH system in different racial groups. *Vox Sanguinis*, **69**, 50–2.

Franco-Gamboa, E., Ibarra, B., Zuñiga, P., Figuero-Peña, C.L. and Cantú, J.M. (1981). Abnormal hemoglobins in northwestern Mexico. *Abstracts, 6th International Congress of Human Genetics*, 35.

Freimer, N.B., Reus, V.I., Escamilla, M., *et al.* (1996). An approach to investigating linkage for bipolar disorder using large Costa Rican pedigrees. *American Journal of Medical Genetics*, **67**, 254–63.

Freire-Maia, A. (1969). Human genetics studies in areas of high natural radiation. I. Methodology. *Acta Geneticae Medicae et Gemellologiae*, **18**, 175–212.

Freire-Maia, A. (1970). The handless and footless families of Brazil. *Lancet*, **I**, 519–20.

Freire-Maia, A. (1971*a*). A recessive form of ectrodactyly and its implications in genetic counseling. *Journal of Heredity*, **62**, 53.

Freire-Maia, A. (1971*b*). Human genetics studies in areas of high natural radiation. II. First results of an investigation in Brazil. *Anais da Academia Brasileira de Ciências*, **43**, 457–9.

Freire-Maia, A. (1974*a*). Genética da aquiropodia ('the handless and footless families of Brazil'). I. Mortalidade, seleção e carga genética. *Ciência e Cultura*, **26**, 376–82.

Freire-Maia, A. (1974*b*). Genética da aquiropodia ('the handless and footless families of Brazil'). II. Aquiria e casos 'similares' à aquiropodia. *Revista Paulista de Medicina*, **84**, 107–10.

Freire-Maia, A. (1974*c*). Genética da aquiropodia. III. Averiguação. *Revista Brasileira de Medicina*, **31**, 600–6.

Freire-Maia, A. (1974*d*). Estudos de genética humana em áreas de alta radiação natural. IV. Pesquisas em áreas radioativas. *Anais da Academia Brasileira de Ciências*, **46**, 333–47.

Freire-Maia, A. (1974*e*). Estudos de genética humana em áreas de alta radiação natural. V. Características regionais e populacionais. *Ciência e Cultura*, **26**, 1138–48.

Freire-Maia, A. (1974*f*). Estudos de genética humana em áreas de alta radiação natural. VI. Carga genética e grupo étnico. *Revista Brasileira de Biologia*, **34**, 523–30.

Freire-Maia, A. (1975*a*). Genética da aquiropodia ('the handless and footless families of Brazil'). V. Teste da hipótese genética. *Ciência e Cultura*, **27**, 130–42.

Freire-Maia, A. (1975*b*). Genetics of acheiropodia ('the handless and footless families of Brazil'). VIII. Penetrance and expressivity. *Clinical Genetics*, **7**, 98–102.

Freire-Maia, A. (1975*c*). Estudos de genética humana em áreas de alta radiação natural. VII. Carga genética. *Revista Brasileira de Pesquisas Médicas e Biológicas*, **8**, 287–99.

Freire-Maia, A. (1981). Historical note: the extraordinary handless and footless families of Brazil: 50 years of acheiropodia. *American Journal of Medical Genetics*, **9**, 31–41.

Freire-Maia, A. (1984). *Guerra e Paz com Energia Nuclear*. São Paulo: Ática.

Freire-Maia, A. and Chakraborty, R. (1975). Genetics of acheiropodia ('the handless and footless families of Brazil'). IV. Sex ratio, consanguinity and birth order. *Annals of Human Genetics*, **39**, 151–61.

Freire-Maia, A. and Freire-Maia, D.V. (1967). Mortality rates in a Brazilian area of high background radiation (preliminary analysis based on official records). *Anais da Academia Brasileira de Ciências*, **39**, 467–9.

Freire-Maia, A. and Freire-Maia, N. (1982). *Efeitos Genéticos das Radiações no Homem*. São Paulo: Hucitec and Universidade Estadual Paulista.

Freire-Maia, A. and Krieger, H. (1975). Human genetic studies in areas of high natural radiation. VIII. Genetic load not related to radiation. *American Journal of Human Genetics*, **27**, 385–93.

Freire-Maia, A. and Krieger, H. (1978). Human genetic studies in areas of high natural radiation. IX. Effects on mortality, morbidity and sex ratio. *Health Physics*, **34**, 61–5.

Freire-Maia, A., Freire-Maia, D.V. and Morton, N.E. (1982). Epidemiology and genetics of endemic goiter. II. Genetic aspects. *Human Heredity*, **32**, 176–80.

Freire-Maia, A., Freire-Maia, N. and Schull, W.J. (1975*c*). Genetics of acheiropodia ('the handless and footless families of Brazil'). IX. Genetic counseling. *Human Heredity*, **25**, 329–36.

Freire-Maia, A., Freire-Maia, N., Morton, N.E., Azevêdo, E.S. and Quelce-Salgado, A. (1975*a*). Genetics of acheiropodia ('the handless and footless families of Brazil'). VI. Formal genetic analysis. *American Journal of Human Genetics*, **27**, 521–7.

Freire-Maia, A., Laredo-Filho, J. and Freire-Maia, N. (1978). Genetics of acheiropodia ('the handless and footless families of Brazil'). X. Roentgenologic study. *American Journal of Medical Genetics*, **2**, 321–30.

Freire-Maia, A., Li, W-H. and Maruyama, T. (1975*b*). Genetics of acheiropodia ('the handless and footless families of Brazil'). VII. Population dynamics. *American Journal of Human Genetics*, **27**, 665–75.

Freire-Maia, D.V. (1981). Bócio endêmico: Classificação e histórico. *Revista da Associação Médica Brasileira*, **27**, 111–12.

Freire-Maia, D.V. and Freire-Maia, A. (1981*a*). Age effect on goiter in a hyperendemic area of the Mato Grosso Plateau, Brazil. *Nagoya Medical Journal*, **26**, 5–11.

Freire-Maia, D.V. and Freire-Maia, A. (1981*b*). Sex effect on the familial aggregation of endemic goiter. *Revista Brasileira de Genética*, **4**, 449–57.

Freire-Maia, D.V. and Freire-Maia, A. (1982). Epidemiologia do bócio endêmico. *Revista da Associação Médica Brasileira*, **28**, 59–62.

Freire-Maia, L. (1961). Aspectos genéticos, clínicos e radiológicos da osteocondromatose, com estudo de duas famílias. *Revista Brasileira de Medicina*, **18**, 1–8.

Freire-Maia, L. and Freire-Maia, A. (1981). Exostoses cartiloginosas múltiplas. Alguns aspectos clínico-genéticos, com revisão da literatura. *Revista da Associação Médica de Minas Gerais*, **32**, 6–9.

Freire-Maia, N. (1957). Inbreeding in Brazil. *American Journal of Human Genetics*, **9**, 284–98.

Freire-Maia, N. (1968). Inbreeding levels in American and Canadian populations, a comparison with Latin America. *Eugenics Quarterly*, **15**, 22–33.

Freire-Maia, N. (1971). Consanguineous marriages and inbreeding load. In Salzano, F.M., editor, *The Ongoing Evolution of Latin American Populations*. Springfield, IL: C.C. Thomas, pp. 189–220.

Freire-Maia, N. (1972). *Radiogenética Humana.* São Paulo: Edgard Blücher and Editora da Universidade de São Paulo.

Freire-Maia, N. (1975). Adaptation and genetic load. In Salzano, F.M., editor, *The Role of Natural Selection in Human Evolution.* Amsterdam: North-Holland, pp. 295–321.

Freire-Maia, N. (1984). Effects of consanguineous marriages on morbidity and precocious mortality, genetic counseling. *American Journal of Medical Genetics*, **18**, 401–6.

Freire-Maia, N. (1990). Genetic effects in Brazilian populations due to consanguineous marriages. *American Journal of Medical Genetics*, **35**,

115–17.

Freire-Maia, N. and Cavalli, I.J. (1978). Genetic investigations in a northern Brazilian island. I. Population structure. *Human Heredity*, **28**, 386–96.

Freire-Maia, N. and Freire-Maia, A. (1967). Recurrence risks of bone aplasias and hypoplasias of the extremities. *Acta Genetica et Statistica Medica*, **17**, 418–21.

Freire-Maia, N. and Pinheiro, M. (1984*a*). *Ectodermal Dysplasias: a Clinical and Genetic Study*. New York: Alan R. Liss.

Freire-Maia, N. and Pinheiro, M. (1984*b*). *Displasias Ectodérmicas. Manual para Profissionais da Área de Saúde*. Curitiba: Universidade Federal do Paraná.

Freire-Maia, N., Andrade, F.L., Athayde-Neto, A., Cavalli, I.J., Oliveira, J.C., Marçallo, F.A. and Coelho, A. (1978). Genetic investigations in a northern Brazilian island. II. Random drift. *Human Heredity*, **28**, 401–10.

Freire-Maia, N., Braga, J.C., Abdala, H. and Tartuce, N. (1977). Estudo clinicogenético de uma amostra de retardados mentais. II. Aspectos genéticos. *Ciência e Cultura*, **29**, 992–9.

Freire-Maia, N., Freire-Maia, A. and Quelce-Salgado, A. (1965). Studies on the offspring of Brazilian physicians working with ionizing radiations. *Proceedings of the Fifth Inter-American Symposium on the Peaceful Application of Nuclear Energy*, pp. 133–6.

Freire-Maia, N., Maia, N.A. and Pacheco, C.N.A. (1980). Mohr-Wriedt (A2) brachydactyly. Analysis of a large Brazilian kindred. *Human Heredity*, **30**, 225–31.

Freire-Maia, N., Quelce-Salgado, A. and Koehler, R.A. (1959). Hereditary bone aplasias and hypoplasias of the upper extremities. *Acta Genetica et Statistica Medica*, **9**, 33–40.

Freitas-Filho, L. (1956). *Vida e Morte nas Capitais Brasileiras*. Rio de Janeiro: Instituto Brasileiro de Geografia e Estatística.

Freyre, G. (1930). *Casa Grande & Senzala*. Rio de Janeiro: Maia and Schmidt.

Fridman, C., Varela, M.C., Nicholls, R.D. and Koiffmann, C.P. (1998). Unusual clinical features in an Angelman syndrome patient with uniparental disomy due to a translocation 15q15q. *Clinical Genetics*, **54**, 303–8.

Frimm, C.E. 1947. A drepanocitose. Doctoral thesis, Faculdade de Medicina, Universidade de Porto Alegre.

Frisancho, A.R. (1978). Nutritional influences on human growth and maturation. *Yearbook of Physical Anthropology*, **21**, 174–91.

Frisancho, A.R., Farrow, S., Friedenzohn, I., *et al.* (1999*a*). Role of genetic and environmental factors in the increased blood pressures of Bolivian blacks. *American Journal of Human Biology*, **11**, 489–98.

Frisancho, A.R., Frisancho, H.G., Albalak, R., Villain, M., Vargas, E. and Soria, R. (1997). Developmental, genetic, and environmental components of lung volumes at high altitude. *American Journal of Human Biology*, **9**, 191–203.

Frisancho, A.R., Frisancho, H.G., Milotich, M., *et al.* (1995). Developmental, genetic, and environmental components of aerobic capacity at high altitude. *American Journal of Physical Anthropology*, **96**, 431–42.

Frisancho, A.R., Juliao, P.C., Barcelona, V., *et al.* (1999*b*). Developmental components of resting ventilation among high- and low-altitude Andean children and adults. *American Journal of Physical Anthropology*, **109**, 295–301.

Frisancho, A.R., Wainwright, R. and Way, A. (1981). Heritability of components of phenotypic expression in skin reflectance of Mestizos from the Peruvian lowlands. *American Journal of Physical Anthropology*, **55**, 203–8.

Frota, M. and de Paula, A.V. (1960). Freqüência do sistema ABO e do fator Rh entre operários de Varginha, Sul de Minas Gerais. *Revista de Medicina do Sul de Minas*, **6**, 83–93.

Frota-Pessoa, O. (1971). Models of genetic structure and their applications in Brazilian populations. In Salzano, F.M., editor, *The Ongoing Evolution of Latin American Populations.* Springfield, IL: C.C. Thomas, pp. 165–87.

Frota-Pessoa, O. and Aratangy, L.R. (1968). The degeneration of the Y chromosome. *Revista Brasileira de Pesquisas Médicas e Biológicas*, **1**, 241–4.

Frozi, V.M. and Mioranza C. (1975). *Imigração Italiana no Nordeste do Rio Grande do Sul.* Caxias do Sul: Editora Movimento and Universidade de Caxias do Sul.

Fumeron, F., Feingold, N., Bois, E., Mayer, F. and Hors, J. (1981). Studies on an isolated West Indies population. I. Analysis of HLA genotypes. *Tissue Antigens*, **17**, 338–42.

Gadow, E.C., Paz, J.E., Lopez-Camelo, J.S., Dutra, M.G., Queenan, J.T., Simpson, J.L., Jennings, V.H., Castilla, E.E. and the NFP-ECLAMC Group (1998). Unintended pregnancies in women delivering at 18 South American hospitals. *Human Reproduction*, **13**, 1991–5.

Gage, T.B. (1989). Bio-mathematical approaches to the study of human variation in mortality. *Yearbook of Physical Anthropology*, **32**, 185–214.

Gagliardi, A.R.T., Gonzalez, C.H. and Pratesi, R. (1984). GAPO syndrome: report of three affected brothers. *American Journal of Medical Genetics*, **19**, 217–23.

Gaitan-Yanguas, M. (1978). Retinoblastoma: analysis of 235 cases. *Radiation Oncology, Biology, Physics*, **4**, 359–65.

Galacteros, F., Garin, J.D., Monplaisir, N., *et al.* (1984). Two new cases of heterozygosity of Hemoglobin Knossos $\alpha_2\beta_2^{27\ Ala \rightarrow Ser}$ detected in the French West Indies and Algeria. *Hemoglobin*, **8**, 215–28.

Galera, M.F., Patrício, F.R.S., Cernach, M.C.S.P., Lederman, H.M. and Brunoni, D. (1998). Clinical, genetical, radiological, and anatomopathological survey of 17 patients with lethal osteochondrodysplasias. *Genetics and Molecular Biology*, **21**, 267–72.

Gallango, M.L. (1967). Los grupos séricos determinables por métodos inmunológicos. *Acta Científica Venezolana*, (Suppl. 3), 179–89.

Gallardo Velázquez, A. and Pimienta Merlín, M. (1997). Nuevas fórmulas para la reconstrucción de la estatura a partir de los huesos de la mano. *Antropología Física Latinoamericana*, **1**, 33–45.

García, M., Estrada, M., Gutiérrez, A., *et al.* (1982). Glyoxalase I polymorphism and racial admixture in the Cuban population. *Human Genetics*, **61**, 50–1.

García, G., González, B., Kameyama, R., *et al.* (1984). Estudio cefalométrico de los ángulos SNA, SNB y ANB en población mexicana masculina y femenina de 12 años de edad. In Ramos Galván, R. and Ramos Rodríguez, R.M., editors, *Estudios de Antropología Biológica (II Coloquio de Antropologia Física Juan Comas)*. México: Universidad Nacional Autónoma de México, pp. 343–53.

Garcia-Godoy, F., Michelen, A. and Townsend, G. (1985). Crown diameter of the deciduous teeth in Dominican Mulatto children. *Human Biology*, **57**, 27–31.

García-Moro, C. and Hernández, M. (1997). Patrones de mortalidad en la población chilena de Tierra del Fuego. *Revista Española de Antropología Biológica*, **18**, 231–45.

Gargiulo, A., Moisset de Españes, E. and Pozzi, A.R. (1939). Grupos sanguíneos en estudiantes de la Facultad de Medicina. *La Semana Médica*, **46**, 112.

Garsaud, P., Boisseau-Garsaud, A-M., Ossondo, M., *et al.* (1998). Epidemiology of cutaneous melanoma in the French West Indies (Martinique). *American Journal of Epidemiology*, **147**, 66–8.

Garza-Chapa, R., Escobar, M.S., Cerda, R. and Leal-Garza, C.H. (1984). Factors related to the frequency of twinning in the state of Nuevo León, México during 1977 and 1978. *Human Biology*, **56**, 277–90.

Garza-Chapa, R., Gonzalez-Rendon, M.R. and Joffre, G. (1978). Grupos sanguíneos ABO y Rh°(D) en poblaciones del IMSS en el Estado de Nuevo León (cálculo de la frecuencia de matrimonios e hijos com incompatibilidad simple y doble). *Archivos de Investigación Médica* (México), **9**, 541–58.

Geiger, C.J., Salzano, F.M. and da Rocha, F.J. (1985). Who seeks genetic counseling and why: a Brazilian evaluation. *Revista Brasileira de Genética*, **8**, 395–403.

Geiger, C.J., Salzano, F.M., Mattevi, M.S., Erdtmann, B. and da Rocha, F.J. (1987). Chromosome variation and genetic counseling: 20 years of experience in Brazil. *Revista Brasileira de Genética*, **10**, 581–91.

Gené, M., Huguet, E., Sánchez-Garcia, C., Moreno, P., Corbella, J. and Mezquita, J. (1995). Study of the 3' Apo B minisatellite performed by PCR in the population of Catalonia (Northeast Spain). *Human Heredity*, **45**, 70–4.

Genovés, S. and Comas, J. (1964). La antropologia física en México, 1943–64. Inventario bibliográfico. *Cuadernos del Instituto de Investigaciones Históricas, Serie Antropológica*, **17**, 1–55.

Genóves, S. and Messmacher, M. (1959). Valor de los patrones tradicionales para la determinación de la edad por medio de las suturas en craneos mexicanos (indígenas y mestizos). *Cuadernos del Instituto de Historia, Serie Antropológica*, **7**, 1–55.

Gerbase-de-Lima, M. (1997). HLA e Doenças: Estudos na população Brasileira. Privat-Docent thesis, University of São Paulo.

Gerbase-de-Lima, M., de Lima, J.J.G., Persoli, L.B., Silva, H.B., Marcondes, M. and Bellotti, G. (1989). Essential hypertension and histocompatibility antigens. A linkage study. *Hypertension*, **14**, 604–9.

Gerbase-de-Lima, M., Ladalardo, M.A., de Lima, J.J.G., Silva, H.B., Bellotti, G. and Pileggi, F. (1992c). Essential hypertension and histocompatibility antigens. An association study. *Hypertension*, **19**, 400–2.

Gerbase-de-Lima, M., Pereira-Santos, A., Sesso, R., Temin, J., Argão, E.S. and Ajzen, H. (1998a). Idiopathic focal segmental glomerulosclerosis and HLA antigens. *Brazilian Journal of Medical and Biological Research*, **31**, 387–9.

Gerbase-de-Lima, M., Pinto, L.C., Grumach, A. and Carneiro-Sampaio, M.M.S. (1998b). HLA antigens and haplotypes in IgA-deficient Brazilian paediatric patients. *European Journal of Immunogenetics*, **25**, 281–5.

Gerbase-de-Lima, M., Scala, L.C.N., Temin, J., Santos, D.V. and Otto, P.A. (1994). Rheumatic fever and the HLA complex. A cosegregation study. *Circulation*, **89**, 138–41.

Gerbase-de-Lima, M., Stilman, R., Porta, G., Lieber, S.R. and Persoli, L.L. (1992b). HLA antigens in children with autoimmune and cryptogenic chronic active hepatitis. In Gorodezky, C., Sierp, G. and Albert, E., editors, *Proceedings of the Fifth Latin American Histocompatibility Workshop.* München: Immunogenetics Laboratory, pp. 129–32.

Gerbase-de-Lima, M., Torres, E.A., Moraes-Silva, M.R.B. and Pellegrino, J. Jr. (1992a). Possible linkage between HLA and chronic open-angle glaucoma. In Gorodezky, C., Sierp, G. and Albert, E., editors, *Proceedings of the Fifth Latin American Histocompatibility Workshop.* München: Immunogenetics Laboratory, pp. 133–6.

Gerber, L.M. and Halberstein, R.A. (1999). Blood pressure: genetic and environmental influences. *Human Biology*, **71**, 467–708.

Gewehr, W.V., Gonçalves, L.M.K., Vaz, M. and Schwartsmann, L.B. (1984). Estudo antropométrico comparativo na determinação do estado nutricional numa vila periférica de Porto Alegre. *Arquivos de Medicina Preventiva* (Porto Alegre), **6**, 100–5.

Giglioli, G.G. (1968). Malaria in the American Indian. In *Biomedical Challenges Presented by the American Indian.* Washington, D.C.: Pan American Health Organization (Scientific publication No. 165), pp. 104–13.

Giles, E., Hansen, A.T., McCullough, J.M., Metzger, D.G. and Wolpoff, M.H. (1968). Hydrogen cyanide and phenylthiocarbamide sensivity, mid-phalangeal hair and color blindness in Yucatán, México. *American Journal of Physical Anthropology*, **28**, 203–12.

Giorgiutti, E.M.P. (1989). Diagnóstico prenatal. *Actas, XIX Congreso Argentino de Genética*, pp. 53–62.

Giraldo, A., Pino, W., García-Ramírez, L.F., Pineda, M. and Iglesias, A. (1995). Vitamin D-dependent rickets type II and normal vitamin D receptor cDNA sequence. A cluster in a rural area of Cauca, Colombia, with more than 200 affected children. *Clinical Genetics*, **48**, 57–65.

Gispert, H.C., Brisco, P.L. and García, P.E. (1950). Distribución de los grupos sanguineos en 11.704 dadores voluntarios de sangre. *Archivos Uruguaios de Medicina*, **37**, 41–3.

Giugliani, R. and Coelho, J.C. (1997). Diagnóstico de erros inatos do metabolismo na América Latina. *Brazilian Journal of Genetics*, **20** (Suppl.),

147–54.
Giugliani, R. and Ferrari, I. (1980a). Metabolic factors in urolithiasis: a study in Brazil. *Journal of Urology*, **124**, 503–7.
Giugliani, R. and Ferrari, I. (1980b). Ácido oxálico e litíase urinária: uma investigação. *Revista Paulista de Medicina*, **96**, 15–17.
Giugliani, R. and Ferrari, I. (1981). Some observations on genetic factors in urolithiasis. *Urology*, **17**, 33–8.
Giugliani, R. and Ferrari, I. (1987). Metabolic factors in urolithiasis: a study in Brazil. *Journal of Urology*, **137**, 320.
Giugliani, R., Dutra, J.C., Barth, M.L., Dutra-Filho, C.S., Goldenfum, S.L. and Wajner, M. (1991). Seven-year experience of a reference laboratory for detection of inborn errors of metabolism in Brazil. *Journal of Inherited Metabolic Diseases*, **14**, 400–2.
Giugliani, R., Dutra, J.C., Pereira, M.L.S., *et al.* (1985b). GM1 gangliosidosis: Clinical and laboratory findings in eight families. *Human Genetics*, **70**, 347–54.
Giugliani, R., Dutra-Filho, C.S., Barth, M.L., Enk, V. and Alexandre-Netto, C. (1990b). Age-related concentrations of glycosaminoglycans in random urine: a contribution to the laboratorial detection of mucopolysaccharidoses. *Revista Brasileira de Genética*, **13**, 599–605.
Giugliani, R., Dutra-Filho, C.S., Barth, M.L., *et al.* (1989b). Inborn errors of metabolism. Sensitivity of screening tests in high risk patients. *Clinical Pediatrics*, **28**, 494–9.
Giugliani, R., Ferrari, I. and Greene, L.J. (1985a). Heterozygous cystinuria and urinary lithiasis. *American Journal of Medical Genetics*, **22**, 703–15.
Giugliani, R., Ferrari, I. and Greene, L.J. (1986). Frequency of cystinuria among stone-forming patients in region of Brazil. *Urology*, **27**, 38–40.
Giugliani, R., Ferrari, I. and Greene, L.J. (1987). An evaluation of four methods for the detection of heterozygous cystinuria. *Clinica Chimica Acta*, **164**, 227–33.
Giugliani, R., Figueiredo, M.S.R.B., Folberg, A., Giugliani, M.C.K., Padovan, G. and Greene, L.J. (1989a). Cystinuria in the south of Brazil. *Revista Brasileira de Genética*, **12**, 865–70.
Giugliani, R., Jardim, L., Edelweiss, M.I. and Rosa, A. (1990a). Early and unusual presentation of Type I primary hyperoxaluria. *Child Nephrology and Urology*, **10**, 107–8.
Go, R.C.P., Elston, R.C. and Salzano, F.M. (1977). Association and linkage between genetic markers and morphological and behavioral attributes in dizygotic twins. *Social Biology*, **24**, 62–8.
Godoy, O. (1937). Grupos sangüíneos nos criminosos de São Paulo. *Arquivos da Sociedade de Medicina Legal e Criminalística*, **8**, 185–91.
Goldani, L.Z., Monteiro, C.M., Donadi, E.A., Martinez, R. and Voltarelli, J.C. (1991). HLA antigens in Brazilian patients with paracoccidioidomycosis. *Mycopathologia*, **114**, 89–91.
Goldberg, A.C. and Kalil, J. (1989a). DNA analysis of histocompatibility antigens: Fragments from the recently identified DR loci. *Brazilian Journal*

of Medical and Biological Research, **22**, 849–57.
Goldberg, A.C. and Kalil, J. (1989*b*). DNA analysis of histocompatibility antigens: Identification of new DQw specificities and of DPw patterns. *Brazilian Journal of Medical and Biological Research*, **22**, 859–67.
Goldberg, A.C. and Kalil, J. (1989*c*). DNA analysis of histocompatibility antigens of a sample of the São Paulo population. *Brazilian Journal of Medical and Biological Research*, **22**, 869–75.
Goldberg, A.C., Yamamoto, J.H., Chiarella, J.M., *et al.* (1998*a*). HLA-DRB1*0405 is the predominant allele in Brazilian patients with Vogt–Koyanagi–Harada disease. *Human Immunology*, **59**, 183–8.
Goldberg, A.C., Chiarella, J.M., Marin, M.L.C., *et al.* (1998*b*). Molecular typing of HLA Class II antigens in a São Paulo population. *Genetics and Molecular Biology*, **21**, 301–5.
Goldstein, J.R. and Morning, A.J. (2000). The multiple-race population of the United States: issues and estimates. *Proceedings of the National Academy of Sciences, USA*, **97**, 6230–5.
Goldstein, M.S. and Kobyliansky, E. (1984). Anthropometric traits, balanced selection and fertility. *Human Biology*, **56**, 35–46.
Gollop, T.R. and Eigier, A. (1987). Early prenatal ultrasound diagnosis of fetal hydrocephalus. *Revista Brasileira de Genética*, **10**, 575–80.
Gollop, T.R., Eigier, A., Vianna-Morgante, A.M. and Naccache, N. (1986). Chorionic villi sampling for early prenatal genetic diagnosis. *Revista Brasileira de Genética*, **9**, 381–5.
Gollop, T.R., Eigier, A. and Naccache, N. (1987). Amostra de vilo corial (seguimento de 56 casos). *Femina*, **15**, 112–14.
Gollop, T.R., Pieri, P.C., Naccache, N.F., Auler-Bittencourt, E., Eigier, A. and Hauschild, D. (1990). Transabdominal chorionic villus sampling: experience with 70 cases. *Revista Brasileira de Genética*, **13**, 591–7.
Gomes, M.L., Macedo, A.M., Pena, S.D.J. and Chiari, E. (1998). Genetic relationship between *Trypanosoma cruzi* strains isolated from chagasic patients on southern Brazil as revealed by RAPD and SSR-PCR analysis. *Acta Tropica*, **69**, 99–109.
Gomes, M.L., Galvão, L.M.C., Macedo, A.M., Pena, S.D.J. and Chiari, E. (1999). Chagas' disease diagnosis: comparative analysis of parasitologic, molecular, and serologic methods. *American Journal of Tropical Medicine and Hygiene*, **60**, 205–10.
Gomes, R.F., Macedo, A.M., Melo, N.M. and Pena, S.D.J. (1995). *Leishmania* (*Viannia*) *brasiliensis*, genetic relationships between strains isolated from different areas of Brazil as revealed by DNA fingerprinting and RAPD. *Experimental Parasitology*, **80**, 681–7.
Gómez Arbesú, J., Paradoa Pérez, M.L. and Basanta Otero, P. (1984). Inmunización fetomaterna frente a antígenos HLA en mujeres cubanas. *Inmunologia*, **3**, 78–82.
Gonçalves, A. and Gonçalves, N.N.S. (1985). Contribuição dos dermatóglifos para o estudo da infertilidade, a partir de casuística de nosso meio. *Revista da Fundação Serviço Especial de Saúde Pública*, **30**, 63–72.

Gonçalves, M.S., Fahel, S., Figueiredo, M.S., *et al.* (1995). Molecular identification of hereditary persistence of fetal hemoglobin type 2 (HPFH type 2) in patients from Brazil. *Annals of Hematology*, **70**, 159–61.

Gonçalves, M.S., Nechtman, J.F., Figueiredo, *et al.* (1994*b*). Sickle cell disease in a Brazilian population from São Paulo: A study of the β^S haplotypes. *Human Heredity*, **44**, 322–7.

Gonçalves, M.S., Sonati, M.F., Kimura, E.M., *et al.* (1994*a*). Association of Hb Santa Ana [α 2β 2 88(F4)Leu→Pro] and Hb Porto Alegre [$\alpha(2)\beta(2)$ 9(A6)Ser→Cys] in a Brazilian female. *Hemoglobin*, **18**, 235–9.

Gonçalves, P.L. and Beiguelman, B. (1974). Dermatoglifos e esquizofrenia: um estudo de pares de irmãos. *Revista Brasileira de Pesquisas Médicas e Biológicas*, **7**, 293–300.

Gonçalves-Pimentel, M.M., Falcão-Conceição, D.N. and Santos, D.M. (1988). Detection of carriers of Duchenne muscular dystrophy gene using logistic discriminating analysis: creatine-kinase and pyruvate-kinase. *Revista Brasileira de Genética*, **11**, 981–93.

Gonzaga, H.F.S., Torres, E.A., Alchorne, M.M.A. and Gerbase-de-Lima, M. (1996). Both psoriasis and benign migratory glossitis are associated with HLA-Cw6. *British Journal of Dermatology*, **135**, 368–70.

Gonzales, G., Crespo-Retes, I. and Guerra-García, R. (1982). Secular change in growth of native children and adolescents at high altitude. I. Puno, Peru (3800 meters). *American Journal of Physical Anthropology*, **58**, 191–5.

Gonzales, G.F. and Villena, A. (1996). Body mass index and age at menarche in Peruvian children living at high altitude and at sea level. *Human Biology*, **68**, 265–75.

Gonzales, G.F., Villena, A., and Aparicio, R. (1998). Acute mountain sickness: is there a lag period before symptoms? *American Journal of Human Biology*, **10**, 669–77.

Gonzales, G.F., Villena, A., Góñez, C. and Zevallos, M. (1994). Relationship between body mass index, age, and serum adrenal androgen levels in Peruvian children living at high altitude and at sea level. *Human Biology*, **66**, 145–53.

González Martinez, E.E. (1990). Españoles en Brasil, Características generales de un fenómeno emigratorio. *Ciência e Cultura*, **42**, 341–6.

Gonzalez, C.H., Vargas, F.R., Perez, A.B.A., *et al.* (1993). Limb deficiency with or without Möbius sequence in seven Brazilian children associated with misoprostol use in the first trimester of pregnancy. *American Journal of Medical Genetics*, **47**, 59–64.

Gorodezky, C., Escobar-Gutiérrez, A. and Salazer-Mallén, M. (1972). Distribution of some of the HL-A system lymphocyte antigens in Mexicans. *Vox Sanguinis*, **23**, 439–43.

Goulart, M. (1975). *A Escravidão Africana no Brasil. Das Origens à Extinção do Tráfico*. São Paulo: Editora Alfa-Ômega.

Goza, F. (1992). A imigração brasileira na América do Norte. *Revista Brasileira de Estudos de População*, **9**(1), 65–82.

Graham, D.H. and Merrick, T.W. (1976). Dois séculos de crescimento populacional brasileiro: suas tendências e seus componentes demográficos. In IBGE. Rio de Janeiro: Fundação Instituto Brasileiro de Geografia e Estatística, pp. 494–503.

Granda Ibarra, H. and Hernández Fernández, R. (1984). Pesquisaje de hemoglobinas anormales en el recien nacido por la tecnica de enfoque isoelectrico. *Memorias, VI Congreso Latinoamericano de Genética*, 318.

Granda Ibarra, H., Martín Ruíz, M., Dorticós Balea, A., Silva Ballester, H., Ojeda Sosa, S. and Cuadra Brown, Y. (1984). Resultados del programa para la prevención de la anemia por hematies falciformes en Cuba. *Memorias, VI Congreso Latinoamericano de Genética*, 300–1.

Gray, R.H., Ferraz, E.M., Amorim, M.S. and Melo, L.F. (1991). Levels and determinants of early neonatal mortality in Natal, northeastern Brazil: results of a surveillance and case–control study. *International Journal of Epidemiology*, **20**, 467–73.

Green, L.D., Derr, J.N. and Knight, A. (2000). mtDNA affinities of the peoples of north-central Mexico. *American Journal of Human Genetics*, **66**, 989–98.

Greenberg, J.H. (1987). *Language in the Americas*. Stanford, CA: Stanford University Press.

Greenberg, J.H., Turner, C.G.II and Zegura, S.L. (1986). The settlement of the Americas: a comparison of the linguistic, dental, and genetic evidence. *Current Anthropology*, **27**, 477–97.

Greene, L.S. (1973). Physical growth and development, neurological maturation and behavioral functioning in two Ecuadorian Andean communities in which goiter is endemic. I. Outline of the problem of endemic goiter and cretinism. Physical growth and neurological maturation in the adult population of La Esperanza. *American Journal of Physical Anthropology*, **38**, 119–34.

Greene, L.S. (1974). Physical growth and development, neurological maturation, and behavioral functioning in two Ecuadorian Andean communities in which goiter is endemic. II. PTC taste sensitivity and neurological maturation. *American Journal of Physical Anthropology*, **41**, 139–51.

Greenwald, I. (1957). The history of goiter in the Inca Empire: Peru, Chile, and the Argentine Republic. Its significance for the etiology of the disease. *Texas Report on Biology and Medicine*, **15**, 874–89.

Greenwald, I. (1969). The history of goiter in Bolivia, Paraguay and Brazil. *Texas Reports on Biology and Medicine*, **27**, 7–26.

Greksa, L.P. (1998*a*). Skin reflectance of lowland Bolivian youths of European ancestry. *American Journal of Human Biology*, **10**, 559–65.

Greksa, L.P. (1998*b*). Comparison of skin reflectances between Bolivian lowlanders and highlanders of European of ancestry. *Human Biology*, **70**, 889–900.

Greksa, L.P., Spielvogel, H., Paredes-Fernandez, L., Paz-Zamora, M. and Caceres, E. (1984). The physical growth of urban children at high altitude. *American Journal of Physical Anthropology*, **65**, 315–22.

Grignoli, C.R.E., Wenning, M.R.S.C., Sonati, M.F., *et al.* (1999). Hb Rio Claro [*β*34(B16) Val→Met]: a novel electrophoretically silent variant found in association with Hb Hasharon [α47(CE5)Asp→His] and α-thalassemia 2 (-α$^{3.7}$). *Hemoglobin*, **23**, 177–88.

Grumach, A.S., Vilela, M.M.S., Gonzalez, C.H., *et al.* (1988). Inherited C3 deficiency of the complement system. *Brazilian Journal of Medical and Biological Research*, **21**, 247–57.

Grunbaum, B.W., Selvin, S., Myhre, B.A. and Pace, N. (1980). Distribution of gene frequencies and discrimination probabilities for 22 human blood genetic systems in four racial groups. *Journal of Forensic Sciences*, **25**, 428–44.

Guadagna, M., Migliano, M., Herrera, J., *et al.* (1998). Diagnóstico molecular de neoplasia endócrina múltiple tipo 1 (MEN–1) en una familia argentina afectada. *Medicina* (Buenos Aires), **58**, 441–5.

Guaraciaba, M.A. (1966). O elemento nipônico de uma comunidade brasileira. *Arquivos do Instituto de Antropologia, Universidade Federal do Rio Grande do Norte*, **2**, 21–38.

Guerra Cedeño, F. (1996). Perfil histórico-antropológico de la población urbana de Caracas. In Municio, A.M. and Barreno, P.G., editors, *Polimorfismo Génico (HLA) en Poblaciones Hispanoamericanas*. Madrid: Real Academia de Ciencias Exactas, Físicas y Naturales, pp. 129–47.

Guerreiro, J.F. and Chautard-Freire-Maia, E. (1988). ABO and Rh blood groups, migration and estimates of racial admixture for the population of Belém, state of Pará, Brazil. *Revista Brasileira de Genética*, **11**, 171–86.

Guerreiro, J.F. and Schneider, H. (1981). Estudo dermatoglífico de uma comunidade negróide de Curiau, Amapá. *Raízes – Ciências Biológicas e Ciências da Saúde* (Belém), **1**, 41–6.

Guerreiro, J.F., Figueiredo, M.S. and Zago, M.A. (1994). Beta-globin gene cluster haplotypes of Amerindian populations from the Brazilian Amazon Region. *Human Heredity*, **44**, 142–9.

Guerreiro, J.F., Ribeiro-dos-Santos, A.K.C., Santos, E.J.M., Cayres, I.M.V. and Santos, S.E.B. (1993). Genetic structure and demography of the human population of Óbidos, in the Brazilian Amazon. *Revista Brasileira de Genética*, **16**, 1075–84.

Guerreiro, J.F., Ribeiro-dos-Santos, A.K.C., Santos, E.J.M., *et al.* (1999). Genetical–demographic data from two Amazonian populations composed of descendants of African slaves: Pacoval and Curiau. *Genetics and Molecular Biology*, **22**, 163–7.

Guevara, J.M., González, M.E. and Arends, T. (1963). Búsqueda de hemoglobinas anormales en el Estado Trujillo. *Acta Científica Venezolana*, **14**, 153–5.

Guilherme, L., Weidebach, W., Kiss, M.H., Snitcowsky, R. and Kalil, J. (1991). Association of human leukocyte Class II antigens with rheumatic fever or rheumatic heart disease in a Brazilian population. *Circulation*, **83**, 1995–8.

Guimarães, E.M., Brasileiro Filho, G. and Pena, S.D.J. (1992). Human papillomavirus detection in cervical dysplasias or neoplasias and in condylomata acuminata by *in situ* hybridization with biotinylated DNA probes. *Revista do Instituto de Medicina Tropical de São Paulo*, **34**, 309–14.

Guimarães, L.V., Latorre, M.R.D.O. and Barros, M.B.A. (1999). Fatores de risco para a ocorrência de déficit estatural em pré-escolares. *Cadernos de Saúde Pública*, **15**, 605–15.

Guion-Almeida, M.L. (1995). Apparent Malpuech syndrome: report on three Brazilian patients with additional signs. *American Journal of Medical Genetics*, **58**, 13–17.

Guion-Almeida, M.L., Kokitsu-Nakata, N.M. and Zechi, R.M. (1998a). Ring-shaped skin creases syndrome: report of a Brazilian family. *Brazilian Journal of Dysmorphology and Speech–Hearing Disorders*, **2**, 13–18.

Guion-Almeida, M.L., Kokitsu-Nakata, N.M. and Zechi, R.M. (1998b). Say syndrome: a new Brazilian case. *Genetics and Molecular Biology*, **21**, 449–51.

Guion-Almeida, M.L., Richieri-Costa, A., Saavedra, D. and Cohen, M.M. Jr. (1996). Cerebrofaciothoracic syndrome. *American Journal of Medical Genetics*, **61**, 152–3.

Gurjão, I. and Macedo, L.M. (1963). Pesquisa genético–antropológica. *VI Reunião Brasileira de Antropologia*, pp. 1–9.

Gusmão, R. d'A. (1998). Brasil: Un ejemplo de eficiencia. *Salud Mundial*, **51**(3), 30.

Haas, J.D., Baker, P.T. and Hunt, E.E. Jr. (1977). The effects of high altitude on body size and composition of the newborn infant in southern Peru. *Human Biology*, **49**, 611–28.

Habicht, J-P., Martorell, R., Yarbrough, C., Malina, R.M. and Klein, R.E. (1974). Height and weight standards for preschool children. *The Lancet,*, **I**, 611–15.

Haddad, L.A. and Pena, S.D.J. (1993). CAT repeat polymorphism in a human expressed sequence tag (EST00444) (D13S308). *Human Molecular Genetics*, **2**, 1748.

Haddad, L.A., Aguiar, M.J.B., Costa, S.S., Mingroni-Netto, R.C., Vianna-Morgante, A.M. and Pena, S.D.J. (1999). Fully mutated and gray-zone *FRAXA* alleles in Brazilian mentally retarded boys. *American Journal of Medical Genetics*, **97**, 808–12.

Halberstein, R.A. (1980). Population regulation in an island community. *Human Biology*, **52**, 479–98.

Halberstein, R.A. (1997). Health and disease in the Caribbean: An historical perspective. In Halberstein, R.A., editor, *Health and Disease in the Caribbean.* Lexington, KY: Association of Caribbean Studies, pp. 1–5.

Halberstein, R.A. (1999). Blood pressure in the Caribbean. *Human Biology*, **71**, 659–84.

Halberstein, R.A. and Crawford, M.H. (1973). Human biology in Tlaxcala, México: demography. *American Journal of Physical Anthropology*, **36**, 199–212.

Halberstein, R.A. and Crawford, M.H. (1975). Demographic structure of a transplanted Tlaxcalan population in the Valley of México. *Human Biology*, **47**, 201–32.

Halberstein, R.A., Davies, J.E. and Mack, A.K. (1981). Hemoglobin variations on a small Bahamian island. *American Journal of Physical Anthropology*, **55**,

217–21.

Haldane, J.B.S. (1932). *The Causes of Evolution.* London: Longman, Green.

Hall, M.S. (1964). Independência do Reino de Guatemala, hoje América Central. In: Levene, R., editor, *História das Américas,* Vol. 7. Rio de Janeiro: W.M. Jackson, pp. 97–262.

Hamel, M.L.H., Salzano, F.M. and Melo e Freitas, M.J. de (1984). The Gm polymorphism and racial admixture in six Amazonian populations. *Journal of Human Evolution,* **13**, 517–29.

Hammer, M.F., Spurdle, A.B., Karafet, T., *et al.* (1997). The geographic distribution of human Y chromosome variation. *Genetics,* **145**, 787–805.

Harb, Z., Llop, E., Moreno, R. and Queiroz, D. (1998). Poblaciones costeras de Chile: marcadores genéticos en cuatro localidades. *Revista Médica de Chile,* **126**, 753–60.

Hardy, E.E. (1998). Saúde reprodutiva na América Latina. *Cadernos de Saúde Pública,* **14** (special suppl 1.), 1–151.

Harrison, G.A. (1971). The application of spectrophotometry to studies of skin color in Latin American populations. In Salzano, F.M., editor, *The Ongoing Evolution of Latin American Populations.* Springfield, IL: C.C. Thomas, pp. 455–69.

Harrison, G.A. (1977). *Population Structure and Human Variation.* Cambridge: Cambridge University Press.

Harrison, G.A. and Boyce, A.J. (1972). *The Structure of Human Populations.* Oxford: Clarendon Press.

Harrison, G.A., Owen, J.J.T., da Rocha, F.J. and Salzano, F.M. (1967). Skin colour in southern Brazilian populations. *Human Biology,* **39**, 21–31.

Hatagima, A., Cabello, P.H. and Krieger, H. (1999). Causal analysis of the variability of IgA, IgG, and IgM immunoglobulin levels. *Human Biology,* **71**, 219–29.

Hayes, J. and Ferraroni, J.J. (1981). Malaria along pioneer highways in the Brazilian Amazon. *Ciência e Cultura,* **33**, 924–8.

Hazelwood, S., Shotelersuk, V., Wildenberg, S.C., *et al.* (1997). Evidence for locus heterogeneity in Puerto Ricans with Hermansky–Pudlak syndrome. *American Journal of Human Genetics,* **61**, 1088–94.

Hedges, H.B. (1992). The number of replicates needed for accurate estimation of the bootstrap P values in phylogenetic studies. *Molecular Biology and Evolution,* **9**, 366–9.

Hegg, R.V. and Bonjardim, E. (1988). The menarche influence on the variation of the bi-acromial and illium crests diameters in female students, age ranging 10.0 to 19.5 in three social economic groups. *Actas, 5th Congresso da Sociedade Européia de Antropologia,* **1**, 253–9.

Heidrich, E.M., Hutz, M.H., Salzano, F.M., Coimbra, C.E.A. Jr. and Santos, R.V. (1995). D1S80 locus variability in three Brazilian ethnic groups. *Human Biology,* **67**, 311–19.

Heit, J.A., Thorland, E.C., Ketterling, R.P., *et al.* (1998). Germline mutations in Peruvian patients with hemophilia B: pattern of mutation in Amerindians is similar to the putative endogenous germline pattern. *Human Mutation,* **11**,

372–6.

Hellwig, D.J. (1992). *African American Reflections on Brazil's Racial Paradise.* Philadelphia, PA: Temple University Press.

Henckel, C. (1977). Estudio seroantropológico de la población hospitalaria de Concepción (Chile). *Boletin de la Sociedad de Biologia de Concepción,* **51,** 113–18.

Henriques, M.H., Silva, N.V., Singh, S. and Wulf, D. (1989). *Adolescentes de Hoje, Pais do Amanhã: Brasil.* New York: The Alan Guttmacher Institute.

Henry, M.U. 1963. The haemoglobinopathies in Trinidad. *Caribbean Medical Journal,* **25,** 26–40.

Heras, C. (1964*a*). O processo da independência no Uruguai. In: Levene, R., editor, *História das Américas,* Vol. 5. Rio de Janeiro: W.M. Jackson, pp. 211–94.

Heras, C. (1964*b*). Independência do Alto Peru. Criação da República da Bolivia. In: Levene, R., editor, *História das Américas,* Vol. 6. Rio de Janeiro: W.M. Jackson, pp. 3–78.

Heredero, L., Granda, I. and Altland, K. (1978). Methods of screening for hemoglobin S for genetic counseling in Cuba: results of analysis of 24 000 blood samples. *Soviet Genetics,* **14,** 766–70.

Heredero-Baute, L. (1984). La interrupción del embarazo en los países iberoamericanos dentro del contexto de la genética humana. Situación en Cuba. *Actas, VI Congreso Latinoamericano de Genética,* pp. 230–1.

Hernández, M. and García-Moro, C. (1997). El poblamiento de Tierra del Fuego (Chile): estudio de los matrimonios. *Revista Española de Antropología Biológica,* **18,** 131–47.

Hernández Castellón, R. (1986). *El Proceso de la Revolución Demográfica en Cuba.* Havana: Centro de Estudios Demográficos, Universidad de la Habana.

Herrera, M., Theiler, G., Augustovski, F., *et al.* (1994). Molecular characterization of HLA class II genes in celiac disease patients of Latin American Caucasian origin. *Tissue Antigens,* **43,** 83–7.

Herrera Cabral, J.M. (1950). La anemia falciforme en la evolución de la tuberculosis pulmonar. *Revista Médica Dominicana,* **5,** 336–45.

Herrero, B.F. (1996). América y la modernidad europea. *Cuadernos Hispanoamericanos,* **547,** 7–25.

Hickling, F.W. (1997). Mental health in the Caribbean. In Halberstein, R.A., editor, *Health and Disease in the Caribbean.* Lexington, KY: Association of Caribbean Studies, pp. 57–71.

Hidalgo, P.C. and Heredero, L. (1985). Consideraciones sobre la genetica de la población de la region central de Cuba. Comunicación al VII Congreso Latinoamericano de Genética, Bogotá (unpublished).

Hidalgo, P.C., Fernández-Estrada, J., Castellanos, T., Méndez, A., Carrazana, M. and Ortega, R. (1995). Distribution of ABO, Rh, MN, and Kell blood groups in Central Cuba. *Brazilian Journal of Genetics,* **18,** 475–8.

Hidalgo, P.C., Soto, G., Machado, M.J., Ortega, R. and García, A. (1974). Frecuencia de portadores de hemoglobina-S y otras hemoglobinas anormales en una muestra de población infantil. *Centro, Série Ciencias Médicas*, **1**, 85–91.

Higman, B.W. (1979). Growth in Afro-Caribbean slave populations. *American Journal of Physical Anthropology*, **50**, 373–85.

Himes, J.H. and Malina, R.M. (1975). Age and secular factors in the stature of adult Zapotec males. *American Journal of Physical Anthropology*, **43**, 367–9.

Himes, J.H. and Malina, R.M. (1977). Sexual dimorphism in metacarpal dimensions and body size of Mexican school children. *Acta Anatomica*, **99**, 15–20.

Himes, J.H. and Mueller, W.H. (1977). Aging and secular change in adult stature in rural Colombia. *American Journal of Physical Anthropology*, **46**, 275–9.

Himes, J.H., Malina, R.M. and Stepick, C.D. (1976*b*). Relationships between body size and second metacarpal dimensions in Oaxaca, México, school children 6 to 14 years of age. *Human Biology*, **48**, 677–92.

Himes, J.H., Martorell, R., Habicht, J-P., Yarbrough, C., Malina, R.M. and Klein, R.E. (1975). Patterns of cortical bone growth in moderately malnourished preschool children. *Human Biology*, **47**, 337–50.

Himes, J.H., Martorell, R., Habicht, J-P., Yarbrough, C., Malina, R.M. and Klein, R.E. (1976*a*). Sexual dimorphism in bone growth as a function of body size in moderately malnourished Guatemalan preschool age children. *American Journal of Physical Anthropology*, **45**, 331–5.

Hirano, M., Garcia-de-Yebenes, J., Jones, A.C., *et al.* (1998). Mitochondrial neurogastrointestinal encephalomyopathy syndrome maps to chromosome 22q13.32-qter. *American Journal of Human Genetics*, **63**, 526–33.

Hladik, C.M. (1993). Fruits of the rain forest and taste perception as a result of evolutionary interactions. In Hladik, C.M., Hladik, A., Linares, O.F., Pagezy, H., Semple, A. and Hadley, M., editors, *Tropical Forests, People and Food. Biocultural Interactions and Applications to Development.* Paris: UNESCO and Parthenon Publishing, pp. 73–82.

Hladik, C.M., Hladik, A., Linares, O.F., Pagezy, H., Semple, A. and Hadley, M. (1993). *Tropical Forests, People and Food. Biocultural Interactions and Applications to Development.* Paris: UNESCO, and New York: Parthenon.

Ho, L., Chan, S-Y., Chow, V., Chong, T., Tay, S-K., Villa, L.L. and Bernard, H-U. (1991). Sequence variants of human papillomavirus type 16 in clinical samples permit verification and extension of epidemiological studies and construction of a phylogenetic tree. *Journal of Clinical Microbiology*, **29**, 1765–72.

Hoetink, H. (1973). *Slavery and Race Relations in the Americas.* New York: Harper and Row.

Hoetink, H. (1997). 'Raça' e cor no Caribe. *Estudos Afro-Asiáticos*, **31**, 7–36.

Hoffecker, J.F., Powers, W.R. and Goebel, T. (1993). The colonization of Beringia and the peopling of the New World. *Science*, **259**, 46–53.

Hoffstetter, R. (1949). Las caracteristicas serologicas de la poblacion urbana de Quito (Ecuador). *Boletin Informativo Científico Nacional*, **2**, 47–73.

Hojas-Bernal, R., McNab-Martin, P., Fairbanks, V.F., *et al.* (1999). Hb Chile [β28(B10)Leu→Met]: an unstable hemoglobin associated with chronic methemoglobinemia and sulfonamide or methylene blue-induced hemolytic anemia. *Hemoglobin*, **23**, 125–34.

Holman, D.J. and Jones, R.E. (1998). Longitudinal analysis of deciduous tooth emergence. II. Parametric survival analysis in Bangladeshi, Guatemalan, Japanese, and Javanese children. *American Journal of Physical Anthropology*, **105**, 209–30.

Horai, S., Kondo, R., Nakagawa-Hattori, Y., Hayashi, S., Sonoda, S. and Tajima, K. (1993). Peopling of the Americas, founded by four major lineages of mitochondrial DNA. *Molecular Biology and Evolution*, **10**, 23–47.

Hulse, F.S. (1979). Migration et selection de groupe: Le cas de Cuba. *Bulletins et Mémoirs de la Societé d'Anthropologie de Paris*, **6**, série, **13**, 137–46.

Huntington, S.P. (1997). *O Choque de Civilizações e a Recomposição da Ordem mundial.* Rio de Janeiro: Editora Objetiva.

Hunziker, J.H., von der Pahlen, A., Magnelli, N. and Bianchi, N.O. (1980). *Actas, IV Congreso Latinoamericano de Genética y X Congreso Argentino de Genética.* Buenos Aires: Asociación Latinoamericana de Genética.

Hurault, J-M. (1989). *Français et Indiens en Guyane: 1604–1972.* Cayenne: Guyane Presse Diffusion.

Hutchinson, J. (1986). Association between stress and blood pressure variation in a Caribbean population. *American Journal of Physical Anthropology*, **71**, 69–79.

Hutchinson, J. and Byard, P.J. (1987). Family resemblance for anthropometric and blood pressure measurements in Black Caribs and Creoles from St. Vincent island. *American Journal of Physical Anthropology*, **73**, 33–9.

Hutchinson, J. and Crawford, M.H. (1981). Genetic determinants of blood pressure level among the Black Caribs of St Vincent. *Human Biology*, **53**, 453–66.

Huttly, S.R.A., Victora, C.G., Barros, F.C. and Vaughan, J.P. (1992). Birth spacing and child health in urban Brazilian children. *Pediatrics*, **89**, 1049–54.

Hutz, M.H. and Salzano, F.M. (1983*a*). Sickle cell anemia in Rio de Janeiro, Brazil: demographic, clinical and laboratory data. *Brazilian Journal of Medical and Biological Research*, **16**, 219–26.

Hutz, M.H. and Salzano, F.M. (1983*b*). Fecundidade em uma amostra brasileira de mulheres com anemia falciforme. *Revista da Associação Médica Brasileira*, **29**, 66–8.

Hutz, M.H., Mattevi, V.S., Callegari-Jacques, S.M., *et al.* (1997). D1S80 locus variability in South American Indians. *Annals of Human Biology*, **24**, 249–55.

Hutz, M.H., Salzano, F.M. and Adams, J. (1983). Hb F levels, longevity of homozygotes and clinical course of sickle cell anemia in Brazil. *American Journal of Medical Genetics*, **14**, 669–76.

Ibarra, B., Franco-Gamboa, E., Ramírez, M.L., Cantú, J.M., Wilson, J.B., Lam, H. and Huisman, T.H.J. (1981). Hb Chiapas $\alpha_2 114$ PRO→ARGβ_2, identification by high pressure liquid chromatography. *Hemoglobin*, **5**, 605–8.

Ibarra, B., Javier-Perea, F. and Villalobos-Arámbula, A.R. (1995). Alelos talasémicos en mestizos mexicanos. *Revista de Investigación Clínica*, **47**, 127–31.

IBGE (Instituto Brasileiro de Geografia e Estatística) (1976). *Encontro Brasileiro de Estudos Populacionais.* Rio de Janeiro: Fundação Instituto Brasileiro de Geografia e Estatística.

IBGE (1984). *Perfil Estatístico de Crianças e Mães no Brasil.* Rio de Janeiro, Fundação Instituto Brasileiro de Geografia e Estatística.

IBGE (1987). *Estatísticas Históricas do Brasil. Séries Econômicas, Demográficas e Sociais de 1550 a 1985.* Rio de Janeiro: Fundação Instituto Brasileiro de Geografia e Estatística.

Iglesias, D.M., Manrique, M., Arrizurieta, E.E., *et al.* (1999). Poliquistosis renal autosomica dominante. Detección de una nueva mutación en el gen *PKD1. Medicina* (Buenos Aires), **59**, 133–7.

Ilha, D.O. and Salzano, F.M. (1961). A roentgenologic and genetic study of a rare osseous dystrophy. *Acta Geneticae Medicae et Gemellologiae*, **10**, 340–52.

Inglehart, R. and Carballo, M. (1997). Does Latin America exist? (And is there a Confucian culture?): a global analysis of cross-cultural differences. *PS: Political Science & Politics*, **30**, 34–47.

Itskan, S.B. and Saldanha, P.H. (1975). Atividade da glucose-6-fosfato desidrogenase eritrocitária em população de área malarígena de São Paulo (Iguape). *Revista do Instituto de Medicina Tropical*, **17**, 83–91.

Ivens-de-Araujo, M.E., Fandinho, F.C., Werneck-Barreto, A.M., *et al.* (1998). DNA fingerprinting of *Mycobacterium tuberculosis* from patients with and without AIDS in Rio de Janeiro. *Brazilian Journal of Medical and Biological Research*, **31**, 369–72.

James, P.E. (1969). *Latin America.* 4th edn. Indianapolis, IN: Odyssey Press.

Jara, L., Aspillaga, M., Avendaño, I., Obreque, V., Blanco, R. and Valenzuela, C.Y. (1998). Distribution of (CGG)n and FMR–1 associated microsatellite alleles in a normal Chilean population. *American Journal of Medical Genetics*, **75**, 277–82.

Jaramillo-Rangel, G., Cerda-Flores, R.M., Cardenas-Ibarra, L., Tamayo-Orozco, J., Morrison, N. and Barrera-Saldaña, H.A. (1999). Vitamin D receptor polymorphisms and bone mineral density in Mexican women without osteoporosis. *American Journal of Human Biology*, **11**, 793–7.

Jardim, L.B., Leite, J.C.L., Silveira, E.L., Barth, M.L. and Giugliani, R. (1992*c*). Resultados preliminares de um programa de detecção precoce para o hipotireoidismo congênito no Rio Grande do Sul. *Jornal de Pediatria*, **68**, 239–42.

Jardim, L.B., Silveira, E.L., Leite, J.C.L., Barth, M.L. and Giugliani, R. (1992*b*). Resultados preliminares de um programa de detecção precoce para aminoacidopatias no Rio Grande do Sul. *Jornal de Pediatria*, **68**, 189–91.

Jardim, L.B., Carneiro, A., Hansel, S., Rieder, C.R.M. and Giugliani, R. (1991). CT hypodensity on cerebral white matter in Wilson's disease. *Arquivos de Neuro-Psiquiatria*, **49**, 211–14.

Jardim, L.B., Giugliani, R. and Fensom, A.H. (1992*a*). Thalamic and basal ganglia hyperdensities: CT marker for globoid cell leukodystrophy? *Neuropediatrics*, **23**, 30–1.

Jardim, L.B., Giugliani, R., Coelho, J.C., Dutra-Filho, C.S. and Blau, N. (1994). Possible high frequency of tetrahydropterin deficiency in South Brazil. *Journal of Inherited Metabolic Diseases*, **17**, 223–9.

Jardim, L.B., Martins, C.S., Pires, R.F., *et al.* (1995). Uma experiência terapêutica no manejo da doença da urina do xarope do bordo. *Jornal de Pediatria*, **71**, 279–84.

Jenkins, C.L. (1981). Patterns of growth and malnutrition among preschoolers in Belize. *American Journal of Physical Anthropology*, **56**, 169–78.

Jiménez, R., Bacallao, J., Molina, J.R. and Valladares, B. (1987). Relación entre el grosor de los pliegues cutaneos de los padres y el neonato: Utilidad en la evaluación nutricional. In Faulhaber, M.E.S. and Lizárraga Cruchaga, X., editors, *Estudios de Antropología Biológica (III Coloquio de Antropologia Física Juan Comas)*. México: Universidad Nacional Autónoma de México, pp. 29–45.

Jiménez, S., Martínez, B., Hernández, M., Munera, A. and Caraballo, L. (1996). Análisis inmunogenético y antropológico de la población del Palenque de San Basilio (Colombia). In Martin Municio, A. and García Barreno, P., editors, *Polimorfismos Génicos (HLA) en Poblaciones Hispanoamericanas*. Madrid: Real Academia de Ciencias, pp. 245–69.

Johnson, H.B. (1997). A colonização portuguesa do Brasil, 1500–1580. In: Bethell, L., editor, *História da América Latina. Vol. 1. América Latina Colonial*, São Paulo: Editora da Universidade de São Paulo, pp. 241–81.

Johnston, F.E. and MacVean, R.B. (1995). Growth faltering and catch-up growth in relation to environmental change in children of a disadvantaged community from Guatemala City. *American Journal of Human Biology*, **7**, 731–40.

Johnston, F.E., Low, S.M., Baessa, Y. and MacVean, R.B. (1985). Growth status of disadvantaged urban Guatemalan children of a resettled community. *American Journal of Physical Anthropology*, **68**, 215–24.

Johnston, F.E., Reid, W., de Baessa, Y. and Macvean, R.B. (1989). Socieconomic correlates of fertility, mortality, and child survival in mothers from disadvantaged urban Guatemalan community. *American Journal of Human Biology*, **1**, 25–30.

Jones, D. (1995). Sexual selection, physical attractiveness, and facial neoteny. *Current Anthropology*, **36**, 723–48.

Jones, D. (1996). An evolutionary perspective on physical attractiveness. *Evolutionary Anthropology*, **5**, 97–109.

Jonxis, J.H.P. (1959). The frequency of haemoglobin S and haemoglobin C carriers in Curaçao and Surinam. In Jonxis, J.H.P. and Delafresnaye, D.F., editors, *Abnormal Haemoglobins*. Oxford: Blackwell, pp. 300–6.

Jorquera, A., Ledezma, E., Sousa, L., *et al.* (1998). Epidemiologic characterization of American cutaneous leishmaniasis in an endemic region of eastern Venezuela. *American Journal of Tropical Medicine and Hygiene*,

58, 589–93.

Junqueira, P.C. and Wishart, P.J. (1958). Distribuição dos grupos sangüíneos ABO em brancos, mulatos e pretos do Rio de Janeiro, de acordo com a presença ou ausência de sobrenome. *Revista Clínica* (São Paulo), **34**(4), 79–83.

Junqueira, P.C., Ottensooser, F., Montenegro, L., Junqueira, N.C. and Cunha, A.B. (1962). Grupos sangüíneos de nordestinos. *Anais da Academia Brasileira de Ciências*, **34**, 143–52.

Kac, G. (1998). Tendência secular em estatura em recrutas da marinha do Brasil nascidos entre 1940 e 1965. *Cadernos de Saúde Pública*, **14**, 565–73.

Kac, G. and Santos, R.V. (1996). Crescimento físico em estatura de escolares de ascendência japonesa na cidade de São Paulo, Brasil. *Cadernos de Saúde Pública*, **12**, 253–7.

Kac, G. and Santos, R.V. (1997). Secular trend in height in enlisted men and recruits from the Brazilian Navy born from 1970 to 1977. *Cadernos de Saúde Pública*, **13**, 479–87.

Kádasi, L., Gécz, J., Feráková, I., *et al.* (1994). Distribution of ApoBII, MCT118(D1S80), YNZ22(D17S30), and COL2A1) Amp-FLPs (Amplified Fragment Length Polymorphisms) in Caucasoid population of Slovakia. *Gene Geography*, **8**, 121–7.

Kahn, M.C. (1936). Blood grouping of 336 Upper Aucaner Bush Negroes and 70 Alukuyana Indians in Dutch Guiana. *Journal of Immunology*, **31**, 377–85.

Kaku, M. and Freire-Maia, N. (1992). Inbreeding effects on morbidity. 4. Further data on Brazilian populations. *American Journal of Medical Genetics*, **42**, 467–9.

Kanter, N. (1963). A disciplinação da pesquisa monodactilar. Conclusões do levantamento estatístico realizado no arquivo monodactilar do Instituto de Polícia Técnica do Rio Grande do Sul. *Revista de Criminalística do Rio Grande do Sul*, **1**, 12–32.

Karlin, S. and McGregor, J. (1967). The number of mutant forms maintained in a population. In Le Cam, L.M. and Neyman, J., editors, *Proceedings of the Fifth Berkeley Symposium on Mathematics, Statistics, and Probability*, Vol. 4. Berkeley, CA: University of California Press, pp. 415–438.

Katsuyama, Y., Inodo, H., Imanishi, T., Misuki, N., Gojobori, T. and Ota, M. (1998). Genetic relationships among Japanese, Northern Han, Hui, Uygur, Kazakh, Greek, Saudi Arabian, and Italian populations based on allelic frequencies at four VNTR (D1S80), D4S43, COL2A1, D17S5) and one STR (ACTBP2) loci. *Human Heredity*, **48**, 126–37.

Kaufman, L., Carnese, F.R., Goicoechea, A., Dejean, C., Salzano, F.M. and Hutz, M.H. (1998). Beta-globin gene cluster haplotypes in the Mapuche Indians of Argentina. *Genetics and Molecular Biology*, **21**, 435–7.

Kéclard, L., Ollendorf, V., Berchel, C., Loret, H. and Mérault, G. (1996). β^S haplotypes, α-globin gene status, and hematological data of sickle cell disease patients in Guadeloupe (F.W.I). *Hemoglobin*, **20**, 63–74.

Kéclard, L., Romana, M., Lavocat, E., Saint-Martin, C., Berchel, C. and Mérault, G. (1997). Sickle cell disorder, β-globin gene cluster haplotypes and α-thalassemia in neonates and adults from Guadeloupe. *American Journal of Hematology*, **55**, 24–7.

Kelso, A.J., Siffert, T. and Maggi, W. (1992). Association of ABO phenotypes and body weight in a sample of Brazilian infants. *American Journal of Human Biology*, **4**, 607–11.

Keyeux, G. (1993). Poblaciones negras de Colombia: una primera aproximación a su estructura molecular. *America Negra*, **5**, 21–33.

Keyeux, G. and Bernal, J.E. (1996). New allele variants of the immunoglobulin switch (Sα) regions. *Human Genetics*, **97**, 695–6.

Khan, A.D., Schroeder, D.K., Martorell, R., Haas, J.D. and Rivera, J. (1996). Early childhood determinants of age at menarche in rural Guatemala. *American Journal of Human Biology*, **8**, 717–23.

Kim, C.A., Passos-Bueno, M.R., Marie, S.K., *et al.* (1999). Clinical and molecular analysis of spinal muscular atrophy in Brazilian patients. *Genetics and Molecular Biology*, **22**, 487–92.

Kimmel, M., Chakraborty, R., Stivers, D.N. and Deka, R. (1996). Dynamics of repeat polymorphisms under a forward-backward mutation model: within- and between-population variability at microsatellite loci. *Genetics*, **143**, 549–55.

Klein, H.S. (1989). A integração social e econômica dos imigrantes portugueses no Brasil no fim do século XIX e no século XX. *Revista Brasileira de Estudos de População*, **6**(2), 17–37.

Knijff, P., Kayser, M., Cáglia, A., *et al.* (1997). Chromosome Y microsatellites: population genetic and evolutionary aspects. *International Journal of Legal Medicine*, **110**, 134–40.

Kobyliansky, E. (1984). Heritability of some morphological traits in man: regression and segregation analysis of familial resemblance. *Zeitschrift für Morphologie und Anthropologie*, **75**, 77–84.

Kobyliansky, E. and Goldstein, M.S. (1987). Changes in anthropometric traits of humans in the process of migration: comparative analysis of two generations of Mexicans in México and the U.S.A. *Anthropologie*, **25**, 63–80.

Koch, W. (1980). *III Colóquio de Estudos Teuto-Brasileiros*. Porto Alegre: Editora da Universidade Federal do Rio Grande do Sul.

Koifman, S. (1997). Epidemiologia do câncer do estômago no Brasil. *Cadernos de Saúde Pública*, **13** (Suppl.), 1–112.

Koifman, S. (1998). Environmental and occupational cancer in Latin America. *Cadernos de Saúde Pública*, **14** (Suppl. 3), 1–198.

Kokitsu-Nakata, N.M. and Richieri-Costa, A. (1998). Blepharophimosis, ptosis, epicanthus inversus syndrome (BPES) and cleft lip and palate. Report of two Brazilian families. *Genetics and Molecular Biology*, **21**, 259–62.

Kolski, R. and Cordido, P. (1974). Heredabilidad para numero de triradii. *Ciência e Cultura*, **26**, 1033–4.

Kolski, R. and Nunes de Langguth, E. (1971). Diametro de cabellos, otro procedimiento para el diagnóstico de zigosidad. *Ciência e Cultura*, **23**, 237–9.

Kolski, R. and Salvat, G. (1977). Valores de 'trc' en dermatoglifos de hermanos y de mellizos dizigóticos. *Ciência e Cultura*, **29**, 1035–8.

Kolski, R., Lazaro, C. and Olaizola, L. (1965). Número de deltas en impresiones digitales. Comparaciones por sexo. *Revista de la Facultad de Humanidades y Ciencias, Universidad de la República* (Montevideo), **22**, 221–6.

Körner, H., Rodriguez, L., Yero, J.L.F., *et al.* (1986). Maternal serum alpha-fetoprotein screening for neural tube defects and other disorders using an ultramicro-ELISA. Collaborative study in Cuba and in the German Democratic Republic. *Human Genetics*, **73**, 60–3.

Kraemer, M.H., Donadi, E.A., Tambascia, M.A., Magna, L.A. and Prigenzi, L.S. (1998). Relationship between HLA antigens and infectious agents in contributing towards the development of Graves' disease. *Immunological Investigations*, **27**, 17–29.

Kragh-Hansen, U., Pedersen, A.O., Galliano, M., *et al.* (1996). High-affinity binding of laurate to naturally occurring mutants of human serum albumin and proalbumin. *Biochemical Journal*, **320**, 911–16.

Krieger, H. (1969). Inbreeding effects on metrical traits in northeastern Brazil. *American Journal of Human Genetics*, **21**, 537–46.

Krieger, H. and Feitosa, M.F. (1999). Genetic epidemiology of infectious disease. *Ciência e Cultura*, **51**, 191–8.

Krieger, H., Morton, N.E., Mi, M.P., Azevêdo, E., Freire-Maia, A. and Yasuda, N. (1965). Racial admixture in north-eastern Brazil. *Annals of Human Genetics*, **29**, 113–25.

Krieger, H., Morton, N.E., Rao, D.C. and Azevêdo, E.S. (1980). Familial determinants of blood pressure in northeastern Brazil. *Human Genetics*, **53**, 415–18.

Krings, M., Stone, A., Schmitz, R.W., Krainitzki, H., Stoneking, M. and Pääbo, S. (1997). Neandertal DNA sequences and the origin of modern humans. *Cell*, **90**, 19–30.

Kulozik, A.E., Wainscoat, J.S., Serjeant, G.R., *et al.* (1986). Geographical survey of β^S-globin gene haplotypes: Evidence for an independent Asian origin of the sickle-cell mutation. *American Journal of Human Genetics*, **39**, 239–44.

Kunter, M. (1987). Population and racial history of Argentina, Uruguay and Paraguay. In Schwidetzky, I., editor, *Rassengeschichte der Menschheit*. München: R. Oldenbourg Verlag, pp. 237–92.

Kuwano, S.T., Chiba, A.K., Figueiredo, M.S., Vieira-Filho, J.P.B. and Bordin, J.O. (1998). Expression of NA1, NA2, and SH alleles of Fcγ receptor IIIB in South American Indians and Brazilian blood donors. *Transfusion*, **38** (Suppl.), 28S.

Kuzawa, C.W. (1998). Adipose tissue in human infancy and childhood: An evolutionary perspective. *Yearbook of Physical Anthropology*, **41**, 177–209.

Kvitko, K. and Weimer, T.A. (1988). Placental phosphoglucomutase polymorphism in Porto Alegre, Brazil. *Revista Brasileira de Genética*, **11**, 949–55.

Labie D., Srinivas, R., Dunda, O., *et al.* (1989). Haplotypes in tribal Indians bearing the sickle gene: evidence for the unicentric origin of the β^S mutation and unicentric origin of the tribal populations of India. *Human Biology*, **61**, 479–91.

Lacaz, C.S. (1939). Índice bioquímico racial. *Folia Clinica et Biologica*, **11**, 168–74.
Lacaz, C.S. and Maspes, V. (1951). Acidente pós-transfusional com imunização a diversos aglutinógenos. Incidência do fator P na cidade de São Paulo. *Resenha Clínica Científica*, **20** (2), 47–50.
Lacaz, C.S., Ferreira, H.C., Mellone, O. and Yahn, O. (1946). Novos dados estatísticos sobre o fator Rh em São Paulo (Brasil). *Resenha Clínica Científica*, **15**, 279–83.
Lacaz, C.S., Mellone, O. and Yahn, Q. (1951). Diagnóstico, profilaxia e tratamento da doença hemolítica do recém-nascido. *Anais do Instituto Pinheiros*, **14**(28), 161–283.
Lacaz, C.S., Pinto, D.O., Borges, S.O., Mellone, O. and Yahn, O. (1955). The incidence of the Kell factor in São Paulo, Brazil. *American Journal of Physical Anthropology*, **13**, 349–50.
Lacerda, G.B., Arce-Gomez, B. and Telles-Filho, F.Q. (1988). Increased frequency of HLA-B40 in patients with paracoccidioidomycosis. *Journal of Medical and Veterinary Mycology*, **26**, 253–6.
Lahr, M.M. (1995). Patterns of modern human diversification: implications for Amerindian origins. *Yearbook of Physical Anthropology*, **38**, 163–98.
Lando, A.M. and Barros, E.C. (1976). *A Colonização Alemã no Rio Grande do Sul. Uma Interpretação Sociológica.* Porto Alegre: Editora Movimento and Instituto Estadual do Livro.
Languillon, J. (1951). Recherches sur la répartition des groupes sanguins chez les Guadeloupéens de race noire. *Bulletin de la Societé de Pathologie Exotique*, **44**, 181–4.
Lapouméroulie, C., Dunda, O., Ducrocq, R., *et al.* (1992). A novel sickle mutation of yet another origin in Africa: The Cameroon type. *Human Genetics*, **89**, 333–7.
Lara, M.L., Layrisse, Z., Scorza, J.V., Garcia, E., Stoikow, Z., Granados, J. and Bias, W. (1991). Immunogenetics of human American cutaneous leishmaniasis. Study of HLA haplotypes in 24 families from Venezuela. *Human Immunology*, **30**, 129–35.
Laredo-Filho, J., Carneiro-Filho, M., Rangel, J.P.A. and Carrasco, M.J.M. (1987). Disostose clidocraniana hereditária. Aspectos clínicos, radiológicos e genéticos. *A Folha Médica*, **95**, 161–8.
Larrandaburu, M., Schüler, L., Ehlers, J.A., Reis, A.M. and Silveira, E.L. (1999). The occurrence of Poland and Poland–Moebius syndromes in the same family: further evidence of their genetic component. *Clinical Dysmorphology*, **8**, 93–9.
Larrauri, S. and Rodriguez-Larralde, A. (1984). Dermatoglifos en una muestra de población venezolana. *Acta Cientifica Venezolana*, **35**, 253–64.
Laska-Mierzejewska, T. (1967). Desarrollo y maduración de los niños y jóvenes habaneros. *Materialy i Prace Antropologiczne*, **74**, 9–64.
Lasker, G.W. (1954). Photoelectric measurement of skin color in a Mexican Mestizo population. *American Journal of Physical Anthropology*, **12**, 115–22.
Lasker, G.W. (1960). Variances of bodily measurements in the offspring of

natives of and immigrants to three Peruvian towns. *American Journal of Physical Anthropology*, **18**, 257–61.

Lasker, G.W. (1968). The occurrence of identical (isonymous) surnames in various relationships in pedigrees: a preliminary analysis of the relation of surname combinations to inbreeding. *American Journal of Human Genetics*, **20**, 250–7.

Lasker, G.W. (1969). Isonymy (recurrence of the same surnames in affinal relatives): a comparison of rates calculated from pedigrees, grave markers and death and birth registers. *Human Biology*, **41**, 309–21.

Lasker, G.W. (1978). Increments through migration to the coefficient of relationship between communities estimated by isonymy. *Human Biology*, **50**, 235–40.

Lasker, G.W. (1999). *Happenings and Hearsay. Experiences of a Biological Anthropologist.* Detroit, MI: Savoyard Books.

Lasker, G.W. and Evans, F.G. (1961). Age, environment and migration: further anthropometric findings on migrant and non-migrant Mexicans. *American Journal of Physical Anthropology*, **19**, 203–11.

Lasker, G.W. and Kaplan, B. (1964). The coefficient of breeding isolation: population size, migration rates, and the possibilities for random genetic drift in six human communities in northern Peru. *Human Biology*, **36**, 327–38.

Lasker, G.W. and Kaplan, B. (1974). Anthropometric variables in the offspring of isonymous matings. *Human Biology*, **46**, 713–17.

Lasker, G.W. and Thomas, R. (1976). Relationship between reproductive fitness and anthropometric dimensions in a Mexican population. *Human Biology*, **48**, 775–91.

Lasker, G.W. and Thomas, R.B. (1978). The relationship between size and shape of human head and reproductive fitness. *Studies in Physical Anthropology*, **4**, 3–9.

Lasker, G.W., Wetherington, R.K., Kaplan, B.A. and Kemper, R.V. (1984). Isonymy between two towns in Michoacan, México. In Galvan, R.R. and Rodriguez, R.M.R., editors, *Estudios de Antropología Biológica (II Coloquio de Antropología Física Juan Comas, 1982)*. México: Universidad Nacional Autónoma de México, pp. 159–163.

Laurenti, R. (1978). Mortalidade infantil e desenvolvimento sócio-econômico. O caso de São Paulo. *Revista Brasileira de Pesquisas Médicas e Biológicas*, **11**, 297–302.

Laurenti, R., Silveira, M.H. and Gotlieb, S.L. (1976). Estudos de mortalidade. In IBGE (1976). Rio de Janeiro: Fundação Instituto Brasileiro de Geografia e Estatística, pp. 49–58.

Lautenberger, J.A., Stephens, J.C., O'Brien, S.J. and Smith, M.W. (2000). Significant admixture linkage disequilibrium across 30 cM around the FY locus in African Americans. *American Journal of Human Genetics*, **66**, 969–78.

Lavinha, J., Gonçalves, J., Faustino, P., *et al.* (1992). Importation route of the sickle cell trait into Portugal: A contribution of molecular epidemiology.

Human Biology, **64**, 891–901.

Lavrin, A. (1989). *Marriage and Sexuality in Colonial Latin America*. Lincoln, NE: University of Nebraska Press.

Layrisse, M., Wilbert, J. and Arends, T. (1958). Frequency of blood group antigens in the descendants of Guayquerí Indians. *American Journal of Physical Anthropology*, **16**, 307–18.

Lazo, B., Figueroa, H., Salinas, C., Campusano, C. and Pinto-Cisternas, J. (1970). Consanguinity in the Province of Valparaiso, Chile, 1917–1966. *Social Biology*, **17**, 167–79.

Lazo, B., Ballesteros, S. and Pinto-Cisternas, J. (1973). Utilización de datos demográficos en Genética Humana. *Revista Médica de Chile*, **101**, 257–62.

Lazo, B., Campusano, C., Figueroa, H., Pinto-Cisternas, J. and Zambra, E. (1978). Inbreeding and immigration in urban and rural zones of Chile, with an endogamy index. *Social Biology*, **25**, 228–34.

Leal, M.C. and Szwarcwald, C.L. (1996). Evolução da mortalidade neonatal no Estado do Rio de Janeiro, Brasil (1979–1993): Análise por causa segundo grupo de idade e região de residência. *Cadernos de Saúde Pública*, **12**, 243–52.

Leatherman, T.L. (1998). Illness, social relations, and household production and reproduction in the Andes of southern Peru. In Goodman, A.H. and Leatherman, T.L., editors, *Building a New Biocultural Synthesis. Political-Economic Perspectives on Human Biology*. Ann Arbor, MI: University of Michigan Press, pp. 245–67.

Leatherman, T.L., Carey, J.W. and Thomas, R.B. (1995). Socioeconomic change and patterns of growth in the Andes. *American Journal of Physical Anthropology*, **97**, 307–21.

Lebrón, E.J. and Bonatti, A.A. (1949). Grupos sanguineos y factor Rh en Tucuman. *Revista de la Sociedad Argentina de Hematologia e Hemoterapia*, **1**, 41–8.

Lee, H.S., Sambuughin, N., Cervenakova, L., *et al.* (1999). Ancestral origins and worldwide distribution of the *PRNP 200K* mutation causing familial Creutzfeldt–Jakob disease. *American Journal of Human Genetics*, **64**, 1063–70.

Lee, K., Préhu, C., Mérault, G., *et al.* (1998). Genetic and hematological studies in a group of 114 adult patients with SC sickle cell disease. *American Journal of Hematology*, **59**, 15–21.

Lee, S., Naime, D.S., Reid, M.E. and Redman, C.M. (1997). Molecular basis for the high-incidence antigens of the Kell blood group system. *Transfusion*, **37**, 1117–22.

Lei, D.L.M., Chaves, S.P., Lerner, B.R. and Stefanini, M.L.R. (1995). Retardo do crescimento físico e aproveitamento escolar em crianças do município de Osasco, área metropolitana de São Paulo, Brasil. *Cadernos de Saúde Pública*, **11**, 238–45.

Lei, D.L.M., Freitas, I.C., Chaves, S.P., Lerner, B.R. and Stefanini, M.L.R. (1997). Retardo do crescimento e condições sociais em escolares de Osasco, São Paulo, Brasil. *Cadernos de Saúde Pública*, **13**, 277–83.

Leistner, S. and Giugliani, R. (1998). A useful routine for biochemical detection and diagnosis of mucopolysaccharidoses. *Genetics and Molecular Biology*, **21**, 163–7.

Lejarraga, H., Sanchirico, F. and Cusminsky, M. (1980). Age of menarche in urban Argentinian girls. *Annals of Human Biology*, **7**, 579–81.

Leme Lopes, M.B. and Lopes da Costa, H. (1951). Rotina de classificação do fator Rh em doadores do Banco de Sangue da Prefeitura do Distrito Federal. Resultados estatísticos em quatro meses de trabalho. *Seara Médica*, **6**, 487–9.

Lemle, A., Hazan, E.M., Mandel, M.B. and Magalhães-Pinto, J.C. (1976). The defect in pulmonary gas transfer in patients with sickle cell disease. *Revista Brasileira de Pesquisas Médicas e Biológicas*, **9**, 279–91.

Lemmen, D.S., Duk-Rodkin, A. and Bednarski, J.M. (1994). Late glacial drainage systems along the northwestern margin of the Laurentide ice sheet. *Quaternary Science Reviews*, **13**, 805–28.

Lemoine, V.R. (1984). *Genética. VI Congreso Latinoamericano de Genética. I Congreso Venezolano de Genética*. Maracaibo: Asociación Venezolana de Genética.

Leonard, W.R., de Walt, K., Stansbury, J.P. and McCaston, M.K. (1995). Growth differences between children of Highland and Coastal Ecuador. *American Journal of Physical Anthropology*, **98**, 47–57.

Leslie, P.W. (1983). Cohorts, overlapping generations and consanguinity estimates. *Annals of Human Biology*, **10**, 257–64.

Leslie, P.W., MacCluer, J.W. and Dyke, B. (1978). Consanguinity avoidance and genotype frequencies in human populations. *Human Biology*, **50**, 281–99.

Leslie, P.W., Dyke, B. and Morrill, W.T. (1980). Celibacy, emigration, and genetic structure in small populations. *Human Biology*, **52**, 115–30.

Leslie, P.W., Morrill, W.T. and Dyke, B. (1981). Genetic implications of mating structure in a Caribbean isolate. *American Journal of Human Genetics*, **33**, 90–104.

Levene, R. (1964). A Argentina até a constituição de 1853. In: Levene, R., editor, *História das Américas*, Vol. 5. Rio de Janeiro: W.M. Jackson, pp. 3–210.

Levin, B.R., Lipsitch, M. and Bonhoeffer, S. (1999). Population biology, evolution, and infectious disease: convergence and synthesis. *Science*, **283**, 806–9.

Levisky, R.B., Thomaz, C., Roisenberg, I., *et al.* (1983). Evaluation of techniques for detection of hemophilia A heterozygotes. *Revista Brasileira de Genética*, **6**, 327–35.

Levy, M.S.F. (1974). O papel da migração internacional na evolução da população brasileira (1872 a 1972). *Revista de Saúde Pública*, **8** (Suppl.), 49–90.

Levy, M.S.F. (1975). Condicionantes sociais e medidas antropométricas. Ph.D. thesis, Faculdade de Saúde Pública, Universidade de São Paulo, São Paulo.

Levy, M.S.F. (1996). Cem anos de movimentos populacionais: São Paulo em destaque. *Revista Brasileira de Estudos de População*, **13**(1), 15–41.

Liachowitz, C., Elderkin, J., Guichirit, I., Brown, H.W. and Ranney, H.M.

428   *References*

(1958). Abnormal hemoglobins in the Negroes of Surinam. *American Journal of Medicine*, **24**, 19–24.

Lima, A.M.V.M.D., Azevêdo, E., Krieger, H., Cabello, P.H. and Pollitzer, W.S. (1996). Admixture and relationships of the population of Jacobina, Bahia, Brazil. *American Journal of Human Biology*, **8**, 483–8.

Lima, A.M.V.M.D., Aguiar, M.E., Azevêdo, E.S. and Silva, M.C.B.O. (1981). Liver total protein in relation to cause of death in man. *Biochemical Medicine*, **25**, 92–7.

Lima, D. and Walter, N. (1955). Novos dados estatísticos sobre a distribuição do fator Rh na cidade do Recife. *Revista Clínica* (São Paulo), **31**(3/4), 63.

Lima, G.B., Capra, M.E.Z., Frantz, B.C., Leite, J.C.L. and Giugliani, R. (1996). Síndrome de Down: características clínicas, perfil epidemiológico e citogenético em recém-nascidos no Hospital de Clínicas de Porto Alegre. *Revista da Associação Médica do Rio Grande do Sul*, **40**, 8–13.

Lima, M.L.C. and Ximenes, R. (1998). Violência e morte: Diferenciais da mortalidade por causas externas no espaço urbano do Recife, 1991. *Cadernos de Saúde Pública*, **14**, 829–40.

Lisker, R. (1971). Distribution of abnormal hemoglobins in México. In Arends, T., Bemski, G. and Nagel, R.L., editors, *Genetical, Functional, and Physical Studies of Hemoglobins*. Basel: S. Karger, pp. 8–14.

Lisker, R. (1981). *Estructura Genética de la Población Mexicana. Aspectos Médicos y Antropológicos*. México: Salvat Mexicana de Ediciones.

Lisker, R. and Babinsky, V. (1986). Admixture estimates in nine Mexican Indian groups and five east coast localities. *Revista de Investigaciones Clínicas*, **38**, 145–9.

Lisker, R., Carnevale, A., Villa, J.A., Armendares, S. and Wertz, D.C. (1998). Mexican geneticists' opinions on disclosure issues. *Clinical Genetics*, **54**, 321–9.

Lisker, R., Casas, L., Mutchinick, O., López-Ariza, B. and Labardini, J. (1982a). Patient with chronic myelogenous leukemia and late appearing Philadelphia chromosome. *Cancer Genetics and Cytogenetics*, **6**, 275–7.

Lisker, R., Cervantes, G., Pérez-Briceño, R. and Alva, G. (1988a). Lack of relationship between lactose absorption and senile cataracts. *Annals of Ophthalmology*, **20**, 436–8.

Lisker, R., Cobo de Gutiérrez, A. and Velázquez-Ferrari, M. (1973). Longitudinal bone marrow chromosome studies in potential leukemic myeloid disorders. *Cancer*, **31**, 509–15.

Lisker, R., Córdova, M.S. and Zárate, Q.B.P.G. (1969). Studies on several genetic hematological traits of the Mexican population. XVI. Hemoglobin S and glucose-6-phosphate dehydrogenase deficiency in the East Coast. *American Journal of Physical Anthropology*, **30**, 349–54.

Lisker, R., López, E.T., Loría, Q.B.P.A. and Sánchez Medal, L. (1962). Drepanocitosis en 4 familias mexicanas. *Revista de Investigaciones Clínicas*, **14**, 375–90.

Lisker, R., Loría, A. and Zárate, G. (1967). Studies on several genetic hematological traits of the Mexican population. XIII. Red cell and serum polymorphisms in Spanish immigrants. *Acta Genetica et Statistica Medica*, **17**, 524–9.

Lisker, R., Mutchinick, O., Pérez-Briceño, R., *et al.* (1982*b*). Distribution of ABO blood groups and other genetic markers in mothers of infants with congenital malformations. *Human Heredity*, **32**, 166–9.

Lisker, R., Mutchinick, O., Santos, M.A., Rodarte, E. and Luft, M. (1983). Type A2 brachydactyly associated to zygodactyly in several members of a Mexican family. *Annales de Génétique*, **26**, 177–9.

Lisker, R., Noguerón, A. and Sánchez-Medal, L. (1960). Plasma thromboplastin component deficiency in the Ehlers–Danlos syndrome. *Annals of Internal Medicine*, **53**, 388–95.

Lisker, R., Pérez-Briceño, R., Granados, J. and Babinsky, V. (1988*b*). Gene frequencies and admixture estimates in the State of Puebla, Mexico. *American Journal of Physical Anthropology*, **76**, 331–5.

Lisker, R., Pérez-Briceño, R., Granados, J., *et al.* (1986). Gene frequencies and admixture estimates in a Mexico City population. *American Journal of Physical Anthropology*, **71**, 203–7.

Lisker, R., Ramirez, E., Perez Briceño, R., Granados, J. and Babinsky, V. (1990). Gene frequencies and admixture estimates in four Mexican urban centers. *Human Biology*, **62**, 791–801.

Lisker, R., Ramirez, E., Perez, G., Diaz, R., Siperstein, M. and Mutchinick, O. (1995). Genotypes of alcohol-metabolizing enzymes in Mexicans with alcoholic liver cirrhosis. *Archives of Medical Research*, **26** (Suppl.), S63–7.

Lisker, R., Ruz, L. and Mutchinick, O. (1978). 45,X/47,XYY mosaicism in a patient with Turner's syndrome. *Human Genetics*, **41**, 231–3.

Lisker, R., Solomons, N.W., Pérez-Briceño, R. and Mata, M.R. (1989). Lactase and placebo in the management of the irritable bowel syndrome: a double-blind, cross-over study. *American Journal of Gastroenterology*, **84**, 756–62.

Lisker, R., Taboada, C. and Reyes, J.L. (1964). Distribution of the ABO blood groups in peptic ulcer, gastric carcinoma and liver cirrhosis in a Mexican population. *Vox Sanguinis*, **9**, 202–3.

Little, B.B. and Malina, R.M. (1989). Genetic drift and natural selection in an isolated Zapotec-speaking community in the Valley of Oaxaca, southern México. *Human Heredity*, **39**, 99–106.

Little, B.B., Buschang, P.H. and Malina, R.M. (1988*a*). Socioeconomic variation in estimated growth velocity of schoolchildren from a rural, subsistence agricultural community in southern México. *American Journal of Physical Anthropology*, **76**, 443–8.

Little, B.B., Buschang, P.H. and Malina, R.M. (1991). Heterozygosity and craniofacial dimensions of Zapotec school children from a subsistence community in the Valley of Oaxaca, southern México. *Journal of Craniofacial Genetics and Developmental Biology*, **11**, 18–23.

Little, B.B., Malina R.M. and Buschang, P.H. (1988*b*). Increased heterozygosity and child growth in an isolated subsistence agricultural community in the Valley of Oaxaca, México. *American Journal of Physical Anthropology*, **77**,

85–90.
Little, B.B., Malina, R.M. and Buschang, P.H. (1990). Sibling similarity in annual growth increments in schoolchildren from a rural community in Oaxaca, México. *Annals of Human Biology*, **17**, 41–7.
Little, B.B., Malina, R.M., Buschang, P.H. and DeMoss, J.H. (1987). Sibling correlations for growth status in schoolchildren from a rural community in Oaxaca, México. *Annals of Human Biology*, **14**, 11–21.
Little, B.B., Malina, R.M., Buschang, P.H. and Little, L.R. (1989). Natural selection is not related to reduced body size in a rural subsistence agricultural community in southern México. *Human Biology*, **61**, 287–96.
Little, B.B., Malina, R.M., Buschang, P.H., DeMoss, J.H. and Little, L.R. (1986). Genetic and environmental effects on growth of children from a subsistence agricultural community in southern México. *American Journal of Physical Anthropology*, **71**, 81–7.
Livi-Bacci, M. (1990). Macro versus micro. In Adams, J., Lam, D.A., Hermalin, A.I. and Smouse, P.E., editors, *Convergent Issues in Genetics and Demography*. Oxford, Oxford University Press, pp. 15–25.
Livshits, G. and Kobyliansky, E. (1984). Changes in the heritability components of anthropometric characters due to preselection and environment during migration. *Human Heredity*, **34**, 348–57.
Llerena, J.C. Jr. and Cabral de Almeida, J.C. (1998). Cytogenetic and molecular contributions to the study of mental retardation. *Genetics and Molecular Biology*, **21**, 273–9.
Llop, E., Rothhammer, F. and Depetris, M. (1975). Genes amerindios y rendimiento intelectual. *Revista Médica de Chile*, **103**, 312–16.
Llop, E., Rothhammer, F., Acuña, M. and Apt, W. (1988). HLA antigens in cardiomyopathic Chilean chagasics. *American Journal of Human Genetics*, **43**, 770–3.
Loiola, G.R., Fujita, M.N. and Moura, S.B.P. (1986/87). Levantamento estatístico de parturientes na Maternidade-Escola Assis Chateaubriand (MEAC) expostas a isoimunização Rho (D). *Revista Médica da Universidade Federal do Ceará* (Fortaleza), **26/27**, 23–8.
Londoño, J.L. (1996). Pobreza, desigualdade política social e democracia. In Langoni, C.G., editor, *A Nova América Latina*. Rio de Janeiro: Fundação Getulio Vargas, pp. 127–210.
Long, J.C. (1991). The genetic structure of admixed populations. *Genetics*, **127**, 417–28.
Long, J.C. and Smouse, P.E. (1983). Intertribal gene flow between the Ye'cuana and Yanomama: genetic analysis of an admixed village. *American Journal of Physical Anthropology*, **61**, 411–22.
Long, J.C., Chakravarty, A., Boehm, C.D., Antonarakis, S. and Kazazian, H.H. (1990). Phylogeny of human β-globin haplotypes and its implications for recent human evolution. *American Journal of Physical Anthropology*, **81**, 113–30.
Long, J.C., Williams, R.C., McAuley, J.E., *et al.* (1991). Genetic variation in Arizona Mexican Americans: estimation and interpretation of admixture

proportions. *American Journal of Physical Anthropology*, **84**, 141–57.

Lopez Blanco, M., Izaguirre-Espinoza, I., Macias-Tomei, C. and Saab-Verardi, L. (1995). Growth in stature in early, average, and late maturing children of the Caracas mixed-longitudinal study. *American Journal of Human Biology*, **7**, 517–27.

Lopez Blanco, M., Jimenez, M.L. and Castellano, H.M. (1992). Urban–rural differences in the growth status of Venezuelan children. *American Journal of Human Biology*, **4**, 105–13.

Lopez-Camelo, J.S., Cabello, P.H. and Dutra, M.G. (1996). A simple model for the estimation of congenital malformation frequency in racially mixed populations. *Brazilian Journal of Genetics*, **19**, 659–63.

Lopez-Cendes, I., Teive, H.G.A., Cardoso, F., *et al.* (1997). Molecular characteristics of Machado–Joseph disease mutation in 25 newly described Brazilian families. *Brazilian Journal of Genetics*, **20**, 717–24.

Lorente, M., Lorente, J.A., Wilson, M.R., Budowle, B. and Villanueva, E. (1997). Spanish population data on seven loci: D1S80, D17S5, HUMTH01, HUMVWA, ACTBP2, D21S11, and HLA-DQA1. *Forensic Science International*, **86**, 163–71.

Lorey, F.W., Ahlfors, C.E., Smith, D.G. and Neel, J.V. (1984). Bilirubin binding by variant albumins in Yanomama Indians. *American Journal of Human Genetics*, **36**, 1112–20.

Losekoot, M., Fodde, R., van Heeren, H., Harteveld, C.L., Giordano, P.C. and Bernini, L.F. (1990). A novel frameshift mutation [FSC 47(+A)] causing β-thalassemia in a Surinam patient. *Hemoglobin*, **14**, 467–70.

Luft, S., Mello, E.D., Silveira, T.R., Dutra, J.C. and Giugliani, R. (1990). Colestase neonatal por galactosemia: relato de quatro casos. *Revista do Hospital de Clínicas de Porto Alegre*, **10**, 33–7.

Luna, F.V. and Costa, I.N. (1982). *Sinopse de Alguns Trabalhos de Demografia Histórica Referentes a Minas Gerais.* Vitória, III Encontro Nacional de Estudos Populacionais, Associação Brasileira de Estudos Populacionais.

Luna, M.C., Granados, P.A., Olek, K. and Pivetta, O.H. (1996). Cystic fibrosis in Argentina: the frequency of the ΔF508 mutation. *Human Genetics*, **97**, 314.

Macedo, A.M. and Pena, S.D.J. (1998). Genetic variability of *Trypanosoma cruzi*, implications for the pathogenesis of Chagas disease. *Parasitology Today*, **14**, 119–24.

Macedo, A.M., Martins, M.S., Chiari, E. and Pena, S.D.J. (1992*a*). DNA fingerprinting of *Trypanosoma cruzi*, a new tool for characterization of strains and clones. *Molecular and Biochemical Parasitology*, **55**, 147–54.

Macedo, A.M., Melo, M.N. and Pena, S.D.J. (1990). Identification of species of *Lishmania* by DNA fingerprinting. *Fingerprinting News*, **2**(1), 11.

Macedo, A.M., Melo, M.N., Gomes, R.F. and Pena, S.D.J. (1992*b*). DNA fingerprints. A tool for identification and determination of the relationships between species and strains of *Leishmania*. *Molecular and Biochemical Parasitology*, **53**, 63–70.

Macedo, A.M., Vallejo, G.A., Chiari, E. and Pena, S.D.J. (1993). DNA fingerprinting reveals relationships between strains of *Trypanosoma rangeli* and *Trypanosoma cruzi.* In Pena, S.D.J., Chakraborty, R., Epplen, J.T. and Jeffreys, A.J., editors, *DNA Fingerprinting: State of the Science.* Basel: Birkhäuser Verlag, pp. 321–30.

Machado, C.C. and Hakkert, R. (1988). Uma análise exploratória da informação sobre a migração inter e intramunicipal no censo demográfico de 1980. *Revista Brasileira de Estudos de População,* **5**(2), 1–20.

Machado, M.D.J. (1998). Dengue, ameaça ontem, desafio hoje. *Ciência Hoje,* **24**(139), 28–34.

Macías-Tomei, C., López-Blanco, M., Espinoza, I. and Vasquez-Ramirez, M. (2000). Pubertal development in Caracas upper-middle-class boys and girls in a longitudinal context. *American Journal of Human Biology,* **12**, 88–96.

Maciel, A., Furtado, A.F. and Marzochi, K.B.F. (1999). Perspectivas da municipalização do controle da filariose linfática na região metropolitana do Recife. *Cadernos de Saúde Pública,* **15**, 195–203.

Maciel, P., Gaspar, C., De Stefano, A.L., *et al.* (1995). Correlation between CAG repeat length and clinical features in Machado-Joseph disease. *American Journal of Human Genetics,* **57**, 54–61.

Madrigal, L. (1989). Hemoglobin genotype, fertility, and the malaria hypothesis. *Human Biology,* **61**, 311–25.

Madrigal, L. (1992). Differential sex mortality in a rural nineteenth-century population, Escazú, Costa Rica. *Human Biology,* **64**, 199–213.

Madrigal, L. (1993). Lack of birth seasonality in a nineteenth-century agricultural population, Escazú, Costa Rica. *Human Biology,* **65**, 255–71.

Madrigal, L. (1996). Sex ratio in Escazú, Costa Rica, 1851–1901. *Human Biology,* **68**, 427–36.

Madrigal, L. (1997). Twinning trend in Escazú, Costa Rica, 1851–1901. *Human Biology,* **69**, 269–76.

Madrigal, L. and Ware, B. (1997). Inbreeding in Escazú, Costa Rica (1800–1840, 1850–1899), Isonymy and ecclesiastical dispensations. *Human Biology,* **69**, 703–14.

Maertens, R., de Canck, I., Moraes, M.E., *et al.* (1998). Characterization of a new HLA-B39 allele, B*3913, in a Brazilian Caucasian. *Tissue Antigens,* **52**, 583–6.

Maestri-Filho, M. (1994). *Os Senhores do Litoral: Conquista Portuguesa e a Agonia Tupinambá.* Porto Alegre: Editora da Universidade Federal do Rio Grande do Sul.

Magalhães, A.F.N., Beiguelman, B., Pereira-Filho, R.A. and Pereira, A.S. (1980). Hereditariedade da síndrome de Gilbert. *Revista Paulista de Medicina,* **95**, 27–33.

Maio, M.C. and Santos, R.V. (1996). *Raça, Ciência e Sociedade.* Rio de Janeiro: Editora FIOCRUZ.

Malina, R.M. (1978). Adolescent growth and maturation: selected aspects of current research. *Yearbook of Physical Anthropology,* **21**, 63–94.

Malina, R.M. (1982). Human biology of urban and rural communities in the Valley of Oaxaca, México. In Wolanski, N. and Siniarska, A., editors, *Ekologia Populacji Ludzkich.* Cracow: Ossolineum, pp. 469–503.

Malina, R.M. (1983). Growth and maturity profile of primary school children in the Valley of Oaxaca, México. *Revista do Instituto de Investigação Científica Tropical* (Lisboa), **2**, 153–157.

Malina, R.M. (1985). Secular comparisons of the statures of Mexican and Mexican American children, youth and adults. *Acta Medica Auxologica*, **17**, 21–34.

Malina, R.M. (1986). Child growth and health studies in Oaxaca. *Practicing Anthropology*, **8**, 13–15.

Malina, R.M. (1989). Growth and maturation. In Serrano, C. and Salas, M.E., editors, *Estudios de Antropologia Biológica (IV Coloquio de Antropologia Física Juan Comas)*. México: Universidad Nacional Autónoma de México, pp. 55–73.

Malina, R.M. (1996). Situación de la salud cardiovascular en los niños y jóvenes latinoamericanos. *Anales de Pediatria de México* Enero-Marzo: 17–29.

Malina, R.M. (1999). Body composition research in human biology. Special issue. *American Journal of Human Biology*, **11**, 141–266.

Malina, R.M. and Buschang, P.H. (1985). Growth, strength and motor performance of Zapotec children, Oaxaca, México. *Human Biology*, **57**, 163–81.

Malina, R.M. and Himes, J.H. (1977a). Differential age effects in seasonal variation of mortality in a rural Zapotec-speaking municipio, 1945–70. *Human Biology*, **49**, 415–28.

Malina, R.M. and Himes, J.H. (1977b). Seasonality of births in a rural Zapotec municipio, 1945–70. *Human Biology*, **49**, 125–37.

Malina, R.M. and Himes, J.H. (1978). Patterns of childhood mortality and growth status in a rural Zapotec community. *Annals of Human Biology*, **5**, 517–31.

Malina, R.M. and Little, B.B. (1981). Comparison of TW1 and TW2 skeletal age differences in American Black and White and in Mexican children, 6–13 years. *Annals of Human Biology*, **8**, 543–8.

Malina, R.M. and Little, B.B. (1985). Body composition, strength, and motor performance in undernourished boys. In Binkhorst, R.A., Kemper, H.C.G. and Saris, W.H.M., editors, *Children and Exercise XI*. Champaign, IL: Human Kinetics, pp. 293–300.

Malina, R.M., Buschang, P.H., Aronson, W.L. and Selby, H.A. (1982a). Aging in selected anthropometric dimensions in a rural Zapotec-speaking community in the Valley of Oaxaca, México. *Social Science and Medicine*, **16**, 217–22.

Malina, R.M., Buschang, P.H., Aronson, W.L. and Selby, H.A. (1982b). Childhood growth status of eventual migrants and sedentes in a rural Zapotec community in the Valley of Oaxaca, México. *Human Biology*, **54**, 709–16.

Malina, R.M., Chumlea, C., Stepick, C.D. and Gutierrez Lopez, F. (1977). Age at menarche in Oaxaca, México, schoolgirls, with comparative data for other areas of México. *Annals of Human Biology*, **4**, 551–8.

Malina, R.M., Habicht, J-P., Martorell, R., Lechtig, A., Yarbrough, C. and

Klein, R.E. (1975). Head and chest circumferences in rural Guatemalan
Ladino children, birth to seven years of age. *American Journal of Clinical
Nutrition*, **28**, 1061–70.

Malina, R.M., Habicht, J-P., Yarbrough, C., Martorell, R. and Klein, R.E.
(1974). Skinfold thicknesses at seven sites in rural Guatemalan Ladino
children, birth through seven years of age. *Human Biology*, **46**, 453–69.

Malina, R.M., Himes, J.H. and Stepick, C.D. (1976). Skeletal maturity of the
hand and wrist in Oaxaca school children. *Annals of Human Biology*, **3**,
211–19.

Malina, R.M., Himes, J.H., Stepick, C.D., Gutierrez Lopez, F. and Buschang,
P.H. (1981). Growth of rural and urban children in the Valley of Oaxaca,
México. *American Journal of Physical Anthropology*, **55**, 269–80.

Malina, R.M., Katzmarzyk, P.T. and Siegel, S.R. (1998). Overnutrition,
undernutrition and the body mass index: implications for strength and
motor fitness. In Parizkova, J. and Hills, A.P., editors, *Physical Fitness and
Nutrition during Growth*. Basel: Karger, pp. 13–26.

Malina, R.M., Little, B.B. and Buschang, P.H. (1986). Sibling similarities in the
strength and motor performance of undernourished school children. *Human
Biology*, **58**, 945–53.

Malina, R.M., Little, B.B. and Buschang, P.H. (1991). Estimated body
composition and strength of chronically mild to moderately undernourished
boys in southern México. In Shephard, R.J. and Parizkova, J., editors,
Human Growth, Physical Fitness and Nutrition. Basel: Karger, pp. 119–32.

Malina, R.M., Little, B.B., Buschang, P.H., DeMoss, J. and Selby, H.A. (1985).
Socioeconomic variation in the growth status of children in a subsistence
agricultural community *American Journal of Physical Anthropology*, **68**,
385–91.

Malina, R.M., Little, B.B., Shoup, R.F. and Buschang, P.H. (1987). Adaptive
significance of small body size: strength and motor performance of school
children in México and Papua New Guinea. *American Journal of Physical
Anthropology*, **73**, 489–99.

Malina, R.M., Selby, H.A. and Swartz, L.J. (1972). Estatura, peso y
circunferencia del brazo en una muestra transversal de niños zapotecos de 6
a 14 años. *Anales de Antropología*, **9**, 143–55.

Malina, R.M., Selby, H.A., Aronson, W.L., Buschang, P.H. and Chumlea, C.
(1980a). Re-examination of the age at menarche in Oaxaca, México. *Annals
of Human Biology*, **7**, 281–2.

Malina, R.M., Selby, H.A., Buschang, P.H. and Aronson, W.L. (1980b). Growth
status of school children in a rural Zapotec community in the Valley of
Oaxaca, México, in 1968 and 1978. *Annals of Human Biology*, **7**, 367–74.

Malina, R.M., Selby, H.A., Buschang, P.H., Aronson, W.L. and Little, B.B.
(1983b). Assortative mating for phenotypic characteristics in a Zapotec
community in Oaxaca, México. *Journal of Biosocial Science*, **15**, 273–80.

Malina, R.M., Selby, H.A., Buschang, P.H., Aronson, W.L. and Wilkinson,
R.G. (1983a). Adult stature and age at menarche in Zapotec-speaking
communities in the Valley of Oaxaca, México, in a secular perspective.

American Journal of Physical Anthropology, **60**, 437–49.

Mallmann, M.C., Pinheiro, F.P., Salzano, F.M. and Ayres, M. (1970). Cytogenetic, growth rate and virus susceptibility differences in sublines of H.Ep.2 cells. *Cytologia*, **35**, 262–70.

Mansilla, J. and Pijoan, C.M. (1995). A case of congenital syphilis during the Colonial Period in México City. *American Journal of Physical Anthropology*, **97**, 187–95.

Mantovani, M.S., Ziembar, M.I., Barbieri Neto, J. and Casartelli, C. (1995). Chromosome alterations in two cases of thyroid papillary adenocarcinoma. *Revista Brasileira de Genética*, **18**, 333–7.

Maranhão, E.C. (1968). Grupos sangüíneos ABO e Rh em Curitiba (unpublished).

Marçano, A.C.B. and Richieri-Costa, A. (1998). A newly recognized autosomal dominant mandibulofacial dysostosis (Bauru type): report on a Brazilian family. *Brazilian Journal of Dysmorphology and Speech–Hearing Disorders*, **2**, 37–41.

Marcellino, A.J., Ringuelet, S., Rossi, G. and Seisdedos, L. (1980). Estudio dermatoglífico de una muestra de San Antonio de los Cobres (Provincia de Salta, República Argentina). *Publicaciones del Instituto de Antropología* (Córdoba), **35**, 25–40.

Marcílio, M.L. (1980). A população da América Latina de 1900 a 1975. *Ciência e Cultura*, **32**, 1155–76.

Marcílio, M.L. (1990). Introdução. In Nadalin, S.O., Marcílio, M.L. and Balhana, A.P., editors, *História e População. Estudos sobre a América Latina*. São Paulo: Associação Brasileira de Estudos Populacionais, pp. 1–4.

Marcílio, M.L. (1997). A demografia histórica brasileira nesse final de milênio. *Revista Brasileira de Estudos de População*, **14**(1/2), 125–43.

Marcondes, E. (1989). *Crescimento Normal e Deficiente*, 3rd edn. São Paulo: Sarvier.

Marinho, H.M. (1970). *Hemoglobinopatia S*. Rio de Janeiro: Centro de Estudos, Treinamento e Aperfeiçoamento, Secretaria de Saúde.

Marinho, H.M., Frota, M.T.T. and Coelho, M.C.S. (1972/1973). Hemofilia: distribuição racial no Estado da Guanabara. *Boletim do Instituto Estadual de Hematologia Arthur de Siqueira Cavalcanti*, **2/3**, 5–6.

Marins, V.M.R.V., Coelho, M.A.S.C., Matos, H.J., *et al.* (1995). Perfil antropométrico de crianças de 0 a 5 anos do município de Niterói, Rio de Janeiro, Brasil. *Cadernos de Saúde Pública*, **11**, 246–53.

Marks, J. (1995). *Human Biodiversity. Genes, Race, and History*. New York: Aldine de Gruyter.

Marmitt, C.R., Hutz, M.H. and Salzano, F.M. (1986). Clinical and hematological features of Hemoglobin SC disease in Rio de Janeiro, Brazil. *Brazilian Journal of Medical and Biological Research*, **19**, 731–4.

Marostica, P.J.C., Raskin, S. and Abreu-e-Silva, F.A. (1998). Analysis of the ΔF508 mutation in a Brazilian cystic fibrosis population: Comparison of pulmonary status of homozygotes with other patients. *Brazilian Journal of Medical and Biological Research*, **31**, 529–32.

Marques, A.A., Watanabe, M., Matioli, F., *et al.* (1998). DGGE (Denaturing Gradient Gel Electrophoresis) como método de 'screening' das variações do gene da G6PD. *Genetics and Molecular Biology*, **21** (Suppl.), 315.

Marques, M.N.T. and Marques, J. (1977). Dermatoglyphics in color blindness. *Acta Geneticae Medicae et Gemellologiae*, **26**, 291–2.

Marques, S.B.D., Volpini, W., Caillat-Zucman, S., Lieber, S.R., Pavin, E.J. and Persoli, L.B. (1998). Distribution of HLA-DRB1 alleles in a mixed population with insulin-dependent diabetes mellitus from the southeast of Brazil. *Brazilian Journal of Medical and Biological Research*, **31**, 365–8.

Marques Ferreira, C.S., Souza, M.M.M. and Azevêdo, E.S. (1973). ABO gene frequencies in a mixed sample of 9391 blood donors in Bahia, Brazil. *Ciência e Cultura*, **25**, 573–4.

Márquez, G.G. (1989). *O General em seu Labirinto*. Rio de Janeiro: Editora Record.

Márquez-Morfín, L. (1998). Unequal in death as in life: a sociopolitical analysis of the 1813 México City typhus epidemic. In Goodman, A.H. and Leatherman, T.L., editors, *Building a New Biocultural Synthesis. Political-Economic Perspectives on Human Biology*. Ann Arbor, MI: University of Michigan Press, pp. 229–42.

Marrelli, M.T., Malafronte, R.S. and Kloetzel, J.K. (1997). Seasonal variation of anti-*Plasmodium falciparum* antibodies directed against a repetitive peptide of gametocyte antigen Pfs2400 in inhabitants in the State of Amapá, Brazil. *Acta Tropica*, **63**, 167–77.

Martello, N., Santos, J.L.F. and Frota-Pessoa, O. (1984). Down syndrome in the different physiographic regions of Brazil. *Revista Brasileira de Genética*, **7**, 157–73.

Martine, G. (1984). Os dados censitários sobre migrações internas, Evolução e utilização. In Melo da Silva, L., editor, *Censos, Consensos, Contra-Sensos*. São Paulo: Associação Brasileira de Estudos Populacionais, pp. 183–211.

Martine, G. (1994). Estado, economia e mobilidade geográfica. Retrospectiva e perspectivas para o fim do século. *Revista Brasileira de Estudos de População*, **11**(1), 41–60.

Martine, G. and Camargo, L. (1997/1998). Growth and distribution of the Brazilian population, Recent trends. *Brazilian Journal of Population Studies*, **1**, 59–83.

Martinez, A.J., Carmenate, M.M., Bello, O., Coyula, R. and Gonzalez, O. (1989*a*). Composición corporal, somatotipo y proporcionalidad en bailarines del Ballet Nacional de Cuba. In Serrano, C. and Salas, M.E., editors, *Estudios de Antropología Biológica (IV Coloquio de Antropología Física Juan Comas)*. México, Universidad Nacional Autónoma de México, pp. 377–94.

Martinez, A.J., Carmenate, M.M. and Coyula, R. (1989*b*). Distribución anatómica de la grasa subcutanea. In Serrano, C. and Salas, M.E., editors, *Estudios de Antropología Biológica (IV Coloquio de Antropología Física Juan Comas)*. México: Universidad Nacional Autónoma de México, pp. 399–406.

Martínez, E., Bacallao, J., Devesa, M. and Amador, M. (1995). Relationship between frame size and fatness in children and adolescents. *American*

Journal of Human Biology, **7**, 1–6.

Martínez, G. and Cañizares, M.E. (1982). Genetic hemoglobin abnormalities in 2363 Cuban newborns. *Human Genetics*, **62**, 250–1.

Martínez, G., Carnot, J. and Hernandez, P. (1983). Hb Genova in a Cuban family. Clinical differences between carriers. *Hemoglobin*, **7**, 591–4.

Martínez, G., Ferreira, R., Hernandez, A., Di Rienzo, A., Felicetti, L. and Colombo, B. (1986). Molecular characterization of Hb H disease in the Cuban population. *Human Genetics*, **72**, 318–19.

Martínez, G., Ferreira, R., Hernandez, A., Di Rienzo, A., Felicetti, L. and Colombo, B. (1990). Frequency of the $-\alpha_2^{3.7}$ thalassemia deletion in the non-White Cuban population. *Gene Geography*, **4**, 65–9.

Martínez, G., Lima, F. and Colombo, B. (1977*b*). Haemoglobin J Guantanamo $(\alpha_2\beta_2 128 \text{ (H6) Ala} \rightarrow \text{Asp})$ a new fast unstable haemoglobin found in a Cuban family. *Biochimica et Biophysica Acta*, **491**, 1–6.

Martínez, G., Lima, F., Residenti, C. and Colombo, B. (1978). Hb J Camaguey α_2 [141 (HC3) Arg→Gly] β_2, a new abnormal human hemoglobin. *Hemoglobin*, **2**, 47–52.

Martínez, G., Lima, F., Wade, M., Estrada, M., Colombo, B., Heredero, L. and Granda, H. (1977*a*). Haemoglobin Porto Alegre in a Cuban family. *Journal of Medical Genetics*, **14**, 422–5.

Martinez-Fuentes, A. (1971). Croissance comparée d'Européens, d'Africains et de Cubains. *Bulletin de la Societé Royale Belge d'Anthropologie et de Préhistoire*, **82**, 121–46.

Martínez-Labarga, C., Rickards, O., Scacchi, R., *et al.* (1999). Genetic population structure of two African-Ecuadorian communities of Esmeraldas. *American Journal of Physical Anthropology*, **109**, 159–74.

Martinho, P.S., Otto, P.G., Gonzalez, C.H., Kok, F. and Otto, P.A. (1989). Dermatoglyphic studies in Rett syndrome. *Revista Brasileira de Genética*, **12**, 859–63.

Martins, C.C., Meyer, M., Seben, R.L. and Marsiglia, G.S. (1982). Estudo epidemiológico sobre pressão arterial em comunidades urbana e rural no Rio Grande do Sul. *Arquivos de Medicina Preventiva* (Porto Alegre), **5**, 30–43.

Martins, C.S.B., Ramalho, A.S. and Pinto, W. Jr. (1987). Hb S heterozigótica e tuberculose pulmonar. *Revista Brasileira de Genética*, **10**, 769–76.

Martins, C.S.B., Ramalho, A.S., Sonati, M.F., Gonçalves, M.S. and Costa, F.F. (1993). Molecular characterization of β thalassaemia heterozygotes in Brazil. *Journal of Medical Genetics*, **30**, 797–800.

Martins, C.S.B., Ribeiro, F. and Costa, F.F. (1993). Frequency of the cystic fibrosis ΔF508 mutation in a population from São Paulo State, Brazil. *Brazilian Journal of Medical and Biological Research*, **26**, 1037–40.

Martins, I.S., Velásquez-Meléndez, G. and Cervato, A.M. (1999). Estado nutricional de grupamentos sociais da área metropolitana de São Paulo, Brasil. *Cadernos de Saúde Pública*, **15**, 71–8.

Martorell, R. and Gonzáles-Cossío, T. (1987). Maternal nutrition and birth weight. *Yearbook of Physical Anthropology*, **30**, 195–220.

Martorell, R., Yarbrough, C., Himes, J.H. and Klein, R.E. (1978). Sibling similarities in number of ossification centers of the hand and wrist in a malnourished population. *Human Biology*, **50**, 73–81.

Martorell, R., Yarbrough, C., Klein, R.E. and Lechtig, A. (1979). Malnutrition, body size, and skeletal maturation: interrelationships and implications for catch-up growth. *Human Biology*, **51**, 371–89.

Martorell, R., Yarbrough, C., Malina, R.M., Habicht, J-P., Lechtig, A. and Klein, R.E. (1975). The head circumference/chest circumference ratio in mild to moderate protein-caloric malnutrition. *Environmental Child Health*, **21**, 203–7.

Massud, M., Andrade, S.M.G.M. and Uribe, M.R.N. (1978). Genética e tratamento da tuberculose. *Revista Médica*, **7**, 18–26.

Mastana, S.S., Bernal, J.E., Onyemelukwe, G.C. and Papiha, S.S. (1994). Haptoglobin subtypes among four different populations. *Human Heredity*, **44**, 10–13.

Mateu, E., Comas, D., Calafell, F., Pérez-Lezaun, A., Abade, A. and Bertranpetit, J. (1997). A tale of two islands. Population history and mitochondrial DNA sequence variation of Bioko and São Tomé, Gulf of Guinea. *Annals of Human Genetics*, **61**, 507–18.

Matiotti, M.L.V., Salzano, F.M., Ruiz, C.A. and Pereira, J.M. (1992). Congenital fascial dystrophy: further evidence of autosome recessive inheritance. *Revista Brasileira de Genética*, **15**, 695–701.

Matsudo, S.M.M. and Matsudo, V.K.R. (1994). Self-assessment and physician assessment of sexual maturation in Brazilian boys and girls: concordance and reproducibility. *American Journal of Human Biology*, **6**, 451–5.

Matsumoto, E. (1995). White sand soils in North-East Brazil. In: Nishizawa, T. and Uitto, J.I., editors, *The Fragile Tropics of Latin America. Sustainable Management and Changing Environments.* Tokyo: United Nations University Press, pp. 253–67.

Matsumoto, W.K., Vicente, M.G., Silva, M.A. and Castro, L.L.C. (1998). Comportamento epidemiológico da malária nos municípios que compõem a Bacia do Alto Paraguai, Mato Grosso do Sul, no período de 1990 a 1996. *Cadernos de Saúde Pública*, **14**, 797–802.

Mattevi, M.S. and Salzano, F.M. (1975*a*). Senescence and human chromosome changes. *Humangenetik*, **27**, 1–8.

Mattevi, M.S. and Salzano, F.M. (1975*b*). Effect of sex, age and cultivation time on number of satellites and acrocentric associations in man. *Humangenetik*, **29**, 265–70.

Mattevi, M.S. and Salzano, F.M. (1982). Effect of chromosome changes on body and mind development. *Advances in the Study of Birth Defects*, **5**, 67–87.

Mattevi, M.S., Flores, R.Z., Erdtmann, B. and Salzano, F.M. (1982). Translocation, t(4; 13) in fertile and infertile members of a family. *Revista Brasileira de Genética*, **5**, 201–7.

Mattevi, M.S., Pinheiro, C.E.A., Erdtmann, B., Flores, R.Z. and Salzano, F.M. (1981). Familial pericentric inversion of chromosome 2. *Journal de Génétique Humaine*, **29**, 161–9.

Mattevi, M.S., Wolff, H., Salzano, F.M. and Mallmann, M.C. (1971). Cytogenetic, clinical and genealogical analyses in a series of gonadal dysgenesis patients and their families. *Humangenetik*, **13**, 126–43.

Mattos, T.C., Giugliani, R. and Haase, H.B. (1987). Congenital malformations detected in 731 autopsies of children aged 0 to 14 years. *Teratology*, **35**, 305–7.

Mattos-Fiore, M.A.B. and Saldanha, P.H. (1996). Dermatoglyphics in juvenile epilepsy. I. Finger patterns and ridge counts. *Brazilian Journal of Genetics*, **19**, 151–63.

Matznetter, T. (1977). Mittelphalangalhaar bei Mischpopulationen and der Brasilküste. *Mitteilungen der Anthropologischen Gesellschaft in Wien*, **107**, 115–29.

Matznetter, T. (1978). Das Papillarsystem von Mischpopulationen der Brasilküste. *Mitteilungen der Anthropologischen Gesellschaft in Wien*, **108**, 74–114.

Maynard Smith, J. and Szathmáry, E. (1995). *The Major Transitions in Evolution.* New York: W.H. Freeman.

Mayr, E. (1942). *Systematics and the Origin of Species.* New York: Columbia University Press.

Mayr, E. (1980). Prologue: Some thoughts on the history of the evolutionary synthesis. In Mayr, E. and Provine, W.B., editors, *The Evolutionary Synthesis. Perspectives on the Unification of Biology.* Cambridge, MA: Harvard University Press, pp. 1–48.

Mayr, E. (1982). *The Growth of Biological Thought. Diversity, Evolution, and Inheritance.* Cambridge, MA: Harvard University Press.

Mazza, S., Iraeta, D. and Franke, I. (1928). Observaciones de grupo sanguíneo de recién nacidos en relación al de los padres. *Instituto de Clínica Quirúrgica*, **4**, 442–7.

Mazzella, H. (1949). Contribución al estudio de la fisiopatologia de los fenómenos por incompatibilidad sanguínea. *Anales de la Facultad de Medicina de Montevideo*, **34**, 377–442.

McBryde, F.W. and Costales, A.S. (1969). Human ecology of northwestern Colombia (The Chocó). *BioScience*, **19**, 432–6.

McCullough, J.M. and McCullough, C.S. (1974). Las creencias del síndrome de 'calor-frío' en Yucatan y su importancia para la antropologia aplicada. *Anales de Antropología*, **11**, 295–305.

McInnes, L.A., Escamilla, M.A., Service, S.K., *et al.* (1996). A complete genome screen for genes predisposing to severe bipolar disorder in two Costa Rican pedigrees. *Proceedings of the National Academy of Sciences, USA*, **93**, 13060–5.

McKeigue, P.M. (1998). Mapping genes that underlie ethnic differences in disease risk: methods for detecting linkage in admixed populations, by conditioning on parental admixture. *American Journal of Human Genetics*, **63**, 241–51.

McNally, E.M., Passos-Bueno, M.R., Bönnemann, C.G., *et al.* (1996). Mild and severe muscular dystrophy caused by a single γ-sarcoglican mutation. *American Journal of Human Genetics*, **59**, 1040–7.

Medici, A.C., Machado, M.H., Nogueira, R.P. and Girardi, S.N. (1992). *O Mercado de Trabalho em Saúde no Brasil: Estrutura e Conjuntura.* Rio de Janeiro: Fundação Oswaldo Cruz.

Medina, M.D., Vaca, G., Lopez-Guido, B., Westwood, B. and Beutler, E. (1997). Molecular genetics of glucose-6-phosphate dehydrogenase deficiency in Mexico. *Blood Cells, Molecules, and Diseases,* **23**, 88–94.

Meier, R.J. (1995). Dermatoglyphic development among Colombian Indians. *South African Journal of Science,* **91**, 466–8.

Meira, D.A., Curi, P.R. and Barraviera, B. (1984*a*). Malaria in Humaitá County, Amazon State, Brazil. XXIX. Some comparative epidemiologic aspects in 1976, 1979 and 1983. *Proceedings, IV Japan–Brazil Symposium on Science and Technology,* **2**, 136–43.

Meira, D.A., Pellegrino-Junior, J., Marcondes, J., *et al* (1984*b*). Malaria in Humaitá County, Amazonas State, Brazil. XXXII: frequency of the HLA antigen in the general population and patients. *Proceedings, IV Japan–Brazil Symposium on Science and Technology,* **2**, 162–71.

Meira, D.A., Pellegrino-Junior, J., Marcondes, J., *et al.* (1984*c*). Malaria in Humaitá County, Amazonas State, Brazil. XXXVIII: determination of HLA antigens in the family of patients prone to cerebral malaria. *Proceedings, IV Japan–Brazil Symposium on Science and Technology,* **2**, 172–6.

Meira, D.A., Pita, H.J., Barraviera, B., *et al.* (1980). Malária no município de Humaitá, Estado do Amazonas. I. Alguns aspectos epidemiológicos e clínicos. *Revista do Instituto de Medicina Tropical* (São Paulo), **22**, 124–34.

Meirelles-Nasser, C.M. (1983). Variáveis antropométricas, endocruzamento, etnia e raça em escolares de Curitiba. M.Sc. dissertation, Federal University of Paraná, Curitiba.

Meissner, R.V., Roisenberg, I. and Nardi, N.B. (1992). Production of monoclonal antibodies specific to human von Willebrand factor. *Revista Brasileira de Genética,* **15**, 935–44.

Mejia Sanchez, M. and Rosales Lopez, A. (1989). Brote dental secundario en dos poblaciones de diferentes niveles socioeconomicos de la Ciudad de Méjico. In Serrano, C. and Salas, M.E., editors, *Estudios de Antropología Biológica (IV Coloquio de Antropología Física Juan Comas).* México: Universidad Nacional Autónoma de México, pp. 203–18.

Melendez, C. (1982). *Conquistadores y Pobladores. Orígenes Histórico-Sociales de los Costarricenses.* San José: Editorial Universidad Estatal a Distancia.

Melendez, C. and Duncan, Q. (1974). *El Negro en Costa Rica.* San José: Editorial Costa Rica.

Melo e Freitas, M.J. and Salzano, F.M. (1975). Eruption of permanent teeth in Brazilian Whites and Blacks. *American Journal of Physical Anthropology,* **42**, 145–50.

Melo-dos-Santos, E.J., Ribeiro-dos-Santos, A.K.C., Guerreiro, J.F., Aguiar, G.F.S., and Santos, S.E.B. (1996). Migration and ethnic change in an admixed population from the Amazon region (Santarém, Pará). *Brazilian Journal of Genetics,* **19**, 511–15.

Meltzer, D.J. (1997). Monte Verde and the Pleistocene peopling of the Americas.

Science, **276**, 754–5.

Memoria, J.M.P. and Barbosa, M. (1957). Análise estatística dos grupos sangüíneos em Belo Horizonte. *Arquivos da Escola Superior de Veterinária* (UREMG), **10**, 299–313.

Mendes, C.P.O., Hackel, C., Arruda, V.R. and Annichino-Bizzacchi, J.M. (1999). Determination of the allele frequencies of three polymorphisms in the promoter region of the human protein C gene in three Brazilian ethnic groups. *Human Heredity*, **49**, 27–30.

Mendes de Oliveira, J.R., Otto, P.A., Vallada, H., *et al.* (1998). Analysis of a novel functional polymorphism within the promoter region of the serotonin transporter gene (5-HTT) in Brazilian patients affected by bipolar disorder and schizophrenia. *American Journal of Medical Genetics*, **81**, 225–7.

Mendez, H.M.M., Breda, D.J., Souto, C.A.V. and Salzano, F.M. (1982). Genetic and cytogenetic studies in patients with intersexuality or infertility. *Journal de Génétique Humaine*, **30**, 5–16.

Mendez, M., Sorkin, L., Rossetti, M.V., *et al.* (1998). Familial porphyria cutanea tarda: characterization of seven novel uroporphyrinogen decarboxylase mutations and frequency of common hemochromatosis alleles. *American Journal of Human Genetics*, **63**, 1363–75.

Mendez de Araújo, H.M., Salzano, F.M. and Wolff, H. (1972). New data on the association between PTC and thyroid diseases. *Humangenetik*, **15**, 136–44.

Méndez de Pérez, B. (1982). Características somatotípicas de los atletas venezolanos de alta competencia determinadas por el método de Heath-Carter. In: Villanueva, M. and Serrano, C., editors, *Estudios de Antropología Biológica (I Coloquio de Antropología Física Juan Comas)*. México: Universidad Nacional Autónoma de México, pp. 471–93.

Menegotto, B.G. and Salzano, F.M. (1990). New study on the relationship between oral clefts and fetal loss. *American Journal of Medical Genetics*, **37**, 539–42.

Menegotto, B.G. and Salzano, F.M. (1991*a*). Epidemiology of oral clefts in a large South American sample. *Cleft Palate-Craniofacial Journal*, **28**, 373–6.

Menegotto, B.G. and Salzano, F.M. (1991*b*). Clustering of malformations in the families of South American oral cleft neonates. *Journal of Medical Genetics*, **28**, 110–13.

Menéndez, C.G. (1984). Somatotipo y composición corporal de los basquetbolistas, volibolistas y sofbolistas: Juegos deportivos bolivarianos, Barquisimeto 1981. In Galván, R.R. and Ramos Rodriguez, R.M., editors, *Estudios de Antropología Biológica (II Coloquio de Antropología Física Juan Comas)*. México: Universidad Nacional Autónoma de México, pp. 467–85.

Merault, G., Keclard, L., Saint-Martin, C., *et al.* (1985). Hemoglobin Roseau-Pointe a Pitre $\alpha_2\beta_2$ 90 (F6) Glu→Gly: a new hemoglobin variant with slight instability and low oxygen affinity. *Federation of Experimental Biological Sciences (FEBS) Letters*, **184**, 10–13.

Meredith, H.V. (1971). Worldwide somatic comparisons among contemporary human groups of adult females. *American Journal of Physical Anthropology*, **34**, 89–132.

Merriwether, D.A. and Ferrell, R.E. (1996). The four founding lineage hypotheses for the New World: a critical reevaluation. *Molecular Phylogenetics and Evolution*, **5**, 241–6.

Mesa, N.R., Mondragón, M.C., Soto, I.D., *et al.* (2000). Autosomal, mtDNA and Y-chromosome diversity in Amerindians: pre- and post-Columbian patterns of gene flow in South America. *American Journal of Human Genetics*, **67**, 1277–86.

Mesquita, M.P. and Leite-Ribeiro, V. (1947). Pesquisas sobre o fator Rh na cidade do Rio de Janeiro. *O Hospital*, **32**, 505–22.

Mestriner, M.A. (1976). Estudo populacional de dois polimorfismos enzimáticos no homem: esterase D e anidrase carbônica CA II. Free docent thesis, Faculdade de Medicina, Ribeirão Preto, SP.

Miall, W.E., Milner, P.F., Lovell, H.G. and Standard, K.L. (1967). Hematological investigations of population samples in Jamaica. *British Journal of Preventive and Social Medicine*, **21**, 45–55.

Michalakis, Y. and Excoffier, L. (1996). A genetic estimation of population subdivision using distances between alleles with special reference for microsatellite loci. *Genetics*, **142**, 1061–4.

Milone, C.A., Galli, A.L., Chalar, E.M. and Pintos, L.B. (1966). Frecuencia de los grupos ABO y factor Rh D realizados sobre 3.801 investigaciones y de los antigenos Rh-Hr sobre 1.652 determinaciones en dadores de la Ciudad de La Plata. *Sumários, II Congreso Argentino de Hematologia y Hemoterapia (Córdoba, Argentina)*, 47.

Minayo, M.C.S. (1999). *Os Muitos Brasis. Saúde e População na Década de 80*. São Paulo and Rio de Janeiro: Hucitec–Abrasco.

Minayo, M.C.S. and Souza, E.R. (1993). Violência para todos. *Cadernos de Saúde Pública*, **9**, 65–78.

Minella, L.S. (1998). A produção científica sobre esterilização feminina no Brasil nos anos 80 e no início dos 90, um debate em aberto. *Revista Brasileira de Estudos de População*, **15**(1), 3–22.

Mingroni-Netto, R.C., Pavanello, R.C.M., Otto, P.A. and Vianna-Morgante, A.M. (1997). Experience with molecular and cytogenetic diagnosis of fragile X syndrome in Brazilian families. *Brazilian Journal of Genetics*, **20**, 731–9.

Mingroni-Netto, R.C., Rosenberg, C., Vianna-Morgante, A.M. and Pavanello, R.C.M. (1990). Fragile X frequency in a mentally retarded population in Brazil. *American Journal of Medical Genetics*, **35**, 22–7.

Miranda, K., Sucharov, C.C., Carvalho, A.C.C., *et al.* (1998). Distribution of angiotensin converting enzyme (ACE) I and D allele frequencies in a sample of 200 subjects of Rio de Janeiro-Brazil. *Abstracts, XXVII Reunião Anual, Sociedade Brasileira de Bioquímica e Biologia Molecular*, 15.

Miranda, S.R.P., Figueiredo, M.S., Kerbauy, J., Grotto, H.Z.W., Saad, S.T.O. and Costa, F.F. (1994*b*). Hb Lepore (Baltimore) (δ 50 Ser β 86 Ala) identified by DNA analysis in a Brazilian family. *Acta Haematologica*, **91**, 7–10.

Miranda, S.R.P., Fonseca, S.F., Figueiredo, M.S., *et al.* (1997). Hb Köln [$\alpha_2\beta_2$ 98 (FG5) val-met] identified by DNA analysis in a Brazilian family. *Brazilian*

Journal of Genetics, **20**, 745–8.

Miranda, S.R.P., Kimura, E.M., Saad, S.T.O. and Costa, F.F. (1994*a*). Identification of Hb Zurich [α(2)β(2) 63(E7) His→Arg] by DNA analysis in a Brazilian family. *Hemoglobin*, **18**, 337–41.

Miranda, S.R.P., Kimura, E.M., Teixeira, R.C., *et al.* (1996). Hb Camperdown [α₂β2104(G6) Arg→Ser] identified by DNA analysis in a Brazilian family. *Hemoglobin*, **20**, 147–54.

Misquiatti, A.R.N., Abramides, D.V.M., Giacheti, C.M., Feniman, M.R., Passos-Bueno, M.R. and Richieri-Costa, A. (1998). Speech/language findings in patients with Apert, Crouzon and Pfeiffer syndromes. *Brazilian Journal of Dysmorphology and Speech–Hearing Disorders*, **2**, 29–35.

Molina, I. and Palmer, S. (1998). *The History of Costa Rica*. San José, Editorial de la Universidad de Costa Rica.

Monplaisir, N., Valette, I. and Bach, J-F. (1986*a*). HLA antigens in 88 cases of rheumatic fever observed in Martinique. *Tissue Antigens*, **28**, 209–13.

Monplaisir, N., Valette, I., Pierre-Louis, S., *et al.* (1988). Study of HLA antigens in systemic lupus erythematosus in the French West Indies. *Tissue Antigens*, **31**, 238–42.

Monplaisir, N., Cassius de Linval, J., Sellaye, M., *et al.* (1981). Dépistage des hemoglobinopathies à la naissance par isoélectrofocalisation. Étude de la population de la Martinique. *Nouvelle Presse Médicale*, **10**, 3127–30.

Monplaisir, N., Galacteros, F., Arous, N., *et al.* (1985). Hémoglobines anormales identifiées à la Martinique. *Nouvelle Revue Française d'Hématologie*, **27**, 11–14.

Monplaisir, N., Merault, G., Poyart, C., *et al.* (1986*b*). Hemoglobin S Antilles: a variant with lower solubility than hemoglobin S and producing sickle cell disease in heterozygotes. *Proceedings of the National Academy of Sciences, USA*, **83**, 9363–7.

Monsalve, M.V. and Hagelberg, E. (1997). Mitochondrial DNA polymorphisms in Carib people of Belize. *Proceedings of the Royal Society of London, Series B*, **264**, 1217–24.

Monsalve, M.V., Erdtmann, B., Otto, P.A. and Frota-Pessoa, O. (1980). The human Y chromosome: racial variation and evolution. *Revista Brasileira de Genética*, **3**, 433–46.

Monsalve, M.V., Espinel, A., Groot de Restrepo, H., Calvo, M., Suarez, M.C. and Rodriguez, A. (1987). Frequency of five genetic polymorphisms in two populations of Colombia. *Revista Brasileira de Genética*, **10**, 247–51.

Montagu, M.F.A. (1960). *An Introduction to Physical Anthropology*. Springfield, IL: C.C. Thomas.

Monteiro, C., Rueff, J., Falcão, A.B., Portugal, S., Weatherall, D.J. and Kulozik, A.E. (1989). The frequency and origin of the sickle cell mutation in the district of Coruche/Portugal. *Human Genetics*, **82**, 255–8.

Monteiro, C.A. (1995). *Velhos e Novos Males da Saúde no Brasil. A Evolução do País e de suas Doenças*. São Paulo: Hucitec-Nupens/USP.

Monteiro, C.A. (1997). Cai o índice de desnutrição infantil em São Paulo. *Notícias FAPESP*, **24**, 6–7.

Monteiro, C.A., Benício, M.H.D., Zuñiga, H.P.P. and Szarfarc, S.C. (1986*b*). Estudo das condições de saúde das crianças do município de São Paulo, SP (Brasil), 1984–85. II. Antropometria nutricional. *Revista de Saúde Pública*, **20**, 446–53.

Monteiro, C.A., D'Aquino Benicio, M.H. and Gouveia, N.C. (1994). Secular growth trends in Brazil over three decades. *Annals of Human Biology*, **21**, 381–90.

Monteiro, C.A., Zuñiga, H.P.P., Benício, M.H.D. and Szarfarc, S.C. (1986*a*). Estudos das condições de saúde das crianças do município de São Paulo, SP (Brasil), 1984–85. I. Aspectos metodológicos, características sócio-econômicas e ambiente físico. *Revista de Saúde Pública*, **20**, 435–45.

Montemayor, F. (1961). Intento de apreciación del mestizaje en algunos grupos mexicanos. *Anales del Instituto Nacional de Antropología y Historia*, **13**, 149–76.

Montemayor, F. (1984). Los datos antropométricos de grupos masculinos mexicanos. In Ramos Galván, R. and Ramos Rodriguez, R.M., editors, *Estudios de Antropología Biológica (II Coloquio de Antropologia Física Juan Comas)*. México: Universidad Nacional Autónoma de México, pp. 87–144.

Montemayor, F. (1987). Afinación del analisis de los datos antropométricos de 35 grupos mexicanos. In Faulhaber, M.E.S. and Cruchaga, X.L., editors, *Estudios de Antropología Biológica (III Coloquio de Antropología Física Juan Comas)*. México: Universidad Nacional Autónoma de México, pp. 225–47.

Montenegro, J. (1925). Classificação e transfusão de sangues. *Brasil Médico*, **39**, 109–13.

Montenegro, L. (1960). Frecuencia de los grupos sanguíneos del sistema ABO y del factor D (Rh_0) en Manaus. *Sangre*, **5**, 191–6.

Montestruc, E. and Ragusin, E. (1944). Classification des groupes sanguins en Martinique et en Guadeloupe. *Comptes Rendus, Societé de Biologie de Paris*, **139**, 764–5.

Montestruc, E., Berdonneau, R., Benoist, J. and Collet, A. (1959). Hémoglobines anormales et groupes sanguins A, B, O, chez las Martiniquais. *Bulletin de la Societé de Pathologie Exotique*, **52**, 156–8.

Moore, L.G., Niermeyer, S. and Zamudio, S. (1998). Human adaptation to high altitude: regional and life-cycle perspectives. *Yearbook of Physical Anthropology*, **41**, 25–64.

Moraes, J.R., Moraes, M.E., Fernandez-Viña, M., *et al.* (1991). HLA antigens and risk for development of pemphigus foliaceus (fogo selvagem) in endemic areas of Brazil. *Immunogenetics*, **33**, 388–91.

Moraes, M.E., Fernandez-Viña, M., Lazaro, A., *et al.* (1997). An epitope in the third hypervariable region of the DRB1 gene is involved in the susceptibility to endemic pemphigus foliaceus (fogo selvagem) in three different Brazilian populations. *Tissue Antigens*, **49**, 35–40.

Moraes, M.E., Fernandez-Viña, M., Salatiel, I., Tsai, S., Moraes, J.R. and Stastny, P. (1993). HLA Class II DNA typing in two Brazilian populations. *Tissue Antigens*, **41**, 238–42.

Moreira, E.S., Vainzof, M., Mari, S.K., Nigro, V., Zatz, M. and Passos-Bueno,

M.R. (1998). A first missense mutation in the δ sarcoglican gene associated with a severe phenotype and frequency of limb-girdle muscular dystrophy type 2F (LGMD2F) in Brazilian sarcoglycanopathies. *Journal of Medical Genetics*, **35**, 951–3.

Moreira, L.M.A., Freitas, L.M., Peixoto, L.I.S., Sousa, M.G.F. and Souza, M.L.B. (1995). A case of cri-du-chat syndrome with karyotype del(5)(p14),+mar. *Revista Brasileira de Genética*, **18**, 125–8.

Moreira, M.M. (1998). O envelhecimento da população brasileira, intensidade, feminização e dependência. *Revista Brasileira de Estudos de População*, **15**(1), 79–94.

Morera, B. (1995). Caracterización étnica de la población Costarricense mediante Marcadores Genéticos. M.Sc. dissertation, San José: Universidad de Costa Rica.

Morgan, O.St C., Rodgers-Johnson, P., Mora, C. and Char, G. (1989). HTLV–1 and polymyositis in Jamaica. *The Lancet*, **II**, 1184–7.

Morris, L., Lewis, G., Powell, D.L., Anderson, J., Way, A., Cushing, J. and Lawless, G. (1981). Pesquisas sobre saúde materno-infantil e planejamento familiar, Uma nova fonte de dados sobre o planejamento familiar. *Population Reports, (Series M)*, **5**, 1–42.

Mortara, G. (1947). Os fatores demográficos do crescimento das populações americanas nos últimos cem anos. In Mortara, G., editor, *Pesquisas Sobre Populações Americanas*. Rio de Janeiro: Fundação Getúlio Vargas, pp. 51–70.

Mortara, G. (1951). O aumento da população do Brasil entre 1872 e 1940. In IBGE. *Estudos de Estatística Teórica e Aplicada. Pesquisas sobre o Desenvolvimento da População do Brasil*. Rio de Janeiro: Instituto Brasileiro de Geografia e Estatística, pp. 50–62.

Morton, N.E. and Barbosa, C.A.A. (1981). Age, area, and acheiropody. *Human Genetics*, **57**, 420–2.

Mota, T.M.B., Pasin, I.P. and Gaida, F. (1963). Grupos sanguíneos e fator Rh em Santa Maria, Rio Grande do Sul. *Revista da Faculdade de Farmária de Santa Maria*, **9**, 17–20.

Moura, M.M.F., Feitosa, M.F., Laredo-Filho, J. and Krieger, H. (1996). The inheritance of factors associated with joint mobility. *Genetic Epidemiology*, **13**, 403–9.

Mourant, A.E. (1954). *The Distribution of the Human Blood Groups*. Springfield, IL: C.C. Thomas.

Mourant, A.E., Kopec, A.C. and Domaniewska-Sobczak, K. (1976). *The Distribution of the Human Blood Groups and Other Polymorphisms*. London: Oxford University Press.

Mourão, L.A.C.B. and Salzano, F.M. (1978). New data on the association between PTC tasting and tuberculosis. *Revista Brasileira de Biologia*, **38**, 475–9.

Mueller, W.H. (1979). Fertility and physique in a malnourished population. *Human Biology*, **51**, 153–66.

Mueller, W.H. (1982). The changes with age of the anatomical distribution of fat.

Social Science and Medicine, **16**, 191–6.

Mueller, W.H. (1983). The genetics of human fatness. *Yearbook of Physical Anthropology*, **26**, 215–30.

Mueller, W.H. and Reid, R.M. (1979). A multivariate analysis of fatness and relative fat patterning. *American Journal of Physical Anthropology*, **50**, 199–208.

Mueller, W.H., Lasker, G.W. and Evans, F.G. (1981). Anthropometric measurements and Darwinian fitness. *Journal of Biosocial Science*, **13**, 309–16.

Muller, H.J. (1914). A gene for the fourth chromosome of *Drosophila*. *Journal of Experimental Zoology*, **17**, 325–36.

Munford, D., Zanini, M.C. and Neves, W.A. (1995). Human cranial variation in South America: implications for the settlement of the New World. *Brazilian Journal of Genetics*, **18**, 673–88.

Muniz, A., Corral, L., Alaez, C., *et al.* (1995). Sickle cell anemia and β-gene cluster haplotypes in Cuba. *American Journal of Hematology*, **49**, 163–84.

Muniz, A., Martinez, G., Lavinha, J. and Pacheco, P. (2000). β-thalassaemia in Cubans: novel allele increases the genetic diversity at the HBB locus in the Caribbean. *American Journal of Hematology*, **64**, 7–14.

Muniz, Y.C.N., Rosa, F.C., Maegawa, F.B. and Souza, I.R. (1998). Novos dados referentes às comunidades isoladas da ilha de Santa Catarina: sul do Brasil. *Genetics and Molecular Biology*, **21** (Suppl. 3), 355.

Murguia, R., Dickinson, F., Cervera, M. and Uc, L. (1991). Socio-economic activities, ecology and somatic differences in Yucatan, México. *Studies in Human Ecology*, **9**, 111–34.

Mutchinick, O.M., Shaffer, L.G., Kashork, C.D. and Cervantes, E.I. (1999). Miller–Dieker syndrome and trisomy 5p in a child carrying a derivative chromosome with a microdeletion in 17p13.3 telomeric to the *LIS1* and *D17S379* loci. *American Journal of Medical Genetics*, **85**, 99–104.

Nadalin, S.O., Marcílio, M.L. and Balhana, A.P. (1990). *História e População. Estudos sobre a América Latina*. São Paulo: Fundação Sistema Estadual de Análise de Dados.

Naoum, P.C. (1987). *Diagnóstico das Hemoglobinopatias*. São Paulo: Sarvier.

Naoum, P.C. (1997). *Hemoglobinopatias e Talassemias*. São Paulo: Sarvier.

Naoum, P.C., Domingos, C.R.B., Mazziero, P.A., Castilho, E.M. and Gomes, C.T. (1986). Hemoglobinopatias no Brasil. *Boletim da Sociedade Brasileira de Hematologia*, **8**, 180–8.

Naoum, P.C., Machado, P.E.A. and Michelim, O.C. (1976). Detection of haemoglobin D Punjab in patient with Sydenham's Chorea. *Revista Brasileira de Pesquisas Médicas e Biológicas*, **9**, 273–8.

Naoum, P.C., Mattos, L.C. and Curi, P.R. (1984). Prevalence and geographic distribution of abnormal hemoglobins in the State of São Paulo, Brazil. *Bulletin of the Pan American Health Organization*, **18**, 127–38.

Navajas, L. and Vianna-Morgante, A.M. (1989). Relationship between age and mental status and the expression of the fragile (X) in heterozygotes for the Martin–Bell syndrome. *Revista Brasileira de Genética*, **12**, 391–404.

Navarrete, J.I., Lisker, R. and Pérez-Briceño, R. (1979). Serum atypical pseudocholinesterase and leprosy. *International Journal of Dermatology*, **18**, 822–3.

Nazer, J. and Valenzuela, C.Y. (1973). Posibles efectos biológicos de los anticonceptivos. *Revista Médica de Chile*, **101**, 234–6.

Neel, J.V. (1949). The inheritance of sickle cell anemia. *Science*, **110**, 64–6.

Neel, J.V., Biggar, R.J. and Sukernik, R.I. (1994). Virologic and genetic studies relate Amerind origins to the indigenous people of the Mongolia/Manchuria/southeastern Siberia region. *Proceedings of the National Academy of Sciences, USA*, **91**, 10737–41.

Nei, M. (1973). Analysis of gene diversity in subdivided populations. *Proceedings of the National Academy of Sciences, USA*, **70**, 3321–3.

Nei, M. (1986). Definition and estimation of fixation indices. *Evolution*, **40**, 643–5.

Nei, M. (1987). *Molecular Evolutionary Genetics*. New York: Columbia University Press.

Nei, M., Tajima, F. and Tateno, Y. (1983). Accuracy of estimated phylogenetic trees from molecular data. *Journal of Molecular Evolution*, **19**, 153–70.

Neves, W.A. and Costa, M.A. (1998). Adult stature and standard of living in the prehistoric Atacama desert. *Current Anthropology*, **39**, 278–81.

Neves, W.A., Meyer, D. and Pucciarelli, H.M. (1996). Early skeletal remains and the peopling of the Americas. *Revista de Antropologia*, **39**, 121–39.

Niddrie, D.L. (1985). The Caribbean. In Blakemore H. and Smith, C.T., editors, *Latin America. Geographical Perspectives*. London: Methuen, pp. 77–132.

Niederman, J.C., Henderson, J.R., Opton, E.M., Black, F.L. and Skvrnova, K. (1967). A nationwide serum survey of Brazilian military recruits, 1964. II. Antibody patterns with arboviruses, polioviruses, measles and mumps. *American Journal of Epidemiology*, **86**, 319–29.

Nielsen, H., Kragh-Hansen, U., Minchiotti, L., *et al.* (1997). Effect of genetic variation on the fatty acid-binding properties of human serum albumin and proalbumin. *Biochimica et Biophysica Acta*, **1342**, 191–204.

Nijenhuis, L.E. and Gemser-Runia, J. (1965). Hereditary and acquired blood factors in the negroid population of Surinam. III. Blood group studies. *Tropical and Geographical Medicine*, **17**, 69–79.

Nishizawa, T. and Uitto, J.I. (1995). *The Fragile Tropics of Latin America: Sustainable Management of Changing Environments*. Tokyo: United Nations University Press.

Novaes, H.M. (1997). Los hospitales en América Latina y el Caribe. *Foro Mundial de la Salud*, **18**, 54–8.

Novais, M. (1953). Grupos sangüíneos na população de Salvador (Bahia). *O Hospital*, **43**, 471–80.

Nunes Moreira, J.A. (1963). Análise estatística dos grupos sangüíneos humanos em Fortaleza. *Revista Brasileira de Biologia*, **23**, 355–60.

Nunes, V., Gasparini, P., Novelli, G., *et al.* (1991). Analysis of 14 cystic fibrosis mutations in five South European populations. *Human Genetics*, **87**, 737–8.

Nunesmaia, H.G. and Azevêdo, E.S. (1973). Unusual frequency of Negro admixture in necropsies of Chagas disease cases in Bahia, Brazil. *Revista do Instituto de Medicina Tropical*, **15**, 10–13.

Nunesmaia, H.G., Azevêdo, E.S., Arandas, E.A. and Widmer, C.G. (1975). Composição racial e anaptoglobinemia em portadores de esquistossomose mansônica forma hepatesplênica. *Revista do Instituto de Medicina Tropical*, **17**, 160–3.

Nuñez Montiel, A., Perez, K.A., Montilla, L.G.P. and Ferrer, A. (1962). Estudios hematológicos sobre la población de la isla de Toas (Estado Zulia). *Acta Científica Venezolana*, **13**, 94–6.

O'Gorman, E. (1986). *A Invenção da América. Reflexão a Respeito da Estrutura Histórica do Seu Devir*. São Paulo: Editora da UNESP.

Ochoa-Díaz López, H., Sánchez-Pérez, H.J., Ruíz-Flores, M. and Fuller, M. (1999). Social inequalities and health in rural Chiapas, México: agricultural economy, nutrition, and child health in La Fraylesca Region. *Cadernos de Saúde Pública*, **15**, 261–70.

Oda, L.S., Iyomasa, M.M. and Watanabe, I-S. (1977). Morphologic analysis of the 'spina mentalis' in adult mandibles of Brazilian whites and negroes. *Revista Brasileira de Pesquisas Médicas e Biológicas*, **10**, 357–60.

Oh, J., Ho, L., Ala-Mello, S., et al. (1998). Mutation analysis of patients with Hermansky – Pudlak syndrome: A frameshift hot spot in the *HPS* gene and apparent locus heterogeneity. *American Journal of Human Genetics*, **62**, 593–8.

Oldeman, R.A.A., Clement, C.R. and Kabala, D.M. (1993). Food and the future of the tropical forest: Management alternatives: background. In Hladik, C.M., Hladik, A., Linares, O.F., Pagezy, H., Semple, A. and Hadley, M., editors, *Tropical Forests, People and Food. Biocultural Interactions and Applications to Development*. Paris: UNESCO, and New York: Parthenon, pp. 685–90.

Oliveira, E.K. (1954). Fissuras Congênitas. Thesis submitted to obtain the Chair of Buccal-Facial Protesis and Surgery, School of Odontology, Pelotas, Universidade Federal do Rio Grande do Sul.

Oliveira, J.M. (1977). Grupos sangüíneos e doença coronária. *Medicina de Hoje*, **3**, 766–76.

Oliveira, J.R.M. and Zatz, M. (1999). The study of genetic polymorphisms related to serotonin in Alzheimer's disease: a new perspective in a heterogenic disorder. *Brazilian Journal of Medical and Biological Research*, **32**, 463–7.

Oliveira, L.A.P. and Simões, C.C.S. (1988). As informações sobre fecundidade, mortalidade e anticoncepção nas PNADs. In Sawyer, D.O., editor, PNADs em Foco: Anos 80. Belo Horizonte, Associação Brasileira de Estudos Populacionais, pp. 183–225.

Oliveira, M.A., Bermudez, J.A.Z. and Souza, A.C.M. (1999). Talidomida no Brasil: vigilância com responsabilidade compartilhada? *Cadernos de Saúde Pública*, **15**, 99–112.

Oliveira, M.C.V.C. (1983). Incidência de imunização eritrocitária em mulheres, na Região Metropolitana do Recife-PE. *Revista de Pediatria de Pernambuco*,

1, 165–9.

Oliveira, M.C.V.C., Oliveira, A.M.A. and Salzano, F.M. (1983). ABO and Rh$_o$ gene frequencies in the Metropolitan Region of Recife, State of Pernambuco, Brazil. *Revista Brasileira de Genética*, **6**, 375–80.

Oliveira, M.P.M.S. and Azevêdo, E.S. (1977). Racial differences in anthropometric traits in school children of Bahia, Brazil. *American Journal of Physical Anthropology*, **46**, 471–5.

Oliveira, R.C. and Castro Faria, L. (1971). Interethnic contact and the study of populations. In Salzano, F.M., editor, *The Ongoing Evolution of Latin American Populations*. Springfield, IL: C.C. Thomas, pp. 41–59.

Oliveira, R.P., Broude, N.E., Macedo, A.M., Cantor, C.R., Smith, C.L. and Pena, S.D.J. (1998). Probing the genetic population structure of *Trypanosoma cruzi* with polymorphic microsatellites. *Proceedings of the National Academy of Sciences, USA*, **95**, 3776–80.

Olivo, E.A. and Ablan, E. (1999). Principales características de la disponibilidad de energía alimentaria en Venezuela en el período 1970–96. *Interciencia*, **24**, 308–16.

Olsson, M.L., Guerreiro, J.F., Zago, M.A. and Chester, M.A. (1997). Molecular analysis of the 0 alleles at the blood group ABO locus in populations of different ethnic origin reveals novel crossing-over events and point mutations. *Biochemical and Biophysical Research Communications*, **234**, 779–82.

Olsson, M.L., Santos, S.E.B., Guerreiro, J.F., Zago, M.A. and Chester, M.A. (1998). Heterogeneity of the 0 alleles at the blood group ABO locus in Amerindians. *Vox Sanguinis*, **74**, 46–50.

Omran, H., Fernandez, C., Jung, M., *et al.* (2000). Identification of a new gene locus for nephronophthisis, on chromosome 3q22 in a large Venezuelan pedigree. *American Journal of Human Genetics*, **66**, 118–27.

Öner, P.D., Dimovski, A.J., Olivieri, N.F., *et al.* (1992). β^S haplotypes in various world populations. *Human Genetics*, **89**, 99–104.

Orioli, I.M. (1995). Segregation distortion in the offspring of Afro-American fathers with postaxial polydactyly. *American Journal of Human Genetics*, **56**, 1207–11.

Orioli, I.M., Castilla, E.E. and Carvalho, W.P. (1982). Inbreeding in a South American newborn series. *Acta Anthropogenetica*, **6**, 45–55.

Orrillo, M., Descailleaux, J., Pereda, J., Hermoza, P. and Alvarado, E. (1976). Numerical X-chromosome anomalies at the Maternity Hospital of Lima. *Journal de Génétique Humaine*, **24**, 221–6.

Ortiz, L.P. (1990). As condições de saúde de mães e crianças do Ceará (Avaliação através de um sistema de informações de maternidades). *Revista Brasileira de Estudos de População*, **7**, 54–73.

Osorio, F., Amaral, M. and Maulaz, P. (1957). Indice bioquimico racial em Pelotas. *Revista da Faculdade de Odontologia de Pelotas*, **3**, 101–8.

Osório, H. (1996). Estruturas socio-econômicas coloniais. In: Wasserman, C., editor, *História da América Latina: Cinco Séculos (Temas e Problemas)*. Porto Alegre: Editora da Universidade Federal do Rio Grande do Sul, pp.

38–76.

Ottensooser, F. (1944). Cálculo do grau de mistura racial através dos grupos sangüíneos. *Revista Brasileira de Biologia*, **4**, 531–7.

Ottensooser, F. (1962). Analysis of trihybrid populations. *American Journal of Human Genetics*, **14**, 278–80.

Ottensooser, F. and Pasqualin, R. (1946). O fator Rh na mortalidade fetal e neonatal. *Arquivos de Biologia* (São Paulo), **30** (271), 20–8.

Ottensooser, F., Leon, N. and Almeida, T.V. (1975). ABO blood groups and isoagglutinins in systemic lupus erythematosus. *Revista Brasileira de Pesquisas Médicas e Biológicas*, **8**, 421–6.

Ottensooser, F., Nakamizo, Y., Sato, M., Miyamoto, Y. and Takizawa, K. (1974). Lectins detecting group C streptococci. *Infection and Immunity*, **9**, 971–3.

Otto, P.A. and Otto, P.G. (1980). The importance of A'-d ridge count in dermatoglyphic diagnosis of the Ulrich–Turner syndrome. *American Journal of Medical Genetics*, **6**, 145–52.

Otto, P.A., Vieira, J., Filho and Marques, S.A. (1989). Comparative analysis of dermatoglyphic indices used for diagnosis of Down's syndrome. *Revista Brasileira de Genética*, **12**, 145–59.

Paes-Alves, A.F., Azevêdo, E.S., Sousa, M.G.F., Almeida-Melo, N. and Oliveira-Filho, O.J. (1991). Autosomal recessive malformation syndrome with minor manifestation in the heterozygotes: a preliminary report of a possible new syndrome. *American Journal of Medical Genetics*, **41**, 141–52.

Paez Perez, C. and Freudenthal, K. (1944). Grupos sanguineos de los indios Sibundoy, Santiaguenos, Kuaiker e indios y mestizos de los alrededores de Pasto. *Revista del Instituto Etnológico Nacional*, **1**, 411–15.

Pagni, D. and Melaragno, M.I. (1998). Cytogenetic study of women with premature ovarian failure. *Genetics and Molecular Biology*, **21**, 263–6.

Pagnier, J., Mears, J.G., Dunda-Belkhodja, O., *et al.* (1984). Evidence for the multicentric origin of the sickle cell hemoglobin gene in Africa. *Proceedings of the National Academy of Sciences, USA*, **81**, 1771–3.

Paiva e Silva, R.B. and Ramalho, A.S. (1997). Riscos e benefícios da triagem genética: o traço falciforme como modelo de estudo em uma população brasileira. *Cadernos de Saúde Pública*, **13**, 285–94.

Palatnik, M. (1966). Seroantropología argentina. *Sangre*, **11**, 395–412.

Palatnik, M. (1976). Genética demográfica de una comunidad criolla (Fortín Lavalle, El Chaco, Argentina). *Mendeliana*, **1**, 17–25.

Palatnik, M., Soares, M.B.M., Junqueira, P.C., Corvelo, T.C.O., Soares, V.M.F.C. and Faria, G.M.F. (1982). Phenotype A_3B in two generations of a Brazilian Negroid family. *Revista Brasileira de Genética*, **5**, 165–74.

Palavecino, E.A., Mota, A.H., Awad, J., *et al.* (1990). HLA and celiac disease in Argentina: Involvement of the DQ subregion. *Disease Markers*, **8**, 5–10.

Palazzo, R., Tenconi, J. (1939). Estadística sobre 15 000 clasificaciones de grupo sanguíneo, realizadas en Buenos Aires. *La Semana Médica*, **46**, 459–60.

Palomino, H., Cerda-Flores, R.M., Blanco, R., *et al.* (1997*b*). Complex segregation analysis of facial clefting in Chile. *Journal of Craniofacial*

Genetics and Developmental Biology, **17**, 57–64.

Palomino, H.M., Palomino, H., Cauvi, D., Barton, S. and Chakraborty, R. (1997*a*). Facial clefting and Amerindian admixture in populations of Santiago, Chile. *American Journal of Human Biology*, **9**, 225–32.

Pang, S. and Clark, A. (1993). Congenital hyperplasia due to 21-hydroxylase deficiency: newborn screening and its relationship to the diagnosis and treatment of the disorder. *Screening*, **2**, 105–39.

Pantaleão, S.M., Conceição, M.M., Silva, G.F. and Lopes, C.R.F.R. (1998). Análise de hemoglobinopatias estruturais em doadores de sangue do HEMOSE. *Anais, XIII Encontro de Genética do Nordeste*, 412.

Pante-de-Souza, G., Mousinho-Ribeiro, R.C., Santos, E.J.M. and Guerreiro, J.F. (1999). *β*-globin haplotypes analysis in Afro-Brazilians from the Amazon region: evidence for a significant gene flow from Atlantic West Africa. *Annals of Human Biology*, **26**, 365–73.

Pante-de-Souza, G., Mousinho-Ribeiro, R.C., Santos, E.J.M., Zago, M.A. and Guerreiro, J.F. (1998). Origin of the hemoglobin S gene in a northern Brazilian population: the combined effects of slave trade and internal migrations. *Genetics and Molecular Biology*, **21**, 427–30.

Papiha, S.S. (1996). Genetic variation in India. *Human Biology*, **68**, 607–28.

Pareira, K.S., Goulart, L.R. and Rezende, W.L. (1998). Frequency of the CKR–5 mutant allele in the Brazilian population. *Abstracts, XXVII Reunião Anual, Sociedade Brasileira de Bioquímica e Biologia Molecular*, 16.

Parra, E.J., Marcini, A., Akey, J., *et al.* (1998). Estimating African American admixture proportions by use of population-specific alleles. *American Journal of Human Genetics*, **63**, 1839–51.

Passos, L.N.M., Marques, H.O., Naoum, P.C., Carrel, R.W. and Williamson, R.T. (1998). Hb Mainz [Beta 98 (FG5) Val-Glu]: hemoglobina instável identificada em paciente com anemia hemolítica em Manaus-AM. *Boletim da Sociedade Brasileira de Hematologia e Hemoterapia*, **20** (Suppl.), 98.

Passos, V.M.A., Calazans, F.F. and Carneiro-Proietti, A.B.F. (1998). Counseling blood donors seropositive for human T-lymphotropic virus types I and II in a developing country. *Cadernos de Saúde Pública*, **14**, 417–20.

Passos-Bueno, M.R. (1999). Molecular studies in Brazilian patients with muscular dystrophies and craniofacial disorders. *Ciência e Cultura*, **51**, 218–25.

Passos-Bueno, M.R., Cerqueira, A., Vainzof, M., Marie, S.K. and Zatz, M. (1995). Myotonic dystrophy: genetic, clinical, and molecular analysis of patients from 41 Brazilian families. *Journal of Medical Genetics*, **32**, 14–18.

Passos-Bueno, M.R., Moreira, E.S., Marie, S.K., *et al.* (1996*a*). Main clinical features of the three mapped autosomal recessive limb-girdle muscular dystrophies and estimated proportion of each form in 13 Brazilian families. *Journal of Medical Genetics*, **33**, 97–102.

Passos-Bueno, M.R., Moreira, E.S., Vainzof, M., Marie, S.K. and Zatz, M. (1996*b*). Linkage analysis in autosomal recessive limb-girdle muscular dystrophy (AR LGMD) maps a sixth form to 5q33–34 (LGMD2F) and indicates that there is at least one more subtype of AR LGMD. *Human*

Molecular Genetics, **5**, 815–20.

Passos-Bueno, M.R., Sertié, A.L., Richieri-Costa, A., *et al.* (1998). Description of a new mutation and characterization of *FGFR1*, *FGFR2*, and *FGFR3* mutations among Brazilian patients with syndromic craniosynostoses. *American Journal of Medical Genetics*, **78**, 237–41.

Passos-Bueno, M.R., Vainzof, M., Moreira, E.S. and Zatz, M. (1999*b*). Seven autosomal recessive limb-girdle muscular dystrophies in the Brazilian population: From LGMD2A to LGMD2G. *American Journal of Medical Genetics*, **82**, 392–8.

Passos-Bueno, M.R., Wilcox, W.R., Jabs, E.W., Sertié, A.L., Alonso, L.G. and Kitoh, H. (1999*a*). Clinical spectrum of fibroblast growth factor receptor mutations. *Human Mutation*, **14**, 115–25.

Pastuszak, A.L., Schüler, L., Speck-Martins, C.E., *et al.* (1998). Use of misoprostol during pregnancy and Möbius syndrome in infants. *New England Journal of Medicine*, **338**, 1881–5.

Patarra, N.L. (1996). Migrações internacionais, Uma nova questão demográfica. *Revista Brasileira de Estudos de População*, **13**(1), 111–13.

Paterniani, E. and Malavolta, E. (1999). La conquista del 'cerrado' en el Brazil. Victoria de la investigación científica. *Interciencia*, **24**, 173–6.

Patiño Salazar, R. and de Mello, R.P. (1948). Determinação dos grupos sangüíneos e Rh em brancos, pretos e mulatos do Estado de Minas Gerais. *Revista Brasileira de Biologia*, **8**, 169–72.

Pauling, L., Itano, H.A., Singer, S.J. and Wells, I.C. (1949). Sickle cell anemia: a molecular disease. *Science*, **110**, 543–8.

Paulino, L.C., Araujo, M., Guerra, G. Jr., Marini, S.H.V.L. and de Mello, M.P. (1999). Mutation distribution and CYP21/C4 locus variability in Brazilian families with the classical form of the 21-hydroxylase deficiency. *Acta Paediatrica*, **88**, 275–83.

Pavan, C. and Brito da Cunha, A. (1968). *A Energia Atômica e o Futuro do Homem*. São Paulo: Companhia Editora Nacional and Editora da Universidade de São Paulo.

Pavelka, M.S.M. and Fedigan, L.M. (1991). Menopause: a comparative life history perspective. *Yearbook of Physical Anthropology*, **34**, 13–38.

Paz y Miño, C. (1991). Patologia genética y cromosómica en el Hospital de Niños Baca Ortiz. *Boletin Científico, Fundación Simon Bolivar*, **3**, 2–11.

Paz y Miño, C. (1997). La genética médica en el Ecuador: situación y perspectivas de desarrollo. *Brazilian Journal of Genetics*, **20** (Suppl. 1), 55–62.

Peceguini, M.C. and Cabello, P.H. (1986). Alguns aspectos demográficos de uma população rural do nordeste brasileiro. *Ciência e Cultura*, **38**, 776–82.

Pedreira, C.M. (1954). Fatores Rh-Hr (Aspectos de Sua Pesquisa na Bahia). Ph.D. thesis, Faculdade de Medicina, Universidade da Bahia, Salvador, BA.

Pedrollo, E., Hutz, M.H. and Salzano, F.M. (1990). Alpha thalassemia frequency in newborn children from Porto Alegre, Brazil. *Revista Brasileira de Genética*, **13**, 573–81.

Pedron, C.G., Gaspar, P.A., Giugliani, R. and Pereira, M.L.S. (1999). Arylsulfatase A pseudodeficiency in healthy Brazilian individuals. *Brazilian*

Journal of Medical and Biological Research, **32**, 941–5.

Pedrosa, M.P., Salzano, F.M., Mattevi, M.S. and Viégas, J. (1983). Quantitative analysis of C-bands in chromosomes 1, 9, 16, and Y of twins. *Acta Geneticae Medicae et Gemellologiae*, **32**, 257–60.

Pedrosa, M.P., Silva, R., Vasconcelos, A.P.C., Oliveira, R.M.A., Cavalcante, C.N. and Guerreiro, J.F. (1995*b*). Variabilidade genética em isolados negróides do Estado de Alagoas. II. Quilombo. *Revista Brasileira de Genética*, **18** (Suppl. 3), 346.

Pedrosa, M.P., Souza, M.C.G., Vasconcelos, A.P.C., Silva, R. and Guerreiro, J.F. (1995*a*). Variabilidade genética em isolados negróides do Estado de Alagoas. I. Muquén. *Revista Brasileira de Genética*, **18** (Suppl.), 346.

Peetoom, F., Crommelin, S., Fontijn, A. and Prins, H.K. (1965). Hereditary and acquired blood factors in the negroid population of Surinam. VI. Serum group-specific components and transferrin types. *Tropical and Geographic Medicine*, **17**, 243–5.

Pena, H.B., Souza, C.P., Simpson, A.J.G. and Pena, S.D.J. (1995). Intracellular promiscuity in *Schistosoma mansoni*, Nuclear transcribed DNA sequences are part of a mitochondrial minisatellite region. *Proceedings of the National Academy of Sciences, USA*, **92**, 915–19.

Peña, H.F., Salzano, F.M. and Callegari, S.M. (1973). Dermatoglyphics in twins. *Acta Geneticae Medicae et Gemellologiae*, **22**, 91–8.

Peña, M.E., Cárdenas, E. and Olmo, J.L. (1984). Crecimiento y maduración osea en deportistas preadolescentes y adolescentes. In Galván, R.R. and Ramos Rodríguez, R.M., editors, *Estudios de Antropología Biológica (II Coloquio de Antropología Física Juan Comas)*. México: Universidad Nacional Autónoma de México, pp. 452–66.

Pena, S.D.J., Carvalho-Silva, D.R., Alves-Silva, J. and Prado, V.F. (2000). Retrato molecular do Brasil. *Ciência Hoje*, **27**(159), 16–25.

Pena, S.D.J., Souza, K.T., Andrade, M. and Chakraborty, R. (1994). Allelic associations of two polymorphic microsallites in intron 40 of the human von Willebrand factor gene. *Proceedings of the National Academy of Sciences, USA*, **91**, 723–7.

Peñaloza, R., García-Carranca, A., Ceras, T., *et al.* (1995). Frequency of haplotypes in the beta globin gene cluster in a selected sample of the Mexican population. *American Journal of Human Biology*, **7**, 45–9.

Peñalver, J.A. and de Miani, M.S.A. (1974). Variantes de hemoglobinopatias observadas en nuestro medio. *Medicina* (Buenos Aires), **34**, 287–92.

Penchaszadeh, V.B. and Beiguelman, B. (1997). Serviços de genética médica na América Latina: estado atual e perspectivas. *Brazilian Journal of Genetics,*, **20** (Suppl.), 1–174.

Penhalber, E.F., Barco, L.D., Maestrelli, S.R.P. and Otto, P.A. (1994). Dermatoglyphics in a large normal sample of Caucasoids from southern Brazil. *Revista Brasileira de Genética*, **17**, 197–214.

Penney, J.B. Jr., Young, A.B., Shoulson, I., *et al.* (1990). Huntington's disease in Venezuela: 7 years of follow-up on symptomatic and asymptomatic individuals. *Movement Disorders*, **5**, 93–9.

Perea, F.J., Casas-Castañeda, M., Villalobos-Arámbula, A.R., et al. (1999b). HbD-Los Angeles associated with HbS or β-thalassemia in four Mexican Mestizo families. Hemoglobin, 23, 231–7.

Perea, F.J., Zamudio, G., Meillon, L.A. and Ibarra, B. (1999a). The Hb Tarrant [α126(H9)Asp→Asn] mutation is localized in the α2-globin gene. Hemoglobin, 23, 295–7.

Pereira, C.A.S., Priore, S.E., Costa, T.M.B., Fernandes, R.M.G. and Del Nero Fragoso, S.H. (1985). Avaliação antropométrica de crianças do Ciclo Básico do Município de Bauru, SP. Salusvita, 4, 58–63.

Pereira, E.T., Cabral de Almeida, J.C., Cunha, A.C.Y.R.G., Patton, M., Taylor, R. and Jeffery, S. (1991). Use of probes for ZFY, SRY and the Y pseudoautosomal boundary in XX males, XX true hermaphrodites, and an XY female. Journal of Medical Genetics, 28, 591–5.

Pereira, E.T., Cabral de Almeida, J.C., Patton, M. and Jeffery, S. (1996). Clinical and molecular studies in a mother and son with Xp22.3 deletion. Brazilian Journal of Genetics, 19, 359–64.

Pereira, J.M., Constancio, W.F., Callado, A.N., Lajchter, D. and Póvoa, H. (1971). Serum levels of immunoglobulins in Brazilian people. Acta Biologica Medica Germanica, 27, 1005–6.

Pereira, L., Raskin, S., Freund, A.A. et al. (1999). Cystic fibrosis mutations R1162X and 2183AA→G in two southern Brazilian states. Genetics and Molecular Biology, 22, 291–4.

Pereira, M.B., Conceição, G.C., Coelho, J.C., Wajner, M. and Giugliani, R. (1997). Detection of metabolic disorders in high-risk patients. Arquivos de Neuropsiquiatria, 55, 209–12.

Pereira, M.G. and Albuquerque, Z.P. (1983). Características da mortalidade na infância no Distrito Federal. Revista da Associação Médica Brasileira, 29, 47–51.

Pereira, R.A., Sichieri, R. and Marins, V.M.R. (1999). Razão cintura/quadril como preditor de hipertensão arterial. Cadernos de Saúde Pública, 15, 333–44.

Pereira-Lima, J.E., Utz, E. and Roisenberg, I. (1966). Hereditary nonhemolytic conjugated hyperbilirubinemia without abnormal liver cell pigmentation. A family study. American Journal of Medicine, 40, 628–33.

Pérez, B., Desviat, L.R., de Lucca, M., Schmidt, B., Loghin-Grosso, N., Giugliani, R., Pires, R.F. and Ugarte, M. (1996). Mutation analysis of phenylketonuria in South Brazil. Human Mutation, 8, 262–4.

Perez de Gianella, T. (1997). La genética médica en el Perú. Brazilian Journal of Genetics, 20 (Suppl. 1), 91–9.

Pérez-Bravo, F., Carrasco, E., Gutierrez-López, M.D., Martínez, M.T., López, G. and García de los Rios, M. (1996). Genetic predisposition and environmental factors leading to the development of insulin-dependent diabetes mellitus in Chilean children. Journal of Molecular Medicine, 74, 105–9.

Pérez-Rojas, G.E., Paul-Moya, H., Bianco, N.E. and Abadí, I. (1984). Seronegative spondyloarthropathies and HLA antigens in a Mestizo population. *Tissue Antigens*, **23**, 107–11.

Perlingeiro, R.C.R., Costa, F.F., Saad, S.T.O., Arruda, V.R. and Queiroz, M.L.S. (1999). Spontaneous erythroid colony formation in Brazilian patients with sickle cell disease. *American Journal of Hematology*, **61**, 40–5.

Pettener, D., Fiorini, S. and Tarazona-Santos, E. (1997). Within-lineage repeated pair isonymy (RPw) in a high altitude Quechua community in the Peruvian Central Andes, 1825–1914. *Revista Española de Antropología Biológica*, **18**, 25–37.

Pettener, D., Pastor, S. and Tarazona-Santos, E. (1998). Surnames and genetic structure of a high-altitude Quechua community from the Ichu River Valley, Peruvian Central Andes, 1825–1914. *Human Biology*, **70**, 865–87.

Petzl-Erler, M.L. (1999). Genetics of the immune responses and disease susceptibility. *Ciência e Cultura*, **51**, 199–211.

Petzl-Erler, M.L. and Santamaria, J. (1989). Are HLA Class II genes controlling susceptibility and resistance to Brazilian pemphigus foliaceus (fogo selvagem)? *Tissue Antigens*, **33**, 408–14.

Petzl-Erler, M.L., Belich, M.P. and Queiroz-Telles, F. (1991). Association of mucosal leishmaniasis with HLA. *Human Immunology*, **32**, 254–60.

Pezzi, A.R., Mazzieri, R. and Martinez, J.A. (1951). Estatisticas de grupos sanguíneos. *Archivos Clínicos* (Rio de Janeiro), **12**, 564–5.

Piccini, R.X. and Victora, C.G. (1997). How well is hypertension managed in the community? A population-based survey in a Brazilian city. *Cadernos de Saúde Pública*, **13**, 595–600.

Pik, C., Loos, J.A., Jonxis, J.H.P. and Prins, H.K. (1965). Hereditary and acquired blood factors in the negroid population of Surinam. II. The incidence of haemoglobin anomalies and the deficiency of glucose-6-phosphate dehydrogenase. *Tropical and Geographical Medicine*, **17**, 61–8.

Pilotto, R.F. (1978). Estudo Genético-Clínico da Síndrome de Ellis-van Creveld. M.Sc. dissertation. Universidade Federal do Paraná, Curitiba.

Pina-Neto, J.M. and Petean, E.B.L. (1999). Genetic counseling follow-up: a retrospective study with a quantitative approach. *Genetics and Molecular Biology*, **22**, 295–307.

Pina-Neto, J.M., Defino, H.L.A., Guedes, M.L. and Jorge, S.M. (1996). Spondyloepimetaphyseal dysplasia with joint laxity (SEMDJL): a Brazilian case. *American Journal of Medical Genetics*, **61**, 131–3.

Pina-Neto, J.M., Vieira de Souza, N., Velludo, M.A.S.L., Perosa, G.B.D., Freitas, M.M.S. and Colafêmina, J.F. (1998). Retinal changes and tumorigenesis in Ramon syndrome: follow-up of a Brazilian family. *American Journal of Medical Genetics*, **77**, 43–6.

Pineda, L., Pinto-Cisternas, J. and Arias, A. (1985). Consanguinity in Colonia Tovar, a Venezuela isolate of German origin (1843–1977). *Journal of Human Evolution*, **14**, 587–96.

Pineda, L., Pinto-Cisternas, J. and Barrai, I. (1984). Consanguinidad por isonimia en poblaciones iberoamericanas, Extension del método de Crow y Mange. b) Su aplicación a una poblacion venezolana. In Rodríguez

Lemoine, V., editor, *Genética. VI Congreso Latinoamericano de Genética. I Congreso Venezolano de Genética.* Maracaibo: Asociación Latinoamericana de Genética, pp. 315–16.

Pinheiro, M. and Freire-Maia, N. (1985). Atrichias and hypotrichoses: a brief review with description of a recessive atrichia in two brothers. *Human Heredity*, **35**, 53–5.

Pinheiro, M. and Freire-Maia, N. (1994). Ectodermal dysplasias: a clinical classification and a causal review. *American Journal of Medical Genetics*, **53**, 153–62.

Pinheiro, M.F., Pontes, M.L., Hughet, E., Gené, M., Pinto da Costa, J. and Moreno, P. (1996). Study of three AMPFLPs (D1S80, 3' ApoB and YNZ22) in the population of North Portugal. *Forensic Science International*, **79**, 23–9.

Pinto, D.O., Borges, S.O. and Lacaz, C.S. (1955). Distribuição dos fatores M e N em São Paulo. *Arquivos de Biologia* (São Paulo), **39**(323), 85–8.

Pinto, F., González, A.M., Hernández, M., Larruga, J.M. and Cabrera, V.M. (1996). Genetic relationship between the Canary Islanders and their African and Spanish ancestors inferred from mitochondrial sequences. *Annals of Human Genetics*, **60**, 321–30.

Pinto, L.I.B., Paskulin, G.A., Graziadio, C. and Mendez, H.M.M. (1996). The frequency of genetic diseases in a high risk ward in a pediatric hospital. *Brazilian Journal of Genetics*, **19**, 145–9.

Pinto, W., Jr. and Beiguelman, B. (1971). Twin pair analysis of some palm indexes. *Acta Geneticae Medicae et Gemellologiae*, **20**, 295–300.

Pinto-Cisternas, J. and Figueroa, H. (1968). Genetic structure of a population of Valparaiso. II. Distribution of two dental traits with anthropological importance. *American Journal of Physical Anthropology*, **29**, 339–48.

Pinto-Cisternas, J., Arvelo, H., Martínez, R. and Castro de Guerra, D. (1997). Coefficient of relationship (RI) within and between four Black Venezuelan populations. *International Journal of Anthropology*, **12**, 55–62.

Pinto-Cisternas, J., Castelli, M.C. and Pineda, L. (1981). La consanguinidad en la parroquia de Los Teques, Venezuela desde 1790 a 1869. *Acta Científica Venezolana*, **32**, 262–8.

Pinto-Cisternas, J., Figueroa, H., Lazo, B., Salinas, C. and Campusano, C. (1971*a*). Genetic structure of the population of Valparaiso. V. ABO blood groups, color vision deficiency and their relationship to other variables. *Human Heredity*, **21**, 431–9.

Pinto-Cisternas, J., Lazo, B., Campusano, C. and Ballesteros, S. (1977). Some determinants of mating structure in a rural zone of Chile, 1810–1959. *Social Biology*, **24**, 234–44.

Pinto-Cisternas, J., Pineda, L. and Barrai, I. (1985). Estimation of inbreeding by isonymy in Iberoamerican populations: an extension of the method of Crow and Mange. *American Journal of Human Genetics*, **37**, 373–85.

Pinto-Cisternas, J., Rodriguez-Larralde, A. and Castro de Guerra, D. (1990*b*). Comparison of two Venezuelan populations using the coefficient of relationship by isonymy. *Human Biology*, **62**, 413–19.

Pinto-Cisternas, J., Salinas, C., Campusano, C., Figueroa, H. and Lazo, B. (1971*b*). Preliminary migration data on a population of Valparaiso, Chile. *Social Biology*, **18**, 305–10.

Pinto-Cisternas, J., Zimmer, E. and Barrai, I. (1990*a*). Comparisons of Lasker's coefficient of relationship in a Venezuelan town in two different periods. *Annals of Human Biology*, **17**, 305–14.

Plaseska-Karanfilska, D., Weinstein, B.I. and Efremov, G.D. (2000). Hb Rambam [β69(E13)Gly→Asp]/β⁰-thalassemia [codon 5(-CT)] in a family from Argentina. *Hemoglobin*, **24**, 157–61.

PNAD. (1976). *Pesquisa Nacional por Amostra de Domicílios*. Rio de Janeiro: Instituto Brasileiro de Geografia e Estatística.

Pollitzer, W.S., Azevêdo, E.S., Barefoot, J., *et al.* (1982). Characteristics of a population sample of Jacobina, Bahia, Brazil. *Human Biology*, **54**, 697–707.

Pollitzer, W.S., Barefoot, J., Azevêdo, E.S. and Lima, A.M.V.D. (1990). Populations, pigmentation, physique and surnames in northeastern Brazil: a factor analysis. In Lara Tapia, L., editor, *Para Conocer el Hombre. Homenaje a Santiago Genovés*. México: Universidad Autónoma de México, pp. 317–21.

Pospísil, M.F. (1969). El peso y la talla de los escolares de la Ciudad de La Habana. *Memorias de la Facultad de Ciencias, Universidad de La Habana, Serie Ciencias Biológicas,*, **3**, 59–87.

Pospisil, M.F. (1977). Dermatoglyphics of the Cuban population. *Acta Facultatis Rerum Naturalium Universitatis Comenianae Anthropologia*, **25**, 139–63.

Pospisil, M.F. (1980). Palmar dermatoglyphics of the Cuban population. *Acta Facultatis Rerum Naturalium Universitatis Comenianae Anthropologia*, **27**, 149–69.

Pospisil, M.F. (1984). Correlation analysis of the finger ridge count in Cuban population. *Acta Facultatis Rerum Naturalinm Universitatis Comenianae Anthropologia*, **32/33**, 65–79.

Post, G.B., Kemper, H.C.G., Welten, D.C. and Coudert, J. (1997). Dietary pattern and growth of 10–12-year-old Bolivian girls and boys: Relation between altitude and socioeconomic status. *American Journal of Human Biology*, **9**, 51–62.

Potter, J.D. (1997). *Food, Nutrition and the Prevention of Cancer: a Global Perspective*. Washington, D.C.: American Institute for Cancer Research.

Pourchet, M.J. (1957/1958). Antropologia física de escolares. *Revista de Educação Pública*, Janeiro-Dezembro, 11–29.

Prado, R.A.R. (1992). *Curriculum Vitae del Estudio Colaborativo Latinoamericano de Malformaciones Congénitas (ECLAMC), 1967–1992.* Rio de Janeiro: Universidade Federal do Rio de Janeiro.

Prata, P.R. (1992). A transição epidemiológica no Brasil. *Cadernos de Saúde Pública*, **8**, 168–75.

Pratesi, R., Santos, M. and Ferrari, I. (1998). Costello syndrome in two Brazilian children. *Journal of Medical Genetics*, **35**, 54–7.

Praxedes, H. and Lehmann, H. (1972). Haemoglobin Niteroi: a new unstable variant. *Abstracts, XIV International Congress of Hematology*, 400.

Prolla, J.C. and Dietz, J. (1985). Dados epidemiológicos do câncer de mama feminina no Rio Grande do Sul, Brasil. Mortalidade no período 1970–1982, incidência em 1980–1981. *Revista da Associação Médica do Rio Grande do Sul*, **29**, 217–23.

Prolla, J.C., Achutti, A.C. and Dietz, J. (1985). Freqüência relativa da mortalidade por neoplasias malignas no Rio Grande do Sul, Brasil; 1970–1980. *Revista da Associação Médica do Rio Grande do Sul*, **29**, 208–13.

Prous, A. (1995). Archeological analysis of the oldest settlements in the Americas. *Brazilian Journal of Genetics*, **18**, 689–99.

Province, M.A. and Rao, D.C. (1985). Path analysis of family resemblance with temporal trends: applications to height, weight, and Quetelet index in northeastern Brazil. *American Journal of Human Genetics*, **37**, 178–92.

Pucciarelli, H.M., Carnese, F.R., Pinotti, L.V., Guimarey, L.M. and Goicoechea, A.S. (1993). Sexual dimorphism in schoolchildren of Villa IAPI neighborhood (Quilmes, Buenos Aires, Argentina). *American Journal of Physical Anthropology*, **92**, 165–72.

Puffer, R.R. and Griffith, G.W. (1967). *Patterns of Urban Mortality.* Washington, DC: Pan American Health Organization.

Puffer, R.R. and Serrano, C.V. (1973). *Patterns of Mortality in Childhood.* Washington, DC: Pan American Health Organization.

Pugliese, L., Arruda, V.R. and Annichino-Bizzacchi, J.M. (1999). A novel nonsense mutation 6,E-X in the protein S gene causes type I deficiency. *Human Heredity*, **49**, 121–2.

Quarleri, J.F., Robertson, B.H., Mathet, V., *et al.* (1998). Genomic and phylogenetic analysis of hepatitis C virus strains from Argentina. *Medicina* (Buenos Aires), **58**, 153–9.

Quezada Díaz, J.E., Roisenberg, I. and Fischer, R.R. (1988). A study of VIIIC and VIIICAg in normals and hemophilia A patients in a Brazilian population. *Revista Brasileira de Genética*, **11**, 193–202.

Quezada, M. and Barrantes, R. (1973). Analisis de la endogamia en una población chilena. *Revista Médica de Chile*, **101**, 227–30.

Rabb, L. and Serjeant, G. (1982). Sickle-cell-hemoglobin Caribbean: a benign syndrome. *Hemoglobin*, **6**, 403–5.

Rabbi-Bortolini, E., Bernardino, A.L.F., Lopes, A.L., Ferri, A.S., Passos-Bueno, M.R. and Zatz, M. (1998). Sweat electrolyte and cystic fibrosis mutation analysis allows early diagnosis in Brazilian children with clinical signs compatible with cystic fibrosis. *American Journal of Medical Genetics*, **76**, 288–90.

Raff, R.A. and Kaufman, T.C. (1983). *Embryos, Genes and Evolution.* New York: MacMillan.

Raffin, C.N., Giugliani, R. and Armbrust-Figueiredo, J. (1980). Genetic predisposition to febrile convulsions. *Arquivos de Neuro-Psiquiatria*, **38**, 327–30.

Rahm, G. (1931). Blutgruppen in Südamerika. Eine Reise nach dem Feuerland. *Die Umschau*, **35**, 777–81.

Rainha de Souza, I., Creplive, S., Susin, M.F., Barros, M.S. and Alves, L. Jr.

(1995). Diversidade genética da comunidade isolada de Witmarsum, no sul do Brasil. *Revista Brasileira de Genética*, **18** (Suppl.), 352.

Ramalho, A.S. (1986*a*). *As Hemoglobinopatias Hereditárias. Um Problema de Saúde Pública no Brasil.* Ribeirão Preto: Sociedade Brasileira de Genética.

Ramalho, A.S. (1986*b*). A talassemia minor como causa de anemia no Estado de São Paulo. *Revista Brasileira de Patologia Clínica*, **22**, 32–8.

Ramalho, A.S. and Beiguelman, B. (1977). Sickle cell trait and tuberculosis. *Ciência e Cultura*, **29**, 1149–51.

Ramalho, A.S., Jorge, R.N., Oliveira, J.A. and Pereira, D.A. (1976). Hemoglobina 'S' em recém-nascidos brasileiros. *Jornal de Pediatria*, **41**, 22–4.

Ramalho, A.S., Magna, L.A., Costa, F.F. and Grotto, H.Z.W. (1985*b*). Talassemia menor: um problema de saúde pública no Brasil? *Revista Brasileira de Genética*, **8**, 747–54.

Ramalho, A.S., Pinto, W. Jr., Magna, L.A. and Beiguelman, B. (1983). Talassemia e hanseníase. *A Folha Médica*, **86**, 461–4.

Ramalho, A.S., Silva, R.B.P., Teixeira, R.C. and Compri, M.B. (1999). Hemoglobin screening: Response of a Brazilian community to optional programs. *Cadernos de Saúde Pública*, **15**, 591–5.

Ramalho, A.S., Velloso, L.A. and Diniz, M. (1985*a*). Síndromes falcêmicas e úlceras de membros inferiores. *Anais Brasileiros de Dermatologia*, **60**, 307–10.

Ramalho, A.T. and Nascimento, A.C.H. (1991). The fate of chromosomal aberrations in ^{137}Cs-exposed individuals in the Goiânia radiation accident. *Health Physics*, **60**, 67–70.

Ramalho, A.T., Nascimento, A.C.H. and Natarajan, A.T. (1988). Dose assessments by cytogenetic analysis in the Goiânia (Brazil) radiation accident. *Radiation Protection Dosimetry*, **25**, 97–100.

Ramalho, J.P. (1975). Edição e coordenação geral. *Arquivos de Anatomia e Antropologia*, **1**, 1–658.

Ramalho, J.P. (1977). Edição e coordenação geral. *Arquivos de Anatomia e Antropologia*, **2**, 1–168.

Ramalho, J.P. (1978). Edição e coordenação geral. *Arquivos de Anatomia e Antropologia*, **3**, 1–535.

Ramalho, J.P. (1979/80). Edição e coordenação geral. *Arquivos de Anatomia e Antropologia*, **4/5**, 1–604.

Ramalho, R.A., Anjos, L.A. and Flores, H. (1998). Hipovitaminose A em recém-nascidos em duas maternidades públicas no Rio de Janeiro, Brasil. *Cadernos de Saúde Pública*, **14**, 821–7.

Ramos, E.S., Moreira-Filho, C.A., Vicente, Y.A.M.V.A., *et al.* (1996). SRY-negative true hermaphrodites and an XX male in two generations of the same family. *Human Genetics*, **97**, 596–8.

Ramos Galván, R. (1982). Dimorfismo sexual en la composición corporal. Un analisis somatometrico. In Villanueva, M. and Serrano, C., editors, *Estudios de Antropología Biológica (I Coloquio de Antropología Física Juan Comas)*. México: Universidad Nacional Autónoma de México, pp. 433–60.

Ramos Rodriguez, R.M. (1987). Valor predictivo de los segmentos de la talla. Estudio en Cuentepec, Morelos. In Faulhaber, M.E.S. and Lizárraga Cruchaga, X., editors, *Estudios de Antropología Biológica (III Coloquio de Antropología Física Juan Comas)*. México: Universidad Nacional Autónoma de México, pp. 57–84.

Ramos Rodriguez, R.M. (1989). Lo biológico y lo social en el crecimiento físico. In Serrano, C. and Salas, M.E., editors, *Estudios de Antropología Biológica (IV Coloquio de Antropología Física Juan Comas)*. México: Universidad Nacional Autónoma de México, pp. 107–14.

Ramsay, M. and Jenkins, T. (1987). Globin gene-associated restriction-fragment-length polymorphisms in Southern African peoples. *American Journal of Human Genetics*, **41**, 1132–44.

Rao, D.C., MacLean, C.J., Morton, N.E. and Yee, S. (1975). Analysis of family resemblance. V. Height and weight in northeastern Brazil. *American Journal of Human Genetics*, **27**, 509–20.

Rapaport, D., Colletto, G.M.D., Vainzof, M. and Zatz, M. (1991). Estimates of genetic and environmental components of serum isocitrate dehydrogenase (ICDH) in normal twins. *Acta Geneticae Medicae et Gemellologiae*, **40**, 77–82.

Raposo-do-Amaral, C.M., Krieger, H., Cabello, P.H. and Beiguelman, B. (1989). Heritability of quantitative orbital traits. *Human Biology*, **61**, 551–7.

Raskin, S., Phillips, J.A. III, Kaplan, G., *et al.* (1999). Geographic heterogeneity of 4 common worldwide cystic fibrosis non-DF508 mutations in Brazil. *Human Biology*, **71**, 111–21.

Raskin, S., Phillips, J.A. III, Krishnamani, M.R.S., *et al.* (1997*a*). Regional distribution of cystic fibrosis-linked DNA haplotypes in Brazil: Multicenter study. *Human Biology*, **69**, 75–88.

Raskin, S., Phillips, J.A. III, Krishnamani, M.R.S., *et al.* (1997*b*). Cystic fibrosis in the Brazilian population: DF508 mutation and KM–19/XV–2C haplotype distribution. *Human Biology*, **69**, 499–508.

Raskin, S., Phillips, J.A.III, Krishnamani, M.R.S., *et al.* (1993). DNA analysis of cystic fibrosis in Brazil by direct PCR amplification from Guthrie cards. *American Journal of Medical Genetics*, **46**, 665–9.

Reader, J. (1998). *Africa. A Biography of the Continent*. New York: Knopf.

Rebelatto, C.L.K. (1996). Estudo de Associação entre os Antígenos HLA-A, B, C, DR e DQ e a Paracoccidioide Infecção e a Forma Crônica de Paracoccidioide Doença. M.Sc. thesis, Federal University of Paraná, Caritiba.

Reher, D.S. (1997). Desafios e conquistas da demografia histórica no final do século. *Revista Brasileira de Estudos de População*, **14**(1/2), 101–24.

Restrepo, A. (1965). Distribucion geográfica de las hemoglobinas anormales. *Boletin del Instituto de Antropologia, Universidad de Antioquia*, **3**(9), 63–6.

Restrepo, A. (1971). Frequency and distribution of abnormal hemoglobins and thalassemia in Colombia, South America. In Arends, T., Bemski, G. and Nagel, R.L., editors, *Genetical, Functional, and Physical Studies of Hemoglobins*. Basel: S. Karger, pp. 39–52.

References   461

Restrepo, A., Palacio, S. and Forero, J.M. (1965). Frecuencia de los grupos sanguineos ABO y RHO en poblacion mixta de la ciudad de Medellin (Antioquia) y, en negros de la ciudad de Quibdo (Choco) y revision de la literatura colombiana. *Boletin del Instituto de Antropologia, Universidad de Antioquia*, **3**, 53–62.
Restrepo, A.M. and Gutierrez, E. (1968). The frequency of glucose-6-phosphate dehydrogenase deficiency in Colombia. *American Journal of Human Genetics*, **20**, 82–5.
Restrepo, M., Rojas, W., Montoya, F., Montoya, A.E. and Dawson, D.V. (1988). HLA and malaria in four Colombian ethnic groups. *Revista do Instituto de Medicina Tropical*, **30**, 323–31.
Reyes, M.E.P., Malina, R.M., Little, B.B. and Buschang, P. (1995). Consumo de alimentos en una comunidad rural zapoteca en el Valle de Oaxaca. In Rodríguez, R.M.R. and Alonso, S.L., editors, *Estudios de Antropología Biológica*, Vol. 5. México: Universidad Nacional Autónoma de México, pp. 407–14.
Ribeiro, D. (1970). The culture-historical configurations of the American peoples. *Current Anthropology*, **11**, 403–34.
Ribeiro, D. (1977). *As Américas e a Civilização*. Petrópolis: Editora Vozes.
Ribeiro, D. (1997). *O Povo Brasileiro. A Formação e o Sentido do Brasil*. São Paulo: Companhia das Letras.
Ribeiro, E., Pinheiro, H.C.F. and Kessler, R.G. (1998). Nonimmune hydrops fetalis associated with chromosomal trisomy 21: a case report. *Brazilian Journal of Dysmorphology and Speech-Hearing Disorders*, **2**, 3–6.
Ribeiro, E.B. (1963). Os grupos sangüíneos ABO e o câncer do estômago em São Paulo. *Anais Paulistas de Medicina e Cirurgia*, **86**, 87–95.
Ribeiro, L.R., Cavalli, I.J., Fontoura, E., Jr., Sbalqueiro, I.J., Maia, N.A. and Muniz, E.C.N. (1982). The human Y chromosome: racial variation. *Revista Brasileira de Genética*, **5**, 217–20.
Ribeiro, R.C., Sandrini Neto, R., Schell, M.J., Sambaio, G.A. and Cat, I. (1990). Adrenocortical carcinoma in children: a study of 40 cases. *Journal of Clinical Oncology*, **8**, 67–74.
Ribeiro-dos-Santos, A.K.C. (1993). A influência de marcadores genéticos na epidemiologia do virus da hepatite B. M.Sc. dissertation, Federal University of Pará, Belém.
Ribeiro-dos-Santos, A.K.C., Santos, E.J.M., Guerreiro, J.F. and Santos, S.E.B. (1995). Demographic and genetic structure of the population of Castanhal, in the Amazon region of Brazil. *Brazilian Journal of Genetics*, **18**, 469–74.
Richard, I., Roudaut, C., Saenz, A., *et al.* (1999). Calpainopathy: a survey of mutations and polymorphisms. *American Journal of Human Genetics*, **64**, 1524–40.
Richards, M., Côrte-Real, H., Forster, P., *et al.* (1996). Paleolithic and neolithic lineages in the European mitochondrial gene pool. *American Journal of Human Genetics*, **59**, 185–203.
Richieri-Costa, A. and Gorlin, R.J. (1994). Oblique facial clefts: report on 4 Brazilian patients. Evidence for clinical variability and genetic heterogeneity.

American Journal of Medical Genetics, **53**, 222–6.

Richieri-Costa, A., Miranda, E., Kamiya, T.Y. and Freire-Maia, D.V. (1990). Autosomal dominant tibial hemimelia-polysyndactyly-triphalangeal thumbs syndrome: report of a Brazilian family. *American Journal of Medical Genetics*, **36**, 1–6.

Rife, D.C. (1972). Genetic variability among peoples of Aruba and Curaçao. *American Journal of Physical Anthropology*, **36**, 21–9.

Ringuelet, S. (1978). Investigación auxológica diferencial de dos poblaciones argentinas. *Revista del Instituto de Antropología, Universidad Nacional de Córdoba*, **6**, 135–48.

Rittler, M., Paz, J.E. and Castilla, E.E. (1997). VATERL: an epidemiologic analysis of risk factors. *American Journal of Medical Genetics*, **73**, 162–9.

Roberts, D.F. and Hiorns, R.W. (1962). The dynamics of racial admixture. *American Journal of Human Genetics*, **14**, 261–77.

Roberts, D.F. and Hiorns, R.W. (1965). Methods of analysis of the genetic composition of a hybrid population. *Human Biology*, **37**, 38–43.

Robinson, D.J. and Gilbert, A. (1985). Colombia and Venezuela. In Blakemore, H. and Smith, C.T., editors, *Latin America. Geographical Perspectives*, 2nd edn. London: Methuen, pp. 187–240.

Robinson, D.J., Blakemore, H. and Smith, C.T. (1985). The Guianas. In Blakemore, H. and Smith, C.T., editors, *Latin America. Geographical Perspectives*, 2nd edn. London: Methuen, pp. 241–52.

Robinson, W.M. and Roisenberg, I. (1975). Asociación entre trombosis venosa, sistema sanguineo ABO y parámetros hemostáticos. *Revista Médica de Chile*, **103**, 317–21.

Robinson, W.M. and Roisenberg, I. (1980). Venous thromboembolism and ABO blood groups in a Brazilian population. *Human Genetics*, **55**, 129–31.

Robinson, W.M., Borges-Osório, M.R., Callegari-Jacques, S.M., *et al.* (1991). Genetic and nongenetic determinants of blood pressure in a southern Brazilian population. *Genetic Epidemiology*, **8**, 55–67.

Robinson, W.M., Salzano, F.M., Achutti, A.C. and Franco, M.H.L.P. (1984). Blood groups, salivary secretion and other immunologic variables in rheumatic fever and rheumatic heart disease. *Acta Anthropogenetica*, **8**, 217–21.

Robles, A. (1990). Medición de la migración interna en Costa Rica, 1883–1950. In Nadalin, S.O., Marcílio, M.L. and Balhana, A.P., editors, *História e População. Estudos sobre a América Latina*. São Paulo: Fundação Sistema Estadual de Análise de Dados, pp. 7–15.

Rocha, A.M.A., Fragoso, S.C. and Junqueira, P.C. (1973). Tipos de haptoglobinas: sua incidência em doadores de sangue do Estado da Guanabara. *Boletim do Instituto Estadual de Hematologia Arthur de Siqueira Cavalcanti* (Rio de Janeiro) **2/3**, 13–14.

Rocha, S. (1996). Renda e pobreza, os impactos do Plano Real. *Revista Brasileira de Estudos de População*, **13**(2), 117–33.

Rocha-Ferreira, M.B., Malina, R.M. and Rocha, L.L. (1991). Anthropometric, functional and psychological characteristics of 8-year-old Brazilian children from low socioeconomic status. In Shephard, R.J. and Parizkova, J., editors, *Human Growth, Physical Fitness and Nutrition*. Basel: Karger, pp. 109–18.

Roche, A.F. (1992). *Growth, Maturation and Body Composition. The Fels Longitudinal Study 1929–91*. Cambridge: Cambridge University Press.

Rodgers-Johnson, P., Morgan, O.St C., Mora, C., *et al.* (1988). The role of HTLV-I in tropical spastic paraparesis in Jamaica. *Annals of Neurology*, **23** (Suppl.), S121–6.

Rodini, E.S.O. and Bortolozzi, J. (1986). Estimativa das freqüências gênicas dos sistemas sangüíneos ABO e Rh em Bauru, SP. *Salusvita*, **5**, 83–8.

Rodini, E.S.O. and Richieri-Costa, A. (1990*a*). EEC syndrome: report on 20 new patients, clinical and genetic considerations. *American Journal of Medical Genetics*, **37**, 42–53.

Rodini, E.S.O. and Richieri-Costa, A. (1990*b*). Autosomal recessive ectodermal dysplasia, cleft lip/palate, mental retardation, and syndactyly: the Zlotogora–Ogur syndrome. *American Journal of Medical Genetics*, **36**, 473–6.

Rodini, E.S.O., Freitas, J.A.S. and Richieri-Costa, A. (1990). Rapp–Hodgkin syndrome: report of a Brazilian family. *American Journal of Medical Genetics*, **36**, 463–6.

Rodrigues, J.D. (1999). Análise da estrutura genética da população de Belém utilizando polimorfismos de DNA autossômicos, mitocondriais e do cromossomo Y. Master's dissertation, Universidade Federal do Pará, Belém.

Rodrigues, N. (1988). *Os Africanos no Brasil*. Brasília: Editora da Universidade de Brasília.

Rodrigues, V. Jr., Abel, L., Piper, K. and Dessein, A.J. (1996). Segregation analysis indicates a major gene in the control of interleukine–5 production in humans infected with *Schistosoma mansoni*. *American Journal of Human Genetics*, **59**, 453–61.

Rodriguez Romero, W.E., Castillo, M., Chaves, M.A., *et al.* (1996). Hb Costa Rica or $\alpha_2\beta_2 77$ (EF1) His→Arg: The first example of a somatic cell mutation in a globin gene. *Human Genetics*, **97**, 829–33.

Rodriguez-Delfin, L., Santos, S.E.B. and Zago, M.A. (1997). Diversity of the human Y chromosome of South American Amerindians: a comparison with Blacks, Whites, and Japanese from Brazil. *Annals of Human Genetics*, **61**, 439–48.

Rodríguez-Larralde, A. (1989). Relationship between 17 Venezuelan counties estimated through communality of surnames. *Human Biology*, **61**, 31–44.

Rodríguez-Larralde, A. (1993). Genetic distance estimated through surname frequencies of 37 counties from the state of Lara, Venezuela. *Journal of Biosocial Science*, **25**, 101–10.

Rodríguez-Larralde, A. and Barrai, I. (1997*a*). Isonymy structure of Sucre and Táchira, two Venezuelan states. *Human Biology*, **69**, 715–31.

Rodríguez-Larralde, A. and Barrai, I. (1997*b*). Estructura genética por isonimia, de los estados Anzoátegui y Trujillo, Venezuela. *Revista Española de Antropología Biológica*, **18**, 39–56.

Rodríguez-Larralde, A., Barrai, I. and Alfonzo, J.C. (1993). Isonymy structure of four Venezuelan states. *Annals of Human Biology*, **20**, 131–45.

Rodríguez-Pombo, P., Hoenicka, J., Muro, S., *et al.* (1998). Human propionyl-CoA carboxylase β subunit gene: exon-intron definition and mutation spectrum in Spanish and Latin American propionic acidemia patients. *American Journal of Human Genetics*, **63**, 360–9.

Roisenberg, I. (1964). Hemophilia B in a pair of monozygotic Negro twins. *Acta Geneticae Medicae et Gemellologiae*, **13**, 240–52.

Roisenberg, I. (1995). Comparative study between biochemical techniques (FVIII/FvW) and DNA analysis (RFLPs) in the diagnosis of hemophilia A carriers. *Revista Iberoamericana de Trombosis y Hemostasia*, **8**, 24–7.

Roisenberg, I. and Morton, N.E. (1971). Genetic aspects of hemophilias A and B in Rio Grande do Sul, Brazil. *Human Heredity*, **21**, 97–107.

Roisenberg, I., Palombini, B.C. and Petersen, N. (1962). Simultaneous occurrence of spherocytosis and polydactyly in a Brazilian family. *Acta Geneticae Medicae et Gemellologiae*, **11**, 55–70.

Rojas-Alvarado, M.A. and Garza-Chapa, R. (1994). Relationships by isonymy between persons with monophyletic and polyphyletic surnames from the Monterrey Metropolitan Area, México. *Human Biology*, **66**, 1021–36.

Roldán, A., Gutiérrez, M., Cygler, A., Bonduel, M., Sciuccati, G. and Torres, A.F. (1997). Molecular characterization of β-thalassemia genes in an Argentine population. *American Journal of Hematology*, **54**, 179–82.

Roman, T., Bau, C.H.D., Almeida, S. and Hutz, M.H. (1999). Lack of association of the Dopamine D4 receptor gene with alcoholism in a Brazilian population. *Addiction Biology*, **4**, 203–7.

Romana, M., Kéclard, L., Froger, A., *et al.* (2000). Diverse genetic mechanisms operate to generate atypical β^S haplotypes in the population of Guadeloupe. *Hemoglobin*, **24**, 77–87.

Romana, M., Keclard, L., Guillemin, G., *et al.* (1996). Molecular characterization of β-thalassemia mutations in Guadeloupe. *American Journal of Hematology*, **53**, 228–33.

Romero Molina, J., de Garay, A.L., Faulhauber, J. and Comas, J. (1976). *Antropología Física. Época Moderna y Contemporánea.* México: Instituto Nacional de Antropología e História.

Romiti, R., Abrucio-Neto, L., Rivitti, E.A. and Giugliani, R. (1996). Arilsulfatase C e o diagnóstico de ictiose recessiva ligada ao X. *Anais Brasileiros de Dermatologia*, **71**, 201–3.

Romney, A.K. (1999). Cultural consensus as a statistical model. *Current Anthropology*, **40** (Suppl.), S103–15.

Rona, R. (1975). Secular trend of pubertal development in Chile. *Journal of Human Evolution*, **4**, 251–7.

Rona, R. and Pereira, G. (1974). Factors that influence age of menarche in girls in Santiago, Chile. *Human Biology*, **46**, 33–42.

Rona, R. and Pierret, T. (1973). Genótipo y estatura en niñas adolescentes de Santiago. *Revista Médica de Chile*, **101**, 207–11.

Rona, R., Rozovski, J. and Stekel, A. (1973). Sickle-talasemia: estudio de una genealogia chilena. *Revista Médica de Chile*, **101**, 237–9.

Roosevelt, A.C., Lima da Costa, M., Machado, C.L., *et al* (1996). Paleoindian cave dwellers in the Amazon: the peopling of the Americas. *Science*, **272**, 373–84.

Rosa, V.L., Salzano, F.M., Franco, M.H.L.P. and Freitas, M.J.M. (1984). Blood genetic studies in five Amazonian populations. *Revista Brasileira de Genética*, **7**, 569–82.

Rosales, T., Guilherme, L., Chiarella, J., *et al.* (1992). Human leukocyte A and B antigen, gene and haplotype frequencies in the population of the city of São Paulo in Brazil. *Brazilian Journal of Medical and Biological Research* 25, 39–47.

Rosé, G., DeLuca, M., Falcone, E., Spadafora, P., Garrieri, G. and De Benedictis, G. (1996). Allele frequency distributions at seven DNA hypervariable loci in a population sample from Calabria (Southern Italy). *Gene Geography*, **10**, 135–45.

Rothhammer, F. (1987). Biological population history of Continental Chile. In Schwidetzky, I., editor, *Rassengeschichte der Menschheit. 12. Lieferung. Amerika II: Mittel und Südamerika*. Munich: R. Oldenbourg Verlag, pp. 219–36.

Rothhammer, F., Llop, E. and Neel, J.V. (1982). Dermatoglyphic characters and physique: a correlation study. *American Journal of Physical Anthropology*, **57**, 99–101.

Rothhammer, F., Pereira, G., Camousseight, A. and Benado, M. (1971). Dermatoglyphics in schizophrenic patients. *Human Heredity*, **21**, 198–202.

Rouquayrol, M.Z. (1966). Indicadores de saúde no município de Fortaleza: 1920–1965. *Revista da Faculdade de Medicina da Universidade Federal do Ceará*, **6**, 59–71.

Rouquayrol, M.Z. and Carneiro, P.C.A. (1976). Indicadores de saúde no município de Fortaleza-II. *Revista da Faculdade de Medicina da Universidade Federal do Ceará*, **16**, 35–44.

Rouquayrol, M.Z., Alencar, J.E. and Bernardo e Sá, L.C. (1967). Causas de óbito no município de Fortaleza, Nomenclatura local e causas de erro. *Revista da Faculdade de Medicina da Universidade Federal do Ceará*, **7**, 45–54.

Roychoudhury, A.K. and Nei, M. (1988). *Human Polymorphic Genes. World Distribution.* New York: Oxford University Press.

Ruffié, J., Quilici, J-C. and Lacoste, M-C. (1977). *Anthropologie des Populations Andines.* Paris: INSERM.

Ruiz Reyes, G. (1983). Hemoglobin variants in Mexico. *Hemoglobin*, **7**, 603–10.

Ruiz-Linares, A., Nayar, K., Goldstein, D.B., *et al.* (1996). Geographic clustering of human Y-chromosome haplotypes. *Annals of Human Genetics*, **60**, 401–8.

Rummler, M.C.O., Santos, J.B., Araújo, M.C.S., Ferro, N.P.B. and Duarte, P.C.T. (1985). Cronologia da erupção dos dentes permanentes em escolares baianos, melanodermos, nível sócio-econômico baixo. *Universitas, Ciência* (Salvador), **34**, 31–42.

Russo, E. (1964). Estudo de 4390 classificações de fatores Rh e de grupos sangüíneos com relação ao sexo e à cor. 4 casos de incompatibilidade sangüínea materno-fetal. *O Hospital*, **65**, 141–60.

Ryan, A.S. (1997). Iron-deficiency anemia in infant development: implications for growth, cognitive development, resistance to infection, and iron supplementation. *Yearbook of Physical Anthropology*, **40**, 25–62.

Saad, P.M. (1991). Tendências e conseqüências do envelhecimento populacional no Brasil. In Branco, P.P.M., editor, *A População Idosa e o Apoio Familiar.* São Paulo: Fundação Sistema Estadual de Análise de Dados, pp. 3–10.

Saad, S.T.O. and Costa, F.F. (1992). Glucose-6-phosphate dehydrogenase deficiency and sickle cell disease in Brazil. *Human Heredity*, **42**, 125–8.

Saad, S.T.O. and Zago, M.A. (1991). Leg ulceration and abnormalities of calf blood flow in sickle-cell anemia. *European Journal of Haematology*, **46**, 188–90.

Saad, S.T.O., Salles, T.S.I., Arruda, V.R., Sonati, M.F. and Costa, F.F. (1997*b*). G6PD Sumaré: a novel mutation in the G6PD gene (1292T→G) associated with chronic nonspherocytic anemia. *Human Mutation*, **10**, 245–7.

Saad, S.T.O., Salles, T.S.I., Carvalho, M.H.M. and Costa, F.F. (1997*a*). Molecular characterization of glucose-6-phosphate dehydrogenase deficiency in Brazil. *Human Heredity*, **47**, 17–21.

Saavedra, D., Richieri-Costa, A., Guion-Almeida, M.L. and Cohen, M.M. Jr. (1996). Craniofrontonasal syndrome: study of 41 patients. *American Journal of Medical Genetics*, **61**, 147–51.

Saddy, E.M.V., Obadia, I., Saddy, J.C. and Azulay, R.D. (1985). Neurofibromatose. 1. Dados epidemiológicos e etiológicos. Aspectos genéticos e dermatológicos. *Anais Brasileiros de Dermatologia*, **60**, 371–8.

Sáenz, G.F. (1986). Hemoglobinopathies in Central America. In Winter, W.P., editor, *Hemoglobin Variants in Human Populations,* Vol. I. Boca Raton, FL: CRC Press, pp. 101–16.

Sáenz, G.F. (1988). Hemoglobinopatías en los países de la Cuenca del Caribe. *Revista de Biologia Tropical*, **36**, 361–72.

Sáenz, G.F., Arroyo, G., Jiménez, J., *et al.* (1971). Investigación de hemoglobinas anormales en población de raza negra costarricense. *Revista de Biología Tropical*, **19**, 251–6.

Sáenz, G.F., Romero, W.R. and Villalobos, M.C. (1993). Variantes estructurales de la hemoglobina en Iberoamérica. *Revista de Biologia Tropical*, **41**, 393–403.

Saitou, N. and Nei, M. (1987). The neighbor-joining method. A new method for reconstructing phylogenetic trees. *Molecular Biology and Evolution*, **4**, 406–25.

Sajantila, A., Budowle, B., Ström, M., *et al.* (1992). PCR amplification of alleles at the D1S80 locus: Comparison of a Finnish and a North American Caucasian population sample, and forensic casework evaluation. *American Journal of Human Genetics*, **50**, 816–25.

Sala, A., Penacino, G. and Corach, D. (1997). VNTR polymorphism in the Buenos Aires, Argentina, metropolitan population. *Human Biology*, **69**, 777–83.

Sala, A., Penacino, G. and Corach, D. (1998). Comparison of allele frequencies of eight STR loci from Argentinian Amerindian and European populations. *Human Biology*, **70**, 937–47.

Sala, A., Penacino, G., Carnese, R. and Corach, D. (1999). Reference database of hypervariable genetic markers of Argentina: application for molecular anthropology and forensic casework. *Electrophoresis*, **20**, 1733–9.

Salaru, N.N. and Otto, P.A. (1989). Blood groups in a large sample from the city of São Paulo (Brazil): allele and haplotype frequencies for MNSs, Kell-Cellano, Rh and ABO systems. *Revista Brasileira de Genética*, **12**, 625–43.

Salazar Mallen, M. and Arteaga, C. (1951). Estudio de los grupos sanguíneos de los mexicanos. Consecuencias desde el punto de vista etnológico. *Revista Mexicana de Estudios Antropológicos*, **12**, 9–29.

Salazar Mallen, M. and Portilla, R.H. (1944). Existencia del aglutinogeno Rh en los hematies de 250 individuos mexicanos. *Revista de la Sociedad Mexicana de Historia Natural*, **5**(3–4), 183–5.

Saldanha, G.M., Alves, I. Jr., Rosa, F.C., Maegawa, F.B., Muniz, Y.N. and Souza, I.R. (1998). Identificação do grau de mistura racial nas comunidades da Costa da Lagoa e de São João do Rio Vermelho na ilha de Santa Catarina. *Genetics and Molecular Biology*, **21** (Suppl.), 359.

Saldanha, P.H. (1956a). ABO blood groups, age and disease. *Revista Brasileira de Biologia*, **16**, 349–53.

Saldanha, P.H. (1956b). Apparent pleiotropic effect of genes determining taste thresholds for phenylthiourea. *The Lancet*, **II**, 74.

Saldanha, P.H. (1956c). Distribuição dos grupos sangüíneos ABO e tipos sangüíneos Rh no nordeste de São Paulo (Vale do Paraíba). *Revista Brasileira de Biologia*, **16**, 243–50.

Saldanha, P.H. (1957). Gene flow from White into Negro populations in Brazil. *American Journal of Human Genetics*, **9**, 299–309.

Saldanha, P.H. (1960). Frequencies of consanguineous marriages in northeast of São Paulo, Brazil. *Acta Genetica et Statistica Medica*, **10**, 71–88.

Saldanha, P.H. (1962a). The genetic effects of immigration in a rural community of São Paulo, Brazil. *Acta Geneticae Medicae et Gemellologiae*, **11**, 158–224.

Saldanha, P.H. (1962b). Race mixture among northeastern Brazilian populations. *American Anthropologist*, **64**, 751–9.

Saldanha, P.H. (1994). A tragédia da talidomida e o advento da teratologia experimental. *Revista Brasileira de Genética*, **17**, 449–64.

Saldanha, P.H., Frota-Pessoa, O., Eveleth, P.B., Ottensooser, F., Cunha, A.B. and Cavalcanti, M.A.A. (1960). Estudo genético e antropológico de uma colônia de holandeses do Brasil. *Revista de Antropologia* (São Paulo), **8**, 1–42.

Saldanha, S.G. (1982). ABO blood groups and salivary secretion of ABH substances among three racial groups in São Paulo city. *Revista Brasileira de Genética*, **5**, 175–86.

Saleh, A.W. Jr., Velvis, H.J.R., Gu, L.H., Hillen, H.F.P. and Huisman, T.H.J. (1997). Hydroxyurea therapy in sickle cell anemia patients in Curaçao, The Netherlands Antilles. *Acta Haematologica*, **98**, 125–9.

468 References

Saleh, M.C., Botelli, A., Melano de Botelli, M., Rezzonico, C.A. and Argaraña, C.E. (1996). Cystic fibrosis: frequency of ΔF508 and G542X mutations in Córdoba, Argentina. *Medicina* (Buenos Aires), **56**, 14–16.

Sales, M.M., de Lucca, E.J., Yamashita, S. and Saad, L.H.C. (1999). Bleomycin sensitivity in patients with familial and sporadic polyposis: a pilot study. *Genetics and Molecular Biology*, **22**, 17–20.

Salle, C., Blanco, A., Baas, M-J. and Cazenave, J.P. (1990). Ethnic polymorphism of DX52 (ST14) locus linked to coagulation factor VIII gene in Argentina. *Annales de Génétique*, **33**, 24–8.

Salles, M.R.R. (1996). Os médicos italianos em São Paulo (1890–1930): um projeto de ascenção social. *Revista Brasileira de Estudos de População*, **13**(1), 43–65.

Salles, V. (1988). *O Negro no Pará. Sob o Regime da Escravidão.* Belém: Fundação Cultural do Pará Tancredo Neves.

Salzano, F.M. (1961). Osteopetrosis: review of dominant cases and frequency in a Brazilian state. *Acta Geneticae Medicae et Gemellologiae*, **10**, 353–8.

Salzano, F.M. (1963). Blood groups and gene flow in Negroes from southern Brazil. *Acta Genetica et Statistica Medica*, **13**, 9–20.

Salzano, F.M. (1965). Genética de populações humanas brasileiras. In Genovés, S., editor, *Homenaje a Juan Comas en su 65 Aniversario. II. Antropología Física.* México: Editorial Libros de México, pp. 253–318.

Salzano, F.M. (1967). Blood groups and leprosy. *Journal of Medical Genetics*, **4**, 102–6.

Salzano, F.M. (1971*a*). Genetic polymorphisms in Brazilian populations. In Salzano, F.M., editor, *The Ongoing Evolution of Latin American Populations.* Springfield, IL: Charles C. Thomas, pp. 631–59.

Salzano, F.M. (1971*b*). *The Ongoing Evolution of Latin American Populations.* Springfield, IL: C.C. Thomas.

Salzano, F.M. (1979). Abnormal hemoglobin studies and counseling in Brazil. *Proceedings, First International Conference on Sickle Cell Disease: A World Health Problem.* Washington, D.C.: Center for Sickle Cell Disease, pp. 67–9.

Salzano, F.M. (1985). Incidence, effects and management of sickle cell disease in Brazil. *American Journal of Pediatric Hematology/Oncology*, **7**, 240–4.

Salzano, F.M. (1986). Hemoglobin variants in Brazil. In Winter, W.P., editor, *Hemoglobin Variants in Human Populations.* Vol. I. Boca Raton, FL: CRC Press, pp. 117–30.

Salzano, F.M. (1987). Brazil. In Schwidetzky, I., editor, *Rassengeschichte der Menschheit*, 12. *Lieferung. Amerika II: Mittel- und Südamerika.* Munich: R. Oldenbourg Verlag, pp. 137–75.

Salzano, F.M. (1991). Interdisciplinary approaches to the human biology of South Amerindians. *Human Biology*, **63**, 875–82.

Salzano, F.M. (1995). *Evolução do Mundo e do Homem. Liberdade ou Organização?* Porto Alegre: Editora da Universidade Federal do Rio Grande do Sul.

Salzano, F.M. (1997). Human races: myth, invention, or reality? *Interciencia*, **22**, 221–7.

Salzano, F.M. and Benevides, F.R.S., Jr., (1974). Fingerprint quantitative variation and asymmetry in Brazilian Whites and Blacks. *American Journal of Physical Anthropology*, **40**, 325–8.

Salzano, F.M. and Blumberg, B.S. (1970). The Australia antigen in Brazilian healthy persons and in leprosy and leukaemia patients. *Journal of Clinical Pathology*, **23**, 39–42.

Salzano, F.M. and Callegari-Jacques, S.M. (1988). *South American Indians. A Case Study in Evolution.* Oxford: Clarendon Press.

Salzano, F.M. and Ebling, H. (1966). Cherubism in a Brazilian kindred. *Acta Geneticae Medicae et Gemellologiae*, **15**, 296–301.

Salzano, F.M. and Freire-Maia, N. (1970). *Problems in Human Biology. A Study of Brazilian Populations.* Detroit, MI: Wayne State University Press.

Salzano, F.M. and Hirschfeld, J. (1965). The dynamics of the Gc polymorphism in a Brazilian population. *Acta Genetica et Statistica Medica*, **15**, 116–25.

Salzano, F.M. and Pena, S.D.J. (1989). Ethics and medical genetics in Brazil. In Wertz, D.C. and Fletcher, J.C., editors, *Ethics and Human Genetics. A Cross-Cultural Perspective.* Berlin: Springer-Verlag, pp. 100–18.

Salzano, F.M. and Rao, D.C. (1976). Path analysis of aptitude, personality, and achievement scores in Brazilian twins. *Behavior Genetics*, **6**, 461–6.

Salzano, F.M. and Schmidt, M. (1988). New case of an EEC-like syndrome in twins. *Acta Geneticae Medicae et Gemellologiae*, **37**, 347–50.

Salzano, F.M. and Schüler, L. (1998). Questões éticas em genética humana. In De Boni, L.A., Jacob, G. and Salzano, F.M., editors, *Ética e Genética*. Porto Alegre: Editora da Pontifícia Universidade Católica do Rio Grande do Sul, pp. 193–210.

Salzano, F.M. and Tondo, C.V. (1982). Hemoglobin types in Brazilian populations. *Hemoglobin*, **6**, 85–97.

Salzano, F.M., Suñé, M.V. and Ferlauto, M. (1967). New studies on the relationship between blood groups and leprosy. *Acta Genetica et Statistica Medica*, **17**, 530–44.

Samara, E.M. and Costa, I.N. (1984). *Demografia Histórica. Bibliografia Brasileira.* São Paulo: Instituto de Pesquisas Econômicas.

Sampaio, D.A., Mattevi, M.S., Cavalli, I.J. and Erdtmann, B. (1989). Densitometric measurements of C bands of chromosomes 1, 9, 16, and Y in leukemic and preleukemic disorders. *Cancer Genetics and Cytogenetics*, **41**, 71–8.

Sampaio, Z.A. (1987). Polimorfismo de hemoglobinas humanas no Estado do Piauí: distribuição, prevalência, relações históricas e antropológicas. Master's dissertation, Universidade Estadual Paulista, São José do Rio Preto.

Sampaio, Z.A., Mourão, L.A.C.B., Nunesmaia, H.G., *et al.* (1986). Perfil da consangüinidade no Brasil. *Ciência e Cultura*, **38** (Suppl. 7), 867.

Sandoval, L. (1961). Grupos sanguineos en la Provincia de Cautin. *Antropologia Chilena* (Santiago, Chile), **2**, 9–15.

Sandoval, L. and Hidalgo, M. (1961). El sistema de grupos sanguineos Duffy en la población de Santiago. *Antropologia Chilena* (Santiago, Chile), **2**, 19–25.

Sans, M. (1992). Genética e historia, Hacia una revisión de nuestra identidad como 'país de inmigrantes'. In Pi Hugarte, R., editor, *Ediciones del Quinto Centenario*. Vol. 1. Montevideo: Universidad de la República, pp. 19–42.

Sans, M., Alvarez, I., Bentancor, N., *et al.* (1995). Blood protein genetic markers in a northeastern Uruguayan population. *Revista Brasileira de Genética*, **18**, 317–20.

Sans, M., Barreto, I. and Portas, M. (1996). The evolution of the Uruguayan population. *International Journal of Anthropology*, **11**(2–4), 19–32.

Sans, M., Mañé-Garzón, F. and Kolski, R. (1986). Presencia de mancha mongólica en recien nacidos de Montevideo. *Archivos de Pediatria*, **57**, 149–56.

Sans, M., Mañé-Garzon, F. and Kolski, R. (1991). Utilización de marcadores bioantropológicos para el estudio del mestizaje en la población uruguaya. *Antropología Biológica*, **1**, 71–86.

Sans, M., Salzano, F.M. and Chakraborty, R. (1997). Historical genetics in Uruguay: Estimates of biological origins and their problems. *Human Biology*, **69**, 161–70.

Sans, M., Sosa, M., Alvarez, I., Toledo, R., Bengochea, M. and Salzano, F.M. (1993). Blood group frequencies and the question of race admixture in Uruguay. *Interciencia*, **18**, 29–32.

Sans, M., Weimer, T.A., Franczak, S.C., *et al.* (2001). Unequal contributions of male and female gene pools from parental populations in the African descendants of the city of Melo, Uruguay. *American Journal of Physical Anthropology*, (in press).

Santiago-Delpin, E.A. (1991). Histocompatibility profile of selected Latin American countries. *Transplantation Proceedings*, **23**, 1861–4.

Santos, B.R., Monteiro, M.G. and Thomasson, H.R. (1997). Allele frequency of ADH2 and ALDH2 among Brazilians of different ethnic groups. *Alcohol*, **14**, 205–7.

Santos, B.R., Monteiro, M.G., Walzer, C., Turler, H., Balant, L. and von-Wartburg, J.P. (1995a). Alcohol flushing, patch test, and ADH and ALDH genotypes in Brazilian ethnic groups. *Brazilian Journal of Medical and Biological Research*, **28**, 513–18.

Santos, C.M.C., Correia, P.S., Santa Rosa, A.A., *et al.* (1998). Early infantile form of galactosialidosis in a female baby with a prenatal diagnosis of fetal ascites: first case in Brazil. *Genetics and Molecular Biology*, **21**, 443–6.

Santos, E.J.M., Ribeiro-dos-Santos, A.K.C., Guerreiro, J.F., Aguiar, G.F.S. and Santos, S.E.B. (1996). Migration and ethnic change in an admixed population from the Amazon Region (Santarém, Pará). *Brazilian Journal of Genetics*, **19**, 511–15.

Santos, E.O., Loureiro, E.C.B., Jesus, I.M., *et al.* (1995). Diagnóstico das condições de saúde de uma comunidade garimpeira na região do Rio Tapajós, Itaituba, Pará, Brasil, 1992. *Cadernos de Saúde Pública*, **11**, 212–25.

Santos, F.R., Bianchi, N.O. and Pena, S.D.J. (1996). Worldwide distribution of human Y-chromosome haplotypes. *Genome Research*, **6**, 601–11.

Santos, F.R., Gerelsaikhan, T., Munkhtuja, B., Oyunsuren, T., Epplen, J.T. and

References   471

Pena, S.D.J. (1995b). Geographic differences in the allele frequencies of the human Y-linked tetranucleotide polymorphism DYS19. *Human Genetics*, **97**, 309–13.

Santos, F.R., Pena, S.D.J. and Epplen, J.T. (1993). Genetic and population study of a Y-linked tetranucleotide repeat DNA polymorphism with a simple non-isotopic technique. *Human Genetics*, **90**, 655–6.

Santos, L.C., Filho (1979). A medicina no Brasil. In Ferri, M.G. and Motoyama, S., editors, *História das Ciências no Brasil*. São Paulo: Editora Pedagógica e Universitária e Editora da Universidade de São Paulo, pp. 191–217.

Santos, M., Kuzmin, A.I., Eisensmith, R.C., *et al.* (1996). Phenylketonuria in Costa Rica: preliminary spectrum of PAH mutations and their associations with highly polymorphic haplotypes. *Human Heredity*, **46**, 128–31.

Santos, M.C.N. and Azevêdo, E.S. (1981). Generalized joint hypermobility and Black admixture in school children of Bahia, Brazil. *American Journal of Physical Anthropology*, **55**, 43–6.

Santos, M.S., Alves-Silva, J., Bandelt, H-J., Pena, S.D.J. and Prado, V.F. (1998). mtDNA sequencing in the Brazilian Caucasian population. *Abstracts, XXVII Reunião Anual, Sociedade Brasileira de Bioquímica e Biologia Molecular*, 13.

Santos, R.C.S., Barreto, O.C.O., Nonoyama, K., *et al.* (1991). X-linked syndrome: Mental retardation, hip luxation, and G6PD variant (Gd(+)Butantan]. *American Journal of Medical Genetics*, **39**, 133–6.

Santos, R.V. and Anjos, L.A. (1993). Crescimento e desenvolvimento físico da criança brasileira. *Cadernos de Saúde Pública*, **9** (Suppl. 1), 5–19.

Santos, S.E.B. and Guerreiro, J.F. (1995). The indigenous contribution to the formation of the population of the Brazilian Amazon region. *Revista Brasileira de Genética*, **18**, 311–15.

Santos, S.E.B., Guerreiro, J.F., Salzano, F.M., Weimer, T.A., Hutz, M.H. and Franco, M.H.L.P. (1987). Mobility, blood genetic traits and race mixture in the Amazonian population of Oriximiná. *Revista Brasileira de Genética*, **10**, 745–59.

Santos, S.E.B., Rodrigues, J.D., Ribeiro-dos-Santos, A.K.C. and Zago, M.A. (1999). Differential contribution of indigenous men and women to the formation of an urban population in the Amazon Region as revealed by mtDNA and Y-DNA. *American Journal of Physical Anthropology*, **109**, 175–80.

Santos, S.E.B., Salzano, F.M., Franco, M.H.L.P. and Freitas, M.J.M. (1983). Mobility, genetic markers, susceptibility to malaria and race mixture in Manaus, Brazil. *Journal of Human Evolution*, **12**, 373–81.

Sanz, L., Beraún, Y., Nieto, A., Martín, J., Vilches, C. and De Pablo, R. (1999). A new HLA-Cw*15 allele, CW*1508, identified in the Peruvian population. *Tissue Antigens*, **53**, 391–3.

Satz, M.L., Chertokoff, L.P., Herrera, M., *et al.* (1989). HLA y enfermedad en la Argentina. Polimorfismo serológico y genómico. *Medicina* (Buenos Aires), **49**, 119–24.

Satz, M.L., Fernandez-Viña, M., Theiler, G.C., *et al.* (1995). Allelic heterogeneity of HLA-B35 subtypes in different populations as assessed by DNA typing. *Tissue Antigens*, **46**, 196–203.

Sauvain-Dugerdil, C. (1987). The biological diversity of the human populations of Central America. In Schwidetzky, I., editor, *Rassengeschichte der Menschheit. 12. Lieferung. Amerika II: Mittel- und Südamerika.* Munich: R. Oldenbourg Verlag, pp. 23–80.

Sauvain-Dugerdil, C. (1989). Diversidad biológica de las poblaciones humanas de Centroamérica: Elementos de discusión. In Serrano, C. and Salas, M.E., editors, *Estudios de Antropología Biológica (IV Coloquio de Antropología Física Juan Comas).* México: Universidad Nacional Autónoma de México, pp. 269–95.

Sawaya, A.L. (1997). *Desnutrição Urbana no Brasil.* São Paulo: Cortez.

Scalco, F.B., Giugliani, R., Tobo, P. and Coelho, J.C. (1999). Effect of dimethylsulfoxide on sphingomyelinase activity and cholesterol metabolism in Niemann–Pick type C fibroblasts. *Brazilian Journal of Medical and Biological Research*, **32**, 23–8.

Schächter, F. (1998). Causes, effects, and constraints in the genetics of human longevity. *American Journal of Human Genetics*, **62**, 1008–14.

Schaffar-Deshayes, L., Chavance, M., Monplaisir, N., *et al.* (1984). Antibodies to HTLV-I p24 in sera of blood donors, elderly people and patients with hemopoietic diseases in France and in French West Indies. *International Journal of Cancer*, **34**, 667–70.

Schanfield, M.S. (1992). Immunoglobulin allotypes (Gm and Km) indicate multiple founding populations of native Americans: evidence of at least four migrations to the New World. *Human Biology*, **64**, 381–402.

Schanfield, M.S., Brown, R. and Crawford, M.H. (1984). Immunoglobulin allotypes in the Black Caribs and Creoles of Belize and St. Vincent. In Crawford, M.H., editor, *Current Developments in Anthropological Genetics*, Vol. 3. *Black Caribs.* New York: Plenum Press, pp. 345–63.

Schepeler, M. and von Dessauer, R. (1945). Factor Rh y transfusion sanguinea en el niño. *Revista Médica de Chile*, **73**, 816–18.

Schilirò, G., Spena, M., Giambelluca, E. and Maggio, A. (1990). Sickle hemoglobinopathies in Sicily. *American Journal of Hematology*, **33**, 81–5.

Schlaepfer, L. (1987). Modelos matematicos y estudios longitudinales de crecimiento. In Faulhaber, M.E.S. and Lizárraga Cruchaga, X., editors, *Estudios de Antropología Biológica (I Coloquio de Antropología Física Juan Comas).* México: Universidad Nacional Autónoma de México, pp. 193–224.

Schmidt, M. and Salzano, F.M. (1980). Dissimilar effects of thalidomide in dizygotic twins. *Acta Geneticae Medicae et Gemellologiae*, **29**, 295–7.

Schmidt, M. and Salzano, F.M. (1983). Clinical studies on teenage Brazilian victims of thalidomide. *Brazilian Journal of Medical and Biological Research*, **16**, 105–9.

Schmidt, M., Salzano, F.M. and Simões, G.V. (1981). Padrões de morbidade e mortalidade em Porto Alegre e suas implicações biológicas. *Ciência e Cultura*, **33**, 1618–22.

Schnee-Greise, J., Bläss, G., Herrmann, S., Schneider, H.R., Förster, R., Bässler,

G. and Pflug, W. (1993). Frequency distribution of D1S80 alleles in the German population. *Forensic Science International*, **59**, 131–6.

Schneider, H., and Salzano, F.M. (1979). Gm allotypes and racial admixture in two Brazilian populations. *Human Genetics*, **53**, 101–5.

Schneider, H., Guerreiro, J.F., Santos, S.E.B., Weimer, T.A., Schneider, M.P.C. and Salzano, F.M. (1987). Isolate breakdown in Amazonia: the Blacks of the Trombetas river. *Revista Brasileira de Genética*, **10**, 565–74.

Schramm, J.M.A., Sanches, O. and Szwarcwald, C.L. (1996). Análise da mortalidade por tétano neonatal no Brasil (1979–1987). *Cadernos de Saúde Pública*, **12**, 217–24.

Schroeder, D.G., Martorell, R. and Flores, R. (1999). Infant and child growth and fatness and fat distribution in Guatemalan adults. *American Journal of Epidemiology*, **149**, 177–85.

Schüler, L. and Salzano, F.M. (1994). Patterns in multimalformed babies and the question of the relationship between sirenomelia and VACTERL. *American Journal of Medical Genetics*, **49**, 29–35.

Schüler, L., Pastuszak, A., Sanseverino, M.T.V., *et al.* (1999). Pregnancy outcome after exposure to misoprostol in Brazil: a prospective, controlled study. *Reproductive Toxicology*, **13**, 147–51.

Schüler, L., Salzano, F.M., Franco, M.H.L.P., Freitas, M.J.M., Mestriner, M.A. and Simões, A.L. (1982). Demographic and blood genetic characteristics in an Amazonian population. *Journal of Human Evolution*, **11**, 549–58.

Schüler, L., Sanseverino, M.T., Clavijo, H.A., *et al.* (1993). Preliminary report on the First Brazilian Teratogen Information Service (SIAT). *Revista Brasileira de Genética*, **16**, 1085–95.

Schull, W.J. and Rothhammer, F. (1990). *The Aymara Strategies in Human Adaptation to a Rigorous Environment*. Dordrecht: Kluwer.

Schwantes, A.R., Salzano, F.M., Castro, I.V. and Tondo, C.V. (1967). Haptoglobins and leprosy. *Acta Genetica et Statistica Medica*, **17**, 127–36.

Schwarcz, L.M. (1993). *O Espetáculo das Raças. Cientistas, Instituições e Questão Racial no Brasil, 1870–1930*. São Paulo: Companhia das Letras.

Schwartzman, J.S., Zatz, M., Vasquez, L.R., *et al.* (1999). Rett syndrome in a boy with a 47, XXY karyotype. *American Journal of Human Genetics*, **64**, 1781–5.

Schwidetzky, I. (1987). The Caribbean islands. In Schwidetzky, I., editor, *Rassengeschichte der Menschheit. 12. Lieferung. Amerika II: Mittel- und Südamerika*. Munich: R. Oldenbourg Verlag, pp. 81–103.

Sciarratta, G.V., Ivaldi, G. and Moruzzi, F. (1990). HB J-Guantanamo in a Chilean baby. *Hemoglobin*, **14**, 115–17.

Scrimshaw, N.S., Méndez, J., Flores, M., Guzmán, M.A. and De León, R. (1961). Diet and serum cholesterol levels among the 'Black Caribs' of Guatemala. *American Journal of Clinical Nutrition*, **9**, 206–10.

Secor, W.E., del Corral, H., dos Reis, M.G., *et al.* (1996). Association of hepatosplenic schistosomiasis with HLA-DQB1*0201. *Journal of Infectious Diseases*, **174**, 1131–5.

Seid-Akhavan, M., Ayres, M., Salzano, F.M., Winter, W.P. and Rucknagel, D.L.

(1973). Two more examples of Hb Porto Alegre, $\alpha_2\beta_2^{Ser \to Cys}$ in Belém, Brazil. *Human Heredity*, **23**, 175–81.

Seielstad, M.T., Minch, E. and Cavalli-Sforza, L.L. (1998). Genetic evidence for a higher female migration rate in humans. *Nature Genetics*, **20**, 278–80.

Seiler, E. and Bier, O.G. (1935). Distribution des agglutinogènes M et N de Landsteiner et Levine dans la population de São Paulo. *Comptes Rendus, Societé de Biologie*, **120**, 1111–2.

Sepich, L.F. (1965). Inmunización al factor Kell. Reacción transfusional. Frecuencia del factor en la población de Mendoza. *Revista de la Asociación Bioquímica Argentina*, **157**, 25–31.

Serjeant, G.R. (1974). *The Clinical Features of Sickle Cell Disease*. Amsterdam: North-Holland.

Serjeant, G.R., Serjeant, B.E., Forbes, M., Hayes, R.J., Higgs, D.R. and Lehmann, H. (1986). Haemoglobin gene frequencies in the Jamaican population: A study in 100 000 newborns. *British Journal of Haematology*, **64**, 253–62.

Serra de Castro, A. (1934). A anemia de hematias falciformes. *Jornal de Pediatria*, **1**, 427–45.

Serra, J.P., Silva, N.A. and Mundim, T.L. (1964). Grupos sangüíneos em Goiânia. *Ciência e Cultura*, **16**, 127–8.

Serrano, C. (1975). El surco palmar transverso en la población del Valle Poblano-Tlaxcalteca, México. *Anales de Antropología*, **12**, 103–15.

Serrano, C. (1982). Dermatoglifos de Coras, Huicholes y mestizos de la Sierra de Nayarit, México. In Villanueva, M. and Serrano, C., editors, *Estudios de Antropología Biológica (I Coloquio de Antropología Física Juan Comas)*. México: Universidad Nacional Autónoma de México, pp. 155–63.

Sertié, A.L., Quimby, M., Moreira, E.S., *et al.* (1996). A gene which causes severe ocular alterations and occipital encephalocele (Knobloch syndrome) is mapped to 21q22.3. *Human Molecular Genetics*, **5**, 843–7.

Sertié, A.L., Sousa, A.V., Steman, S., Pavanello, R.C. and Passos-Bueno, M.R. (1999). Linkage analysis in a large Brazilian family with van der Woude syndrome suggests the existence of a susceptibility locus for cleft palate at 17p11.2–11.1. *American Journal of Human Genetics*, **65**, 433–40.

Sesso, R., Almeida, M.A., Figueiredo, M.S. and Bordin, J.O. (1998). Renal dysfunction in patients with sickle cell anemia or sickle cell trait. *Brazilian Journal of Medical and Biological Research*, **31**, 1257–62.

Settineri, W.M.F., Salzano, F.M. and Freitas, M.J.M. (1976). X-linked anhidrotic ectodermal dysplasia with some unusual features. *Journal of Medical Genetics*, **13**, 212–16.

Shansis, M. and Carpilovsky, J.C. (1956). Freqüência dos grupos sanguineos A, B, O e do fator Rh na população de Porto Alegre. *Revista Brasileira de Biologia*, **16**, 483–9.

Shephard, R.J. (1985). Factors associated with population variation in physiological working capacity. *Yearbook of Physical Anthropology*, **28**, 97–122.

Silva, A.E. and Varella-Garcia, M. (1989). Plasma folate and vitamin B_{12} levels in thalassemia heterozygotes. *Brazilian Journal of Medical and Biological Research*, **22**, 1225–8.

Silva, C.M.D., Severini, M.H., Sopelsa, A., *et al.* (1999). Six novel β–galactosidase gene mutations in Brazilian patients with Gm1-gangliosidosis. *Human Mutation*, **13**, 401–9.

Silva, E.O. (1991). Waardenburg I syndrome: a clinical and genetic study of two large Brazilian kindreds, and literature review. *American Journal of Medical Genetics*, **40**, 65–74.

Silva, E.O. and Batista, J.E.M. (1994). Craniofacial disproportions in Waardenburg syndrome type I. *Revista Brasileira de Genética*, **17**, 97–9.

Silva, E.O., Batista, J.E.M., Medeiros, M.A.B. and Fonteles, S.M.S. (1993). Craniofacial anthropometric studies in Waardenburg syndrome type I. *Clinical Genetics*, **44**, 20–5.

Silva, E.O., Janovitz, D. and Albuquerque, S.C. (1980). Ellis van Creveld syndrome: report of 15 cases in an inbred kindred. *Journal of Medical Genetics*, **17**, 349–56.

Silva, E.O., Salzano, F.M. and Weimer, T.A. (1990). Is the Waardenburg I syndrome gene located on chromosome 9? *Revista Brasileira de Genética*, **13**, 551–6.

Silva, H.P., Crews, D.E. and Neves, W.A. (1995). Subsistence patterns and blood pressure variation in two rural caboclo communities of Marajó Island, Pará, Brazil. *American Journal of Human Biology*, **7**, 535–42.

Silva, L.C.S., Pires, R.F., Coelho, J.C., Jardim, L.B. and Giugliani, R. (1997). Evaluation of an aspartame loading test for the detection of heterozygotes for classical phenylketonuria. *Clinical Genetics*, **51**, 231–5.

Silva, L.M. and Donadi, E.A. (1996). Is immunogenetic susceptibility to neuropsychiatric systemic lupus erythematosus (SLE) different from nonneuropsychiatric SLE? *Annals of Rheumatic Diseases*, **55**, 544–7.

Silva, M.C.B.O., Passos-Bueno, M.R., Sousa, M.G.F., Carvalho, R.D.S. and Azevêdo, E.S. (1985). Níveis da fosfatase alcalina placentária no soro de gestantes: estudo em Salvador, Bahia. *Revista Paulista de Medicina*, **103**, 280–3.

Silva, M.I.A.F. (1979). Estudos genéticos na população de Natal e em anêmicos de Porto Alegre. M.Sc. dissertation, Federal University of Rio Grande do Sul.

Silva, M.I.A.F., Salzano, F.M. and Lima, F.A.M. (1981). Migration, inbreeding, blood groups and hemoglobin types in Natal, Brazil. *Studies in Physical Anthropology*, **7**, 3–11.

Silva, N.V. (1987). Distância social e casamento inter-racial no Brasil. *Estudos Afro-Asiáticos*, **14**, 54–83.

Silva, N.V. (1994). Uma nota sobre 'raça social' no Brasil. *Estudos Afro-Asiáticos*, **26**, 67–80.

Silva, R. (1983). *Evaluación Genética y Estudio de Malformaciones Congénitas*. Popayan: Universidad del Cauca.

Silva, W.A. Jr. and Figueiredo, M.S. (1994). Analysis of Factor VIII gene polymorphisms in Brazilian Blacks reveals further differences in the Black population. *Human Heredity*, **44**, 252–60.

Silva, W.A. Jr., Bortolini, M.C., Meyer, D., *et al.* (1999). Genetic diversity of two African and sixteen South American populations determined on the basis of six hypervariable loci. *American Journal of Physical Anthropology*, **109**, 425–37.

Silva, W.A. Jr., Targino de Araújo, J. and Zago, M.A. (1992). Hemoglobin J Rovigo (α^{53} Ala→Asp) is not associated with an α-globin gene deletion. *Revista Brasileira de Genética*, **13**, 667–73.

Silvestre, N., Gonzales, G.F. and Villena, A. (1997). Delayed visuomotor development in children born to adolescent mothers. *American Journal of Human Biology*, **9**, 717–23.

Simonney, N., de Bosch, N., Argueyo, A., Garcia, E. and Layrisse, Z. (1985). HLA antigens in hemophiliacs with or without factor VIII antibodies in a Venezuelan Mestizo population. *Tissue Antigens*, **25**, 216–19.

Simpson, A.J.G., Dias Neto, E., Pena, H.B. and Pena, S.D.J. (1995). The detection of DNA polymorphism in schistosomes and their snail hosts. *Memórias do Instituto Oswaldo Cruz*, **90**, 211–13.

Simpson, G.G. (1944). *Tempo and Mode in Evolution.* New York: Columbia University Press.

Simpson, G.G. (1949). *The Meaning of Evolution.* New Haven, CT: Yale University Press.

Siniarska, A. and Dickinson, F. (1996). *Annotated Bibliography in Human Ecology.* Vol. 1. Delhi: Kamla-Raj.

Skidmore, T. (1993). *Black into White: Race and Nationalism in Brazilian Thought.* Durham, NC: Duke University Press.

Slatkin, M. (1995). A measure of population subdivision based on microsatellite allele frequencies. *Genetics*, **139**, 457–62.

Smink, D.A. and Prins, H.K. (1965). Hereditary and acquired blood factors in the negroid population of Surinam. V. Electrophoretic heterogeneity of glucose-6-phosphate dehydrogenase. *Tropical and Geographic Medicine*, **17**, 236–42.

Smith, C.T. (1985). The Central Andes. In Blakemore, H. and Smith, C.T., editors, *Latin America. Geographical Perspectives.* 2nd edn. London: Methuen, pp. 253–324.

Smolka, M.O. (1992). Mobilidade intra-urbana no Rio de Janeiro, Da estratificação social à segregação residencial no espaço. *Revista Brasileira de Estudos de População*, **9**(2), 97–114.

Snyder, L.H. (1929). The blood groups of the Jamaicans. In Davenport, C.B. and Steggerda, M. editors, *Race Crossing in Jamaica.* New York: Carnegie Institute, publication no. 395, pp. 277–81.

Soares, I.S., Oliveira, S.G., Souza, J.M. and Rodrigues, M.M. (1999). Antibody response to the N- and C-terminal regions of the *Plasmodium vivax* merozoite surface protein 1 in individuals living in an area of exclusive transmission of *P. vivax* malaria in the north of Brazil. *Acta Tropica*, **72**, 13–24.

Solá, J.E. (1931). Los grupos sanguineos en las diferentes formas de alienación mental. *La Semana Médica*, **38**, 699–706.

Solaun, M. and Kronus, S. (1973). *Discrimination Without Violence*. New York: Wiley.

Sommer, M., Gathof, B.S., Podskarbi, T., Giugliani, R., Kleinlein, B. and Shin, Y.S. (1995). Mutations in the galactose-1-phosphate uridyltransferase gene of two families with mild galactosaemia variants. *Journal of Inherited Metabolic Diseases*, **18**, 567–76.

Sonati, M.F., Farah, S.B., Ramalho, A.S. and Costa, F.F. (1991). High prevalence of α-thalassemia in a Black population of Brazil. *Hemoglobin*, **15**, 309–11.

Sonati, M.F., Kaeda, J., Kimura, E.M., Costa, F.F. and Luzzatto, L. (1998). Mild clinical expression of S-β thalassemia in a Brazilian patient with the β+IVS-I-6 (T→C) mutation. *Genetics and Molecular Biology*, **21**, 431–3.

Sonati, M.F., Kimura, E.M. and Costa, F.F. (1992b). Red cell indices and alpha-thalassemia. *Revista Brasileira de Genética*, **15**, 687–93.

Sonati, M.F., Kimura, E.M., Grotto, H.Z.W., Gervasio, S.A. and Costa, F.F. (1996). Hereditary hemoglobinopathies in a population from southeast Brazil. *Hemoglobin*, **20**, 175–9.

Sonati, M.F., Kimura, E.M., Grotto, H.Z.W., Tavella, M.H. and Costa, F.F. (1992a). Hb H disease associated with the (–MED) deletion in a Brazilian black woman. *Acta Haematologica*, **87**, 145–7.

Sonati, M.F., Mason, P.J. and Luzzatto, L. (1994). Duas novas mutações causando anemia hemolítica crônica por deficiência de G6PD. *Revista Brasileira de Genética*, **17** (Suppl.), 82.

Soriano Lleras, A. (1954). Grupos sanguineos en los tuberculosos. *Anales de la Sociedad de Biologia de Bogotá*, **6**, 145–6.

Soto Pradera, E., Latour, J.S. and Montalvo, J.R. (1955). Grupos sanguíneos y clasificación del factor Rh en madres y recién nacidos de la Clínica de Maternidad Obrera de la Habana. *Revista Cubana de Pediatria*, **27**, 433–6.

Sousa, M.G.F., Passos-Bueno, M.R., Silva, M.C.B.O., Duarte, A.F.B.G. and Azevêdo, E.S. (1990). The contribution of sex, electrophoretic phenotype, pregnancy and race to the variability of delta-aminolevulinate dehydrase (ALADH) levels in human erythrocytes. A study in Black mixed Brazilians. *Clinica Chimica Acta*, **194**, 229–34.

Souto, F.J.D., Fontes, C.J.F., Oliveira, J.M., Gaspar, A.M.C. and Lyra, L.G.C. (1997). Epidemiological survey of infection with hepatitis B virus in the savannah and wetlands (Pantanal) of central Brazil. *Annals of Tropical Medicine and Parasitology*, **91**, 411–16.

Souza, A.H.O. (1997). Estudo Genético das Crianças de Carretéis. Deficiência Familial Isolada do Hormônio de Crescimento, Itabaianinha, Sergipe. M.Sc. dissertation, Universidade Federal de Alagoas, Aracaju.

Souza, E.R. (1993). Violência velada e revelada, Estudo epidemiológico da mortalidade por causas externas em Duque de Caxias, Rio de Janeiro. *Cadernos de Saúde Pública*, **9**, 48–64.

Souza, I.R. and Culpi, L. (1992). Valongo, an isolated Brazilian black community. I. Structure of the population. *Revista Brasileira de Genética*,

15, 439–47.

Souza, R.L.R., Castro, R.M.V., Pereira, L., Freund, A.A., Culpi, L. and Chautard-Freire-Maia, E. (1998). Frequencies of the butyrylcholinesterase K mutation in Brazilian populations of European and African origin. *Human Biology*, **70**, 965–70.

Sow, A., Peterson, E., Josiforska, O., Fabry, M.E., Krishnamoorthy, R. and Nagel, R. (1995). Linkage disequilibrium of the Senegal haplotype with the β^S gene in the Republic of Guinea. *American Journal of Hematology*, **50**, 301–3.

Spencer, H.C., Miller, L.H., Collins, W.E., *et al.* (1978). The Duffy blood group and resistance to *Plasmodium vivax* in Honduras. *American Journal of Tropical Medicine and Hygiene*, **27**, 664–70.

Spielman, R.S. (1973). Do the natives all look alike? Size and shape components of anthropometric differences among Yanomama Indian villages. *American Naturalist*, **107**, 694–708.

Spuhler, J.N. (1962). Empirical studies on quantitative human genetics. In *Proceedings, UN/WHO Seminar on the Use of Vital and Health Statistics for Genetic and Radiation Studies*, New York: WHO, pp. 241–52.

Spuhler, J.N. (1968). Assortative mating with respect to physical characteristics. *Eugenics Quarterly*, **15**, 128–40.

Spurgeon, J.H. and Meredith, H.V. (1979). Body size and form of Black and White male youths: South Carolina youths compared with youths measured at earlier times and other places. *Human Biology*, **51**, 187–200.

Spurgeon, J.H., Meredith, E.M. and Meredith, H.V. (1978). Body size and form of children of predominantly black ancestry living in West and Central Africa, North and South America, and the West Indies. *Annals of Human Biology*, **5**, 229–46.

Spurr, G.B. (1983). Nutritional status and physical work capacity. *Yearbook of Physical Anthropology*, **26**, 1–35.

Spurr, G.B., Dufour, D.L. and Reina, J.C. (1996). Energy expenditures of urban Colombian girls and women. *American Journal of Human Biology*, **8**, 237–49.

Spurr, G.B., Reina, J.C. and Dufour, D.L. (1997). Comparative study of flex heart rate in Colombian children and in pregnant, lactating, and non-pregnant, nonlactating women. *American Journal of Human Biology*, **9**, 647–57.

Spurr, G.B., Reina, J.C., Dufour, D.L. and Narváez, J.V. (1994). $\dot{V}_{O2max}$ and nutritional status in urban Colombian girls and women. *American Journal of Human Biology*, **6**, 641–9.

Spurr, G.B., Reina, J.C., Narváez, J.V. and Dufour, D.L. (1992). Maximal oxygen consumption of Colombian women of differing socioeconomic status. *American Journal of Human Biology*, **4**, 625–33.

Stanbury, J.B. (1969). *Endemic Goiter*. Washington, D.C.: Pan American Health Organization.

Staten, L.K., Dufour, D.L., Reina, J.C. and Spurr, G.B. (1998). Household headship and nutritional status: female-headed versus male/dual-headed households. *American Journal of Human Biology*, **10**, 699–709.

Stearns, S.C. (1992). *The Evolution of Life Histories.* Oxford: Oxford University Press.

Stebbins, G.L. (1950). *Variation and Evolution in Plants.* New York: Columbia University Press.

Steele, D.G. and Powell, J.F. (1992). Peopling of the Americas: paleobiological evidence. *Human Biology*, **64**, 303–36.

Stepan, N.L. (1991). *The Hour of Eugenics: Race, Gender, and Nation in Latin America.* Ithaca, NY: Cornell University Press.

Stevenson, A.C., Johnston, H.A., Golding, D.R. and Stewart, M.I.P. (1966*b*). *Basic Tabulations in Respect of Consecutive Post 28-Week Births Recorded in the Co-operating Centres. World Health Organization Comparative Study of Congenital Malformations.* Oxford: Medical Research Council of Great Britain.

Stevenson, A.C., Johnston, H.A., Stewart, M.I.P. and Golding, D.R. (1966*a*). *Congenital Malformations. A Report of a Study of Series of Consecutive Births in 24 Centres.* Geneva: World Health Organization.

Stine, O.C., Dover, G.J., Zhu, D. and Smith, K.D. (1992). The evolution of two West African populations. *Journal of Molecular Evolution*, **34**, 336–44.

Stinson, S. (1982). The effect of high altitude on the growth of children of high socioeconomic status in Bolivia. *American Journal of Physical Anthropology*, **59**, 61–71.

Stinson, S. (1985). Sex differences in environmental sensitivity during growth and development. *Yearbook of Physical Anthropology*, **28**, 123–47.

Stoneking, M., Hedgecock, D., Higuchi, R.G., Vigilant, L. and Erlich, H.A. (1991). Population variation of human mtDNA control region sequences detected by enzymatic amplification and sequence-specific oligonucleotide probes. *American Journal of Human Genetics*, **48**, 370–82.

Stringer, C.B. and Gamble, C. (1993). *In search of the Neanderthals.* New York: Thames and Hudson.

Stueber-Odebrecht, N., Chautard-Freire-Maia, E.A., Carrenho, M.X. and Primo-Parmo, S.L. (1984). Distribuição dos sistemas sangüíneos ABO e RH com relação à origem dos sobrenomes, em uma amostra de Blumenau, SC. *Ciência e Cultura* **36** (Suppl. 7), 873.

Suarez, R.M., Buso, R., Meyer, L.M. and Olavarrieta, S.T. (1959). Distribution of abnormal hemoglobins in Puerto Rico and survival studies of red blood cells using Cr^{51}. *Blood*, **14**, 255–61.

Sugimura, H., Hamada, G.S., Suzuki, I., *et al.* (1995). *CYP1A1* and *CYP2E1* polymorphism and lung cancer, case-control study in Rio de Janeiro, Brazil. *Pharmacogenetics*, **5** (Suppl.), S145–8.

Sumita, D.R., Vainzof, M., Campiotto, S., *et al.* (1998). Absence of correlation between skewed X inactivation in blood and serum creatine-kinase levels in Duchenne/Becker female carriers. *American Journal of Medical Genetics*, **80**, 356–61.

Suñé, M.V., Centeno, J.V. and Salzano, F.M. (1970). Gonadoblastoma in a phenotypic female with 45,X/47XYY mosaicism. *Journal of Medical*

Genetics, **7**, 410–12.

Suzuki, K., Bustos, T. and Spritz, R.A. (1998). Linkage disequilibrium mapping of the gene for Margarita island ectodermal dysplasia (ED4) to 11q23. *American Journal of Human Genetics*, **63**, 1102–7.

Szathmáry, E.J.E. and Reed, T.E. (1978). Calculation of the maximum amount of gene admixture in a hybrid population. *American Journal of Physical Anthropology*, **48**, 29–34.

Szwarcwald, C.L., Leal, M.C., Castilho, E.A. and Andrade, C.L.T. (1997). Mortalidade infantil no Brasil, Belíndia ou Bulgária? *Cadernos de Saúde Pública*, **13**, 503–16.

Tajara, E.H., Gagliardi, A.R.T. and Varella-Garcia, M. (1982*b*). The Prader–Willi syndrome and mosaicism of an extra chromosome. *Revista Brasileira de Genética*, **5**, 209–16.

Tajara, E.H., Varella-Garcia, M., Gagliardi, A.R.T. and Beiguelman, B. (1982*a*). Cytogenetical observations in mental deficiency. *Revista Brasileira de Genética*, **5**, 195–200.

Tanaka, M., Hirai, H., LoVerde, P.T., *et al.* (1995). Yeast artificial chromosome (YAC)-based genome mapping of *Schistosoma mansoni*. *Molecular and Biochemical Parasitology*, **69**, 41–51.

Targino de Araújo, J. (1971). Geographical distribution and incidence of hemoglobin variants in Brazil. In Arends, T., Bemski, G. and Nagel, R.L., editors, *Genetical, Functional, and Physical Studies of Hemoglobins*. Basel: S. Karger, pp. 26–31.

Targino de Araújo, J., Lehmann, H., Plowman, D., *et al.* (1977). Hemoglobina J Rovigo, alfa 53 (E–2) alanina-aspártico: primeiro caso identificado em São Paulo, Brasil. *Anais, VI Congresso do Colégio Brasileiro de Hematologia*, 81.

Targino de Araújo, J., Plowman, D., Targino de Araújo, R.A., de Souza, L.F. and Lehmann, H. (1980). Haemoglobin J Rovigo 53 alpha (E–2) aspartic acid alanin. *Revista Brasileira de Pesquisas Médicas e Biológicas*, **13**, 37–9.

Targino de Araújo, J., Ribeiro, V.S. and Targino de Araújo, R.A. (1987). Formas de talassemia identificadas na cidade de São Paulo. *Revista do Hospital de Clínicas da Faculdade de Medicina de São Paulo*, **42**, 267–72.

Targino de Araújo, J., Ribeiro, V.S., Costa e Silva, M.S., de Bonis, J.C. and Targino de Araújo, R.A. (1982). Hemoglobina E $\alpha_2\beta_2^{26}$ glutâmico-lisina (Hb E $\alpha_2\beta_2^{26\ Glu\rightarrow Lys}$). *Revista do Instituto de Medicina Tropical*, **24**, 229–33.

Taschner, S.P. and Bógus, L.M.M. (1986). Mobilidade espacial da população brasileira: aspectos e tendências. *Revista Brasileira de Estudos de População*, **3(2)**, 87–129.

Teixeira, R.C. and Ramalho, A.S. (1994). Genetics and public health: response of a Brazilian population to an optional hemoglobinopathy program. *Revista Brasileira de Genética*, **17**, 435–8.

Tejada, C., de González, N.L.S. and Sánchez, M. (1965). El factor Diego y el gene de celulas falciformes entre los Caribes de raza negra de Livingston, Guatemala. *Revista del Colegio Médico* (Guatemala), **16**, 83–6.

Tejada, C., Strong, J.P., Montenegro, M.R., Restrepo, C. and Solberg, L.A. (1968). Distribution of coronary and aortic atherosclerosis by geographic

References 481

location, race, and sex. *Laboratory Investigation*, **18**, 509–26.

Telatar, M., Teraoka, S., Wang, Z., *et al.* (1998*a*). Ataxia-telangiectasia: identification and detection of founder-effect mutations in the ATM gene in ethnic populations. *American Journal of Human Genetics*, **62**, 86–97.

Telatar, M., Wang, Z., Castellvi-Bel, S., *et al.* (1998*b*). A model for ATM heterozygote identification in a large population: four founder-effect ATM identify most of Costa Rican patients with ataseia telangiectafia. *Molecular Genetics and Metabolism*, **64**, 36–43.

Telles da Silva, B.T., Borges-Osóreo, M.R.L. and Salzano, F.M. (1975). School achievement, intelligence, and personality in twins. *Acta Geneticae Medicae et Genellologiae*, **24**, 213–19.

Teodoro, U., Casavechia, M.T.G., Dias, M.L.G.G., *et al.* (1988). Perfil epidemiológies das parasitosis intestinais no município de Maringá, Paramá, *Ciência e Cultura*, **40**, 698–702.

Theme-Filha, M.M., Silva, R.I. and Noronha, C.P. (1999). Mortalidade materna no municípia do Rio de Janeiro, 1993 a 1996. *Cadernos de Saúde Pública*, **25**, 397–403.

Thomas, R.B. (1998). The evolution of human adaptability paradigms: toward a biology of poverty. In Goodman, A. and Leatherman, T.L., editors, *Building a New Biocultural Synthesis. Political-Economic Perspectives on Human Biology*. Ann Arbor, MI: University of Michigan Press, pp. 43–73.

Thomas, R.B., Winterhalder, B. and McRae, S.D. (1979). An anthropological approach to human ecology and adaptive dynamics. *Yearbook of Physical Anthropology*, **22**, 1–46.

Tiburcio, V., Romero, A. and de Garay, A.L. (1978). Gene frequencies and racial intermixture in a Mestizo population from Mexico City. *Annals of Human Biology*, **5**, 131–8.

Tills, D., Kopec, A.C. and Tills, R.E. (1983). *The Distribution of the Human Blood Groups and Other Polymorphisms*, Supplement 1. Oxford: Oxford University Press.

Timm, E. and Timm, L. (1992). *Culturas em Movimento. A Presença Alemã no Rio Grande do Sul*. Porto Alegre: Riocell.

Tiziani, V., Reichenberger, E., Buzzo, C.L., *et al.* (1999). The gene for cherubism maps to chromosome 4p16. *American Journal of Human Genetics*, **65**, 158–66.

Todorov, T. (1983). *A Conquista da América: a Questão do Outro*. São Paulo: Martins Fontes.

Toledo, S.P.A., Saldanha, S.G., Laurenti, R. and Saldanha, P.H. (1969). Dermatoglifos digitais e palmares de indivíduos da população de São Paulo. *Revista Paulista de Medicina*, **75**(7), 1–10.

Toledo, V.M. and Castillo, A. (1999). La ecología en Latinoamérica: siete tesis para una ciencia pertinente en una región en crisis. *Interciencia*, **24**, 157–68.

Tolentino, H. (1974). *Raza e Historia en Santo Domingo. Los Orígenes del Perjuicio Racial en America*. Santo Domingo, Editorial de la Universidad Autónoma.

Tondo, C.V., Salzano, F.M. and Rucknagel, D.L. (1963). Hemoglobin Porto

Alegre, a possible polymer of normal hemoglobin in a Caucasian Brazilian family. *American Journal of Human Genetics*, **15**, 265–79.

Toplin, R.B. (1974). S*lavery and Race Relations in Latin America.* Westport, CT: Greenwood Press.

Torrealba, J.F. (1956). Algunas consideraciones sobre la enfermedad de hematíes falciformes o enfermedad de Herrick en Venezuela (Addendum). *Gaceta Médica de Caracas*, **63**, 53–73.

Torregrosa, M.V. (1945/46). Incidence of the Rh agglutinogen among Puerto Ricans. *Puerto Rico Journal of Public Health*, **21**, 166–8.

Torres, O. (1930). Contribuição ao estabelecimento dos grupos sangüíneos na Bahia. *Brasil Médico*, **44**, 186–8.

Torres Carmona, M.A. and Márquez Monter, H. (1984). Frecuencia de taurodoncia en una muestra de población clínica de la Facultad de Odontología. In Ramos Galván, R. and Ramos Rodríguez, R.M., editors, *Estudios de Antropología Biológica (II Coloquio de Antropología Física Juan Comas)*. México: Universidad Nacional Autónoma de México, pp. 331–41.

Torroni, A., Brown, M.D., Lott, M.T., Newman, N.J., Wallace, D.C. and the Cuba Neuropathy Field Investigation Team (1995). African Native American, and European mitochondrial DNAs in Cubans from Pinar del Rio Province and implications for the recent epidemic neuropathy in Cuba. *Human Mutation*, **5**, 310–17.

Torroni, A., Cruciani, F., Rengo, C., *et al.* (1999). The A1555G mutation in the 12S rRNA gene of human mtDNA: Recurrent origins and founder events in families affected by sensorineural deafness. *American Journal of Human Genetics*, **65**, 1349–58.

Torroni, A., Neel, J.V., Barrantes, R., Schurr, T.G. and Wallace D.C. (1994). Mitochondrial DNA 'clock' for the Amerinds and its implications for timing their entry into North America. *Proceedings of the National Academy of Sciences, USA*, **91**, 1158–62.

Toupance, B., Godelle, B., Gouyon, P-H. and Schächter, F. (1998). A model for antagonistic pleiotropic gene action for mortality and advanced age. *American Journal of Human Genetics*, **62**, 1525–34.

Tovée, M.J., Maisey, D.S., Emery, I.L. and Cornelissen, P.L. (1999). Visual cues to female physical attractiveness. *Proceedings of the Royal Society of London, Series B*, **266**, 211–18.

Trachtenberg, A., Jobim, L.F.J., Kraemer, E., *et al.* (1988*b*). The HLA polymorphism in five Brazilian populations. *Annals of Human Biology*, **15**, 213–21.

Trachtenberg, A., Salzano, F.M. and Jobim, L.F.J. (1988*a*). Imunização a antígenos HLA em diferentes grupos étnicos brasileiros. *Revista do Hospital de Clínicas de Porto Alegre*, **8**(2), 79–83.

Trachtenberg, A., Salzano, F.M., Kraemer, E. and Jobim, L.F.J. (1993). HLA-A e B e transplantes de rim. Dados de um centro brasileiro. *Jornal Brasileiro de Medicina*, **64**, 221–4.

Trachtenberg, A., Stark, A.E., Salzano, F.M. and da Rocha, F.J. (1985). Canonical correlation analysis of assortative mating in two groups of

Brazilians. *Journal of Biosocial Science*, **17**, 389–403.

Trachtenberg, E.A., Keyeux, G., Bernal, J., Noble, J.A. and Erlich, H.A. (1996). Results of Expedicion Humana. II. Analysis of HLA Class II alleles in three African American populations from Colombia using the PCR/SSOP: Identification of a novel DQB1*02 (*0203) allele. *Tissue Antigens*, **48**, 192–8.

Travassos, L.R. (1992). Protozoology and public health in the Amazon. *Ciência e Cultura*, **44**, 81–163.

Trinkaus, E. and Shipman, P. (1993). *The Neanderthals: Changing the Image of Mankind*. New York: Knopf.

Trivers, R., Manning, J.T., Thornhill, R., Singh, D. and McGuire, M. (1999). Jamaican Symmetry Project: long-term study of fluctuating asymmetry in rural Jamaican children. *Human Biology*, **71**, 417–30.

Tsuneto, L.T., Arce-Gomez, B., Petzl-Erler, M.L. and Queiroz-Telles, F. (1989). HLA-A29 and genetic susceptibility to chromoblastomycosis. *Journal of Medical and Veterinary Mycology*, **27**, 181–5.

Tsuyuoka, R., Bailey, J.W., Guimarães, A.M.D.N., Gurgel, R.Q. and Cuevas, L.E. (1999). Anemia and intestinal parasitic infections in primary school students in Aracaju, Sergipe, Brazil. *Cadernos de Saúde Pública*, **15**, 413–21.

Tyrrell, R.M. (1978). *International Symposium on Current Topics in Radiobiology and Photobiology*. Rio de Janeiro: Academia Brasileira de Ciências.

Uhrhammer, N., Lange, E., Porras, O., *et al.* (1995). Sublocation of an ataxia-telangiectasia gene distal to D11S384 by ancestral haplotyping in Costa Rican families. *American Journal of Human Genetics*, **57**, 103–11.

Ulijaszek, S.J. (1992). Human energetics methods in biological anthropology. *Yearbook of Physical Anthropology*, **35**, 215–42.

Ulijaszek, S.J., Johnston, F.E. and Preece, M.A. (1998). *Cambridge Encyclopedia of Human Growth and Development*. New York: Cambridge University Press.

Urdaneta, M., Prata, A., Struchiner, C.J., Tosta, C.E., Tauil, P. and Boulos, M. (1998). Evaluation of SPf66 malaria vaccine efficacy in Brazil. *American Journal of Tropical Medicine and Hygiene*, **58**, 378–85.

Ureña, P.H. (1964). A emancipação e o primeiro período de vida independente da Ilha de S. Domingos. In: Levene, R., editor, *História das Américas*, Vol. 7. Rio de Janeiro: W.M. Jackson, pp. 375–400.

Utagawa, C.Y., Sugayama, S.M.M., Ribeiro, E.M., *et al.* (1999). Infantile sialic acid storage disease: Report of the first case in South America. *Clinical Genetics*, **55**, 386–7.

Vago, A.R., Macedo, A.M., Oliveira, R. *et al.* (1996). Kinetoplast DNA signatures of *Trypanosoma cruzi* strains obtained directly from infected tissues. *American Journal of Pathology*, **149**, 2153–9.

Vainzof, M., Moreira, E.S., Ferraz, G., Passos-Bueno, A.R., Marie, S.K. and Zatz, M. (1999). Further evidence for the organization of the four sarcoglycans proteins within the dystrophin–glycoprotein complex. *European Journal of Human Genetics*, **7**, 251–4.

Vainzof, M., Passos-Bueno, M.R., Man, N.T. and Zatz, M. (1995). Absence of correlation between utrophin localization and quantity and the clinical severity in Duchenne/Becker dystrophies. *American Journal of Medical*

Genetics, **58**, 305–9.

Valdez, G.A. (1965). Los grupos y factores sanguineos. Estudio estadístico en el Hospital Obrero de Lima. *Revista del Cuerpo Médico* (Lima), **4**, 201–10.

Valenzuela, C.Y. (1975). Dimorfismo sexual pondoestatural en una población chilena. Evidencia de genes para estatura en los cromosomas sexuales? *Revista Médica de Chile*, **103**, 322–6.

Valette, I., Monplaisir, N., Sorel, G., Ribal, C., Dijon, V. and Raffoux, C. (1988). HLA A, B, C and DR association with insulin-dependent diabetes in Martinique. *Tissue Antigens*, **32**, 1–5.

Valiente, M.C., Martinez, A.J., Valdés, T., Carmenate, M.M. and González, O. (1987). Canalización del crecimiento: Estudio longitudinal del peso, la estatura y sus relaciones. In Faulhaber, M.E.S. and Lizárraga Cruchaga, X., editors, *Estudios de Antropología Biológica (III Coloquio de Antropología Física Juan Comas)*. México: Universidad Nacional Autónoma de México, pp. 47–56.

Vallada, E.P., Roseiro, A.M., Moreno, G.M. and Pinhão Filho, M.N. (1967). Grupo sangüíneo e fator Rh em Botucatu e Itapetininga (Estado de São Paulo). *O Hospital*, **71**, 557–61.

Vallejo, G.A., Chiari, E., Macedo, A.M. and Pena, S.D.J. (1993). A simple laboratory method for distinguishing between *Trypanosoma cruzi* and *Trypanosoma rangeli*. *Transactions of the Royal Society of Tropical Medicine and Hygiene*, **87**, 165–6.

Vallejo, G.A., Macedo, A.M., Chiari, E. and Pena, S.D.J. (1994). Kinetoplast DNA from *Trypanosoma rangeli* contains two distinct classes of minicircles with different size and molecular organization. *Molecular Biochemical Parasitology*, **67**, 245–53.

Vallinoto, A.C.R., Cayres-Vallinoto, I.M.V., Ribeiro-dos-Santos, A.K.C., Zago, M.A., Santos, S.E.B. and Guerreiro, J.F. (1999). Heterogeneity of Y chromosome markers among Brazilian Amerindians. *American Journal of Human Biology*, **11**, 481–7.

Van den Ende, J.J., van Bever, Y., Rodini, E.S.O. and Richieri-Costa, A. (1992). Marden–Walker-like syndrome without psychomotor retardation: report of a Brazilian girl born to consanguineous parents. *American Journal of Medical Genetics*, **42**, 467–9.

Van der Sar, A. (1943). Anemia con eritrocitos en forma de hoz en la gestacion. *Revista de la Policlínica de Caracas*, **12**(68), 1–12.

Van der Sar, A. (1959). The occurrence of carriers of abnormal haemoglobin S and C on Curaçao. Ph.D. thesis, Rijksuniversiteit te Groningen.

van Wering, E.R. (1981*a*). The secular growth trend on Aruba between 1954 and 1974. *Human Biology*, **53**, 105–15.

van Wering, E.R. (1981*b*). The anthropometric status of Aruban children: 1974. *Human Biology*, **53**, 117–35.

Van Zanen, G.E. (1962). Expression of the haemoglobin S gene on the island of Curaçao. Ph.D. thesis, Rijksuniversiteit te Groningen.

Vanderborght, B.O.M., Reis, A.M.M., Rouzere, C.D., *et al.* (1993). Prevalence of anti-hepatitis C virus in the blood donor population of Rio de Janeiro. *Vox*

Sanguinis, **65**, 122–5.

Vandermeer, J. (1996). *Reconstructing Biology. Genetics and Ecology in the New World Order.* New York: Wiley.

Vargas, L.A. (1990). Old and new transitions and nutrition in México. In Swedlund, A.C. and Armelagos, G.J., editors, *Disease in Populations in Transition. Anthropological and Epidemiological Perspectives.* New York: Bergin and Garvey, pp. 145–60.

Vargas, L.A. and Casillas, L.E. (1977). La prueba del escalón de Harvard en jóvenes mexicanos. *Anales de Antropología*, **14**, 381–8.

Vargas, L.A., Ramírez, M.E. and Flores, L. (1973). El dimorfismo sexual en fémures mexicanos modernos. *Anales de Antropología*, **10**, 329–36.

Vasconcelos, F.A.G. (1999). Os Arquivos Brasileiros de Nutrição. Uma revisão sobre a produção científica em nutricão no Brasil (1944 a 1988). *Cademos de Saúde Pública*, **15**, 303–16

Vázquez, M.L., Mosquera, M., Cuevas, L.E., *et al.* (1999). Incidência e fatores de risco de diarréia e infecções respiratórias agudas em comunidades urbanas de Pernambuco, Brasil. *Cadernos de Saúde Pública*, **15**, 163–71.

Vela, M., Gamboa, S., Loera-Luna, A., Aguirre, B.E., Pérez-Palacios, G. and Velázquez, A. (1999). Neonatal screening for congenital hypothyroidism in México: experience, obstacles, and strategies. *Journal of Medical Screening*, **6**, 77–9.

Velázquez, A. (1997*a*). Biotin deficiency in protein-energy malnutrition: implications for nutritional homeostasis and individuality. *Nutrition*, **13**, 991–2.

Velázquez, A. (1997*b*). Gene–nutrient interactions in single-gene defects and polygenic diseases: methodologic considerations. In Simopoulos, A.P., editor, *Genetic Variation and Dietary Response.* Basel: Karger, pp. 145–64.

Velázquez, A., Bilbao, G., González-Trujillo, J.L., *et al.* (1996). Apparent higher frequency of phenylketonuria in the Mexican state of Jalisco. *Human Genetics*, **97**, 99–102.

Velázquez, A., Loera-Luna, A., Aguirre, B.E., Gamboa, S., Vargas, H. and Robles, C. (1994). Tamiz neonatal para hipotireoidismo congénito y fenilcetonuria. *Salud Pública de México*, **36**, 249–56.

Vellini Ferreira, F. (1967/68). Distribuição dos valores relativos ao comprimento (glabella-opisthocranion) e largura (bi-euryon) máximas do crânio cerebral, em brancos e negros brasileiros, de ambos os sexos. *Revista de Antropologia*, **15/16**, 91–8.

Venturelli, L.E. and Moraes, M.H.B. (1986). Freqüências gênicas dos sistemas ABO, MNSs e Rh em caucasóides e negróides da cidade de Campinas, SP. *Revista Brasileira de Genética*, **9**, 179–85.

Verano, J.W. and Ubelaker, D.H. (1992). *Disease and Demography in the Americas.* Washington, DC, Smithsonian Institution Press.

Veras, R.P. and Dutra, S. (1993). Envelhecimento da população brasileira: Reflexões e aspectos a considerar quando da definição de desenhos de pesquisas para estudos populacionais. *Physis*, **3**, 107–26.

Vercesi, A.M.L., Carvalho, M.R.S., Aguiar, M.J.B. and Pena, S.D.J. (1999).

486   References

Prevalence of Prader–Willi and Angelman syndromes among mentally retarded boys in Brazil. *Journal of Medical Genetics*, **36**, 498.

Vespucio, A. (1951). *El Nuevo Mundo: Cartas Relativas a sus Viajes y Descubrimientos*. Buenos Aires: Editora Nova.

Viana, M.B., Giugliani, R., Leite, V.H.R., *et al.* (1990). Very low levels of high density lipoprotein cholesterol in four sibs of a family with non-neuropathic Niemann-Pick disease and sea-blue histiocytosis. *Journal of Medical Genetics*, **27**, 499–504.

Viana, M.B., Leite, V.H.R., Giugliani, R. and Fensom, A. (1992). Sea-blue histiocytosis in a family with Niemann–Pick disease. *Sangre*, **37**, 59–67.

Vianna-Morgante, A.M., Costa, S.S., Pavanello, C.M., Otto, P.A. and Mingroni-Netto, R.C. (1999). Premature ovarian failure (POF) in Brazilian fragile X carriers. *Genetics and Molecular Biology*, **22**, 471–4.

Victora, C.G. and Vaughan, J.P. (1985). Land tenure patterns and child health in southern Brazil, The relationship between agricultural production, malnutrition and child mortality. *International Journal of Health Services*, **15**, 253–74.

Victora, C.G., Barros, F.C., Martines, J.C., Béria, J.U. and Vaughan, J.P. (1985). Estudo longitudinal das crianças nascidas em 1982 em Pelotas, RS, Brasil. *Revista de Saúde Pública*, **19**, 58–68.

Viégas, J. and Salzano, F.M. (1978). C-bands in chromosomes 1, 9, and 16 of twins. *Human Genetics*, **45**, 127–30.

Viégas, J., Souza, P.L.R. and Salzano, F.M. (1974). Progeria in twins. *Journal of Medical Genetics*, **11**, 384–6.

Villa, L.L. (1999). Molecular variability of human papillomaviruses and susceptibility to cervical neoplasia. *Ciência e Cultura*, **51**, 212–17.

Villalobos-Arámbula, A.R., Bustos, R., Casas-Castañeda, M., *et al.* (1997). β-thalassemia and βAglobin gene haplotypes in Mexican Mestizos. *Human Genetics*, **99**, 498–500.

Villalobos-Torres, C., Rojas-Martínez, A., Villareal-Castellanos, E., *et al.* (1997). Analysis of 16 cystic fibrosis mutations in Mexican patients. *American Journal of Medical Genetics*, **69**, 380–2.

Villanueva, M. (1982). La antropologia física de los antropólogos físicos en México. Inventario bibliografico (1930–1979). In Villanueva, M. and Serrano, C., editors, *Estudios de Antropología Biológica*, Vol. 1. México: Universidad Nacional Autónoma de México, pp. 75–124.

Villanueva, M. (1989). La somatotipologia. Un recurso viable para evaluar la composición corporal? In Serrano, C. and Salas, M.E., editors, *Estudios de Antropología Biológica (IV Coloquio de Antropología Física Juan Comas)*. México: Universidad Nacional Autónoma de México, pp. 417–21.

Villanueva, M., Sáenz, M.E. and Serrano, C. (1984). Crecimiento y desarrollo en escolares de la Villa de Las Margaritas, Chiapas, México. In Ramos Galván, R. and Ramos Rodríguez, R.M., editors, *Estudios de Antropología Biológica (II Coloquio de Antropología Física Juan Comas)*. México: Universidad Nacional Autónoma de México, pp. 427–52.

Villareal, M.T., Chávez, M., Lezana, J.L., *et al.* (1996). G542X mutation in

Mexican cystic fibrosis patients. *Clinical Genetics*, **49**, 54–6.
Villarroel, H., Lazo, B., Campusano, C., Pinto-Cisternas, J. and Ballesteros, S. (1980). Consanguinidad y migración en Viña del Mar durante 1888 a 1967. *Revista Médica de Chile*, **108**, 407–12.
Villasante, G.S., Hidalgo, P.C., Heredero-Bauté, L. and García, B.B. (1990). Diagrama de flujo para el estudio de enfermedades lisosomales. *Medicentro*, **6**, 231–5.
Visentainer, J.E., Tsuneto, L.T., Serra, M.F., Peixoto, P.R. and Petzl-Erler, M.L. (1997). Association of leprosy with HLA-DR2 in a southern Brazilian population. *Brazilian Journal of Medical and Biological Research*, **30**, 51–9.
Visentainer, J.E.L., Tsuneto, L.T., Moliterno, R.A. and Telles-Filho, F.Q. (1993). Lack of association between paracoccidioidomycosis and HLA histocompatibility antigens. *Revista Brasileira de Genética*, **16**, 1035–41.
Vogel, F. and Motulsky, A.G. (1997). *Human Genetics. Problems and Approaches*. Berlin: Springer-Verlag.
Wade, P. (1993). *Blackness and Race Mixture: The Dynamics of Racial Identity in Colombia*. Baltimore, MD: Johns Hopkins University Press.
Wagley, C. (1952). *Race and Class in Rural Brazil*. Paris: UNESCO.
Wagley, C. (1971). The formation of the American population. In Salzano, F.M. editor, *The Ongoing Evolution of Latin American Populations*. Springfield, IL: C.C. Thomas, pp. 19–39.
Wagner, S.C., Friedrish, J.R., Job, F. and Hutz, M.H. (1996). Caracterização molecular da anemia falciforme em pacientes de Porto Alegre. *Brazilian Journal of Genetics*, **19** (Suppl.), 244.
Wainscoat, J.S., Bell, J.I., Thein, S.L., Higgs, D.R., Serjeant, G.R. and Peto, T.E.A. (1983). Multiple origins of the sickle mutation: Evidence from β^S gene cluster polymorphism. *Molecular Biology and Medicine*, **1**, 191–7.
Wajner, M., Goldbeck, A.S., Lhullier, F., Pires, R.F., Netto, C.B.O. and Giugliani, R. (1994). Neonatal screening for amino acidopathies and congenital hypothyroidism: a pilot program in southern Brazil. *Revista Brasileira de Genética*, **17**, 229–30.
Wajner, M., Sanseverino, M.T., Giugliani, R., *et al.* (1992). Biochemical investigation of a Brazilian patient with a defect in mitochondrial acetoacetyl-coenzyme A thiolase. *Clinical Genetics*, **41**, 202–5.
Wajner, M., Wannmacher, C.M.D., Gaidzinski, D., Dutro-Filho, C.S., Buchalter, M.S. and Giugliani, R. (1986). Detection of inborn errors of metabolism in patients of pediatric intensive care units of Porto Alegre, Brazil: comparison of the prevalence of such disturbances in a selected and an unselected sample. *Revista Brasileira de Genética*, **9**, 331–40.
Waldvogel, B. and Morais, L.C.C. (1998). Mortalidade por AIDS em São Paulo: Dezoito anos de história. *Boletim Epidemiológico*, (São Paulo), **16**(2), 3–22.
Walker, S.P., Grantham-McGregor, S.M., Himes, J.H. and Williams, S. (1996). Anthropometry in adolescent girls in Kingston, Jamaica. *Annals of Human Biology*, **23**, 23–9.
Wallace, D.C. (1995). Mitochondrial DNA variation in human evolution, degenerative disease, and aging. *American Journal of Human Genetics*, **57**,

201–223.

Wannmacher, C.M.D., Wajner, M., Giugliani, R. and Dutra-Filho, C.S. (1982a). An improved specific laboratory test for homocystinuria. *Clinica Chimica Acta*, 125, 367–9.

Wannmacher, C.M.D., Wajner, M., Giugliani, R., Costa, M.G., Giugliani, E.R.J. and Castro, J.Z. (1984). Clinical and biochemical effects of long-term vitamin A administration to a patient with Hurler–Scheie compound. *Brazilian Journal of Medical and Biological Research*, 17, 43–7.

Wannmacher, C.M.D., Wajner, M., Giugliani, R., Giugliani, E.R.J., Costa, M.G. and Giugliani, C.K. (1982b). Detection of metabolic disorders among high-risk patients. *Revista Brasileira de Genética*, 5, 187–94.

Ward, R.H. and Weiss, K.M. (1976). Demographic aspects of the biology of human populations. Special Issue, *Journal of Human Evolution*, 5(1), 1–154.

Wasserman, C. (1996a). *História da América Latina: Cinco Séculos*. Porto Alegre: Editora da Universidade Federal do Rio Grande do Sul.

Wasserman, C. (1996b). A formação do estado nacional na América Latina: As emancipações políticas e o intrincado ordenamento dos novos países. In: Wasserman, C., editor, *História da América Latina: Cinco Séculos (Temas e Problemas)*. Porto Alegre: Editora da Universidade Federal do Rio Grande do Sul, pp. 178–215.

Wasserman, C. and Guazzelli, C.B. (1996). *História da América Latina. Do Descobrimento a 1900*. Porto Alegre: Editora da Universidade Federal do Rio Grande do Sul.

Waterlow, J.C. and Bunjé, H.W. (1966). Observations on mountain sickness in the Colombian Andes. *The Lancet*, II, 655–61.

Watson, E., Forster, P., Richards, M. and Bandelt, H. (1997). Mitochondrial footprints of human expansion in Africa. *American Journal of Human Genetics*, 61, 691–704.

Waynforth, D. (1998). Fluctuating asymmetry and human male life-history traits in rural Belize. *Proceedings of the Royal Society of London, Series B*, 265, 1497–501.

Waynforth, D. (1999). Differences in time use for mating and nepotistic effort as a function of male attractiveness in rural Belize. *Evolution and Human Behavior*, 20, 19–28.

Waynforth, D., Hurtado, A.M. and Hill, K. (1998). Environmentally contingent reproductive strategies in Mayan and Ache males. *Evolution and Human Behavior*, 19, 369–85.

Weckmann, A.L., Vargas-Alarcón, G., López, M., *et al.* (1997). Frequencies of HLA-A and HLA-B alleles in a Mexico City Mestizo sample. *American Journal of Human Biology*, 9, 1–5.

Wehling, A. and Wehling, M.J.C. (1994). *Formação do Brasil Colonial*. Rio de Janeiro: Editora Nova Fronteira.

Weimer, T.A., Franco, M.H.L.P., Salzano, F.M., Layrisse, Z., Méndez-Castellano, H. and Brennan, S.O. (1999). Genetic studies in a semi-isolated Venezuelan population with the molecular identification of an albumin variant. *Interciencia*, 24, 49–53.

Weimer, T.A., Salzano, F.M., Westwood, B. and Beutler, E. (1993). Molecular characterization of glucose-6-phosphate dehydrogenase variants from Brazil. *Human Biology*, **65**, 41–7.

Weimer, T.A., Salzano, F.M., Westwood, B. and Beutler, E. (1998). G6PD variants in three South American ethnic groups: population distribution and description of two new mutations. *Human Heredity*, **48**, 92–6.

Weimer, T.A., Tavares-Neto, J., Franco, M.H.L.P., *et al.* (1991). Genetic aspects of *Schistosoma mansoni* infection severity. *Revista Brasileira de Genética*, **14**, 623–30.

Weiner, J.S. and Lourie, J.A. (1969). *Human Biology. A Guide to Field Methods.* Oxford: Blackwell.

Weinstein, B.I., Kutlar, A., Webber, B.B., Wilson, J.B. and Huisman, T.H.J. (1985). Hemoglobin Daneshgah-Tehran or α_2 72 (EF1) HIS→ARGβ_2 in an Argentinian family. *Hemoglobin*, **9**, 409–14.

Weinstein, B.I., Plaseska-Karanfilska, D. and Efremov, G.D. (2000). Hb Saint Etienne or Hb Instanbul [β92(F8)His→Gln] found in an Argentinean family. *Hemoglobin*, **24**, 149–52.

Wenning, M.R.S.C., Silva, N.M., Jorge, S.B., *et al.* (2000). Hb Campinas [α26(B7)Ala→Val]: A novel, electrophoretically silent, variant. *Hemoglobin*, **24**, 143–8.

Went, L.N. (1957). Incidence of abnormal haemoglobins in Jamaica. *Nature*, **180**, 1131–2.

Went, L.N. and MacIver, J.E. (1958). Distribution of abnormal haemoglobins in Jamaica. *Schweizer Zeitschrift für Allgemeine Pathologie und Bakteriologie*, **21**, 614–17.

Went, L.N. and MacIver, J.E. (1961). Thalassemia in the West Indies. *Blood*, **17**, 166–81.

Wertheimer, F.W., Brewster, R.H. and White, C.L. (1967). Periodontal disease and nutrition in Trinidad. *Journal of Periodontology*, **38**, 100–4.

Wertz, D.C. and Fletcher, J.C. (1989). *Ethics and Human Genetics. A Cross-Cultural Perspective.* Berlin: Springer-Verlag.

Whittle, M.R., Zatz, M. and Reinach, F.C. (1993). The use of chromosome 5q markers for confirming the diagnosis of proximal spinal muscular atrophy. *Brazilian Journal of Medical and Biological Research*, **26**, 1157–73.

WHO (World Health Organization) (1964). *Research in Population Genetics of Primitive Groups.* Geneva: World Health Organization Technical report no. 279.

WHO (1968). *Research on Human Population Genetics.* Geneva: World Health Organization Technical report no. 387.

WHO (1998). *Rapport sur la Santé dans le Monde, 1998. La Vie au 21ᵉ. Siècle. Une Perspective pour Tous.* Geneva: World Health Organization.

Wilkie, A.O.M., Buckle, V.J., Harris, P.C., *et al.* (1990). Clinical features and molecular analysis of the αthalassemia/mental retardation syndromes. I. Cases due to deletions involving chromosome band 16p13.3. *American Journal of Human Genetics*, **46**, 1112–26.

Williams-Blangero, S., Vandeberg, J.L., Blangero, J. and Teixeira, A.R.L. (1997).

Genetic epidemiology of seropositivity for *Trypanosoma cruzi* infection in rural Goiás, Brazil. *American Journal of Tropical Medicine and Hygiene*, **57**, 538–43.

Wilson, W.M., Dufour, D.L., Staten, L.K., Barac-Nieto, M., Reina, J.C. and Spurr, G.B. (1999). Gastrointestinal parasitic infection, anthropometrics, nutritional status, and physical work capacity in Colombian boys. *American Journal of Human Biology*, **11**, 763–71.

Winnie, W.W., Jr. (1965). Estimates of interstate migration in México, 1950–1960: data and methods. *Antropologica*, **14**, 38–60.

Witkop, C.J., Jr. and Barros, L. (1963). Oral and genetic studies of Chileans 1960. I. Oral anomalies. *American Journal of Physical Anthropology*, **21**, 15–24.

Wolanski, N. (1998). Comparison of growth patterns of subcutaneous fat tissue in Mexican and Polish with US and Peruvian populations. *Annals of Human Biology*, **25**, 467–77.

Wolanski, N., Dickinson, F. and Siniarska, A. (1994). Seasonal rhythm of menarche as a sensitive index of living conditions. *Studies in Human Ecology*, **11**, 171–91.

Wong, L.R., Hakkert, R. and Lima, R.A. (1987). *Futuro da População Brasileira. Projeções, Previsões e Técnicas*. São Paulo: Associação Brasileira de Estudos Populacionais.

Wright, S. (1930). The genetical theory of natural selection. *Journal of Heredity*, **21**, 349–56.

Wright, S. (1931). Evolution in Mendelian populations. *Genetics*, **16**, 97–159.

Xavier, A.R., Carvalheiro, C.D.G. and Ferrari, I. (1978). Incidence and prevalence of Down's syndrome in the city of Ribeirão Preto, São Paulo (Brazil), 1972. *Revista Brasileira de Pesquisas Médicas e Biológicas*, **11**, 39–42.

Yarbrough, C., Habicht, J-P., Malina, R.M., Lechtig, A. and Klein, R.E. (1975). Length and weight in rural Guatemalan Ladino children, birth to seven years of age. *American Journal of Physical Anthropology*, **42**, 439–47.

Yarbrough, C., Martorell, R., Klein, R.E., Himes, J., Malina, R.M. and Habicht, J-P. (1977). Stature and age as factors in the growth of second metacarpal cortical bone in moderately malnourished children. *Annals of Human Biology*, **4**, 43–8.

Yasuda, N. (1969). The inbreeding coefficient in northeastern Brazil. *Human Heredity*, **19**, 444–56.

Youlton, R. (1971). Sindrome de Turner. Caracteres clinicos y hallazgos citogenéticos en 11 casos. *Revista Médica de Chile*, **99**, 125–8.

Yunes, J., Somensi, J. and Ronchezel, V.S.C. (1976). Tendência da mortalidade por causas no Brasil. In IBGE. Rio de Janeiro: Fundação Instituto Brasileiro de Geografia e Estatística, pp. 112–25.

Zabaglia, S.F.C., Pedro, A.O., Pinto Neto, A.M., Guarisi, T., Paiva, L.H.S.C. and Lane, E. (1998). Estudo exploratório da associação entre o perfil lipídico e a densidade mineral óssea em mulheres menopausadas, em hospital de referência de Campinas. *Cadernos de Saúde Pública*, **14**, 779–86.

Zago, M.A. (1989). Fetal hemoglobin heterogeneity in Brazilian newborns and β-thalassemia homozygotes. *Haematologica*, **74**, 347–50.

Zago, M.A. and Bottura, C. (1983). Splenic function in sickle-cell diseases. *Clinical Science*, **65**, 297–302.

Zago, M.A. and Costa, F.F. (1985). Hereditary haemoglobin disorders in Brazil. *Transactions of the Royal Society of Tropical Medicine and Hygiene*, **79**, 385–8.

Zago, M.A. and Costa, F.F. (1988). Hb D-Los Angeles in Brazil: simple heterozygotes and associations with β-thalassemia and with Hb S. *Hemoglobin*, **12**, 399–403.

Zago, M.A., Costa, F.F. and Bottura, C. (1981). Beta-thalassemia in Brazil. *Brazilian Journal of Medical and Biological Research*, **14**, 383–8.

Zago, M.A., Costa, F.F. and Bottura, C. (1982). Thalassaemia intermedia in a family with β⁰-thalassaemia and Hb Hasharon. *Journal of Medical Genetics*, **19**, 437–40.

Zago, M.A., Costa, F.F. and Bottura, C. (1983*b*). β⁺-thalassaemia intermedia with low Hb F. *Klinische Wochenschrift*, **61**, 95–8.

Zago, M.A., Costa, F.F. and Bottura, C. (1984). Hemoglobin H disease in three Brazilian families. *Revista Brasileira de Genética*, **7**, 137–47.

Zago, M.A., Costa, F.F. and Bottura, C. (1985). β⁺-thalassemia intermedia resulting from the interaction of the high HbA2 and the silent β-thalassemia genes. *Revista Brasileira de Genética*, **8**, 545–53.

Zago, M.A., Costa, F.F. and Tavella, M.H. (1989). Hb F heterogeneity in sickle cell anemia and sickle cell β-thalassemia from Brazil. In: Schmidt, B.J., Diament, A.J. and Loghin-Grosso, N.S., editors, *Current Trends in Infant Screening*, Amsterdam: Elsevier, pp. 349–51.

Zago, M.A., Costa, F.F., Freitas, T.C. and Bottura, C. (1980). Clinical, hematological and genetic features of sickle-cell anemia and sickle cell-βthalassemia in a Brazilian population. *Clinical Genetics*, **18**, 58–64.

Zago, M.A., Costa, F.F., Ismael, S.J. and Bottura, C. (1983*c*). Enfermedades drepanocíticas en una población brasileña. *Sangre*, **28**, 191–8.

Zago, M.A., Costa, F.F., Tone, L.G. and Bottura, C. (1983*a*). Hereditary hemoglobin disorders in a Brazilian population. *Human Heredity*, **33**, 125–9.

Zago, M.A., Figueiredo, M.S. and Ogo, S.H. (1992*a*). Bantu β^S cluster haplotype predominates among Brazilian Blacks. *American Journal of Physical Anthropology*, **88**, 295–8.

Zago, M.A., Kerbauy, J., Souza, H.M., *et al.* (1992*b*). Growth and sexual maturation of Brazilian patients with sickle cell diseases. *Tropical and Geographical Medicine*, **44**, 317–21.

Zago, M.A., Silva, W.A. Jr. and Franco, R.F. (1999). Hemoglobinopathies and other hereditary hematological diseases in the Brazilian population. *Ciência e Cultura*, **51**, 226–34.

Zago, M.A., Silva, W.A. Jr., Dalle, B., *et al.* (2000). Atypical β^S haplotypes are generated by diverse genetic mechanisms. *American Journal of Hematology*, **63**, 79–84.

Zago, M.A., Tavella, M.H., Simões, B.P., Franco, R.F., Guerreiro, J.F. and

Santos, S.E.B. (1996). Racial heterogeneity of DNA polymorphisms linked to the A and the O alleles of the ABO blood group gene. *Annals of Human Genetics*, **60**, 67–72.

Zalis, M.G., Pang, L., Silveira, M.S., Milhous, W.K. and Wirth, D.F. (1998). Characterization of *Plasmodium falciparum* isolated from the Amazon Region of Brazil: evidence for quinine resistance. *American Journal of Tropical Medicine and Hygiene*, **58**, 630–7.

Zanenga, R., Mattevi, M.S. and Erdtmann, B. (1984). Smaller autosomal C band sizes in Blacks than in Caucasoids. *Human Genetics*, **66**, 286.

Zarama de Martinez, S. (1992). Ceguera. Por herencia o por negligencia? *América Negra*, **3**, 191–6.

Zatz, M., Marie, S.K., Cerqueira, A., Vainzof, M., Pavanello, R.C.M. and Passos-Bueno, M.R. (1998b). The facioscapulohumeral muscular dystrophy (FSHD1) gene affects males more severely and more frequently than females. *American Journal of Medical Genetics*, **77**, 155–61.

Zatz, M., Marie, S.K., Passos-Bueno, M.R., Vainzof, M., *et al.* (1995b). High proportion of new mutations and possible anticipation in Brazilian facioscapulohumeral muscular dystrophy families. *American Journal of Human Genetics*, **56**, 99–105.

Zatz, M., Passos-Bueno, M.R., Cerqueira, A., Marie, S.K., Vainzoff, M. and Pavanello, R.C.M. (1995a). Analysis of the CTG repeat in skeletal muscle of young and adult myotonic dystrophy patients: when does the expansion occur? *Human Molecular Genetics*, **4**, 401–6.

Zatz, M., Sumita, D., Campiotto, S., Canovas, *et al.* (1998a). Paternal inheritance or different mutations in maternally related patients occur in about 3% of Duchenne familial cases. *American Journal of Medical Genetics*, **78**, 361–5.

Zatz, M., Vainzof, M. and Passos-Bueno, M.R. (1993). The impact of molecular biology for the understanding of hereditary myopathies in the Brazilian population. *Ciência e Cultura*, **45**, 241–8.

Zavala, C., Cobo, A., Lisker, R., Madrid, L. and Mendoza, Y. (1972). Alteraciones de los cromosomas en enfermos mentales. Estudio en dos hospitales psiquiátricos. *Revista de Investigación Clínica*, **24**, 77–81.

Zavala, C., Morton, N.E., Rao, D.C., Lalouel, J.M., Gamboa, I.A., Tejeda, A. and Lisker, R. (1979). Complex segregation analysis of diabetes mellitus. *Human Heredity*, **29**, 325–33.

Zavalla, S.A. (1964). México. A revolução. A independência. A constituição de 1824. In: Levene, R., editor, *História das Américas,* Vol. 7. Rio de Janeiro: W.M. Jackson, pp. 3–96.

Zimmerman, P.A., Phadke, P.M., Lee, A., Elson, L.H., Araujo, E., Guderian, R. and Nutman, T.B. (1995). Migration of a novel DAQ1* allele (DQA1*0502) from African origin to North and South America. *Human Immunology*, **42**, 233–40.

Zuiani, T.B.B., Trindade, I.E.K., Yamashita, R.P. and Trindade, A.S. Jr. (1998). The pharyngeal flap surgery in patients with velopharyngeal insufficiency: perceptual and nasometric speech assessment. *Brazilian Journal of*

Dysmorphology and Speech–Hearing Disorders, **1**, 31–42.

Zuñiga Ide, J. (1980). *La Consanguinidad en el Valle de Elqui. Un Estudio de Genética de Poblaciones Humanas.* La Serena: Universidad de Chile.

Zvelebil, M. (1986). Mesolithic prelude and Neolithic revolution. In Zvelebil, M., editor, *Hunters in Transition: Mesolithic Societies of Temperate Eurasia and their Transition to Farming.* Cambridge: Cambridge University Press, pp. 5–15.

Author index

Subject index